SELECTED SOLUTIONS MANUAL

C. ALTON HASSELL
Baylor University

GENERAL CHEMISTRY

FOURTH EDITION

HILL • PETRUCCI • McCREARY • PERRY

PEARSON
Prentice Hall

Upper Saddle River, NJ 07458

Project Manager: Kristen Kaiser
Senior Editor: Kent Porter-Hamann
Editor-in-Chief, Science: John Challice
Vice President of Production & Manufacturing: David W. Riccardi
Executive Managing Editor: Kathleen Schiaparelli
Assistant Managing Editor: Becca Richter
Production Editor: Jeffrey Rydell
Supplement Cover Manager: Paul Gourhan
Supplement Cover Designer: Joanne Alexandris
Manufacturing Buyer: Ilene Kahn
Cover Image Credit: MgO(100) surface/Royce Copenheaver

Pearson Prentice Hall
Pearson Education, Inc.
Upper Saddle River, NJ 07458

Printed in the United States of America

10 9 8 7 6

ISBN 0-13-140346-X

Pearson Education Ltd., *London*
Pearson Education Australia Pty. Ltd., *Sydney*
Pearson Education Singapore, Pte. Ltd.
Pearson Education North Asia Ltd., *Hong Kong*
Pearson Education Canada, Inc., *Toronto*
Pearson Educación de Mexico, S.A. de C.V.
Pearson Education—Japan, *Tokyo*
Pearson Education Malaysia, Pte. Ltd.

Table of Contents

Table of Contents

To the Student

This manual is meant to be an aid in your learning chemistry, especially to your learning to solve chemical problems. It can be misused and not be helpful, but used correctly, the manual can be a great help.

Contained in these pages are the worked out solutions for the in-chapter exercises and the end-of-chapter review questions and problems with blue numbers for General Chemistry, 4nd ed. by John W. Hill, Ralph H. Petrucci, Terry Mcreary, and Scott Perry, Prentice Hall (2005). In working the mathematical problems, values of constants, such as the universal gas constant or the atomic weights, were used that included one more significant figure than the least well known input data. The reason was to insure that the input data would be the limiting factor in the accuracy of the answer. When the answer to an intermediate step was listed, it was usually rounded to the number of significant figures than should be listed in the final answer. You should keep one more than the number in the final answer so that the input data, not the intermediate answer, should be the limiting value. The final answer was written with the correct number of significant figures. The answer in the back of the text book was calculated without intermediate rounding and will occasionally varyfrom the solution manual by 1 or 2 in the final digit of the answer.

Learning problem solving is much like learning to play a sport or a musical instrument. You must do it yourself; you cannot learn by watching. Skill is developed by practice and more practice. Practice time should be planned so that you are at your best. Very little will be gained during practice time when you are very tired or distracted.

Before attempting the problems, read the text chapter carefully, reading the example problems and working out the exercises. If you work on an exercise for ten or fifteen minutes and still do not have an answer, only then should you look in the solutions manual to get an idea of how to start working the problem. After the text is read, work through the review questions and the odd numbered problems. Working a problem may require looking back in the text to find a similar example. Only after ten or fifteen minutes of attempting a problem without result should the solutions manual be consulted. Any problem that requires use of the solutions manual should be reworked a few days later to see if the problem solving skill was really acquired or just the answer read from the manual.

When you refer to the manual, the solution method may be different from yours as there is usually more than one way to work a problem. There are some problems in the manual that actually show two different methods. Sometimes these solutions differ in the last significant figure because of the round off of numbers. If your answer is within 1 or 2 in the last significant figure, then it is probably the same number.

Don't work these problems just to get an answer. Work these problems to develop the skills to be able to solve other problems, such as those on the next test or those in your future job. While practicing these skills, it is better if you work in a quiet, undisturbed atmosphere while you are as fresh as possible. It is also better to work some every day than to cram all of your studying into one day.

Baylor University
Waco, Texas

C. Alton Hassell
Alton_Hassell@baylor.edu

Dedication

I dedicate this manual to the memory of two wonderful men.

The first was my father, Clinton A. "Brit" Hassell (1914–1995). My greatest inheritance was the education that he provided for me.

The second was my first college chemistry professor, Dr. Thomas C. Franklin (1923–1997). He became a mentor and dear friend. He is deeply missed.

Acknowledgements

This manual is the compilation of the work of many people. I owe a great debt to each one. They have combined to keyboard, proof, standardize, edit, double-check, clarify, and even beautify the initial rough draft. The errors that remain are my sole responsibility.

Adonna Cook, Barbara Rauls, Andrew Garner, Erin Saenz, and Phi Doan keyboarded the manuscript. That is, they turned illegible scrawling and crude drawings into printed text and illustrative figures. Lee Ann Marshall more than once gave us the expert guidance that was needed in the use of Word.

Proofreaders and/or accuracy-checkers included Michael Wismer, Tony Yiannakos, Matt Johl, and David Shinn. These wonderful, hard-working people have pored over the manual and found my many mistakes.

The Department of Chemistry at Baylor University and especially the department chairman, Dr. Marianna Busch, have given me great support and encouragement.

John Hill, Ralph Petrucci, Terry McCreary, and Scott Perry wrote a wonderful text without which this manual would be unnecessary. Terry McCreary helped with some of the solutions. Ralph went the extra mile (miles) to proof and edit an earlier edition of the manual. Robert Wismer wrote solutions for some problems in another book which are being reused in this manual.

To work with the editors and marvelous staff at Prentice Hall is to work with the best. Paul Corey signed me to the original project. Mary Hornby has become a dear friend and the glue that held us all together for earlier editions. Kristen Kaiser has done the same for the last two editions.

My wife, Patricia, and my children, Clint and Sharina, put up with me or with my absence during the entire process.

Chapter 1

Chemistry: Matter and Measurement

Exercises

1.1A (a) $2.05 \times 10^{-6}\ \text{m} \times \dfrac{\mu\text{m}}{10^{-6}\ \text{m}} = 2.05\ \mu\text{m}$

(b) $4.03 \times 10^{3}\ \text{g} \times \dfrac{\text{kg}}{10^{3}\ \text{g}} = 4.03\ \text{kg}$

(c) $7.06 \times 10^{-9}\ \text{s} \times \dfrac{\text{ns}}{10^{-9}\ \text{s}} = 7.06\ \text{ns}$

(d) $5.15 \times 10^{-2}\ \text{m} \times \dfrac{\text{cm}}{10^{-2}\ \text{m}} = 5.15\ \text{cm}$

1.1B (a) $6217\ \text{g} = 6.217 \times 10^{3}\ \text{g}$

(b) $0.0016\ \text{s} = 1.6 \times 10^{-3}\ \text{s}$

(c) $0.0717\ \text{g} = 7.17 \times 10^{-2}\ \text{g}$

(d) $387\ \text{m} = 3.87 \times 10^{2}\ \text{m}$

1.2A (a) $355\ \mu\text{s} = 3.55 \times 10^{2}\ \mu\text{s} \times \dfrac{10^{-6}\ \text{s}}{\mu\text{s}} = 3.55 \times 10^{-4}\ \text{s}$

(b) $1885\ \text{km} = 1.885 \times 10^{3}\ \text{km} \times \dfrac{10^{3}\ \text{m}}{\text{km}} = 1.885 \times 10^{6}\ \text{m}$

(c) $1350\ \text{cm} = 1.350 \times 10^{3}\ \text{cm} \times \dfrac{10^{-2}\ \text{m}}{\text{cm}} = 1.350 \times 10^{1}\ \text{m}$

(d) $425\ \text{nm} = 4.25 \times 10^{2}\ \text{nm} \times \dfrac{10^{-9}\ \text{m}}{\text{nm}} = 4.25 \times 10^{-7}\ \text{m}$

1.2B (a) $2.28 \times 10^{5}\ \text{g} \times \dfrac{\text{kg}}{10^{3}\ \text{g}} = 2.28 \times 10^{2}\ \text{kg}$

(b) $0.083\ \text{cm} = 8.3 \times 10^{-2}\ \text{cm} \times \dfrac{10^{-2}\ \text{m}}{\text{cm}} = 8.3 \times 10^{-4}\ \text{m}$

(c) $4.05 \times 10^{2}\ \mu\text{m} \times \dfrac{10^{-6}\ \text{m}}{\mu\text{m}} = 4.05 \times 10^{-4}\ \text{m}$

(d) $20.25\ \text{min} \times \dfrac{60\ \text{s}}{\text{min}} = 1215\ \text{s} = 1.215 \times 10^{3}\ \text{s}$

1.3A (a) $t_F = (1.8 \times 85.0\ °C) + 32 = 185\ °F$

(b) $t_F = (1.8 \times -12.2\ °C) + 32.0 = 10.0\ °F$

(c) $t_C = (335\ °F - 32)/1.8 = 168\ °C$

(d) $t_C = (-20.8\ °F - 32.0)/1.8 = -29.3\ °C$ 32 ° is a defined number. it is good to any number of significant figures.

1.3B $t_F = (1.8 \times -273.15\ °C) + 32.00 = -459.67\ °F$

1.4A $6.4\ mm \times \dfrac{10^{-3}\ m}{mm} = 6.4 \times 10^{-3}\ m$

$1.827\ m \times 0.762\ m \times 6.4 \times 10^{-3}\ m = 8.9 \times 10^{-3}\ m^3$

or

$1.39\ m^2 \times 6.4 \times 10^{-3}\ m = 8.9 \times 10^{-3}\ m^3$

1.4B $6.4\ mm \times \dfrac{10^{-3}\ m}{mm} \times \dfrac{cm}{10^{-2}\ m} = 6.4 \times 10^{-1}\ cm$

$1.827\ m \times \dfrac{cm}{10^{-2}\ m} = 182.7\ cm$

$0.762\ m \times \dfrac{cm}{10^{-2}\ m} = 76.2\ cm$

$182.7\ cm \times 76.2\ cm \times 6.4 \times 10^{-1}\ cm = 8.9 \times 10^3\ cm^3$

or

$1.39 \times 10^{-4}\ cm^2 \times 6.4 \times 10^{-1}\ cm = 8.9 \times 10^3\ cm^3$

1.5A $21.60\ g \times 2.04 \times 21 = 925\ g\ Zn$

1.5B $9.2\ mm \times \dfrac{10^{-3}\ m}{mm} \times \dfrac{cm}{10^{-2}\ m} = 9.2 \times 10^{-1}\ cm = 0.92\ cm$

face area $= 6\ faces \times \dfrac{(0.92\ cm)^2}{face} = 5.1\ cm^2$

1.6A (a)

```
   48.2 m
    3.82 m
   48.4394 m
 ___________
  100.4594 m   rounds to 100.5 m
```

(b)

```
  148 g
    2.39 g
    0.0124 g
 ___________
  150.4024 g   rounds to 1.50 × 10^2 g
```

(c) 451 g
−15.46 g
−20.3 g
415.24 g rounds to 415 g

(d) 15.436 L
5.3 L
−6.24 L
−8.177 L
6.319 L rounds to 6.3 L

1.6B (a) 51.5 + 2.67 = 54.17 m; 33.42 − 0.124 = 33.296 m

$54.2 \text{ m} \times 33.30 \text{ m} = 1804.86$ rounds to $1.80 \times 10^3 \text{ m}^2$

(b) 125.1 − 1.22 = 123.88; 52.5 + 0.63 = 53.13

$$\frac{123.9\text{ g}}{53.1\text{ mL}} = 2.33 \text{ g/mL}$$

(c) 47.5 − 1.44 = 46.06

$$\frac{46.1\text{ kg}}{10.5\text{ m} \times 0.35\text{ m} \times 0.175\text{ m}} = 72 \text{ kg/ m}^3$$

(d)

$14.2 \text{ mg} = 14.2 \times 10^{-3} \text{ g} = 0.0142 \text{ g}$

$3.52 \text{ mg} = 3.52 \times 10^{3} \text{ g} = 0.00352 \text{ g}$

0.307 g
− 0.0142 g
− 0.00352 g
0.28928 g

$0.28 \text{ mm} = 0.28 \times 10^{-3} \text{ m} = 0.28 \times 10^{-3} \times 10^{2} \text{ cm} = 0.028 \text{ cm}$

1.22 cm
− 0.028 cm
1.192 cm

$$\frac{0.289\text{ g}}{1.19\text{ cm} \times 0.752\text{ cm} \times 0.51\text{ cm}} = 0.63 \text{ g/cm}^3$$

1.7A (a) $? \text{ m} = 76.3 \text{ mm} \times \frac{10^{-3}\text{ m}}{\text{mm}} = 0.0763 \text{ m} = 7.63 \times 10^{-2} \text{ m}$

(b) $? \text{ mg} = 0.0856 \text{ kg} \times \frac{10^{3}\text{ g}}{\text{kg}} \times \frac{\text{mg}}{10^{-3}\text{ g}} = 8.56 \times 10^{4} \text{ mg}$

(c) $? \text{ ft} = 0.556 \text{ km} \times \frac{0.6214\text{ mi}}{\text{km}} \times \frac{5280\text{ ft}}{\text{mi}} = 1.82 \times 10^{3} \text{ ft}$

1.7B (a) $? \text{ kg} = 1.95 \times 10^{-3} \text{ oz} \times \frac{28.35\text{ g}}{\text{oz}} \times \frac{\mu\text{g}}{10^{-6}\text{ g}} = 5.53 \times 10^{4}\ \mu\text{g}$

(b) $? \text{ fl oz} = 3.50 \text{ gal} \times \frac{3.785 \text{ L}}{1 \text{ gal}} \times \frac{\text{mL}}{10^{-3} \text{ L}} \times \frac{\text{fl oz}}{29.57 \text{ mL}} = 448 \text{ fl oz}$

(c) $? \text{ km} = 7.75 \times 10^4 \text{ ft} \times \frac{\text{mi}}{5280 \text{ ft}} \times \frac{\text{km}}{0.6214 \text{ mi}} = 23.6 \text{ km}$

1.8A (a) $? \text{ in}^2 = 476 \text{ cm}^2 \times \left(\frac{1 \text{ in.}}{2.54 \text{ cm}}\right)^2 = 73.8 \text{ in.}^2$

(b) $? \text{ m}^3 = 1.56 \times 10^4 \text{ in.}^3 \times \left(\frac{1 \text{ m}}{39.37 \text{ in.}}\right)^3 = 0.256 \text{ m}^3$

1.8B $? \text{ g} = 1.00 \text{ L} \times \frac{\text{mL}}{10^{-3} \text{ L}} \times \frac{\text{cm}^3}{\text{mL}} \times \left(\frac{10^{-2} \text{ m}}{\text{cm}}\right)^3 \times \left(\frac{39.37 \text{ in}}{\text{m}}\right)^3 \times \left(\frac{\text{ft}}{12 \text{ in}}\right)^3 \times \frac{62.4 \text{ lb}}{\text{ft}^3}$

$\times \frac{453.6 \text{ g}}{\text{lb}} = 1.00 \times 10^3 \text{ g}$

1.9A $? \text{ kg/m}^2 = \frac{14.70 \text{ lb}}{\text{in.}^2} \times \left(\frac{1 \text{ in.}}{2.54 \text{ cm}}\right)^2 \times \left(\frac{\text{cm}}{10^{-2} \text{ m}}\right)^2 \times \frac{453.6 \text{ g}}{\text{lb}} \times \frac{\text{kg}}{10^3 \text{ g}} = \frac{1.034 \times 10^4 \text{ kg}}{\text{m}^2}$

1.9B $? \text{ m} = ((5 \text{ ft} \times \frac{12 \text{ in}}{\text{ft}}) + 10.5 \text{ in}) \times \frac{\text{m}}{39.37 \text{ in}} = 1.79 \text{ m}$

$? \text{ lb} = (1.79)^2 \times \frac{25.0 \text{ kg}}{\text{m}^2} \times \frac{2.205 \text{ lb}}{\text{kg}} = 177 \text{ lb}$

1.10A $V = \frac{4}{3}\pi r^3 = \frac{4}{3} \times 3.1416 \times \left(5.00 \text{ mm} \times \frac{10^{-3} \text{ m}}{\text{mm}} \times \frac{\text{cm}}{10^{-2} \text{ m}}\right)^3 = 0.524 \text{ cm}^3$

$d = \frac{m}{V} = \frac{4.085 \text{ g}}{0.524 \text{ cm}^3} = \frac{7.80 \text{ g}}{\text{cm}^3}$

1.10B $d = \frac{m}{V} = \frac{25.0 \text{ lb} \times \frac{\text{kg}}{2.205 \text{ lb}} \times \frac{10^3 \text{ g}}{\text{kg}}}{1.62 \text{ qt} \times \frac{\text{L}}{1.057 \text{ qt}} \times \frac{\text{mL}}{10^{-3} \text{ L}} \times \frac{\text{cm}^3}{\text{mL}}} = \frac{7.40 \text{ g}}{\text{cm}^3}$

1.11A $? \text{ gal} = 10.0 \text{ kg} \times \frac{10^3 \text{ g}}{\text{kg}} \times \frac{\text{mL}}{0.791 \text{ g}} \times \frac{10^{-3} \text{ L}}{\text{mL}} \times \frac{1.057 \text{ qt}}{1.00 \text{ L}} \times \frac{\text{gal}}{4 \text{ qt}} = 3.34 \text{ gal}$

1.11B $? \text{ kg} = 15.0 \text{ kg} \times \dfrac{10^3 \text{ g}}{\text{kg}} \times \dfrac{\text{mL}}{0.690 \text{ g}} \times \dfrac{0.789 \text{ g}}{\text{mL}} \times \dfrac{\text{kg}}{10^3 \text{ g}} = 17.2 \text{ kg}$

This simplifies to: $? \text{ kg} = 15.0 \text{ kg} \times \dfrac{\text{mL}}{0.690 \text{ g}} \times \dfrac{0.789 \text{ g}}{\text{mL}} = 17.2 \text{ kg}$

1.12A Water is about 1 g/mL

$20 \text{ qt} \times \dfrac{\text{L}}{1.06 \text{ qt}} \times \dfrac{\text{mL}}{10^{-3} \text{ L}} \times \dfrac{1 \text{ g}}{\text{mL}} \times \dfrac{\text{kg}}{10^3 \text{ g}}$ simplifies to $20 \text{ qt} \times \dfrac{\text{L}}{1.1 \text{ qt}} \times \dfrac{\text{kg}}{\text{L}}$

$\dfrac{20}{1.1}$ is closer to 20 than 15, so 20 kg is a reasonable answer.

1.12B (a) $? \text{ g balsa wood} = 1 \text{ ft}^3 \times \left(\dfrac{12 \text{ in}}{\text{ft}}\right)^3 \times \left(\dfrac{2.54 \text{ cm}}{\text{in}}\right)^3 \times \dfrac{0.11 \text{ g}}{\text{cm}^3}$

$= 3.1 \times 10^3 \text{ g balsa wood}$

(b) $? \text{ g water} = 2.00 \text{ L} \times \dfrac{\text{mL}}{10^{-3} \text{ L}} \times \dfrac{\text{cm}^3}{\text{mL}} \times \dfrac{1.00 \text{ g}}{\text{cm}^3} = 2.00 \times 10^3 \text{ g water}$

(c) ? g mercury = 1850 g mercury

(d) $? \text{ g hexane} = 1.50 \text{ L} \times \dfrac{\text{mL}}{10^{-3} \text{ L}} \times \dfrac{\text{cm}^3}{\text{mL}} \times \dfrac{0.66 \text{ g}}{\text{cm}^3} = 0.99 \times 10^3 \text{ g hexane}$

$? \text{ g water} = 1.00 \text{ L} \times \dfrac{\text{mL}}{10^{-3} \text{ L}} \times \dfrac{\text{cm}^3}{\text{mL}} \times \dfrac{1.00 \text{ g}}{\text{cm}^3} = 1.00 \times 10^3 \text{ g water}$

$? \text{ g mixture} = 0.99 \times 10^3 \text{ g hexane} + 1.00 \times 10^3 \text{ g water} = 1.99 \times 10^3 \text{ g mixture}$

(a) is the greatest mass.

1.13A If the measurement had been made at 21 °C, we would not have known the correct mass as we don't know the density at 21 °C.

1.13B mass of Saturn $= 95.2 \times 6.0 \times 10^{24} \text{ kg} \times \dfrac{10^3 \text{ g}}{\text{kg}} = 5.7 \times 10^{29} \text{ g}$

volume of Saturn $= \dfrac{4}{3}\pi r^3 = \dfrac{4}{3} \times 3.1416 \times (5.82 \times 10^4 \text{ km})^3 \times \left(\dfrac{10^3 \text{ m}}{\text{km}}\right)^3 \times \left(\dfrac{\text{cm}}{10^{-2} \text{ m}}\right)^3 = 8.26 \times 10^{29} \text{ cm}^3$

density of Saturn $= \dfrac{m}{V} = \dfrac{5.7 \times 10^{29} \text{ g}}{8.26 \times 10^{29} \text{ cm}^3} = 0.69 \text{ g/cm}^3$

1.14A The plastic piece would replace the ebony wood. The plastic piece floats on chloroform but sinks in water.

1.14B (a) The block weighs 75.0 g.

(b) The block sinks to the bottom, thus its density is greater than 0.789 g/mL.

(c) The block floats on the chloroform. The volume displaced is the volume of chloroform that weighs 75.0 g. The volume of chloroform is 50.6 mL. That is about 82 % of the block, so the block floats with 82% submerged.

$$? \text{ mL chloroform} = 75.0 \text{ g} \times \frac{\text{mL}}{1.483\text{g}} = 50.6 \text{ mL}$$

$$\% \text{ block} = \frac{\text{mL chloroform}}{\text{mL block}} = \frac{50.6 \text{ mL}}{61.5 \text{ mL}} = 0.82.$$

(d) The block displaces 61.5 mL of water or 61.5 g of water.

? g scale = mass block in air − mass water displaced = 75.0 g − 61.5 g = 13.5 g

The scale reads 13.5g.

Review Questions

1. Matter is something that has mass and takes up space. (d) heat and (f) music are not matter.

8. (c) is not a physical property. (a), (b), and (d) are physical properties.

9. (c) is a chemical change.

10. C, Cl, and Na are symbols representing elements. CO, $CaCl_2$, and KI are combinations of symbols for two elements (formulas) and represent compounds.

11. (a) is a homogeneous mixture or solution. (b) and (c) are pure substances.

16. (b) The first two zeroes are not significant.

20. $$? \text{ lb} = 1.0 \text{ lb nitrogen} \times \frac{100.0 \text{ lb fertilizer}}{20.0 \text{ lb nitrogen}} = 5.0 \text{ lb fertilizer (d)}$$

22. (a). $d = 1.0 \text{ g/cm}^3$

(b). $d = \dfrac{148.9\text{g}}{100.0\text{cm}^3} = 1.489 \text{ g/cm}^3$

(c). $d = \dfrac{7.72 \text{ g}}{10.0 \text{ cm}^3} = 0.772 \text{ g/cm}^3$

(d) $d = 0.83 \text{ g/cm}^3$

(b) has the greatest density.

Problems

23. (a), (b), and (d) are matter.

25. (a) physical (b) physical (c) chemical (d) physical

27. Helium and salt are substances. Lemon juice and wine are mixtures. Helium and salt are composed of only helium atoms and sodium and chloride ions, respectively. Lemon juice and wine are composed of a number of substances.

29. The person has lost both weight and mass. The person must lose mass to weigh less under the same force of gravity.

31. Yes. A set of measurements may contain a consistent error and thus be precise but inaccurate.
Yes. If the errors in individual measurements cancel each other, then the average may be accurate, although imprecise.

33. The mass measurement will be more consistent than the volume measurement. The mass in a certain volume will depend on how the flour is packed together. In earlier times, flour was sifted into the measurement container to provide a more consistent measure.

35. (a) $? \mu g = 8.01 \times 10^{-6}\ g \times \dfrac{1\ \mu g}{10^{-6}\ g} = 8.01\ \mu g$

(b) $? mL = 7.9 \times 10^{-3}\ L \times \dfrac{1\ mL}{10^{-3}\ L} = 7.9\ mL$

(c) $? km = 1.05 \times 10^{3}\ m \times \dfrac{1\ km}{10^{3}\ m} = 1.05\ km$

37. (a) $t_F = 1.8\ (98.6\ °C) + 32.0 = 209\ °F$

(b) $t_C = \dfrac{1069\ °F - 32}{1.8} = 576.1\ °C$

(c) $t_F = 1.8\ (-52\ °C) + 32 = -62\ °F$

39. $t_F = 1.8\ (156\ °C) + 32 = 313\ °F$
Yes, the indium will melt.

41. (a) $? L = 37.4\ mL \times \dfrac{10^{-3}\ L}{mL} = 3.74 \times 10^{-2}\ L$

(b) $? m = 1.55 \times 10^{2}\ km \times \dfrac{10^{3}\ m}{km} = 1.55 \times 10^{5}\ m$

(c) $? mg = 198\ mg \times \dfrac{10^{-3}\ g}{mg} = 0.198 g$

(d) $? \text{ mm}^2 = 1.19 \text{ m}^2 \times \left(\dfrac{\text{cm}}{10^{-2}\text{ m}}\right)^2 = 1.19 \times 10^4 \text{ cm}^2$

(e) $? \text{ ms} = 78\ \mu\text{s} \times \dfrac{10^{-6}\text{ s}}{\mu\text{s}} \times \dfrac{\text{ms}}{10^{-3}\text{ s}} = 0.078 \text{ ms}$

(f) $?\dfrac{\text{m}}{\text{s}} = \dfrac{88\text{ km}}{\text{h}} \times \dfrac{\text{h}}{3600\text{ s}} \times \dfrac{10^3\text{ m}}{\text{km}} = 24\dfrac{\text{m}}{\text{s}}$

43. (a). $? \text{ kg} = \left(\left(15.0\text{ lb} \times \dfrac{16\text{ oz}}{\text{lb}}\right) + 2\text{ oz}\right) \times \dfrac{28.35\text{ g}}{\text{oz}} \times \dfrac{\text{kg}}{10^3\text{ g}} = 6.86 \text{ kg}$

(b). $? \text{ cm} = 1.5 \text{ in} \times \dfrac{2.54\text{ cm}}{\text{in}} = 3.8 \text{ cm}$

$? \text{ cm} = 3.5 \text{ in} \times \dfrac{2.54\text{ cm}}{\text{in}} = 8.9 \text{ cm}$

$\dfrac{5}{8} = 0.625$ $\qquad ? \text{ cm} = 92.6 \text{ in} \times \dfrac{2.54\text{ cm}}{\text{in}} = 235 \text{ cm}$

(c) $? \text{ m}^3 = 16 \text{ in} \times 24 \text{ in} \times 42 \text{ in} \times \left(\dfrac{2.54\text{ cm}}{\text{in}}\right)^3 \times \left(\dfrac{10^{-2}\text{ m}}{\text{cm}}\right)^3 = 0.26 \text{ m}^3$

45. The yardstick (3 ft.) is shorter than the 3-ft 5-in. rattlesnake. The rattlesnake is 41 in. [(3 × 12) + 5], which is shorter than the chain, because 1 meter is about 40 in., so the chain is about 48 (40 × 1.2) in.. The rope is longest at 75 in.
(4) < (3) < (1) < (2)
yardstick < rattlesnake < chain < rope

47. $? \text{ in.} = 1 \text{ link} \times \dfrac{\text{chain}}{100\text{ links}} \times \dfrac{\text{furlong}}{10\text{ chain}} \times \dfrac{\text{mi}}{8\text{ furlongs}} \times \dfrac{5280\text{ ft}}{\text{mi}} \times \dfrac{12\text{ in.}}{\text{ft}} = 7.92 \text{ in.}$

49. (a) 3 (b) 3 (c) 4 (d) 4 (e) 3 (f) 4

51. (a) 2.804×10^3 m (b) 9.01×10^2 s
(c) 9.0×10^{-4} cm (d) 2.210×10^2 s

53. (a) 505.5 m (b) 2120, zero not significant
(c) 0.00610 (d) 40000 mL, last two zeros are not significant

55. (a)

	36.5 m
	−0.132 m
	9.44 m
	45.808 m => 45.8 m

(b)

	154.44 cm
	− 93 cm
	−105.1 cm
	166.54 cm => 167 cm

(c)
$$\begin{array}{rl} 4.61 & \text{g} \\ 39.9 & \text{g} \\ -0.0220 & \text{g} \\ \hline 44.488 & \text{g} \Rightarrow 44.5\ \text{g} \end{array}$$

(d)
$$\begin{array}{rl} 12.52 & \text{L} \\ +1.5 & \text{L} \\ -3.18 & \text{L} \\ -0.72 & \text{L} \\ \hline 10.12 & \text{L} \Rightarrow 10.1\ \text{L} \end{array}$$

57. (a) $73.0\ \text{mm} \times 1.340\ \text{mm} \times (25.31 - 1.6)\ \text{mm} = 73.0\ \text{mm} \times 1.340\ \text{mm} \times 23.7\ \text{mm}$
$= 2318\ \text{mm}^3 \Rightarrow 2.32 \times 10^3\ \text{mm}^3$

(b) $$\frac{33.58\,\text{cm} \times 1.007\,\text{cm}}{0.00705\,\text{g}} = 4796.4624\ \frac{\text{cm}^2}{\text{g}} \Rightarrow 4.80 \times 10^3 \frac{\text{cm}^2}{\text{g}}$$

(c) $$\frac{418.7\,\text{mm} \times 31.8\,\text{mm}}{(19.27\,\text{mg} - 18.98\,\text{mg})} = \frac{418.7\,\text{mm} \times 31.8\,\text{mm}}{(0.29\,\text{mg})} = 45913\ \frac{\text{mm}^2}{\text{mg}}$$

$$\Rightarrow 4.6 \times 10^4 \frac{\text{mm}^2}{\text{mg}}$$

(d) $$\frac{2.023\,\text{g} - (1.8 \times 10^{-3}\ \text{g})}{1.05 \times 10^4\ \text{mL}} = \frac{2.021\,\text{g}}{1.05 \times 10^4\ \text{mL}} = 1.92495 \times 10^{-4} \frac{\text{g}}{\text{mL}}$$

$$\Rightarrow 1.92 \times 10^{-4} \frac{\text{g}}{\text{mL}}$$

59. $$d = \frac{m}{V} = \frac{57.0\ \text{g}}{50.0\ \text{mL}} = 1.14\ \text{g/mL}$$

61.
$$\begin{array}{rl} 18.432\ \text{g} & \text{metal + paper} \\ -\ 1.214\ \text{g} & \text{paper} \\ \hline 17.218\ \text{g} & \text{metal} \end{array}$$

$$d = \frac{m}{V} = \frac{17.218\ \text{g}}{3.29\ \text{cm}^3} = 5.23\ \text{g/cm}^3$$

63. $?\ \text{g} = (7.6\ \text{cm} \times 7.6\ \text{cm} \times 94\ \text{cm}) \times 0.11\ \text{g/cm}^3 = 6.0 \times 10^2\ \text{g}$

65. $$V = \frac{m}{d} = 5.79\ \text{mg} \times \frac{10^{-3}\ \text{g}}{\text{mg}} \times \frac{\text{cm}^3}{19.3\,\text{g}} = 3.00 \times 10^{-4}\ \text{cm}^3$$

$$l = \frac{V}{A} = \frac{3.00\text{x}10^{-4}\ \text{cm}^3}{44.6\,\text{cm}^2} = 6.73 \times 10^{-6}\ \text{cm}$$

67. $$d = \frac{m}{V} = \frac{3.2\ \text{kg}}{0.80\ \text{m x } 0.80\ \text{m x } 1.20\ \text{m}} \times \frac{10^3\,\text{g}}{\text{kg}} \times \left(\frac{10^{-2}\ \text{m}}{\text{cm}}\right)^3 = \frac{3.2\text{x}10^3\ \text{g}}{7.68\text{x}10^5\ \text{cm}^3}$$

$$= 4.2 \times 10^{-3}\ \text{g/cm}^3$$

69. $15\text{ gal} \times \frac{4\text{ qt}}{\text{gal}} \times \frac{\text{L}}{1.06\text{ qt}} \times \frac{\text{mL}}{10^{-3}\text{ L}} \times \frac{1\text{ g}}{\text{mL}} \times \frac{\text{lb}}{454\text{ g}} = 125\text{ lb water}$

$3.0\text{ L} \times \frac{\text{mL}}{10^{-3}\text{ L}} \times \frac{\text{cm}^3}{\text{mL}} \times \frac{13.6\text{ g}}{\text{cm}^3} \times \frac{\text{lb}}{454\text{ g}} = 90\text{ lb mercury}$

(2) The water is the heaviest and thus, the most difficult to lift.

71. $d = \frac{m}{V} = \frac{30.0\text{ g}}{(22.2 - 18.0)\text{ mL}} = 7.143\text{ g/mL} \Rightarrow 7.1\text{ g/mL}$

Additional Problems

74. $t_C = \frac{t_F - 32}{1.8}$ for $t_C = t_F$

$t_C = \frac{t_C - 32}{1.8}$

$1.8\ t_C = t_C - 32$

$0.8\ t_C = -32$

$t_C = \frac{-32}{0.8} = -40°$

They are the same at only one temperature because the degree sizes are different.

76. The meter stick, with 1000-mm markings, is actually 1005 mm long. The true poster area is $1.827\text{ m} \times 0.763\text{ m} = 1.39\text{ m}^2$. The measured size is $1.827\text{ m} \times \frac{1.005\text{ m}}{1.000\text{ m}} \times 0.762\text{ m} \times \frac{1.005\text{ m}}{1.000\text{ m}} = 1.836\text{ m} \times .766\text{ m} = 1.41\text{ m}^2$.

$1.41\text{ m}^2 - 1.39\text{ m}^2 = 0.02\text{ m}^2$ error.

79. $9\text{ d} \times \frac{24\text{ hr}}{\text{d}} \times \frac{3600\text{ s}}{\text{hr}} = 777600\text{ s}$

$3\text{ min} \times \frac{60\text{ s}}{\text{min}} = 180\text{ s}$

44 s

777824 s speed $= \frac{25102\text{ mi}}{777824\text{ s}} \times \frac{3600\text{ s}}{\text{hr}} = 116.18\text{ mi/hr}$

83. $?\frac{\text{cm}^3}{\text{nail}} = \pi r^2 l = \pi \times \left(\frac{0.300\text{ cm}}{2}\right)^2 \times 5.00\text{ cm} = \frac{0.353\text{ cm}^3}{\text{nail}}$

$?\frac{\text{g}}{\text{nail}} = Vd = 0.353\text{ cm}^3 \times \frac{2.698\text{ g}}{\text{cm}^3} = \frac{0.953\text{ g}}{\text{nail}}$

$$?\,\text{nails} = 225\,\text{kg} \times \frac{\text{nail}}{0.953\,\text{g}} \times \frac{10^3\,\text{g}}{\text{kg}} = 2.36 \times 10^5\ \text{nails}$$

86. $$?\ \text{mg/m}^2\ \text{hr} = \frac{10\ \text{tons}}{\text{mi}^2\ \text{month}} \times \left(\frac{\text{mi}}{5280\ \text{ft}}\right)^2 \times \left(\frac{\text{ft}}{12\ \text{in.}}\right)^2 \times \left(\frac{\text{in.}}{2.54\ \text{cm}}\right)^2 \times \left(\frac{\text{cm}}{10^{-2}\ \text{m}}\right)^2$$
$$\times \left(\frac{2000\ \text{lb}}{\text{tons}}\right) \times \left(\frac{454\ \text{g}}{\text{lb}}\right) \times \left(\frac{\text{mg}}{10^{-3}\ \text{g}}\right) \times \left(\frac{\text{month}}{30\ \text{days}}\right) \times \left(\frac{\text{day}}{24\ \text{hr}}\right) = 4.9\ \text{mg/m}^2\ \text{hr}$$

Density is not needed.

88. $V_{\text{brass}} = (2.0\ \text{cm})^3 = 8.0\ \text{cm}^3$

The brass cube will sink to the bottom, displacing 8.0 cm^3, or 8.0 mL of water.

$V_{\text{cork}} = 5.0\ \text{cm} \times 4.0\ \text{cm} \times 2.0\ \text{cm} = 40.\ \text{cm}^3$

$$m_{\text{cork}} = dV = \frac{0.22\ \text{g}}{\text{cm}^3} \times 40.\ \text{cm}^3 = 8.8\ \text{g}$$

The cork will displace 8.8 g of water, which is about 8.8 mL of water. More water will overflow from the vessel of water in which the cork is floated (right).

92. (a) $$\text{microballoon volume} = 1.00\text{L} \times \frac{68\ \text{L}}{100\ \text{L}} \times \frac{\text{mL}}{10^{-3}\ \text{L}} \times \frac{\text{cm}^3}{\text{mL}} = 6.8 \times 10^2\ \text{cm}^3$$

$$?\frac{\text{volume}}{\text{microballoon}} = \frac{4}{3}\pi r^3 = \frac{4}{3} \times 3.1416 \times \left(\frac{0.070\ \text{mm}}{2}\right)^3 \times \left(\frac{10^{-3}\ \text{m}}{\text{mm}}\right)^3$$
$$\times \left(\frac{\text{cm}}{10^{-2}\ \text{m}}\right)^3 = 1.8 \times 10^{-7}\ \text{cm}^3$$

$$?\ \text{microballoons} = 6.8 \times 10^2\ \text{cm}^3 \times \frac{\text{microballoon}}{1.8 \times 10^{-7}\ \text{cm}^3} = 3.8 \times 10^9\ \text{microballoons}$$

$$?\frac{\text{g}}{\text{microballoons}} = \frac{62\ \text{g}}{3.8 \times 10^9\ \text{microballoons}} = \frac{1.6 \times 10^{-8}\ \text{g}}{\text{microballoon}}$$

surface area = $4\pi r^2$

$$= 4 \times 3.1416 \left(\frac{0.070\ \text{mm}}{2}\right)^2 \times \left(\frac{10^{-3}\ \text{m}}{\text{mm}}\right)^2 \times \left(\frac{\text{cm}}{10^{-2}\ \text{m}}\right)^2 = 1.5 \times 10^{-4}\ \text{cm}^2$$

$$\text{volume of glass} = \frac{1.6 \times 10^{-8}\ \text{g}}{\text{microballoon}} \times \frac{\text{cm}^3}{2.2\ \text{g}} = 7.3 \times 10^{-9}\ \text{cm}^3$$

$$\text{thickness} = \frac{\text{volume}}{\text{surface area}} = \frac{7.3 \times 10^{-9}\ \text{cm}^3}{1.5 \times 10^{-4}\ \text{cm}^2} = 4.8 \times 10^{-5}\ \text{cm}$$

(b) mass glue $= \frac{32\text{ L}}{100\text{ L}} \times 1.00\text{ L} \times \frac{\text{mL}}{10^{-3}\text{ L}} \times \frac{1.67\text{ g}}{\text{mL}} = 5.3 \times 10^2$

total mass = mass glue + mass microballoon $= 5.3 \times 10^2\ \text{g} + 62\ \text{g} = 5.9 \times 10^2\ \text{g}$

density $= \frac{\text{mass}}{\text{volume}} = \frac{5.9\times10^2\text{ g}}{1.00\text{ L}} \times \frac{10^{-3}\text{ L}}{\text{mL}} = \frac{0.59\text{ g}}{\text{mL}}$

93. (a) density of object $= \frac{5.15\text{ kg}}{30.2\text{ cm}\times12.9\text{ cm}\times11.5\text{ cm} - \pi(2.5\text{ cm})^2\times11.5\text{ cm}}$

$$\times\left(\frac{10^3\text{ g}}{\text{kg}}\right) = \frac{5150\text{ g}}{4.25\text{x}10^3\text{ cm}^3} = 1.21\text{ g/cm}^3$$

by estimation $\frac{5\times1000\text{ g}}{30\text{ cm}\times13\text{ cm}\times12\text{ cm} - \pi\times(3\text{ cm})^2\times12\text{ cm}}$

$$= \frac{5000\text{ g}}{4680\text{ cm}^3 - 324\text{ cm}^3} = \frac{5000\text{ g}}{4356\text{ cm}^3}$$

The density is greater than 1 g/cm^3. The object will not float.

(b) ? g of balsa $= \pi\ (2.5\text{ cm})^2 \times 11.5\text{ cm} \times \frac{0.11\text{ g}}{\text{cm}^3} = 25$ g balsa

density of object $= \frac{5150\text{g} + 25\text{g}}{30.2\text{ cm}\times12.9\text{ cm}\times11.5\text{ cm}} = \frac{5175\text{ g}}{4.48\times10^3\text{ cm}^3} = 1.16\text{ g/cm}^3$

The object still will not float.

(c) Weight of material of object removed:

? g $= \pi\ (2.5\text{ cm})^2 \times 11.5\text{ cm} \times 1.21\text{ g/cm}^3 = 2.7 \times 10^2$ g

Weight loss when replaced by balsa:

? g $= 2.7 \times 10^2\ \text{g} - 25\ \text{g} = 2.4 \times 10^2\ \text{g}$

Object must weigh less than 4.48×10^3 g to float.

? g = 5150 g – 4480 g = 670 g

? holes $= 670\text{ g} \times \left(\frac{\text{hole}}{250\text{ g}}\right) = 2.7$ holes

At least three holes must be drilled and filled with balsa.

96. Let X be the mL of chloroform.

(1.483 g/mL) × X + (2.890 g/mL) × (100.0 mL - X) = 100.0 mL × 1.950 g/mL

1.483 g/mL X + 289.0 g – 2.890 g/mL X =195.0 g

94.0 g = 1.407 g/mL X

X = 66.8 mL

66.8 mL of chloroform and 33.2 mL of bromoform

Apply Your Knowledge

100. The Antarctic ice-cap melting will raise the mean sea level. The ice is continental ice, and its melting constitutes the net addition of water to the oceans.
The Arctic ice cap floats on the seawater. The water produced by the complete melting of a chunk of this ice occupies the same volume as the submerged portion of the original chunk of ice. Sea level remains unchanged. (Think of ice cubes floating in a glass of water with the tops of the cubes above the rim of the glass. The water does not overflow the glass as the ice melts.)

101. $?\frac{\text{mg}}{\text{L}} = \frac{10\,\mu\text{g}}{\text{dL}} \times \frac{\text{dL}}{10^{-1}\,\text{L}} \times \frac{10^{-6}\,\text{g}}{\mu\text{g}} \times \frac{\text{mg}}{10^{-3}\,\text{g}} = 0.10\frac{\text{mg}}{\text{L}}$

$?\frac{\text{mg}}{\text{L}} = \frac{10\,\text{mg}}{10\,\text{L}} = 1\frac{\text{mg}}{\text{L}}$

They are not the same. They differ by a factor of 10. The reporter may have used the conversion for cL instead of dL, but probably he used mg for μg and daL (10 L) for dL (0.1 L). That would make a factor of 10 difference.

102. (a) ? g = 42.062 g – 32.105 g = 9.957 g

$\text{V pyc} = 9.957\text{ g} \times \frac{\text{mL}}{0.9982\text{ g}} = 9.975\text{ mL}$

? g = 40.873 g – 32.105 g = 8.768 g

$d = \frac{m}{V} = \frac{8.768\text{ g}}{9.975\text{ mL}} = 0.8790\text{ g/mL}$

(b) ? g = 38.1055 g – 36.2142 g = 1.8913 g Zn mass Zn

$?\text{ mL} = (46.1894\text{ g} - 36.2142\text{ g}) \times \frac{\text{mL}}{0.99823\text{ g}} = 9.9929\text{ mL } H_2O$ in pyc without zinc

$?\text{ mL} = (47.8161\text{ g} - 1.8913\text{ g Zn} - 36.2142\text{ g pyc}) \times \frac{\text{mL}}{0.99823\text{ g}} = 9.7278\text{ mL}$ H_2O in pyc with zinc

? mL = 9.9929 mL – 9.7278 mL = 0.2651 mL Zn

$d = \frac{m}{V} = \frac{1.8913\text{ g}}{0.2651\text{ mL}} = 7.134\text{ g/mL}$

Chapter 2

Atoms, Molecules, and Ions

Exercises

2.1A The bulb will still weigh 45.07 g. All of the reactants and products are sealed inside the bulb; nothing can escape or be added.

2.1B The mass would be less as mass is carried away as smoke.

2.2A $? \text{ g magnesium oxide} = 1.500 \text{ g oxygen} \times \left(\frac{3.317 \text{ g magnesium oxide}}{1.317 \text{ g oxygen}}\right)$

$= 3.778 \text{ g magnesium oxide}$

2.2B $? \text{ g magnesium oxide} = 1.554 \text{ g oxygen} \times \left(\frac{3.317 \text{ g magnesium oxide}}{1.317 \text{ g oxygen}}\right)$

$= 3.914 \text{ g magnesium oxide}$

$? \text{ g magnesium} = 3.914 \text{ g magnesium oxide} - 1.554 \text{ g oxygen} = 2.360 \text{ g magnesium}$

2.3A $A = Z + 66 = 50 + 66 = 116$ $\quad {}^{116}_{50}\text{Sn}$

2.3B Cadmium-116 has 48 protons and 48 electrons

$? \text{ neutrons} = A - Z = 116 - 48 = 68 \text{ neutrons}$ $\quad {}^{116}_{48}\text{Cd}$

2.4A
$19.99244 \times 0.9051 = 18.095$
$20.99395 \times 0.0027 = 0.057$
$21.99138 \times 0.0922 = 2.028$
$20.180 \Rightarrow 20.18$

2.4B X = fractional abundance of copper 63

$X \times 62.9298 + (1.000 - X) \times 64.9278 = 63.546$

$62.9298\, X + 64.9278 - 64.9278\, X = 63.546$

$X = \frac{1.382}{1.9980} = 0.6917$

69.17% copper 63

30.83% copper 65

2.5A The most abundant isotope is ^{24}Mg. Since the average atomic mass is 24.3050, the atomic mass is closer to the mass of magnesium 24.
It is difficult to determine the second most abundant unless we know the % abundance of the magnesium 24. There are many combinations of the three percentages that would work. Since both ^{25}Mg and ^{26}Mg are heavier than the

average atomic mass, it is not obvious which is second most abundant. A study of isotopes indicates that the percentage usually decreases the farther away from the average atomic mass [implying ^{25}Mg] but that isotopes with an odd number of neutrons often have a lesser percent abundance [implying ^{26}Mg].
The actual percentages are ^{24}Mg 78.70%, ^{25}Mg 10.13%, and ^{26}Mg 11.17%.

2.5B (a) No. Even if the next lightest isotope were 40.00%, the results would still be heavier than 24.3050 u.
$0.6000 \times 23.98504\ u = 14.391\ u$
$0.4000 \times 24.98584\ u = \underline{9.994\ u}$
$24.385\ u$
(b) For three isotopes, set magnesium-26 at 0.00% and magnesium-24 = X
$23.98504 \times X + 24.98584 \times (1.0000 - X) = 24.3050$
$0.6808 = 1.0008\ X$
$X = 0.6803$
The smallest possible % of magnesium-24 is 68.03%

2.6A N_2F_4 dinitrogen tetrafluoride

2.6B S_8O octasulfur oxide

2.7A (a) P_4O_{10}
(b) heptasulfur dioxide

2.7B SO_2F_2

2.8A (a) K_2S (b) Li_2O (c) AlF_3

2.8B (a) Cr_2O_3 (b) FeS (c) Li_3N

2.9A (a) calcium bromide
(b) lithium sulfide
(c) iron(II) bromide
(d) copper(I) iodide

2.9B (a) copper(I) sulfide Cu_2S
(b) cobalt(III) oxide Co_2O_3
(c) magnesium nitride Mg_3N_2

2.10A (a) $(NH_4)_2CO_3$ (b) $Ca(ClO)_2$ (c) $Cr_2(SO_4)_3$

2.10B (a) $KAl(SO_4)_2$ (b) $MgNH_3PO_4$

2.11A (a) potassium hydrogen carbonate or potassium bicarbonate
(b) iron(III) phosphate

(c) magnesium dihydrogen phosphate

2.11B(a) sodium selenate
(b) iron(III) arsenide
(c) sodium hydrogen phosphite

Review Questions

1. (10.0 g Zn + 8.0 g S) – 14.9 g ZnS = 3.1 g S left over (b)

3. No. The chlorine atoms must go somewhere, but carbon dioxide and water do not contain chlorine, so the report cannot be correct.

7. (a) isobars (b) isotopes (c) identical atoms (d) different elements
(e) different elements

8. (a) $^{8}_{5}B$ (b) $^{14}_{6}C$ (c) $^{235}_{92}U$ (d) $^{60}_{27}Co$

12. (a) Magnesium is an element.
(b) Methyl is a prefix, meaning a CH_3 group.
(c) Chloride is a negatively charged atom of chlorine, the ion Cl^-.
(d) Ammonia is a compound, NH_3.
(e) Ammonium is an ion, NH_4^+.
(f) Ethane is a compound, CH_3CH_3.
The only substances that could be found on a stockroom shelf are (a), (d), and (f).

19. (a) Usually it is organic if it contains C, but there are a few exceptions, so a molecular formula is usually sufficient.
(b) It contains only H and C, so a molecular formula is sufficient.
(c) An alcohol requires a structural formula.
(d) It contains only H and C, and the ratio is 2n + 2 hydrogens for each carbon, so a molecular formula is sufficient.
(e) A carboxylic acid requires a structural formula.

Problems

21. $Zn + S \rightarrow ZnS$
Before reaction: 1.000 g Zn + 0.600 g S = 1.600 g
After reaction: 1.490 g ZnS + 0.110 g Zn = 1.600 g
The mass of substances after the reaction equals the mass of substances before the reaction. The law of conservation of mass is confirmed.

23. (a) 104.3 g before and after reaction.
(b) 40.3 g magnesium oxide
(c) Law of conservation of mass, and law of definite proportions
(d) 80.6 g magnesium oxide is formed. There is twice the amount of

magnesium to react, as well as excess oxygen (more than 32.0 g).

25. $\frac{0.625\text{ g}}{1.000\text{ g}} \times 100\% = 62.5\%\text{ C}$ $\frac{0.0419\text{ g}}{1.000\text{g}} \times 100\% = 4.19\%\text{ H}$
$\frac{0.968\text{ g}}{1.549\text{ g}} \times 100\% = 62.5\%\text{ C}$ $\frac{0.0649\text{ g}}{1.549\text{ g}} \times 100\% = 4.19\%\text{ H}$
$\frac{1.734\text{ g}}{2.774\text{ g}} \times 100\% = 62.51\%\text{ C}$ $\frac{0.116\text{ g}}{2.774\text{ g}} \times 100\% = 4.18\%\text{ H}$
Yes, each sample has the same percent carbon and percent hydrogen.

27. $?\text{ g }SO_2 = 1.305\text{ g S} \times \frac{0.623\text{ g }SO_2}{0.312\text{ g S}} = 2.61\text{ g }SO_2$ produced

29. Using the $\frac{1.142\text{ g oxygen}}{1.000\text{ g nitrogen}}$ ratio but changing the relative number of atoms produces new ratios.
Twice the number of oxygen $\frac{2.284\text{ g oxygen}}{1.000\text{ g nitrogen}}$; (c) is possible.
Twice the number of nitrogen $\frac{1.142\text{ g oxygen}}{2.000\text{ g nitrogen}} = \frac{0.571\text{ g oxygen}}{1.000\text{ g nitrogen}}$; (a) is possible.
(b) and (d) are not possible; no combination of atoms would produce those ratios.

31.

	Protons	Neutrons	Electrons
(a) $^{64}_{30}Zn$	30	34	30
(b) $^{109}_{47}Ag$	47	62	47
(c) $^{14}_{6}C$	6	8	6
(d) $^{243}_{95}Am$	95	148	95

33. #2 $^{40}_{20}Ca$ #6 $^{48}_{22}Ti$ #7 $^{48}_{20}Ca$ #2 and #7 are isotopes, they have the same number of protons but different number of neutrons.

35. There are multiple isotopes whose masses average the atomic mass of 79.904 u.

37. 68.926 × 0.601 = 41.42
70.925 × 0.399 = 28.30
69.72 => 69.7 u

39. 27.97693 × 0.9221 = 25.798
28.97649 × 0.0470 = 1.362
29.97376 × 0.0309 = 0.926
28.086 => 28.09 u

41. $84.91179\,X + (1.000 - X) \times 86.90919 = 85.4678$
$84.91179\,X + 86.90919 - 86.90919\,X = 85.4678$
$X = \dfrac{1.4414}{1.99740} = 0.72164$
72.164% rubidium 85   27.836% rubidium 87

43.

	Group	Period	Type
(a) Ga	3A	4	metal
(b) P	5A	3	nonmetal
(c) I	7A	5	nonmetal
(d) Ra	2A	7	metal
(e) Li	1A	2	metal
(f) La	3B	6	metal
(h) Xe	8A	5	nonmetal

45. (a) N_2 (b) H_2 (c) I_2 (d) P_4

47. $BrCl_3$, NH_3, and HCl are binary molecular compounds. They have two elements and are molecular compounds. MgF_2 is an ionic compound. HIO is a ternary molecular compound.

49. (a) $SiCl_4$
(b) SF_6
(c) CS_2
(d) Cl_2O_3
(e) dinitrogen oxide
(f) diiodine pentoxide
(g) phosphorus pentachloride
(h) diphosphorus trioxide

51. (a) Mg^{2+}
(b) Br^-
(c) Ti^{4+}
(d) iodide ion
(e) sulfide ion
(f) chromium(II) ion

53. (a) ammonium ion
(b) hypochlorite ion
(c) phosphate ion
(d) hydrogen sulfate ion or bisulfate ion
(e) OH^-
(f) SO_3^{2-}
(g) $C_2H_3O_2^-$
(h) CO_3^{2-}

55. (a) potassium iodide
(b) magnesium fluoride
(c) nickel(II) sulfate
(d) titanium(IV) bromide
(e) ammonium hydrogen sulfate or ammonium bisulfate
(f) aluminum oxide
(g) barium chloride dihydrate
(h) lithium oxalate
(i) sodium sulfate decahydrate

57. (a) KCl
    (b) $CaCO_3$
    (c) Cr_2O_3
    (d) $KClO_4$
    (e) $NaClO_3$
    (f) $FeSO_4 \cdot 7H_2O$

59. (a) Na_2O
    (b) CuOH
    (c) BaI_2
    (d) $Al_2(SO_4)_3$
    (e) sodium bromate
    (f) lithium hydrogen sulfate or lithium bisulfate
    (g) ammonium dichromate
    (h) nickel oxalate

61. (a) Ammonium chlorate is NH_4ClO_3; NH_4Cl is ammonium chloride
    (b) Potassium nitrate is KNO_3; KNO_2 is potassium nitrite.
    (c) Sodium sulfate is Na_2SO_4; $NaSO_4$ is not a correct formula.
    (d) Barium hydroxide is $Ba(OH)_2$; BaOH is not a correct formula.
    (e) Zinc oxalate is ZnC_2O_4; ZnO is zinc oxide.
    (f) Manganese(IV) oxide is MnO_2, not MnO_4. MnO_4^- is the permanganate ion.
    (g) Strontium chromate is $SrCrO_4$; $Sr(CrO_4)_2$ is not a correct formula.
    (h) Copper(II) phosphate is $Cu_3(PO_4)_2$; Cu_2PO_4 is not a correct formula.

63. (a) hydrobromic acid
    (b) nitrous acid
    (c) phosphorous acid HNO_2
    (d) $HClO_2$
    (e) KOH
    (f) H_2SO_3
    (g) barium hydroxide
    (h) HI

65. IO_3^- is the iodate ion. HIO_3 is iodic acid.

67.

(a)
$$H-\overset{\overset{\displaystyle H}{|}}{\underset{\underset{\displaystyle H}{|}}{C}}-\overset{\overset{\displaystyle H}{|}}{\underset{\underset{\displaystyle H}{|}}{C}}-\overset{\overset{\displaystyle H}{|}}{\underset{\underset{\displaystyle H}{|}}{C}}-\overset{\overset{\displaystyle H}{|}}{\underset{\underset{\displaystyle H}{|}}{C}}-\overset{\overset{\displaystyle H}{|}}{\underset{\underset{\displaystyle H}{|}}{C}}-H$$
$CH_3(CH_2)_3CH_3$

(b)
$$H-\overset{\overset{\displaystyle H}{|}}{\underset{\underset{\displaystyle H}{|}}{C}}-\overset{\overset{\displaystyle H}{|}}{\underset{\underset{\displaystyle H}{|}}{C}}-\overset{\overset{\displaystyle H}{|}}{\underset{\underset{\displaystyle H}{|}}{C}}-\overset{\overset{\displaystyle O}{\|}}{C}-OH$$
$CH_3(CH_2)_2COOH$

(c)
$$H-O-\overset{\overset{\displaystyle H}{|}}{\underset{\underset{\displaystyle H}{|}}{C}}-\overset{\overset{\displaystyle H}{|}}{\underset{\underset{\displaystyle H}{|}}{C}}-\overset{\overset{\displaystyle H}{|}}{\underset{\underset{\displaystyle H}{|}}{C}}-H$$
$HO(CH_2)_2CH_3$

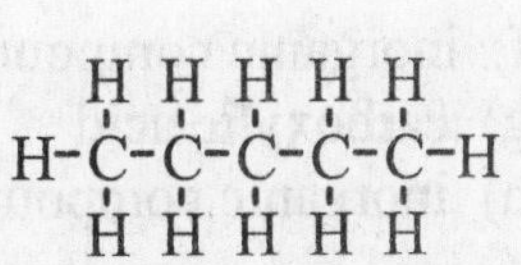

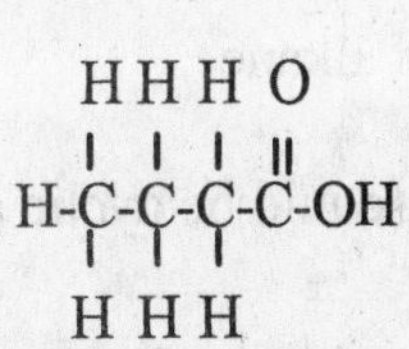

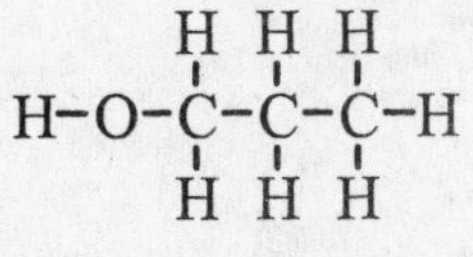

(e) 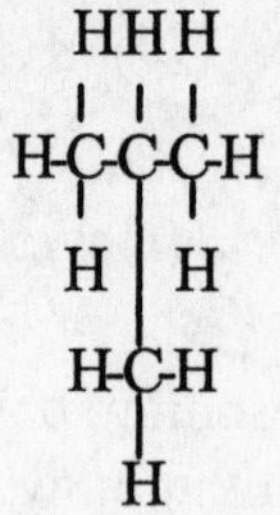

(d) 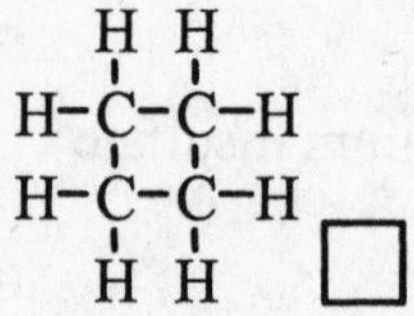

$CH_3CH(CH_3)CH_3$ or $CH(CH_3)_3$

(f)

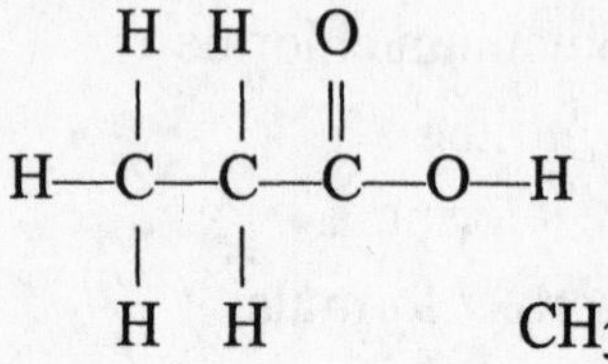

CH_3CH_2COOH could also be called propanoic acid

69. (a)

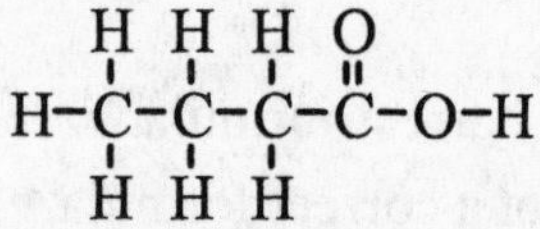

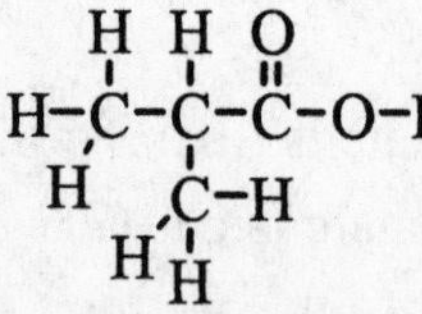

(b) $CH_3CH_2CH_2COOH$ butyric acid

$CH_3CH(CH_3)COOH$ isobutyric acid

(c)

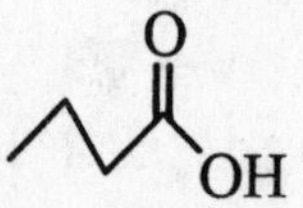

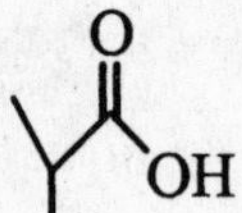

(d) They are isomers.

71. (a) straight-chain alkane
(b) alcohol
(c) hydrocarbon (could be cyclic alkane)
(d) hydrocarbon
(e) carboxylic acid
(f) inorganic compound
(g) carboxylic acid
(h) inorganic compound

73. b: 2,2,4-trimethylpentane

75. (a) dimethyl ether — methyl alcohol (methanol)
(b) ethyl methyl ether — methyl alcohol and ethyl alcohol (methanol and ethanol)
(c) diethyl ether — ethyl alcohol (ethanol)

77.

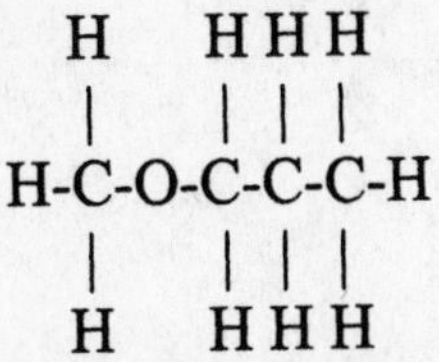

methyl propyl ether

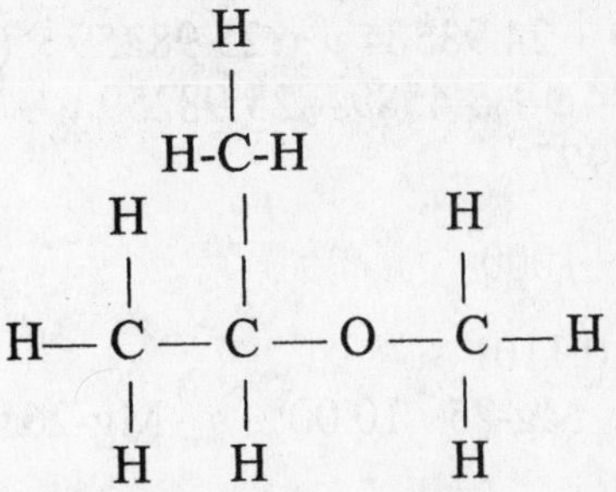

isopropyl methyl ether

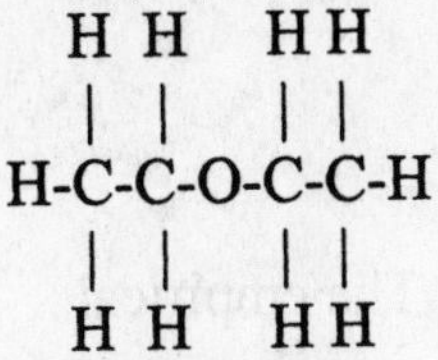

diethyl ether

Additional Problems

79. (a)

(b)

(c)

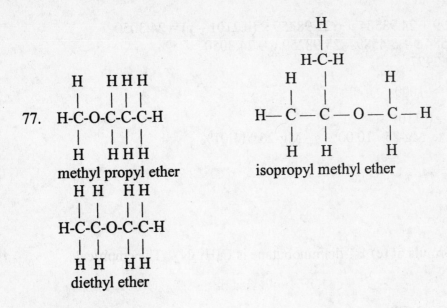

82. (c) Magnesia is produced from a metal strip and oxygen so magnesia, cannot be an element.

86. H_2 is the most likely ($(0.9999)^2 \times 100\% = 99.98\%$), HD is next most likely ($0.9999 \times 0.0001 \times 100\% = 0.009999\%$), and D_2 is the least likely ($(0.0001)^2 \times 100\% = 1 \times 10^{-6}\%$).

89.

	ratio C : sample	ratio H : sample
sample 1	$\frac{0.1141\ g}{0.2450\ g} = 0.4657$	$\frac{0.0216\ g}{0.2450\ g} = 0.0882$
sample 2	$\frac{0.1400\ g}{0.3005\ g} = 0.4659$	$\frac{0.0264\ g}{0.3005\ g} = 0.0879$
sample 3	$\frac{0.0639\ g}{0.1371\ g} = 0.466$	$\frac{0.0121\ g}{0.1371 g} = 0.0883$

The composition of carbon and hydrogen is constant, and the composition of the total oxygen and nitrogen together is consistent, but it is impossible to determine if the ratio of oxygen to nitrogen changes.

92. N_2O_5 would have given a different ratio of mass of oxygen to mass of nitrogen, and thus a different compound produces a different proportion. N_2O_4 would have been the same proportion as NO_2 and would have been difficult for Dalton to explain.

94. y = fractional abundance of Mg 25

$23.98504 \times 0.7899 + 24.98584\,y + 25.98259 \times (0.2101 - y) = 24.3050$
$18.9458 + 24.98584\,y + 5.4589 - 25.98259\,y = 24.3050$
$-0.99675\,y = -0.0997$

$$y = \frac{-0.0997}{-0.99675} = .1000$$

$0.2101 - 0.1000 = 0.1101$
Mg-24 78.99% Mg-25 10.00% Mg-26 11.01%

98. (a) $C_nH_{2n+1}OH$
(b) $C_nH_{2n}O_2$
(c) $C_nH_{2n}O_2$

103. The molecular formula of (e) 1,4-diaminobutane is $C_4H_{12}N_2$. The empirical formula is C_2H_6N.

105. (a) $CH_3(CH_2)_{10}COONH_4$
(b) $(CH_3(CH_2)_{16}COO)_2Ca$
(c) $CH_3(CH_2)_7CH{=}CH(CH_2)_7COOK$

107. (a) methyl acetate from acetic acid and methyl alcohol
(b) propyl acetate from acetic acid and propyl alcohol
(a) isopropyl formate from formic acid and isopropyl alcohol

110.

$CH_3CH_2CH_2CH_2CH_2CH_3$	n-hexane
$CH_3CH(CH_3)CH_2CH_2CH_3$	2-methylpentane
$CH_3CH_2CH(CH_3)CH_2CH_3$	3-methylpentane
$CH_3C(CH_3)_2\ CH_2CH_3$	2,2-dimethylbutane
$CH_3CH(CH_3)CH(CH_3)CH_3$	2,3-dimethylbutane

111. (a) The atomic number is 20, that is, calcium. The mass number is 40, so the isotope is ${}^{40}_{20}$Ca.
(b) $Z + 1.6\,Z = 234$
$2.6\,Z = 234$
$Z = 90$ ${}^{234}_{90}$Th
(c) $A_1 = 16 \times 7.50 = 120$
$A_2 = 4 \times 12 = 48$
$Z_1 = 2\,Z_2$
$A_1 - Z_1 = 3(A_2 - Z_2)$
$120 - Z_1 = 3(48 - Z_2) = 144 - 3\,Z_2$
$120 - 2\,Z_2 = 144 - 3\,Z_2$
$Z_2 = 24$ $Z_1 = 48$

The isotope is ${}^{120}_{48}Cd$. (Isotope 2 is ${}^{48}_{24}Cr$.)

(d) A = (Z + n) × 2
Z = 10, 12, 14, 16, 18, 20, 28, 30, 32, 34, 36, 66, 68, 70, 72, 104, 106, 108, 170, 172
$\#e^- = Z - n$ = 2, 10, 18, 36, 54, 86, 12, 20, 38, 56, 88, 28, 46, 64, 96, 54, 72, 104, 90, 122, 140
for Z 10, 18, 36 are noble gases and could not be an ion.
12, 14, 16, 20 have A = 2Z not 2(Z + n)
170, 172, 104, 106, 108 are too large for common ions.
A is too large for 66, 68, 70, 72.
The only Z – n that is close to 28, 30, 32, or 34 is 28, or if Z – n = 28, then Z can be 30(n = 2) or 32(n = 4). If Z = 30, A = 2 × 32 = 64. One choice is ${}^{64}_{30}Zn^{2+}$. If Z = 32, A = 2 × 36 = 72. Another choice is ${}^{72}_{32}Ge^{4+}$. The 4+ ion is probably less likely than a 2+ ion.

114. (a) 1-propyl bromide
(b) isobutyl chloride
(c) ethyl iodide
(d) tert-butyl fluoride or 1,1-dimethyl ethyl fluoride

Apply Your Knowledge

116. (a) The naturally occurring mixture is slightly heavier than the single oxygen-16. When ratio measurements were made, the atomic masses of the physicists were very slightly larger.
(b) Carbon-12 with an atomic mass of 12.0000 is divisible by 4. The naturally occurring mixture defined as 16.000 is changed to 15.9994. The change is 0.0006 part in 15.9994 or about 0.4 part per 10,000.

120. The atomic weights are averages of the naturally occurring isotopes, most of which have isotopic masses that are very close to integral masses. The atomic mass of chlorine is an average of the two isotopes, mass 35 and mass 37, both of which have masses close to an integral value. Chlorine could be a combination of one atom of mass 37 for each three atoms of mass 35.

Chapter 3

Stoichiometry: Chemical Calculations

Exercises

3.1A (a) P_4 — $4 \times 30.97 = 123.88$ u => 124 u

(b) N_2O_4

N	$2 \times 14.01 =$	28.02
O	$4 \times 16.00 =$	64.00
		92.02 u => 92.0

(c) H_2SO_4

H	$2 \times 1.01 =$	2.02
S	$1 \times 32.07 =$	32.07
O	$4 \times 16.00 =$	64.00
		98.09 u => 98.1

(d) CH_3CH_2COOH

C	$3 \times 12.01 =$	36.03
H	$6 \times 1.01 =$	6.06
O	$2 \times 16.00 =$	32.00
		74.09 u => 74.1 u

3.1B (a) PCl_5

P	$1 \times 30.9738 =$	30.9738
Cl	$5 \times 35.4527 =$	177.2635
		208.2373 u => 208.24 u

(b) N_2O_5

N	$2 \times 14.0067 =$	28.0134
O	$5 \times 15.9994 =$	79.9970
		108.0104 u => 108.01 u

(c) $CH_3CH_2CH_2COOH$

C	$4 \times 12.011 =$	48.044
H	$8 \times 1.0079 =$	8.0632
O	$2 \times 15.9994 =$	31.9988
		88.1060 u => 88.106 u

(d) $CH_3COCH_2CH_2CH_2CH_3$

C	$6 \times 12.011 =$	72.066
H	$12 \times 1.00794 =$	12.0953
O	$1 \times 15.9994 =$	15.9994
		100.1607 u => 100.16 u

3.2A (a)

Li	$2 \times 6.94 =$	13.88
O	$1 \times 16.00 =$	16.00
		29.88 u => 29.9 u

(b)

Mg	$1 \times 24.3 =$	24.3
N	$2 \times 14.0 =$	28.0
O	$6 \times 16.0 =$	96.0
		148.3 u = 148

(c)

Ca	1×40	$= 40$
H	4×1.0	$= 4$
P	2×31	$= 62$
O	8×16	$= \underline{128}$
		234 u

(d)

K	2×39	$= 78$
Sb	1×122	$= 122$
F	5×19	$= \underline{95}$
		295 u

3.2B (a) $NaHSO_3$

Na	1×22.990	$= 22.990$
H	1×1.008	$= 1.008$
S	1×32.066	$= 32.066$
O	3×15.999	$= \underline{47.997}$
		104.061 u => 104.06 u

(b) NH_4ClO_4

N	1×14.007	$= 14.007$
H	4×1.008	$= 4.032$
Cl	1×35.453	$= 35.453$
O	4×15.999	$= \underline{63.996}$
		117.488 => 117.49 u

(c) $Cr_2(SO_4)_3$

Cr	2×51.996	$= 103.992$
S	3×32.066	$= 96.198$
O	12×15.999	$= \underline{191.988}$
		392.178 => 392.18 u

(d) $CuSO_4{\cdot}5H_2O$

Cu	1×63.546	$= 63.546$
S	1×32.066	$= 32.066$
O	9×15.9994	$= 143.995$
H	10×1.0079	$= \underline{10.079}$
		249.686 u => 249.69 u

3.3A (a) $? \text{ mg Ag} = 1.34 \times 10^{-4} \text{ mol Ag} \times \dfrac{107.9 \text{ g Ag}}{\text{mol Ag}} \times \dfrac{\text{mg}}{10^{-3} \text{ g}} = 14.5 \text{ mg Ag}$

(b) $? \text{ atoms} = 20.5 \text{ mol } O_2 \times \dfrac{6.022 \times 10^{23} \text{ molecules}}{\text{mol}} \times \dfrac{2 \text{ atoms}}{\text{molecule}}$

$= 2.47 \times 10^{25} \text{ atoms}$

3.3B (a) $? \text{ g Al} = 5.5 \text{ cm} \times 5.5 \text{ cm} \times 5.5 \text{ cm} \times \dfrac{2.70 \text{ g}}{\text{cm}^3} = 4.49 \times 10^2 \text{ g Al}$

$? \text{ mol Al} = 4.49 \times 10^2 \text{ g Al} \times \dfrac{\text{mol Al}}{27.0 \text{ g Al}} = 17 \text{ mol Al}$

(b) $? \text{ mL} = 4.06 \times 10^{24} \text{ Br atoms} \times \frac{\text{molecule}}{2 \text{ atoms}} \times \frac{\text{mol}}{6.022 \times 10^{23} \text{ molecules}}$

$\times \frac{159.80 \text{ g}}{\text{mol}} \times \frac{\text{mL}}{3.12 \text{ g}} = 173 \text{ mL}$

3.4A (a) ? g/mol = 12.01 + (4 × 35.45) = 153.8 g/mol

$? \text{ atoms Cl} = 125 \text{ mL } CCl_4 \times \frac{1.589 \text{ g}}{\text{mL}} \times \frac{\text{mol } CCl_4}{153.8 \text{ g}} \times \frac{4 \text{ mol Cl}}{\text{mol } CCl_4}$

$\times \frac{6.022 \times 10^{23} \text{ atoms}}{\text{mol}} = 3.11 \times 10^{24} \text{ atoms Cl}$

(b)

C	12 × 12.01 =	144.12
H	22 × 1.01 =	22.22
O	11 × 16.00 =	176.00
	45	342.34 u

$? \text{ g C} = 215 \text{ g } C_{12}H_{22}O_{11} \times \frac{\text{mol } C_{12}H_{22}O_{11}}{342.34 \text{ g } C_{12}H_{22}O_{11}} \times \frac{12 \text{ mol C}}{\text{mol } C_{12}H_{22}O_{11}}$

$\times \frac{12.01 \text{ g C}}{\text{mol C}} = 90.5 \text{ g C}$

3.4B $? \text{ mL} = 1.00 \text{ mol C} \times \frac{\text{mol } C_{12}H_{22}O_{11}}{12 \text{ mol C}} \times \frac{342.3 \text{ g}}{\text{mol } C_{12}H_{22}O_{11}} \times \frac{100 \text{ g solution}}{5.05 \text{ g } C_{12}H_{22}O_{11}}$

$\times \frac{\text{mL}}{1.0181 \text{ g}} = 555 \text{ mL}$

3.5A 6.0×10^{23} atoms is 24 g. Thus, $1.0 \times 10^{23} = 4.0$ g. (d)

$\frac{24.31}{6.022} = 4.04$ actual.

3.5 B(a) $? \text{ atoms} = 1.00 \text{g } CO_2 \times \frac{\text{mol}}{44.01 \text{ g}} \times \frac{6.022 \times 10^{23} \text{ molecules}}{\text{mol}} \times \frac{1 \text{ atom}}{\text{molecule}}$

$= 1.37 \times 10^{22} \text{ atoms}$

? g/mol = 12.01 + (2×16.00) = 44.01 g/mol

(b) $? \text{ atoms} = 1.00 \text{g } C_3H_8 \times \frac{\text{mol}}{44.09 \text{ g}} \times \frac{6.022 \times 10^{23} \text{ molecules}}{\text{mol}} \times \frac{3 \text{ atoms}}{\text{molecule}}$

$= 4.10 \times 10^{22} \text{ atoms}$

? g/mol = (3×12.01) + (8×1.008) = 44.09 g/mol

(c) $? \text{ atoms} = 1.00 \text{g } C_3H_6O_2 \times \frac{\text{mol}}{74.08 \text{ g}} \times \frac{6.022 \times 10^{23} \text{ molecules}}{\text{mol}} \times \frac{3 \text{ atoms}}{\text{molecule}}$

$= 2.44 \times 10^{22} \text{ atoms}$

? g/mol = (3×12.01) + (6×1.008) + (2×16.00) = 74.08 g/mol

(d) $? \text{ atoms} = 1.00\text{g } C_4H_{10}O \times \frac{\text{mol}}{74.12\text{g}} \times \frac{6.022\times10^{23}\text{ molecules}}{\text{mol}} \times \frac{4\text{ atoms}}{\text{molecule}}$

$= 3.25\times10^{22}$ atoms

$? \text{ g/mol} = (4\times12.01) + (10\times1.008) + (1\times16.00) = 74.12 \text{ g/mol}$

These equations simplify to $\frac{\#\text{ atoms C / molecule}}{\text{molar mass}}$

(a) $\frac{1}{44}$ actual 1.37×10^{22} atoms

(b) $\frac{3}{44}$ largest actual 4.10×10^{22} atoms

(c) $\frac{3}{74}$ actual 2.44×10^{22} atoms

(d) $\frac{4}{74}$ actual 3.25×10^{22} atoms

3.6A (a) $(NH_4)_2SO_4$

$2 \times 14.01 + 8 \times 1.008 + 1 \times 32.07 + 4 \times 16.00 = 132.15$ u

$\%N = \frac{28.02 \text{ g N}}{132.15 \text{ g } (NH_4)_2SO_4} \times 100\% = 21.20\%$ N

$\%H = \frac{8.064 \text{ g H}}{132.15 \text{ g } (NH_4)_2SO_4} \times 100\% = 6.10\%$ H

$\%S = \frac{32.07 \text{ g S}}{132.15 \text{ g } (NH_4)_2SO_4} \times 100\% = 24.27\%$ S

$\%O = \frac{64.00 \text{ g O}}{132.15 \text{ g } (NH_4)_2SO_4} \times 100\% = \underline{48.43\%\text{ O}}$

100.00%

(b) $CO(NH_2)_2$ $1 \times 12.01 + 1 \times 16.00 + 2 \times 14.01 + 4 \times 1.008 = 60.06$ u

$\%N = \frac{28.02 \text{ g N}}{60.06 \text{ g } CO(NH_2)_2} \times 100\% = 46.65\%$ N

$\%C = \frac{12.01 \text{ g C}}{60.06 \text{ g } CO(NH_2)_2} \times 100\% = 20.00\%$ C

$\%O = \frac{16.00 \text{ g O}}{60.06 \text{ g } CO(NH_2)_2} \times 100\% = 26.64\%$ O

$\%H = \frac{4.032 \text{ g H}}{60.06 \text{ g } CO(NH_2)_2} \times 100\% = \underline{6.71\%\text{ H}}$

100.00%

$\%N = \frac{28.02 \text{ g N}}{2 \times 14.01 + 4 \times 1.008 + 3 \times 16.00} \times 100\% = 35.00\%$ N in NH_4NO_3

Urea $CO(NH_2)_2$ has the highest %N.

3.6B (a) $N(CH_2CH_2OH)_3$

$(1 \times 14.01) + (6 \times 12.01) + (3 \times 16.00) + (15 \times 1.01) = 149.22$

$$\%N = \frac{14.01}{149.22} \times 100\% = 9.389\%\ N$$

(b) $$\text{mass } \%\ O = \frac{6 \times 16.00 \times 100\%}{(57 \times 12.01) + (110 \times 1.008) + (6 \times 16.00)} = 10.77\%\ O$$

(c) $$\text{mass } \%\ H_2O = \frac{(5 \times 16.00) + (10 \times 1.008) \times 100\%}{63.55 + 32.07 + (4 \times 16.00) + (5 \times 16.00) + (10 \times 1.008)}$$

$$= 36.08\%\ H_2O$$

3.7A $NaHCO_3$ $1 \times 22.99 + 1 \times 1.008 + 1 \times 12.01 + 3 \times 16.00 = 84.01$ u

$$?\ \text{mg Na} = 5.00\text{g } NaHCO_3 \times \frac{\text{mol } NaHCO_3}{84.01\text{ g } NaHCO_3} \times \frac{\text{mol Na}}{\text{mol } NaHCO_3} \times \frac{22.99\text{ g Na}}{\text{mol Na}}$$

$$\times \frac{\text{mg}}{10^{-3}\text{ g}} = 1.37 \times 10^3\text{ mg Na}$$

3.7B NH_4NO_3

$$?\ \frac{\text{molar mass N}}{\text{molar mass } NH_4NO_3} = \frac{2 \times 14.007}{2 \times 14.007 + 3 \times 15.999 + 4 \times 1.0079} = 0.3500$$

$(NH_4)_2SO_4$

$$?\ \frac{\text{molar mass N}}{\text{molar mass } (NH_4)_2SO_4} = \frac{2 \times 14.007}{2 \times 14.007 + 4 \times 15.999 + 8 \times 1.0079 + 1 \times 32.066}$$

$$= 0.2120$$

$CO(NH_2)_2$

$$?\ \frac{\text{molar mass N}}{\text{molar mass } CO(NH_2)_2} = \frac{2 \times 14.007}{2 \times 14.007 + 15.999 + 12.011 + 4 \times 1.0079} = 0.4665$$

$$?\ \text{g N} = 1.00\text{ kg} \times \frac{10^3\text{ g}}{\text{kg}}$$

$$\times \left(\frac{(0.3500 \times 12.5)\text{ g N} + (0.2120 \times 20.3)\text{ g N} + (0.4665 \times 11.4)\text{ g N}}{100\text{ g FERT}} \right) = 1.40 \times 10^2\text{ g N}$$

3.8A LiH_2PO_4 $?\%\ P = \frac{31 \text{ x } 100\%}{7 + 2 + 31 + 64} = \frac{3100}{104} \approx 31\%$ Actual 30%

$Ca(H_2PO_4)_2$ $?\%\ P = \frac{2 \text{ x } 31 \text{ x } 100\%}{40 + 4 + 62 + 128} = \frac{6200}{234} \approx 25\%$ Actual 26%

$(NH_4)_2HPO_4$ $?\%\ P = \frac{31 \text{ x } 100\%}{28 + 8 + 1 + 31 + 64} = \frac{3100}{132} \approx 23\%$ Actual 23%

Once the equations are written out, it is easy to see that the denominator is the smallest for LiH_2PO_4— that is 7 + 2 is less than 28 + 8 + 1 and less than 1/2 of 40 + 4. So the largest percentage is LiH_2PO_4.

3.8B methanol CH_3OH; acetic acid CH_3COOH; butane C_4H_{10}; octane C_8H_{18}

Methane and acetic acid have a lower percent of carbon because an oxygen is included in the formula. Octane has a greater carbon mass percent because octane has 9 hydrogens per 4 carbons compared to 10 hydrogens per 4 carbons for butane.

actual C_8H_{18} $\dfrac{8\times 12.01}{(8\times 12.01)+(18\times 1.008)}\times 100\% = 84.12\%\ C$

C_4H_{10} $\dfrac{4\times 12.01}{(4\times 12.01)+(10\times 1.008)}\times 100\% = 82.66\%\ C$

CH_3COOH $\dfrac{2\times 12.01}{(2\times 12.01)+(2\times 16.00)+(4\times 1.008)}\times 100\% = 40.00\%\ C$

CH_3OH $\dfrac{12.01}{12.01+16.00+(4\times 1.008)}\times 100\% = 37.48\%\ C$

3.9A $? \text{ mol C} = 71.95 \text{ g C} \times \dfrac{\text{mol C}}{12.01 \text{ g C}} = 5.991 \text{ mol C} \times \dfrac{1}{0.9981 \text{ mol O}} = \dfrac{6.002 \text{ mol C}}{\text{mol O}}$

$? \text{ mol H} = 12.08 \text{ g H} \times \dfrac{\text{mol H}}{1.008 \text{ g H}} = 11.98 \text{ mol H} \times \dfrac{1}{0.9981 \text{ mol O}} = \dfrac{12.00 \text{ mol H}}{\text{mol O}}$

$? \text{ mol O} = 15.97 \text{ g O} \times \dfrac{\text{mol O}}{16.00 \text{ g O}} = 0.9981 \text{ mol O} \times \dfrac{1}{0.9981 \text{ mol O}} = \dfrac{1.000 \text{ mol O}}{\text{mol O}}$

$C_6H_{12}O$

3.9B $? \text{ mol C} = 51.70 \text{ g C} \times \dfrac{\text{mol C}}{12.01 \text{ g C}} = 4.305 \text{ mol C} \times \dfrac{1}{0.8608 \text{ mol N}} = \dfrac{5.001 \text{ mol C}}{\text{mol N}}$

$? \text{ mol H} = 8.68 \text{ g H} \times \dfrac{\text{mol H}}{1.008 \text{ g H}} = 8.61 \text{ mol H} \times \dfrac{1}{0.8608 \text{ mol N}} = \dfrac{10.0 \text{ mol H}}{\text{mol N}}$

$? \text{ mol N} = 12.06 \text{ g N} \times \dfrac{\text{mol N}}{14.01 \text{ g N}} = 0.8608 \text{ mol N} \times \dfrac{1}{0.8608 \text{ mol N}} = \dfrac{1.000 \text{ mol N}}{\text{mol N}}$

$? \text{ mol O} = 27.55 \text{ g O} \times \dfrac{\text{mol O}}{16.00 \text{ g O}} = 1.722 \text{ mol O} \times \dfrac{1}{0.8608 \text{ mol N}} = \dfrac{2.001 \text{ mol O}}{\text{mol N}}$

$C_5H_{10}NO_2$

3.10A $? \text{ mol C} = 94.34 \text{ g C} \times \dfrac{\text{mol C}}{12.01 \text{ g C}} = 7.855 \text{ mol C} \times \dfrac{1}{5.62 \text{ mol H}} = \dfrac{1.40 \text{ mol C}}{\text{mol H}}$

$? \text{ mol H} = 5.66 \text{ g H} \times \dfrac{\text{mol H}}{1.008 \text{ g H}} = 5.62 \text{ mol H} \times \dfrac{1}{5.62 \text{ mol H}} = \dfrac{1.00 \text{ mol H}}{\text{mol H}}$

$C_{(1.40\times 5)}H_{(1.00\times 5)} = C_7H_5$

3.10B (a) $? \text{ mol Fe} = 72.4 \text{ g Fe} \times \dfrac{\text{mol Fe}}{55.85 \text{ g Fe}} = 1.30 \text{ mol Fe} \times \dfrac{1}{1.30 \text{ mol Fe}}$

$= 1.00 \dfrac{\text{mol Fe}}{\text{mol Fe}}$

$? \text{ mol O} = 27.6 \text{ g O} \times \dfrac{\text{mol O}}{16.00 \text{ g O}} = 1.73 \text{ mol O} \times \dfrac{1}{1.30 \text{ mol Fe}} = 1.33 \dfrac{\text{mol O}}{\text{mol Fe}}$

$$Fe_{(1.00\times3)}O_{(1.33\times3)} = Fe_3O_4$$

(b) $? \text{ mol C} = 9.93 \text{ g C} \times \dfrac{\text{mol C}}{12.01 \text{ g C}} = 0.827 \text{ mol C} \times \dfrac{1}{0.827 \text{ mol C}}$

$$= 1.00 \frac{\text{mol C}}{\text{mol C}}$$

$? \text{ mol Cl} = 58.6 \text{ g Cl} \times \dfrac{\text{mol Cl}}{35.45 \text{ g Cl}} = 1.65 \text{ mol Cl} \times \dfrac{1}{0.827 \text{ mol C}}$

$$= \frac{2.00 \text{ mol Cl}}{\text{mol C}}$$

$? \text{ mol F} = 31.4 \text{ g F} \times \dfrac{\text{mol F}}{19.00 \text{ g F}} = 1.65 \text{ mol F} \times \dfrac{1}{0.827 \text{ mol C}} = \dfrac{2.00 \text{ mol F}}{\text{mol C}}$

CCl_2F_2

(c) $? \text{ mol C} = 37.01 \text{ g C} \times \dfrac{\text{mol C}}{12.01 \text{ g C}} = 3.082 \text{ mol C} \times \dfrac{1}{1.320 \text{ mol N}}$

$$= \frac{2.335 \text{ mol C}}{\text{mol N}}$$

$? \text{ mol H} = 2.22 \text{ g H} \times \dfrac{\text{mol H}}{1.008 \text{ g H}} = 2.20 \text{ mol H} \times \dfrac{1}{1.320 \text{ mol N}} = \dfrac{1.67 \text{ mol H}}{\text{mol N}}$

$? \text{ mol N} = 18.50 \text{ g N} \times \dfrac{\text{mol N}}{14.01 \text{ g N}} = 1.320 \text{ mol N} \times \dfrac{1}{1.320 \text{ mol N}}$

$$= \frac{1.000 \text{ mol N}}{\text{mol N}}$$

$? \text{ mol O} = 42.27 \text{ g O} \times \dfrac{\text{mol O}}{16.00 \text{ g O}} = 2.642 \text{ mol O} \times \dfrac{1}{1.320 \text{ mol N}}$

$$= \frac{2.002 \text{ mol O}}{\text{mol N}}$$

$C_{2.335}H_{1.67}NO_{2.002}$ (multiplication by 3 to obtain whole numbers) $C_7H_5N_3O_6$

3.11A CH_2 $1 \times 12.01 + 2 \times 1.008 = 14.0 \text{ u}$

ethylene $\dfrac{28.0 \text{ u}}{\text{molecule}} \times \dfrac{\text{formula}}{14.0 \text{ u}} = \dfrac{2 \text{ formula}}{\text{molecule}}$

C_2H_4

cyclohexane $\dfrac{84.0 \text{ u}}{\text{molecule}} \times \dfrac{\text{formula}}{14.0 \text{ u}} = \dfrac{6 \text{ formula}}{\text{molecule}}$

C_6H_{12}

1-pentene $\dfrac{70.0 \text{ u}}{\text{molecule}} \times \dfrac{\text{formula}}{14.0 \text{ u}} = \dfrac{5 \text{ formula}}{\text{molecule}}$

C_5H_{10}

3.11B (a) P_2O_3 (b) C_2H_3 (c) $C_3H_4O_3$ (d) $C_2H_4O_3$ (e) CH_3O (f) $CuCO_2$

3.12A (a) $? \text{ g C} = 0.561 \text{ g } CO_2 \times \dfrac{12.01 \text{ g C}}{44.01 \text{ g } CO_2} = 0.153 \text{ g C}$

$$? \text{ g H} = 0.306 \text{ g } H_2O \times \frac{2 \text{ x } 1.0079 \text{ g H}}{18.015 \text{ g } H_2O} = 0.0342 \text{ g H}$$

$$?\% \text{ C} = \frac{0.153 \text{ g C}}{0.255 \text{ g sample}} \times 100\% = 60.0\% \text{ C}$$

$$?\% \text{ H} = \frac{0.0342 \text{ g H}}{0.255 \text{ g sample}} \times 100\% = 13.4\% \text{ H}$$

$$?\% \text{ O} = 100.0\% - 60.0\% \text{ C} - 13.4\% \text{ H} = 26.6\% \text{ O}$$

(b) $$? \text{ mol C} = 60.0 \text{ g C} \times \frac{\text{mol C}}{12.011 \text{ g C}} = 5.00 \text{ mol C} \times \frac{1}{1.66 \text{ mol O}} = \frac{3.01 \text{ mol C}}{\text{mol C}}$$

$$? \text{ mol H} = 13.4 \text{ g H} \times \frac{\text{mol H}}{1.0079 \text{ g H}} = 13.3 \text{ mol H} \times \frac{1}{1.66 \text{ mol O}} = \frac{8.01 \text{ mol H}}{\text{mol O}}$$

$$? \text{ mol O} = 26.6 \text{ g O} \times \frac{\text{mol O}}{15.999 \text{ g O}} = 1.66 \text{ mol O} \times \frac{1}{1.66 \text{ mol O}} = \frac{1.00 \text{ mol O}}{\text{mol O}}$$

C_3H_8O

3.12B (a) With the information given, it can be assumed that only O in addition to C and H can be present.

$$?\% \text{ C} = 1.0666 \text{ g } CO_2 \times \frac{12.011 \text{ g C}}{44.010 \text{ g } CO_2} \times \frac{100\%}{0.3629 \text{ g sample}} = 80.213\% \text{ C}$$

$$?\% \text{ H} = 0.3120 \text{ g } H_2O \times \frac{2 \text{ x } 1.0079 \text{ g H}}{18.015 \text{ g } H_2O} \times \frac{100\%}{0.3629 \text{ g sample}} = 9.620\% \text{ H}$$

$$100.000\% - 80.213\% \text{ C} - 9.620\% \text{ H} = 10.167\% \text{ O}$$

(b) $$? \text{ mol C} = 80.213 \text{ g C} \times \frac{\text{mol C}}{12.011 \text{ g C}} = 6.6783 \text{ mol C} \times \frac{1}{0.63546 \text{ mol O}} = \frac{10.509 \text{ mol C}}{\text{mol O}}$$

$$? \text{ mol H} = 9.620 \text{ g H} \times \frac{\text{mol H}}{1.0079 \text{ g H}} = 9.545 \text{ mol H} \times \frac{1}{0.63546 \text{ mol O}} = \frac{15.02 \text{ mol H}}{\text{mol O}}$$

$$? \text{ mol O} = 10.167 \text{ g O} \times \frac{\text{mol O}}{15.9994 \text{ g O}} = 0.63546 \text{ mol O} \times \frac{1}{0.63546 \text{ mol O}} = \frac{1.000 \text{ mol O}}{\text{mol O}}$$

$$C_{10.51}H_{15.02}O_{1.00} \Rightarrow C_{21}H_{30}O_2$$

3.13A (a) $SiCl_4 + 2\, H_2O \rightarrow SiO_2 + 4\, HCl$

(b) $PCl_5 + 4\, H_2O \rightarrow H_3PO_4 + 5\, HCl$

(c) $6\, CaO + P_4O_{10} \rightarrow 2\, Ca_3(PO_4)_2$

3.13B (a) $Pb(NO_3)_2\,(aq) + 2\, KI\,(aq) \rightarrow PbI_2\,(s) + 2\, KNO_3\,(aq)$

(b) $4\ HCl\ (g) + O_2\ (g) \rightarrow 2\ H_2O\ (g) + 2\ Cl_2\ (g)$

3.14A (a) $2\ C_4H_{10} + 13\ O_2 \rightarrow 8\ CO_2 + 10\ H_2O$
(b) $C_5H_{12}O_2 + 7\ O_2 \rightarrow 5\ CO_2 + 6\ H_2O$

3.14B $2\ (CH_3)_3COCH_3 + 15\ O_2 \rightarrow 10\ CO_2 + 12\ H_2O$

3.15A (a) $FeCl_3 + 3NaOH \rightarrow Fe(OH)_3 + 3\ NaCl$
(b) $3\ Ba(NO_3)_2 + Al_2(SO_4)_3 \rightarrow 3\ BaSO_4 + 2\ Al(NO_3)_3$

3.15B $3\ Ca(OH)_2(s) + 2\ H_3PO_4(aq) \rightarrow Ca_3(PO_4)_2(s) + 6\ H_2O(l)$

3.16A $? \text{ mol C} = 47.99 \text{ g C} \times \frac{\text{mol C}}{12.011 \text{ g C}} = 3.996 \text{ mol C} \times \frac{1}{2.664 \text{ mol O}} = \frac{1.500 \text{ mol C}}{\text{mol O}}$

$? \text{ mol H} = 9.40 \text{ g H} \times \frac{\text{mol H}}{1.008 \text{ g H}} = 9.325 \text{ mol H} \times \frac{1}{2.664 \text{ mol O}} = \frac{3.500 \text{ mol H}}{\text{mol O}}$

$? \text{ mol O} = 42.62 \text{ g O} \times \frac{\text{mol O}}{15.999 \text{ g O}} = 2.664 \text{ mol O} \times \frac{1}{2.664 \text{ mol O}} = \frac{1.000 \text{ mol O}}{\text{mol O}}$

$C_{(1.5\times2)}H_{(3.5\times2)}O_{(1.0\times2)} \Rightarrow C_3H_7O_2$ empirical formula

formula mass = $(3 \times 12.01 \text{ u}) + (7 \times 1.008 \text{ u}) + (2 \times 16.00 \text{ u}) = 75.09 \text{ u}$

$\frac{? \text{ formula units}}{\text{molecule}} = \frac{150.2 \text{ u}}{\text{molecule}} \times \frac{\text{formula units}}{75.09 \text{ u}} = \frac{2 \text{ formula units}}{\text{molecule}}$

$C_6H_{14}O_4$ molecular formula

$2\ C_6H_{14}O_4 + 15\ O_2 \rightarrow 12\ CO_2 + 14\ H_2O$

3.16B $2\ Pb(NO_3)_2\ (s) \rightarrow 2\ PbO\ (s) + 4\ NO_2\ (g) + O_2\ (g)$

3.17A (a) $C_3H_8 + 5\ O_2 \rightarrow 3\ CO_2 + 4\ H_2O$

$? \text{ mol } CO_2 = 0.529 \text{ mol } C_3H_8 \times \frac{3 \text{ mol } CO_2}{\text{mol } C_3H_8} = 1.59 \text{ mol } CO_2$

(b) $? \text{ mol } H_2O = 76.2 \text{ mol } C_3H_8 \times \frac{4 \text{ mol } H_2O}{\text{mol } C_3H_8} = 305 \text{ mol } H_2O$

(c) $? \ CO_2 = 1.010 \text{ mol } O_2 \times \frac{3 \text{ mol } CO_2}{5 \text{ mol } O_2} = 0.6060 \text{ mol } CO_2$

3.17B $2\ C_4H_{10} + 13\ O_2 \rightarrow 8\ CO_2 + 10\ H_2O$

(a) $? \text{ mol } CO_2 = 55.6 \text{ g } C_4H_{10} \times \frac{\text{mol } C_4H_{10}}{(4\times12.01)+(10\times1.008)} \times \frac{8 \text{ mol } CO_2}{2 \text{ mol } C_4H_{10}}$

$= 3.83 \text{ mol } CO_2$

(b) $? \text{ mol } O_2 = 55.6\text{g } C_4H_{10} \times \dfrac{\text{mol } C_4H_{10}}{(4\times 12.01)+(10\times 1.008)} \times \dfrac{13 \text{ mol } O_2}{2 \text{ mol } C_4H_{10}}$

$= 6.22 \text{ mol } O_2$

3.18A $2\text{ Mg (s)} + TiCl_4 \rightarrow 2\text{ } MgCl_2 + Ti$

$? \text{ g Mg} = 83.6 \text{ g } TiCl_4 \times \dfrac{\text{mol } TiCl_4}{189.68 \text{ g } TiCl_4} \times \dfrac{2 \text{ mol Mg}}{\text{mol } TiCl_4} \times \dfrac{24.31 \text{ g Mg}}{\text{mol Mg}}$

$= 21.4 \text{ g Mg}$

3.18B $2\text{ } NH_4NO_3 \rightarrow 2\text{ } N_2 + O_2 + 4\text{ } H_2O$

$? \text{ g } N_2 = 75.5 \text{ g } NH_4NO_3 \times \dfrac{\text{mol } NH_4NO_3}{80.04 \text{ g } NH_4NO_3} \times \dfrac{2 \text{ mol } N_2}{2 \text{ mol } NH_4NO_3}$

$\times \dfrac{28.02 \text{ g } N_2}{\text{mol } N_2} = 26.4 \text{ g } N_2$

$? \text{ g } O_2 = 75.5 \text{ g } NH_4NO_3 \times \dfrac{\text{mol } NH_4NO_3}{80.04 \text{ g } NH_4NO_3} \times \dfrac{1 \text{ mol } O_2}{2 \text{ mol } NH_4NO_3} \times \dfrac{32.00 \text{ g } O_2}{\text{mol } O_2}$

$= 15.1 \text{ g } O_2$

3.19A $2\text{ } C_8H_{18} + 25\text{ } O_2 \rightarrow 16\text{ } CO_2 + 18\text{ } H_2O$

$? \text{ mL} = 775 \text{ mL } C_8H_{18} \times \dfrac{0.7025 \text{ g } C_8H_{18}}{\text{mL } C_8H_{18}} \times \dfrac{\text{mol } C_8H_{18}}{114.22 \text{ g } C_8H_{18}} \times \dfrac{18 \text{ mol } H_2O}{2 \text{ mol } C_8H_{18}}$

$\times \dfrac{18.015 \text{ g } H_2O}{\text{mol } H_2O} \times \dfrac{\text{mL}}{0.9982 \text{ g}} = 774 \text{ mL}$

3.19B $2\text{ } CH_3OH + 3\text{ } O_2 \rightarrow 2\text{ } CO_2 + 4\text{ } H_2O$

$2\text{ } CH_3CH_2CH_2OH + 9\text{ } O_2 \rightarrow 6\text{ } CO_2 + 8\text{ } H_2O$

$? \text{ g} = 25.0\text{g} \times \dfrac{62.5\text{g } CH_3OH}{100\text{g}} \times \dfrac{\text{mol } CH_3OH}{12.01 + 16.00 + (4\times 1.008) \text{ g } CH_3OH}$

$\times \dfrac{4 \text{ mol } H_2O}{2 \text{ mol } CH_3OH} \times \dfrac{18.02 \text{ g}}{\text{mol } H_2O} = 17.57\text{g}$

$? \text{ g} = 25.0\text{g} \times \dfrac{37.5\text{g } C_3H_8O}{100\text{g}} \times \dfrac{\text{mol } C_3H_8O}{(3\times 12.01) + 16.00 + (8\times 1.008) \text{ g } C_3H_8O}$

$\times \dfrac{8 \text{ mol } H_2O}{2 \text{ mol } C_3H_8O} \times \dfrac{18.02 \text{ g}}{\text{mol } H_2O} = 11.24\text{g}$

$? \text{ mL} = (17.57\text{g} + 11.24\text{g}) \times \dfrac{\text{mL}}{0.9982\text{g}} = 28.9 \text{ mL}$

3.20A $? \text{ g } H_2S = 10.2 \text{ g HCl} \times \dfrac{\text{mol HCl}}{36.46 \text{ g HCl}} \times \dfrac{\text{mol } H_2S}{2 \text{ mol HCl}} \times \dfrac{34.08 \text{ g } H_2S}{\text{mol } H_2S} = 4.77\text{g } H_2S$

$$? \text{ g } H_2S = 13.2 \text{ g FeS} \times \frac{\text{mol FeS}}{87.91 \text{ g FeS}} \times \frac{\text{mol } H_2S}{\text{mol FeS}} \times \frac{34.08 \text{ g } H_2S}{\text{mol } H_2S} = 5.12 \text{ g } H_2S$$

4.77 g H_2S is formed.

$$4.77 \text{ g } H_2S \times \frac{\text{mol } H_2S}{34.08 \text{ g } H_2S} \times \frac{\text{mol FeS}}{\text{mol } H_2S} \times \frac{87.91 \text{ g FeS}}{\text{mol FeS}} = 12.3 \text{ g FeS used}$$

? g FeS excess = 13.2 g – 12.3 g = 0.9 g FeS excess

3.20B $2 \text{ Al} + 6 \text{ HCl} \rightarrow 3 H_2 + 2 \text{ AlCl}_3$

$$? \text{ g } H_2 = 12.5 \text{ g Al} \times \frac{\text{mol Al}}{26.98 \text{ g Al}} \times \frac{3 \text{ mol } H_2}{2 \text{ mol Al}} \times \frac{2.016 \text{ g } H_2}{\text{mol } H_2} = 1.40 \text{ g } H_2$$

$$? \text{ g } H_2 = 250.0 \text{ mL solution} \times \frac{1.13 \text{ g}}{\text{mL}} \times \frac{25.6 \text{ g HCl}}{100 \text{ g solution}} \times \frac{\text{mol HCl}}{36.46 \text{ g HCl}} \times \frac{3 \text{ mol } H_2}{6 \text{ mol HCl}} \times \frac{2.016 \text{ g } H_2}{\text{mol } H_2} = 2.00 \text{ g } H_2$$

1.40 g H_2 are produced.

3.21A $$? \text{ g acet} = 20.0 \text{ g alcohol} \times \frac{\text{mol alch}}{88.15 \text{ g alch}} \times \frac{1 \text{ mol acetate}}{1 \text{ mol alch}} \times \frac{130.18 \text{ g acet}}{\text{mol acet}} = 29.5 \text{ g acet}$$

$$? \text{ g acet} = 25.0 \text{ g acid} \times \frac{\text{mol aa}}{60.05 \text{ g aa}} \times \frac{1 \text{ mol acet}}{1 \text{ mol aa}} \times \frac{130.18 \text{ g acet}}{\text{mol acet}} = 54.20 \text{ g acet}$$

Theoretical yield is 29.5 g isopentyl acetate.

$$\text{Actual Y} = \frac{\text{\% Y theoret. Y}}{100\%} = \frac{29.5 \text{ g} \times 90.0\%}{100\%} = 26.6 \text{ g}$$

3.21B $$? \text{ g} = \text{theor Y} = \frac{\text{actual Y x } 100\%}{\text{\% Y}} = \frac{433 \text{ g} \times 100\%}{78.5\%} = 552 \text{ g}$$

$$? \text{ g} = 552 \text{ g acet} \times \frac{\text{mol acet}}{130.18 \text{ g acet}} \times \frac{\text{mol alch}}{\text{mol acet}} \times \frac{88.15 \text{ g alch}}{\text{mol alch}} = 374 \text{ g isopentyl alcohol}$$

3.22A $H_3PO_4 + 2 \text{ NH}_3 \rightarrow (NH_4)_2HPO_4$

$$? \text{ kg } (NH_4)_2HPO_4 = 1.00 \text{ kg } H_3PO_4 \times \frac{\text{mol } H_3PO_4}{97.99 \text{ g } H_3PO_4} \times \frac{\text{mol } (NH_4)_2HPO_4}{\text{mol } H_3PO_4} \times \frac{132 \text{ g } (NH_4)_2HPO_4}{\text{mol } (NH_4)_2HPO_4} = 1.35 \text{ kg } (NH_4)_2HPO_4$$

3.22B $\text{Ca (s)} + 2 \text{ HCl} \rightarrow H_2 + \text{CaCl}_2$

$\text{Mg (s)} + 2 \text{ HCl} \rightarrow H_2 + \text{MgCl}_2$

$2 \text{ Al (s)} + 6 \text{ HCl} \rightarrow 3 H_2 + 2 \text{ AlCl}_3$

$$1\text{ g Ca} \times \frac{\text{mol Ca}}{40.08\text{g}} \times \frac{\text{mol H}_2}{\text{mol Ca}} = 0.0250\text{ mol}$$

$$1\text{ g Mg} \times \frac{\text{mol Mg}}{24.31\text{g}} \times \frac{\text{mol H}_2}{\text{mol Mg}} = 0.0411\text{ mol}$$

$$1\text{ g Al} \times \frac{\text{mol Al}}{26.98\text{g}} \times \frac{3\text{ mol H}_2}{2\text{ mol Al}} = 0.0556\text{ mol}$$

This simplifies to $\frac{1}{40}$, $\frac{1}{24}$, and $\frac{3}{54} \approx \frac{1}{18}$, so Al produces the most H_2

3.23A (a) $? \text{ M} = \frac{3.00\text{ mol KI}}{2.39\text{ L solution}} = 1.26\text{ M KI}$

(b) $? \text{ M} = \frac{0.522\text{ g HCl}}{0.592\text{ L solution}} \times \frac{\text{mol HCl}}{36.46\text{ g HCl}} = 0.0242\text{ M HCl}$

(c) $? \text{ M} = \frac{2.69\text{ g C}_{12}\text{H}_{22}\text{O}_{11}}{225\text{ mL}} \times \frac{\text{mL}}{10^{-3}\text{ L}} \times \frac{\text{mol C}_{12}\text{H}_{22}\text{O}_{11}}{342.3\text{ g C}_{12}\text{H}_{22}\text{O}_{11}}$

$= 3.49 \times 10^{-2}\text{ M C}_{12}\text{H}_{22}\text{O}_{11}$

3.23B (a) $? \text{ M} = \frac{126\text{ mg}}{100.0\text{ mL}} \times \frac{\text{mol C}_6\text{H}_{12}\text{O}_6}{180.2\text{ g}} = 6.99 \times 10^{-3}\text{ M C}_6\text{H}_{12}\text{O}_6$

(b) $? \text{ M} = \frac{10.5\text{ mL C}_2\text{H}_5\text{OH}}{25.0\text{ mL}} \times \frac{0.789\text{ g C}_2\text{H}_5\text{OH}}{\text{mL C}_2\text{H}_5\text{OH}} \times \frac{\text{mol C}_2\text{H}_5\text{OH}}{46.08\text{ g C}_2\text{H}_5\text{OH}}$

$\times \frac{\text{mL}}{10^{-3}\text{ L}} = 7.19\text{ M C}_2\text{H}_5\text{OH}$

(c) $? \text{ M} = \frac{9.5\text{ mg N}}{\text{mL}} \times \frac{60.07\text{ g CO(NH}_2)_2}{28.02\text{ g N}} \times \frac{\text{mol CO(NH}_2)_2}{60.07\text{ g CO(NH}_2)_2}$

$= 0.34\text{ M CO(NH}_2)_2$

3.24A (a) $? \text{ g} = 2.00\text{ L} \times 6.00\text{ M KOH} \times \frac{56.11\text{ g KOH}}{\text{mol KOH}} = 673\text{ g KOH}$

(b) $? \text{ g} = 10.0\text{ mL} \times \frac{10^{-3}\text{ L}}{\text{mL}} \times 0.100\text{ M KOH} \times \frac{56.11\text{ g KOH}}{\text{mol KOH}} = 0.0561\text{ g KOH}$

(c) $? \text{ g} = 35.0\text{ mL} \times \frac{10^{-3}\text{ L}}{\text{mL}} \times 2.50\text{ M KOH} \times \frac{56.11\text{ g KOH}}{\text{mol KOH}} = 4.91\text{ g KOH}$

3.24B $? \text{ mL} = 725\text{ mL} \times 0.350\text{ M} \times \frac{10^{-3}\text{ mol}}{\text{mmol}} \times \frac{74.12\text{ g C}_4\text{H}_{10}\text{O}}{\text{mol C}_4\text{H}_{10}\text{O}} \times \frac{\text{mL}}{0.810\text{ g}}$

$= 23.2\text{ mL C}_4\text{H}_{10}\text{O}$

3.25A $? \text{ M} = \frac{90.0\text{ g HCOOH}}{100\text{ g solution}} \times \frac{1.20\text{ g}}{\text{mL}} \times \frac{\text{mL}}{10^{-3}\text{ L}} \times \frac{\text{mol HCOOH}}{46.03\text{ g HCOOH}} = 23.5\text{ M HCOOH}$

3.25B $?\%\ HClO_4 = \dfrac{11.7\ \text{mol}\ HClO_4}{\text{L solution}} \times \dfrac{10^{-3}\ \text{L}}{\text{mL}} \times \dfrac{\text{mL}}{1.67\ \text{g}} \times \dfrac{100.5\ \text{g}}{\text{mol}} \times 100\%$

$= 70.4\%\ HClO_4$, by mass

3.26A $?\ \text{mL} = \dfrac{15.0\ \text{L} \times 0.315\ \text{M}}{10.15\ \text{M}} \times \dfrac{\text{mL}}{10^{-3}\ \text{L}} = 466\ \text{mL}$

3.26B $?\ \text{mL} = \dfrac{7.50\ \text{mg}}{\text{mL}} \times 375\ \text{mL} \times \dfrac{\text{mol}\ CH_3OH}{32.05\ \text{g}\ CH_3OH} \times \dfrac{\text{L}}{5.15\ \text{mol}} = 17.0\ \text{mL}\ CH_3OH$

3.27A $?\ \text{mL} = 750.0\ \text{mL} \times 0.0250\ \text{M}\ Na_2CrO_4 \times \dfrac{2\ \text{mol}\ AgNO_3}{\text{mol}\ Na_2CrO_4}$

$\times \dfrac{\text{L}}{0.100\ \text{mol}\ AgNO_3} = 375\ \text{mL}\ AgNO_3$

3.27B (a) $?\ \text{g}\ CO_2 = 175\ \text{mL} \times \dfrac{10^{-3}\ \text{L}}{\text{mL}} \times 1.55\ \text{M}\ NaHCO_3 \times \dfrac{\text{mol}\ CO_2}{\text{mol}\ NaHCO_3}$

$\times \dfrac{44.01\ \text{g}\ CO_2}{\text{mol}\ CO_2} = 11.9\ \text{g}\ CO_2$

$?\ \text{g}\ CO_2 = 235\ \text{mL} \times \dfrac{10^{-3}\ \text{L}}{\text{mL}} \times 1.22\ \text{M}\ HCl \times \dfrac{\text{mol}\ CO_2}{\text{mol}\ HCl}$

$\times \dfrac{44.01\ \text{g}\ CO_2}{\text{mol}\ CO_2} = 12.6\ \text{g}\ CO_2$

$NaHCO_3$ is limiting; 11.9 g CO_2 is produced.

(b) $?\ \text{mol}\ NaCl = 175\ \text{mL} \times \dfrac{10^{-3}\ \text{L}}{\text{mL}} \times 1.55\ \text{M}\ NaHCO_3 \times \dfrac{\text{mol}\ NaCl}{\text{mol}\ NaHCO_3}$

$= 0.271\ \text{mol}\ NaCl$

$?\ \text{M} = \dfrac{0.271\ \text{mol}\ NaCl}{410\ \text{mL}} \times \dfrac{\text{mL}}{10^{-3}\ \text{L}} = 0.661\ \text{M}\ NaCl$

Review Questions

2. $1.00\ \text{mol} \times \dfrac{6.022 \times 10^{23}\ \text{molecules}}{\text{mol}} = 6.02 \times 10^{23}\ \text{molecules}$

$1.00\ \text{mol} \times \dfrac{6.022 \times 10^{23}\ \text{molecules}}{\text{mol}} \times \dfrac{2\ \text{atoms}}{\text{molecule}} = 1.20 \times 10^{24}\ \text{atoms}$

3. (c) $?\ \text{atoms} = 1\ \text{mol}\ F_2 \times \dfrac{6.02 \times 10^{23}\ \text{molecules}}{\text{mol}} \times \dfrac{2\ \text{atoms}}{\text{molecule}} = 1.20 \times 10^{24}\ \text{atoms}$

4. (a) $1.00\ \text{mol Ca(NO}_3)_2 \times \dfrac{6.022 \times 10^{23}\ \text{formula units}}{\text{mol}} \times \dfrac{1\ Ca^{2+}}{\text{formula unit}}$

$= 6.02 \times 10^{23}\ Ca^{2+}$

$1.00\ \text{mol Ca(NO}_3)_2 \times \dfrac{6.022 \times 10^{23}\ \text{formula units}}{\text{mol}} \times \dfrac{2\ NO_3^-}{\text{formula unit}}$

$= 1.20 \times 10^{24}\ NO_3^-$

(b) $1.00\ \text{mol Ca(NO}_3)_2 \times \dfrac{6.022 \times 10^{23}\ \text{formula units}}{\text{mol}} \times \dfrac{2\ \text{N atoms}}{\text{formula unit}}$

$= 1.20 \times 10^{24}\ \text{N atoms}$

$1.00\ \text{mol Ca(NO}_3)_2 \times \dfrac{6.022 \times 10^{23}\ \text{formula units}}{\text{mol}} \times \dfrac{6\ \text{O atoms}}{\text{formula unit}}$

$= 3.61 \times 10^{24}\ \text{O atoms}$

5. (a) HO (b) CH_2 (c) C_5H_4 (d) C_3H_6O

8. (b) $2\ KClO_3(s) \rightarrow 2\ KCl(s) + 3\ O_2(g)$

9. (a) $Hg(NO_3)_2 \xrightarrow{\Delta} Hg(l) + 2\ NO_2(g) + O_2(g)$

(b) $Na_2CO_3\ (aq) + 2\ HCl\ (aq) \rightarrow H_2O\ (l) + CO_2\ (g) + 2\ NaCl\ (aq)$

(c) C $34.62\ \text{g} \times \dfrac{\text{mol}}{12.011\ \text{g}} = 2.882 \times \dfrac{1}{2.882} = 1.00$

H $3.88\ \text{g} \times \dfrac{\text{mol}}{1.008\text{g}} = 3.849 \times \dfrac{1}{2.882} = 1.34$

O $61.50\ \text{g} \times \dfrac{\text{mol}}{15.999\ \text{g}} = 3.844 \times \dfrac{1}{2.882} = 1.33$

$(CH_{1.34}O_{1.33}) \times 3 = C_3H_4O_4$ Since nothing is given about molar mass, it must be assumed that the emperical formula is the molecular formula.

$C_3H_4O_4 + 2\ O_2 \rightarrow 3\ CO_2 + 2\ H_2O$

10. (c) $2.0\ \text{mol CCl}_4 \times \dfrac{\text{mol CCl}_2\text{F}_2}{\text{mol CCl}_4} = 2.0\ \text{mol theoretical yield}$

$\%\ \text{yield} = \dfrac{\text{actual yield}}{\text{theoretical yield}} = \dfrac{1.7\ \text{mol}}{2.0\ \text{mol}} \times 100\% = 85\%$

11. $1.00\ \text{mol } O_2 \times \dfrac{4\ \text{mol } NH_3}{5\ \text{mol } O_2} = 0.80\ \text{mol } NH_3$

$1.00\ \text{mol } O_2 \times \dfrac{4\ \text{mol NO}}{5\ \text{mol } O_2} = 0.80\ \text{mol NO}$

$$1.00 \text{ mol O2} \times \frac{6 \text{ mol } H_2O}{5 \text{ mol } O_2} = 1.2 \text{ mol } H_2O$$

(a) all of the O_2 (g) is consumed.

14. These are the same values. By multiplying both numerators and denominators by 10^{-3}, grams are changed to milligrams, and liters are changed to milliliters.

Problems

17. (a) C_3H_7Br $(3 \times 12.01) + (7 \times 1.008) + 79.90 = 122.99$ u-molecular
(b) $Mg(HCO_3)_2$ $24.31 + (2 \times 1.008) + (2 \times 12.01) + (6 \times 16.00)$
$= 146.35$ u-formula
(c) $Al(ClO_4)_3$ $26.98 + (3 \times 35.45) + (12 \times 16.00) = 325.33$ u-formula
(d) $Fe(NO_3)_3 \cdot 9H_2O$ $55.85 + (3 \times 14.01) + (9 \times 16.00) + (18 \times 1.008)$
$+ (9 \times 16.00) = 404.02$ u-formula
(e) $Ti_2(SO_4)_3$ $(2 \times 47.88) + (3 \times 32.066) + (12 \times 16.00) = 383.96$ u-formula
(f) S_2Cl_2 $(2 \times 32.066) + (2 \times 35.453) = 135.04$ u-molecular
(g) $(CH_3CH_2CH_2)_2O$ $(6 \times 12.011) + (14 \times 1.008) + 16.00 = 102.18$ u-molecular
(h) C_8H_{18} $(8 \times 12.01) + (18 \times 1.008) = 114.22$ u-molecular

19. (a)

C	11 × 12.01 u =	132.11
H	17 × 1.008 u =	17.14
O	5 × 16.00 u =	80.00
P	1 × 30.97 u =	30.97
S	2 × 32.07 u =	64.14
		324.36 u

(b)

C	8 × 12.011 u =	96.088
H	8 × 1.008 u =	8.064
O	3 × 15.999 u =	47.997
		152.15 u

21. (a) ? g/mol = 39.098 + 16.00 + 1.008 = 56.11 g/mol

$$? \text{ g} = 0.773 \text{ mol KOH} \times \frac{56.10\text{g}}{\text{mol}} = 43.4\text{g}$$

(b) ? g/mol = 28.09 + (4 × 35.45) = 169.9 g/mol

$$? \text{ g} = 0.250 \text{ mol } SiCl_4 \times \frac{169.9\text{g}}{\text{mol}} = 42.5\text{g}$$

(c) ? g/mol = 87.62 + (2 × 1.008) + (2 × 32.07) + (8 × 16.00) = 281.8 g/mol

$$? \text{ g} = 0.158 \text{ mol } Sr(HSO_4)_2 \times \frac{281.8\text{g}}{\text{mol}} = 44.5\text{g}$$

(d) ? g/mol = 28.09 + (4 × 12.01) + (12 × 1.008) = 88.23 g/mol

$$? \text{ g} = 3.91 \times 10^{-4} \text{ mol } (CH_3)_4Si \times \frac{88.23\text{g}}{\text{mol}} = 3.45 \times 10^{-2} \text{ g}$$

23. (a) ? g/mol = 137.3 + (2 × 16.00) + (2 × 1.008) = 171.3 g/mol

$$? \text{ mol} = 647\text{g } Ba(OH)_2 \times \frac{\text{mol}}{171.3\text{g}} = 3.78 \text{ mol}$$

(b) ? g/mol = 32.07 + (3 × 16.00) = 80.07 g/mol

$$? \text{ mol} = 16.3\text{g } SO_3 \times \frac{\text{mol}}{80.07\text{g}} = 0.204 \text{ mol}$$

(c) ? g/mol = 58.69 + (2 × 35.45) + (7 × 16.00) +(14 × 1.008) = 255.7 g/mol

$$? \text{ mol} = 35.6\text{g } NiCl_2 \cdot 7H_2O \times \frac{\text{mol}}{255.7\text{g}} = 0.139 \text{ mol}$$

(d) ? g/mol = (2 × 12.01) + (6 × 1.008) + 32.07 = 62.14 g/mol

$$? \text{ mol} = 218\text{mg } CH_3CH_2SH \times \frac{10^{-3}\text{ g}}{\text{mg}} \times \frac{\text{mol}}{62.14\text{g}} = 3.51 \times 10^{-3} \text{ mol}$$

(e) ? g/mol = 55.85 + 32.07 = 87.92 g/mol

$$? \text{ mol} = 3.32 \times 10^4 \text{ kg FS} \times \frac{10^3\text{ g}}{\text{kg}} \times \frac{\text{mol}}{87.92\text{g}} = 3.78 \times 10^5 \text{ mol}$$

25. $$? O^{2-} \text{ ions} = 1.00 \text{ mol } Fe_2O_3 \times \frac{6.022 \times 10^{23} \text{ molecules}}{\text{mol}} \times \frac{3O^{2-}\text{ions}}{\text{molecule}} = 1.81 \times 10^{24}\ O^{2-} \text{ ions}$$

$$? Fe^{3+} \text{ ions} = 1.00 \text{ mol } Fe_2O_3 \times \frac{6.022 \times 10^{23} \text{ molecules}}{\text{mol}} \times \frac{2Fe^{3+}\text{ions}}{\text{molecule}} = 1.20 \times 10^{24}\ Fe^{3+} \text{ ions}$$

27. (a) $$? \text{ molecules} = 4.68 \text{ mol} \times \frac{6.022 \times 10^{23} \text{ molecules}}{\text{mol}} = 2.82 \times 10^{24} \text{ molecules}$$

(b) $$? SO_4^{2-} \text{ ions} = 86.2\text{g } BaSO_4 \times \frac{\text{mol}}{233.4\text{g}} \times \frac{6.022 \times 10^{23}\ SO_4^{2-} \text{ ions}}{\text{mol}} = 2.22 \times 10^{23}\ SO_4^{2-} \text{ ions}$$

(c) $$? \text{ g/atom} = \frac{127.60 \text{ g}}{\text{mol}} \times \frac{\text{mol}}{6.022 \times 10^{23} \text{ atom}} = 2.119 \times 10^{-22} \text{ g/atom}$$

(d) $$? \text{ g/ion} = \frac{54.94 + (4 \times 16.00) \text{ g}}{\text{mol}} \times \frac{\text{mol}}{6.022 \times 10^{23} \text{ ion}} = 1.98 \times 10^{-22} \text{ g/ion}$$

29. 6.1×10^{23} molecules of O_2 is about 2 × Avogadro's number of atoms. 0.76 moles of HCl is about 1.5 times Avogadro's number of atoms. 250.0 g of U is slightly more than Avogadro's number of atoms, and 0.86 mol Al is less than Avogadro's number of atoms.

$Al < U < HCl < O_2$ (a) < (c) < (d) < (b)

actual

(a) $$0.86 \text{ mol Al} \times \frac{6.022 \times 10^{23} \text{ atoms}}{\text{mol}} = 5.2 \times 10^{23} \text{ atoms}$$

(b) $$6.1 \times 10^{23} \text{ molecules } O_2 \times \frac{2 \text{ atoms O}}{\text{molecules } O_2} = 12 \times 10^{23} \text{ atoms}$$

(c) $250.0 \text{ g U} \times \frac{\text{mol}}{238.03 \text{ g}} \times \frac{6.022 \times 10^{23} \text{ atoms}}{\text{mol}} = 6.325 \times 10^{23} \text{ atoms}$

(d) $0.76 \text{ mol HCl} \times \frac{6.022\ 10^{23} \text{ molecules}}{\text{mol}} \times \frac{2 \text{ atoms}}{\text{molecule}} = 9.2 \times 10^{23} \text{ atoms}$

31. (a) $? \% \text{K} = \frac{39.10 \times 100\%}{39.10 + 14.01 + (2 \times 16.00)} = 45.94\% \text{ K}$

$? \% \text{N} = \frac{14.01 \times 100\%}{39.10 + 14.01 + (2 \times 16.00)} = 16.46\% \text{ N}$

$? \% 0 = \frac{(2 \times 16.00) \times 100\%}{39.10 + 14.01 + (2 \times 16.00)} = 37.60\% \text{ O}$

(b) $? \% \text{C} = \frac{(4 \times 12.01) \times 100\%}{(4 \times 12.01) + (10 \times 1.008) + 16.00} = 64.81\% \text{ C}$

$? \% \text{H} = \frac{(10 \times 1.008) \times 100\%}{(4 \times 12.01) + (10 \times 1.008) + 16.00} = 13.60\% \text{ H}$

$? \% \text{O} = \frac{16.00 \times 100\%}{(4 \times 12.01) + (10 \times 1.008) + 16.00} = 21.59\% \text{ O}$

(c) $? \% \text{Al} = \frac{26.98 \times 100\%}{26.98 + (3 \times 30.97) + (9 \times 16.00)} = 10.22\% \text{ Al}$

$? \% \text{P} = \frac{(3 \times 30.97) \times 100\%}{26.98 + (3 \times 30.97) + (9 \times 16.00)} = 35.21\% \text{ P}$

$? \% \text{O} = \frac{(9 \times 16.00) \times 100\%}{26.98 + (3 \times 30.97) + (9 \times 16.00)} = 54.57\% \text{ O}$

(d) $? \% \text{H} = \frac{(6 \times 1.008) \times 100\%}{(6 \times 1.008) + (2 \times 12.01) + (6 \times 16.00)} = 4.80\% \text{ H}$

$? \% \text{C} = \frac{(2 \times 12.01) \times 100\%}{(6 \times 1.008) + (2 \times 12.01) + (6 \times 16.00)} = 19.05\% \text{ C}$

$? \% \text{O} = \frac{(6 \times 16.00) \times 100\%}{(6 \times 1.008) + (2 \times 12.01) + (6 \times 16.00)} = 76.15\% \text{ O}$

33. (a) $?\% \text{O} = \frac{4 \times 16.00}{132.13} \times 100\% = 48.44\% \text{ O}$

(b) $?\% \text{N} = \frac{14.01}{101.15} \times 100\% = 13.85\% \text{ N}$

(c) $?\% \text{Be} = \frac{3 \times 9.012}{537.53} \times 100\% = 5.03\% \text{ Be}$

35. (a) N_2O_5

(b) C_5H_{11}

37. (a)

$$\begin{array}{ll} C & 3 \times 12 = 36 \\ H & 2 \times 1 = 2 \\ Cl & 1 \times 35 = \underline{35} \\ & 73 \end{array}$$

$$\frac{147\ u}{molecule} \times \frac{formula}{73\ u} = \frac{2\ formula}{molecule} \quad C_6H_4Cl_2 \text{ molecular formula}$$

(b)

$$40.00\ g\ C \frac{mol\ C}{12.011\ g\ C} = 3.330\ mol\ C \times \frac{1}{3.330\ mol} = 1 \frac{mol\ C}{mol\ C}$$

$$6.71\ g\ H \times \frac{mol\ H}{1.008\ g\ H} = 6.66\ mol\ H \times \frac{1}{3.330\ mol\ C} = \frac{2\ mol\ H}{mol\ C}$$

$$53.29\ g\ O \times \frac{mol\ O}{16.00\ g\ O} = 3.331\ mol\ O \times \frac{1}{3.330\ mol\ C} = \frac{1\ mol\ O}{mol\ C}$$

CH_2O emperical formula

formula mass = 12.0 + 2 × 1.0 + 16.0 = 30.0

$$\frac{180\ u}{molecule} \times \frac{formula}{30.0\ u} = \frac{6\ formula}{molecule}$$

$C_6H_{12}O_6$ molecular formula

39. (a)

$$? \ mol\ C = 72.22\ g\ C \times \frac{mol\ C}{12.011\ g\ C} = 6.013\ mol\ C \times \frac{1}{0.334\ mol\ N} = \frac{18.00\ mol\ C}{mol\ N}$$

$$? \ mol\ H = 7.07\ g\ H \times \frac{mol\ H}{1.008\ g\ H} = 7.01\ mol\ H \times \frac{1}{0.334\ mol\ N} = \frac{21.0\ mol\ H}{mol\ N}$$

$$? \ mol\ N = 4.68\ g\ N \times \frac{mol\ N}{14.01\ g\ N} = 0.334\ mol\ N \times \frac{1}{0.334\ mol\ N} = \frac{1.00\ mol\ N}{mol\ N}$$

$$? \ mol\ O = 16.03\ g\ O \times \frac{mol\ O}{15.999\ g\ O} = 1.002\ mol\ O \times \frac{1}{0.334\ mol\ N} = \frac{3.00\ mol\ O}{mol\ N}$$

$C_{18}H_{21}NO_3$

(b)

$$21.9\ g\ Mg \times \frac{mol\ Mg}{24.31\ g} = 0.901\ mol\ Mg \times \frac{1}{0.898\ mol\ P} = \frac{1.00\ mol\ Mg}{mol\ P}$$

$$27.8\ g\ P \times \frac{mol\ P}{30.97\ g} = 0.898\ mol\ P \times \frac{1}{0.898\ mol\ P} = \frac{1.00\ mol\ P}{mol\ P}$$

$$50.3\ g\ O \times \frac{mol\ O}{16.00\ g} = 3.14\ mol\ O \times \frac{1}{0.898\ mol\ P} = \frac{3.50\ mol\ O}{mol\ P}$$

$Mg_{(1 \times 2)}P_{(1 \times 2)}O_{(3.5 \times 2)} \rightarrow Mg_2P_2O_7$

41.

$$? \ mol\ C = 65.44\ g\ C \times \frac{mol\ C}{12.011\ g\ C} = 5.448\ mol\ C \times \frac{1}{1.816\ mol\ O} = \frac{3.000\ mol\ C}{mol\ O}$$

$$? \ mol\ H = 5.49\ g\ H \times \frac{mol\ H}{1.008\ g\ H} = 5.446\ mol\ H \times \frac{1}{1.816\ mol\ O} = \frac{2.999\ mol\ H}{mol\ O}$$

$$? \ mol\ O = 29.06\ g\ O \times \frac{mol\ O}{15.999\ g\ O} = 1.816\ mol\ O \times \frac{1}{1.816\ mol\ O} = \frac{1.000\ O}{mol\ O}$$

empirical formula: C_3H_3O

$3 \times 12 + 3 \times 1 + 1 \times 16 = 55$

$$\frac{110\ \text{u}}{\text{molecule}} \times \frac{\text{formula}}{55\ \text{u}} = \frac{2\ \text{formula}}{\text{molecule}} \qquad \text{molecular formula: } C_6H_6O_2$$

43. $? \text{ mol Li} = 4.33\ \text{g Li} \times \dfrac{\text{mol Li}}{6.941\ \text{g Li}} = 0.624\ \text{mol Li} \times \dfrac{1}{0.6234\ \text{mol Cl}} = \dfrac{1.00\ \text{mol Li}}{\text{mol Cl}}$

$$? \text{ mol Cl} = 22.10\ \text{g Cl} \times \frac{\text{mol Cl}}{35.453\ \text{g Cl}} = 0.6234\ \text{mol Cl} \times \frac{1}{0.6234\ \text{mol Cl}} = \frac{1.000\ \text{mol Cl}}{\text{mol Cl}}$$

$$? \text{ mol O} = 39.89\ \text{g O} \times \frac{\text{mol O}}{15.999\ \text{g O}} = 2.493\ \text{mol O} \times \frac{1}{0.6234\ \text{mol Cl}} = \frac{4.000\ \text{mol O}}{\text{mol Cl}}$$

$$? \text{ mol } H_2O = 33.69\ \text{g } H_2O \times \frac{\text{mol } H_2O}{18.015\ \text{g } H_2O} = 1.870\ \text{mol } H_2O \times \frac{1}{0.6234\ \text{mol Cl}} = \frac{3.000\ \text{mol } H_2O}{\text{mol Cl}}$$

$LiClO_4 \cdot 3H_2O$

45.

		Actual
$(NH_4)_2SO_4$	$\% N = \dfrac{2 \times 14}{28 + 8 + 32 + 64}$	21%
NH_4NO_2	$\% N = \dfrac{2 \times 14}{28 + 4 + 32}$	44%
NH_4NO_3	$\% N = \dfrac{2 \times 14}{28 + 4 + 48}$	35%
NH_4Cl	$\% N = \dfrac{14}{14 + 4 + 35}$	26%

The first three all have 2 × 14 in the numerator, so the smallest denominator will be the largest percent. If the 28 + 4 is ignored, then it is easy to see that 32 is less than 48 and less than 4 + 32 + 64. The second one is the lesser of the first three. To compare the second with the fourth, notice that although the numerator is double in the second, only part of the denominator is double, so the second has a smaller denominator and larger percent.

47. $1.278\ \text{g C}\ \dfrac{\text{mol C}}{12.011\ \text{g C}} = 0.1064\ \text{mol C} \times \dfrac{1}{0.1062\ \text{mol S}} = \dfrac{1.002\ \text{mol C}}{\text{mol S}}$

$$0.318\ \text{g H}\ \frac{\text{mol H}}{1.008\ \text{g H}} = 0.315\ \text{mol H} \times \frac{1}{0.1062\ \text{mol S}} = \frac{2.97\ \text{mol H}}{\text{mol S}}$$

$$3.404\ \text{g S}\ \frac{\text{mol S}}{32.066\ \text{g S}} = 0.1062\ \text{mol S} \times \frac{1}{0.1062\ \text{mol S}} = \frac{1.000\ \text{mol S}}{\text{mol S}}$$

CH_3S empirical formula

$$? \frac{\text{formula}}{\text{molecule}} = \frac{94.19\text{ U}}{\text{molecule}} \times \frac{\text{formula}}{47.10\text{ U}} = \frac{2\text{ formula}}{\text{molecule}}$$

$C_2H_6S_2$ molecular formula

49. $$? \text{ mol C} = 1.119 \text{ g } CO_2 \times \frac{\text{mol C}}{44.010 \text{ g } CO_2} = 0.02543 \text{ mol C} \times \frac{1}{0.00635 \text{ mol S}} = \frac{4.00 \text{ mol C}}{\text{mol S}}$$

$$? \text{ mol H} = 0.229 \text{ g } H_2O \times \frac{2 \text{ mol H}}{18.02 \text{ g } H_2O} = 0.0254 \text{ mol H} \times \frac{1}{0.00635 \text{ mol S}} = \frac{4.00 \text{ mol H}}{\text{mol S}}$$

$$? \text{ mol S} = 0.407 \text{ g } SO_2 \times \frac{\text{mol S}}{64.06 \text{ g } SO_2} = 0.00635 \text{ mol S} \times \frac{1}{0.00635 \text{ mol S}} = \frac{1.00 \text{ mol S}}{\text{mol S}}$$

C_4H_4S

51.

		Actual
(a) CH_3OH	$\frac{16}{12+4+16}$	50%
CH_3CH_2OH	$\frac{16}{24+6+16}$	35%
$CH_3OC(CH_3)_3$	$\frac{16}{60+12+16}$	18%

Since the denominator is smaller in CH_3OH, it has the greatest oxygen percent with CH_3CH_2OH next and $CH_3OC(CH_3)_3$ last.

least $CH_3OC(CH_3)_3 < CH_3CH_2OH < CH_3OH$ most

(b) CH_3OH $$\frac{16.0 \times 100\%}{12.0 + 4.0 + 16.0} = 50.0\%$$

$$\frac{?\ \%\text{ O}}{\text{fuel}} = \frac{50.0\%\text{O}}{CH_3OH} \times \frac{10.5 \text{ g } CH_3OH}{100 \text{ g fuel}} = \frac{5.25\%\text{ O}}{\text{fuel}}$$

Yes, methanol does meet the 2.7% O requirement.

(c) $CH_3OC(CH_3)_3$ $$\frac{16.0 \times 100\%}{(5 \times 12.0) + (12 \times 1.01) + 16.0} = 18.2\%$$

$$\frac{?\text{ MTBG}}{\text{gasoline}} = \frac{\frac{2.7\%\text{O}}{\text{gasoline}} \times 100\%}{\frac{18.2\%\text{O}}{\text{MTBG}}} = \frac{15\%\text{ MTBG}}{\text{gasoline}}$$

53. (a) $Cl_2O_5 + H_2O \rightarrow 2\ HClO_3$

(b) $V_2O_5 + 2\ H_2 \rightarrow V_2O_3 + 2\ H_2O$
(c) $4\ Al + 3\ O_2 \rightarrow 2\ Al_2O_3$
(d) $TiCl_4 + 2\ H_2O \rightarrow TiO_2 + 4\ HCl$
(e) $Sn + 2\ NaOH \rightarrow Na_2SnO_2 + H_2$
(f) $PCl_5 + 4\ H_2O \rightarrow H_3PO_4 + 5\ HCl$
(g) $CH_3SH + 3\ O_2 \rightarrow CO_2 + SO_2 + 2\ H_2O$
(h) $3\ Zn(OH)_2 + 2\ H_3PO_4 \rightarrow Zn_3(PO_4)_2 + 6\ H_2O$
(i) $3\ CH_3CH_2OH + PCl_3 \rightarrow 3\ CH_3CH_2Cl + H_3PO_3$

55. (a) $2\ Mg(s) + O_2(g) \rightarrow 2\ MgO(s)$
One mole of oxygen is chemically equivalent to two moles of magnesium.
One mole of oxygen is chemically equivalent to two moles of magnesium oxide.

57. (a) $2\ HCl\ (aq) + Zn\ (s) \rightarrow ZnCl_2\ (aq) + H_2\ (g)$
(b) $C_2H_6\ (g) + 2\ H_2O\ (g) \rightarrow 2\ CO\ (g) + 5\ H_2\ (g)$
(c) $P_4O_{10}\ (s) + 6\ H_2O\ (l) \rightarrow 4\ H_3PO_4\ (aq)$
(d) $Pb(s) + PbO_2(s) + 2\ H_2SO_4(aq) \rightarrow 2\ PbSO_4(s) + 2\ H_2O(l)$

59. $? \text{ mol Fe} = 72.3 \text{ g Fe} \times \frac{\text{mol Fe}}{55.85 \text{ g Fe}} = 1.29 \text{ mol Fe} \times \frac{1}{1.29 \text{ mol Fe}} = \frac{1.00 \text{ mol Fe}}{\text{mol Fe}}$

$? \text{ mol O} = 27.7 \text{ g O} \times \frac{\text{mol O}}{16.00 \text{ g O}} = 1.73 \text{ mol O} \times \frac{1}{1.29 \text{ mol Fe}} = \frac{1.34 \text{ mol O}}{\text{mol Fe}}$

$Fe_{(1 \times 3)}O_{(1.34 \times 3)} \Rightarrow Fe_3O_4$

$3\ Fe_2O_3(s) + H_2(g) \xrightarrow{400\ °C} H_2O(g) + 2\ Fe_3O_4(s)$

61. $4\ NH_3 + 5\ O_2 \rightarrow 4\ NO + 6\ H_2O$

63. (a) $? \text{ mol } CO_2 = 451 \text{ mol } C_8H_{18} \times \frac{16 \text{ mol } CO_2}{2 \text{ mol } C_8H_{18}} = 3.61 \times 10^2 \text{ mol } CO_2$

(b) $? \text{ mol } O_2 = 585 \text{ mol } C_8H_{18} \times \frac{25 \text{ mol } O_2}{2 \text{ mol } C_8H_{18}} = 7.31 \times 10^3 \text{ mol } O_2$

(c) $? \text{ mol } H_2O = 188 \text{ mol } C_8H_{18} \times \frac{18 \text{ mol } H_2O}{2 \text{ mol } C_8H_{18}} = 1.69 \times 10^3 \text{ mol } H_2O$

(d) $? \text{ mol } C_8H_{18} = 2.2 \times 10^4 \text{ mol } O_2 \times \frac{2 \text{ mol } C_8H_{18}}{25 \text{ mol } O_2} = 1.8 \times 10^3 \text{ mol } C_8H_{18}$

65. $CaCN_2 + 3\ H_2O \rightarrow CaCO_3 + 2\ NH_3$
$Mg_3N_2 + 6\ H_2O \rightarrow 3\ Mg(OH)_2 + 2\ NH_3$

$? \text{ kg } NH_3 = 1 \text{ kg } CaCN_2 \times \frac{\text{mol } CaCN_2}{80 \text{ g } CaCN_2} \times \frac{2 \text{ mol } NH_3}{\text{mol } CaCN_2} \times \frac{17 \text{ g } NH_3}{\text{mol } NH_3}$

$$? \text{ kg } NH_3 = 1 \text{ kg } Mg_3N_2 \times \frac{\text{mol } Mg_3N_2}{101 \text{ g } Mg_3N_2} \times \frac{2 \text{ mol } NH_3}{\text{mol } Mg_3N_2} \times \frac{17 \text{ g } NH_3}{\text{mol } NH_3}$$

Since the only difference in the two equations is the molar mass of the compounds $CaCN_2$ and Mg_3N_2, the smaller molar mass will produce the most NH_3; $CaCN_2$ produces more than Mg_3N_2. (Actual 0.43 kg to 0.34 kg)

67. $2\ C_{14}H_{30} + 43\ O_2 \rightarrow 28\ CO_2 + 30\ H_2O$

$$? \text{ g } CO_2 = 7.53 \text{ L } C_{14}H_{30} \times \frac{\text{mL}}{10^{-3} \text{ L}} \times \frac{0.763 \text{ g}}{\text{mL}} \times \frac{\text{mol } C_{14}H_{30}}{198.38 \text{ g } C_{14}H_{30}}$$

$$\times \frac{28 \text{ mol } CO_2}{2 \text{ mol } C_{14}H_{30}} \times \frac{44.01 \text{ g } CO_2}{\text{mol } CO_2} = 1.78 \times 10^4 \text{ g } CO_2$$

69. (a) $CaCO_3 + 2\ HCl \rightarrow CaCl_2 + CO_2 + H_2O$

$$? \text{ g } CO_2 = 12.3 \text{ g chalk} \times \frac{69.7 \text{ g } CaCO_3}{100.0 \text{ g chalk}} \times \frac{\text{mol } CaCO_3}{100.09 \text{ g } CaCO_3} \times \frac{\text{mol } CO_2}{\text{mol } CaCO_3}$$

$$\times \frac{44.01 \text{ g } CO_2}{\text{mol } CO_2} = 3.77 \text{ g } CO_2$$

(b) $$? \text{ g } CaCO_3 = 1.31 \text{ g } CO_2 \times \frac{\text{mol } CO_2}{44.01 \text{ g } CO_2} \times \frac{\text{mol } CaCO_3}{\text{mol } CO_2} \times \frac{100.09 \text{ g } CaCO_3}{\text{mol } CaCO_3}$$

$$= 2.979 \text{ g } CaCO_3$$

$$\frac{2.979 \text{ g } CaCO_3}{4.38 \text{ g chalk}} \times 100\% = 68.0\% \ CaCO_3$$

71. The limiting reactant is the reactant that will be used up in a chemical reaction. Therefore, it limits the amount of product(s) formed.
A reaction can have two limiting reactants if two reactant are in stoiochiometric proportions to each other, that is, both will be used up simultaneously.

73. $2\ LiOH(s) + CO_2(g) \rightarrow Li_2CO_3(s) + H_2O(l)$

$$? \text{ mol } Li_2CO_3 = 4.40 \text{ mol LiOH} \times \frac{\text{mol } Li_2CO_3}{2 \text{ mol LiOH}} = 2.20 \text{ mol } Li_2CO_3$$

$$? \text{ mol } Li_2CO_3 = 3.20 \text{ mol } CO_2 \times \frac{\text{mol } Li_2CO_3}{\text{mol } CO_2} = 3.20 \text{ mol } Li_2CO_3$$

LiOH is limiting and 2.20 mol Li_2CO_3 can be produced.

75. $2\ Hg(l) + O_2\ (g) \rightarrow 2\ HgO(s)$

$$? \text{ mol } O_2 = 4.00\text{g } O_2 \times \frac{\text{mol } O_2}{32.00 \text{ g}} = 0.125 \text{ mol } O_2$$

0.250 mol O, so Hg is limiting.
? mol O excess = 0.250 mol O – 0.200 mol O = 0.050 mol O excess

$$? \text{ g } O_2 = 0.050 \text{ mol O excess} \times \frac{\text{mol } O_2}{2 \text{ mol O}} \times \frac{32.00 \text{ g}}{\text{mol } O_2} = 0.80 \text{ g } O_2$$

(d) is the correct answer.
(a) Some Hg must be used up to make HgO.
(b) The reaction should continue to completion.
(c) The Hg is limiting. The Hg(l) and O_2(g) are *not* in stoichiometric proportions.

77. (a) $HI + KHCO_3 \rightarrow KI + CO_2 + H_2O$

$$? \text{ mol KI} = 398 \text{ g HI} \times \frac{1 \text{ mol HI}}{127.9 \text{ g HI}} \times \frac{1 \text{ mol KI}}{1 \text{ mol HI}} = 3.11 \text{ mol KI}$$

$$? \text{ mol KI} = 318 \text{ g } KHCO_3 \times \frac{1 \text{ mol } KHCO_3}{100.1 \text{ g } KHCO_3} \times \frac{1 \text{ mol KI}}{1 \text{ mol } KHCO_3} = 3.18 \text{ mol KI}$$

HI is the limiting reactant, and the mass of the product is

$$? \text{ g KI} = 3.11 \text{ mol KI} \times \frac{166.0 \text{ g KI}}{1 \text{ mol KI}} = 516 \text{ g KI.}$$

(b) $KHCO_3$ is in excess.

$$? \text{ g } KHCO_{3\,\text{consumed}} = 3.11 \text{ mol KI} \times \frac{1 \text{ mol } KHCO_3}{1 \text{ mol KI}} \times \frac{100.1 \text{ g } KHCO_3}{1 \text{ mol } KHCO_3}$$
$$= 311 \text{ g } KHCO_3 \text{ consumed}$$

$$318 \text{ g } KHCO_{3\,\text{initially}} - 311 \text{ g } KHCO_{3\,\text{consumed}} = 7 \text{ g } KHCO_{3\,\text{in excess}}$$

79. $3\ Cu + 2\ FeCl_3 \rightarrow 3\ CuCl_2 + 2\ Fe$

$$? \text{ g } FeCl_3 = 3.72 \text{ g Cu} \times \frac{\text{mol}}{63.55 \text{ g}} \times \frac{2 \text{ mol } FeCl_3}{3 \text{ mol Cu}} \times \frac{162.2 \text{ g}}{\text{mol}} = 6.33 \text{ g } FeCl_3$$

8.48 g is more than enough.

81. $$? \text{ mol ZnS} = 0.488 \text{ g Zn} \times \frac{\text{mol Zn}}{65.39 \text{ g Zn}} \times \frac{8 \text{ mol ZnS}}{8 \text{ mol Zn}} = 0.00746 \text{ mol ZnS}$$

$$? \text{ mol ZnS} = 0.503 \text{ g } S_8 \times \frac{\text{mol } S_8}{256.53 \text{ g } S_8} \times \frac{8 \text{ mol ZnS}}{\text{mol } S_8} = 0.0157 \text{ mol ZnS}$$

Zn is limiting.

$$? \text{ g ZnS} = 0.00746 \text{ mol ZnS} \times \frac{97.46 \text{ g ZnS}}{\text{mol ZnS}} = 0.727 \text{ g ZnS} \quad \text{theoretical yield}$$

$$? \text{ \% yield} = \frac{0.606 \text{ g ZnS actual}}{0.727 \text{ g ZnS theo.}} \times 100\% = 83.4\%$$

83. $$? \text{ mol } NH_4HCO_3 = 14.8 \text{ g } NH_3 \times \frac{\text{mol } NH_3}{17.03 \text{ g } NH_3} \times \frac{\text{mol } NH_4HCO_3}{\text{mol } NH_3}$$
$$= 0.8691 \text{ mol } NH_4HCO_3$$

$$? \text{ mol } NH_4HCO_3 = 41.3 \text{ g } CO_2 \times \frac{\text{mol } CO_2}{44.01 \text{ g } CO_2} \times \frac{\text{mol } NH_4HCO_3}{\text{mol } CO_2}$$

$= 0.9384 \text{ mol } NH_4HCO_3$

NH_3 is limiting.

$$? \text{ g } NH_4HCO_3 = 0.8691 \text{ mol } NH_4HCO_3 \times \frac{79.06 \text{ g } NH_4HCO_3}{\text{mol } NH_4HCO_3} \times \frac{74.7 \text{ g}}{100.0\text{g}}$$
$$= 51.3 \text{ g } NH_4HCO_3$$

85. (a) $? \text{ M } CaCl_2 = \dfrac{2.60 \text{ mol } CaCl_2}{1.15 \text{ L}} = 2.26 \text{ M } CaCl_2$

(b) $? \text{ M } Li_2CO_3 = \dfrac{0.000700 \text{ mol } Li_2CO_3}{10.0 \text{ mL solution}} \times \dfrac{\text{mL}}{10^{-3} \text{ L}} = 0.0700 \text{ M } Li_2CO_3$

(c) $? \text{ M } NaNO_3 = \dfrac{6.631 \text{ g } NaNO_3}{100.0 \text{ mL}} \times \dfrac{\text{mL}}{10^{-3} \text{ L}} \times \dfrac{\text{mol } NaNO_3}{84.994 \text{ g } NaNO_3} = 0.7802 \text{ M}$

$NaNO_3$

(d) $? \text{ M } C_{12}H_{22}O_{11} = \dfrac{412 \text{ g } C_{12}H_{22}O_{11}}{1.25 \text{ L}} \times \dfrac{\text{mol } C_{12}H_{22}O_{11}}{342.3 \text{ g } C_{12}H_{22}O_{11}}$

$= 0.963 \text{ M } C_{12}H_{22}O_{11}$

(e) $? \text{ M } C_3H_8O_3 = \dfrac{15.50 \text{ mL } C_3H_8O_3}{225.0 \text{ mL}} \times \dfrac{\text{mL}}{10^{-3} \text{ L}} \times \dfrac{1.265 \text{ g } C_3H_8O_3}{\text{mL } C_3H_8O_3}$

$\times \dfrac{\text{mol } C_3H_8O_3}{92.094 \text{ g } C_3H_8O_3} = 0.9463 \text{ M}$

(f) $? \text{ M } C_3H_7OH = \dfrac{35.0 \text{ mL } C_3H_7OH}{250. \text{ mL}} \times \dfrac{\text{mL}}{10^{-3} \text{ L}} \times \dfrac{0.786 \text{ g } C_3H_7OH}{\text{mL } C_3H_7OH}$

$\times \dfrac{\text{mol } C_3H_7OH}{60.10 \text{ g } C_3H_7OH} = 1.83 \text{ M } C_3H_7OH$

87. (a) $? \text{ mol NaOH} = 1.25 \text{ L} \times 0.0235 \text{ M NaOH} = 0.0294 \text{ mol NaOH}$

(b) $? \text{ g } C_6H_{12}O_6 = 10.0 \text{ mL} \times \dfrac{10^{-3} \text{ L}}{\text{mL}} \times 4.25 \text{ M} \times \dfrac{180.2 \text{ g } C_6H_{12}O_6}{\text{mol } C_6H_{12}O_6}$

$= 7.66 \text{ g } C_6H_{12}O_6$

(c) $? \text{ g } CuSO_4 \cdot 5 \, H_2O = 3.00 \text{ L} \times 0.275 \text{ M} \times \dfrac{249.7 \text{ g } CuSO_4 \cdot 5 \, H_2O}{\text{mol } CuSO_4 \cdot 5 \, H_2O}$

$= 206 \text{ g } CuSO_4 \cdot 5 \, H_2O$

(d) $? \text{ mL } C_4H_{10}O = 715 \text{ mL} \times \dfrac{10^{-3} \text{ L}}{\text{mL}} \times 1.34 \text{ M} \times \dfrac{74.12 \text{ g } C_4H_{10}O}{\text{mol } C_4H_{10}O}$

$\times \dfrac{\text{mL } C_4H_{10}O}{0.808 \text{ g } C_4H_{10}O} = 87.9 \text{ mL } C_4H_{10}O$

89. (a) $? \text{ mL} = 0.0867 \text{ mol NaBr} \times \frac{1}{0.215 \text{ M}} \times \frac{\text{mL}}{10^{-3} \text{ L}} = 403 \text{ mL}$

(b) $? \text{ mL} = 32.1 \text{ g } CO(NH_2)_2 \times \frac{\text{mol } CO(NH_2)_2}{60.06 \text{ g } CO(NH_2)_2} \times \frac{1}{0.215 \text{ M}} \times \frac{\text{mL}}{10^{-3} \text{ L}}$

$= 2.49 \times 10^3 \text{ mL}$

(c) $? \text{ mL} = 715 \text{ mg } CH_3OH \times \frac{\text{mol } CH_3OH}{32.04 \text{ g } CH_3OH} \times \frac{1}{0.215 \text{ M}} = 104 \text{ mL}$

91. $? \text{ M } H_3PO_4 = \frac{85.0 \text{ g } H_3PO_4}{100. \text{ g solution}} \times \frac{1.689 \text{ g}}{\text{mL}} \times \frac{\text{mL}}{10^{-3} \text{ L}} \times \frac{\text{mol } H_3PO_4}{97.99 \text{ g } H_3PO_4}$

$= 14.7 \text{ M } H_3PO_4$

93. $? \text{ M} = 14.00 \text{ mL} \times \frac{10^{-3} \text{ L}}{\text{mL}} \times \frac{1.04 \text{ mol}}{\text{L}} \times \frac{1}{0.500 \text{ L}} = 0.0291 \text{ M}$

95. (a) $M_C V_C = M_D V_D$

$V_C = \frac{M_D V_D}{M_C} = \frac{0.5000 \text{ M HCl x } 2.000 \text{ L}}{6.052 \text{ M HCl}} = 0.1652 \text{ L} \times \frac{\text{mL}}{10^{-3} \text{ L}} = 165.2 \text{ mL}$

(b) $? \text{ mL} = \frac{7.150 \text{ mg HCl}}{\text{mL}} \times 500.0 \text{ mL} \times \frac{\text{mol HCl}}{36.46 \text{ g HCl}} \times \frac{1}{6.052 \text{ M}} = 16.20 \text{ mL}$

97. (c) 0.17 M NH_3: The concentration must be greater than the average of 0.10 M and 0.20 M, but less than the greater of the two (0.20M).
The actual calculation would be:

$$C = \frac{(0.100 \text{ L} \times 0.100 \text{ mol/L}) + (0.200 \text{ L} \times 0.200 \text{ mol/L})}{0.100 \text{ L} + 0.200 \text{ L}} = 0.167 \text{ M} \Rightarrow 0.17 \text{ M}$$

99. Measure 25.00 mL of 0.04000 M $AgNO_3$ solution 3 times and 10.00 mL once into a flask. Then add 0.8664 g $AgNO_3$.

$? \text{ mol needed} = (85.00 \text{ ml} \times \frac{10^{-3} \text{ L}}{\text{mL}} \times 0.100 \text{ M})$

$- (85.00 \text{ mL} \times \frac{10^{-3} \text{ L}}{\text{mL}} \times 0.04000 \text{ M}) = 5.10 \times 10^{-3} \text{ mol}$

$? \text{ g } AgNO_3 \text{ added} = 5.10 \times 10^{-3} \text{ mol} \times \frac{169.88 \text{ g } AgNO_3}{\text{mol } AgNO_3} = 0.866 \text{ g } AgNO_3$

(An alternate method is listed below.)
Using a volumetric flask, it is necessary to make 100 mL of solution, as that is the smallest volumetric flask available with a volume greater than 80 mL. Measure 100.0 mL of the 0.04000 M $AgNO_3$ solution in the 100.0 mL volumetric flask, remove 20.00 mL with the 10.00 mL pipet, and add 0.815 g of $AgNO_3$.

$$? \text{ mol needed} = 80.0 \text{ mL} \times \frac{10^{-3}\text{ L}}{\text{mL}} \times 0.100 \text{ M} - 80.0 \text{ mL} \times \frac{10^{-3}\text{ L}}{\text{mL}} \times 0.04000 \text{ M}$$
$$= 4.80 \times 10^{-3} \text{ mol needed}$$

$$? \text{ g } AgNO_3 \text{ added} = 4.80 \times 10^{-3} \text{ mol} \times \frac{169.88 \text{ g } AgNO_3}{\text{mol } AgNO_3} = 0.815 \text{ g } AgNO_3 \text{ added}$$

101. $$? \text{ g } BaSO_4 = 635 \text{ mL} \times \frac{10^{-3}\text{ L}}{\text{mL}} \times 0.314 \text{ M } Na_2SO_4 \times \frac{\text{mol } BaSO_4}{\text{mol } Na_2SO_4}$$
$$\times \frac{233.4 \text{ g } BaSO_4}{\text{mol } BaSO_4} = 46.5 \text{ g } BaSO_4$$

103. $$? \text{ mol HCl used} = 2.02 \text{ g Al} \times \frac{\text{mol Al}}{26.98 \text{ g Al}} \times \frac{6 \text{ mol HCl}}{2 \text{ mol Al}} = 0.225 \text{ mol HCl used}$$
$$? \text{ mol HCl available} = 0.400 \text{ L} \times 2.75 \text{ M HCl} = 1.10 \text{ mol HCl}$$
$$? \text{ M} = \frac{1.10 \text{ mol} - 0.225 \text{ mol}}{0.400 \text{ L}} = 2.2 \text{ M}$$

105. $CaCO_3 + 2HCl \rightarrow CaCl_2 + CO_2 + H_2O$
$$? \text{ g } CO_2 = 4.35 \text{ g } CaCO_3 \times \frac{\text{mol } CaCO_3}{100.1 \text{ g } CaCO_3} \times \frac{\text{mol } CO_2}{\text{mol } CaCO_3} \times \frac{44.01 \text{ g } CO_2}{\text{mol } CO_2}$$
$$= 1.91 \text{ g } CO_2$$
$$? \text{ g } CO_2 = 75.0 \text{ mL} \times \frac{10^{-3}\text{ L}}{\text{mL}} \times 1.50 \text{ M HCl} \times \frac{\text{mol } CO_2}{2 \text{ mol HCl}} \times \frac{44.01 \text{ g } CO_2}{\text{mol } CO_2}$$
$$= 2.48 \text{ g } CO_2$$
$CaCO_3$ is limiting. 1.91 g CO_2 is produced.

107. (a) $2 \text{ Al} + 6 \text{ HCl} \rightarrow 2 \text{ AlCl}_3 + 3 \text{ H}_2$
$$? \text{ cm}^2 = 0.05 \text{ mL} \times \frac{10^{-3}\text{ L}}{\text{mL}} \times 12.0 \text{ M HCl} \times \frac{2 \text{ mol Al}}{6 \text{ mol HCl}} \times \frac{26.98 \text{ g Al}}{\text{mol Al}} \times \frac{\text{cm}^3}{2.70 \text{ g}}$$
$$\times \frac{1}{0.10 \text{ mm}} \times \frac{\text{mm}}{10^{-3} \text{ m}} \times \frac{10^{-2} \text{ m}}{\text{cm}} = 0.2 \text{ cm}^2$$
(b) $$? \text{ drops} = 1.50 \text{ cm}^2 \times 0.065 \text{ mm} \times \frac{10^{-3} \text{ m}}{\text{mm}} \times \frac{\text{cm}}{10^{-2} \text{ m}} \times \frac{8.96 \text{ g}}{\text{cm}^3} \times \frac{\text{mol Cu}}{63.55 \text{ g Cu}}$$
$$\times \frac{8 \text{ mol } HNO_3}{3 \text{ mol Cu}} \times \frac{1}{6.0 \text{ M } HNO_3} \times \frac{\text{mL}}{10^{-3} \text{ L}} \times \frac{\text{drop}}{0.05 \text{ mL}} = 12.2 \text{ drops} \Rightarrow 13 \text{ drops}$$

Additional Problems

110. $$? \text{ tablets} = 550 \text{ mg } Fe^{2+} \times \frac{151.9 \text{ g } FeSO_4}{55.85 \text{ g Fe}} \times \frac{\text{tablet}}{325 \text{ mg } FeSO_4} = 4.6 \text{ tablets}$$

5 tablets would be lethal.

113. $0.210\text{ g HCl} \times \dfrac{\text{mol Cl}}{36.46\text{ g HCl}} = 5.76 \times 10^{-3}\text{ mol Cl} \times \dfrac{1}{1.92\times10^{-3}\text{ mol Br}} = \dfrac{3\text{ mol Cl}}{\text{mol Br}}$

$0.155\text{ g HBr} \times \dfrac{\text{mol Br}}{80.91\text{ HBr}} = 1.92 \times 10^{-3}\text{ mol Br} \times \dfrac{1}{1.92\times10^{-3}\text{ mol Br}} = \dfrac{1\text{ mol Br}}{\text{mol Br}}$

$BrCl_3$

115. $MCl_2 + 2\ AgNO_3 \rightarrow 2\ AgCl + M(NO_3)_2$

$?\text{ mol } MCl_2 = 1.8431\text{ g AgCl} \times \dfrac{\text{mol AgCl}}{143.321\text{ g AgCl}} \times \dfrac{\text{mol } MCl_2}{2\text{ mol AgCl}} = 0.0064300\text{ mol } MCl_2$

$?\text{ molar mass } MCl_2 = \dfrac{0.8150\text{ g } MCl_2}{0.0064300\text{ mol } MCl_2} = 126.75\text{ g/mol}$

$?\text{ atomic mass M} = 126.75\text{ g/mol} - 2 \times 35.45\text{ g/mol} = 55.85\text{ g/mol}$ atomic mass
M is iron.

118. $?\text{ mol } C_4H_9Br = 13.0\text{ g} \times \dfrac{\text{mol } C_4H_9OH}{74.12\text{ g } C_4H_9OH} \times \dfrac{\text{mol } C_4H_9Br}{\text{mol } C_4H_9OH} = 0.175\text{ mol } C_4H_9Br$

$?\text{ mol } C_4H_9Br = 21.6\text{ g} \times \dfrac{\text{mol NaBr}}{102.9\text{ g NaBr}} \times \dfrac{\text{mol } C_4H_9Br}{\text{mol NaBr}} = 0.210\text{ mol } C_4H_9Br$

$?\text{ mol } C_4H_9Br = 33.8\text{ g} \times \dfrac{\text{mol } H_2SO_4}{98.09\text{ g } H_2SO_4} \times \dfrac{\text{mol } C_4H_9Br}{\text{mol } H_2SO_4} = 0.345\text{ mol } C_4H_9Br$

C_4H_9OH limiting

$?\text{ g} = 0.1754\text{ mol } C_4H_9Br \times \dfrac{137.0\text{ g } C_4H_9Br}{\text{mol } C_4H_9Br} = 24.0\text{ g}$ theoretical yield

16.8 g actual yield $\quad ?\%\text{ yield} = \dfrac{16.8\text{ g}}{24.0\text{ g}} \times 100\% = 70.0\%$ yield

120. $37.51\text{ g C} \times \dfrac{\text{mol C}}{12.011\text{ g}} = 3.123\text{ mol C}$

$3.15\text{ g H} \times \dfrac{\text{mol H}}{1.008\text{ g}} = 3.13\text{ mol H}$

$59.34\text{ g F} \times \dfrac{\text{mol F}}{18.998\text{ g}} = 3.123\text{ mol F}$

CHF empirical formula

$$? \frac{\text{formula}}{\text{molecule}} = \frac{96.0524\text{u}}{\text{molecule}} \times \frac{\text{formula}}{32.017\text{u}} = \frac{3\ \text{formula}}{\text{molecule}}$$

$C_3H_3F_3$ molecular formula

$4\ C_3H_3F_3 + 12\ O_2 \rightarrow 3\ CF_4 + 9\ CO_2 + 6\ H_2O$

123. $3\ Mg + N_2 \rightarrow Mg_3N_2$

$$?\ \text{mol}\ Mg_3N_2 = 35.00\ \text{g Mg} \times \frac{\text{mol Mg}}{24.305\ \text{g Mg}} \times \frac{1\ \text{mol}\ Mg_3N_2}{3\ \text{mol Mg}} = 0.4800\ \text{mol}\ Mg_3N_2$$

(a) $$?\ \text{mol}\ Mg_3N_2 = 15.00\ \text{g gas} \times \frac{95\ \text{g}\ N_2}{100\ \text{g gas}} \times \frac{1\ \text{mol}\ N_2}{28.013\ \text{g}\ N_2} \times \frac{\text{mol}\ Mg_3N_2}{\text{mol}\ N_2} = 0.51\ \text{mol}\ Mg_3N_2$$

Mg is still limiting

$$?\ \text{g}\ Mg_3N_2 = 0.4800\ \text{mol} \times \frac{100.93\ \text{g}\ Mg_3N_2}{\text{mol}\ Mg_3N_2} = 48.45\ \text{g}\ Mg_3N_2$$

(b) $$?\ \text{g}\ Mg_3N_2 = 15.00\ \text{g gas} \times \frac{85\ \text{g}\ N_2}{100\ \text{g gas}} \times \frac{1\ \text{mol}\ N_2}{28.013\ \text{g}\ N_2} \times \frac{\text{mol}\ Mg_3N_2}{\text{mol}\ N_2} = 0.46\ \text{mol}\ Mg_3N_2$$

Now N_2 is limiting.

$$?\ \text{g}\ Mg_3N_2 = 0.46\ \text{mol}\ Mg_3N_2 \times \frac{100.93\ \text{g}\ Mg_3N_2}{\text{mol}\ Mg_3N_2} = 46\ \text{g}\ Mg_3N_2$$

(c) $2\ Mg + O_2 \rightarrow 2\ MgO$

$$?\ \text{g MgO} = 15.00\ \text{g gas} \times \frac{25\ \text{g}\ O_2}{100\ \text{g gas}} \times \frac{\text{mol}\ O_2}{32.00\ \text{g}\ O_2} \times \frac{2\ \text{mol MgO}}{\text{mol}\ O_2} \times \frac{40.304\ \text{g MgO}}{\text{mol MgO}} = 9.4\ \text{g MgO}$$

$$?\ \text{g Mg used} = 15.00\ \text{g gas} \times \frac{25\ \text{g}\ O_2}{100\ \text{g gas}} \times \frac{\text{mol}\ O_2}{32.00\ \text{g}\ O_2} \times \frac{2\ \text{mol Mg}}{\text{mol}\ O_2} \times \frac{24.305\ \text{g Mg}}{\text{mol Mg}} = 5.7\ \text{g Mg used}$$

$?\ \text{g Mg left} = 35.00\ \text{g Mg} - 5.7\ \text{g} = 29.3\ \text{g left}$

$$?\ \text{mol}\ Mg_3N_2 = 29.3\ \text{g Mg} \times \frac{\text{mol Mg}}{24.305\ \text{g Mg}} \times \frac{\text{mol}\ Mg_3N_2}{3\ \text{mol Mg}} = 0.402\ \text{mol}\ Mg_3N_2$$

$$?\ \text{mol}\ Mg_3N_2 = 15.00\ \text{g gas} \times \frac{75\ \text{g}\ N_2}{100\ \text{g gas}} \times \frac{1\ \text{mol}\ N_2}{28.013\ \text{g}\ N_2} \times \frac{\text{mol}\ Mg_3N_2}{\text{mol}\ N_2} = 0.40\ \text{mol}\ Mg_3N_2$$

Both are limiting.

$$?\ \text{g}\ Mg_3N_2 = 0.40\ \text{mol}\ Mg_3N_2 \times \frac{100.93\ \text{g}\ Mg_3N_2}{\text{mol}\ Mg_3N_2} = 41\ \text{g}\ Mg_3N_2$$

Total product = 9.4 g MgO + 41 g Mg_3N_2 = 50. g

124. $40.00\text{ g C} \times \dfrac{\text{mol C}}{12.011\text{ g C}} = 3.330\text{ mol C} \times \dfrac{1}{3.330\text{ mol C}} = \dfrac{1\text{ mol C}}{\text{mol C}}$

$6.71\text{ g H} \times \dfrac{\text{mol H}}{1.008\text{ g H}} = 6.66\text{ mol H} \times \dfrac{1}{3.330\text{ mol C}} = \dfrac{2\text{ mol H}}{\text{mol C}}$

$53.29\text{ g O} \times \dfrac{\text{mol O}}{16.00\text{ g O}} = 3.331\text{ mol O} \times \dfrac{1}{3.330\text{ mol C}} = \dfrac{1\text{ mol O}}{\text{mol C}}$

CH_2O

$\dfrac{?\text{ formula}}{\text{molecule}} = \dfrac{60\text{ g}}{\text{molecule}} \times \dfrac{\text{formula}}{30\text{ g}} = \dfrac{2\text{ formula}}{\text{molecule}}$

$C_2H_4O_2$

$22.56\text{ g P} \times \dfrac{\text{mol P}}{30.974\text{ g P}} = 0.7284\text{ mol P} \times \dfrac{1}{0.7284\text{ mol P}} = \dfrac{1\text{ mol P}}{\text{mol P}}$

$77.44\text{ g Cl} \times \dfrac{\text{mol Cl}}{35.453\text{ g Cl}} = 2.184\text{ mol Cl} \times \dfrac{1}{0.7284\text{ mol P}} = \dfrac{3.00\text{ mol Cl}}{\text{mol P}}$

PCl_3

$3.69\text{ g H} \times \dfrac{\text{mol H}}{1.008\text{ g H}} = 3.66\text{ mol H} \times \dfrac{1}{1.219\text{ mol P}} = \dfrac{3\text{ mol H}}{\text{mol P}}$

$37.77\text{ g P} \times \dfrac{\text{mol P}}{30.974\text{ g P}} = 1.219\text{ mol P} \times \dfrac{1}{1.219\text{ mol P}} = \dfrac{1\text{ mol P}}{\text{mol P}}$

$58.53\text{ g O} \times \dfrac{\text{mol O}}{15.999\text{ g O}} = 3.658\text{ mol O} \times \dfrac{1}{1.219\text{ mol O}} = \dfrac{3\text{ mol O}}{\text{mol P}}$

H_3PO_3

$C_2H_4O_2 + PCl_3 \rightarrow H_3PO_3$

$?\text{ mol } C_2H_4O_2 = 10.000\text{ g} \times \dfrac{\text{mol}}{60.052\text{ g}} = 0.16652\text{ mol}$

$?\text{ mol } PCl_3 = 7.621\text{ g} \times \dfrac{\text{mol}}{137.32\text{ g}} = 0.05550\text{ mol}$

$?\text{ mol } H_3PO_3 = 4.552\text{ g} \times \dfrac{\text{mol}}{81.991\text{ g}} = 0.05552\text{ mol}$

$\dfrac{0.16652\text{ mol}}{0.05550\text{ mol}} = 3$

$3\ C_2H_4O_2 + PCl_3 \rightarrow H_3PO_3 + C_6O_3Cl_3H_9$

leads to 6 C + 3 O + 3 Cl + 9 H

$3\ CH_3COOH + PCl_3 \rightarrow H_3PO_3 + 3\ CH_3C(O)Cl$

Apply Your Knowledge

128 (a) $?\ \%\text{ P} = 10\ \%\ P_2O_5 \times \dfrac{\text{mol } P_2O_5}{141.94\text{ g}} \times \dfrac{2\text{ mol P}}{\text{mol } P_2O_5} \times \dfrac{30.97\text{ g P}}{\text{mol P}} = 4.4\ \%\text{ P}$

$?\ \%\text{ K} = 5\ \%\ K_2O \times \dfrac{\text{mol } K_2O}{94.20\text{ g}} \times \dfrac{2\text{ mol K}}{\text{mol } K_2O} \times \dfrac{39.10\text{ g K}}{\text{mol K}} = 4\ \%\text{ K}$

(b) NH_4NO_3

$$? \% N = \frac{2 \times 14.01}{2 \times 14.01 + 4 \times 1.01 + 3 \times 16.00} \times 100\% = 35.00\ \% N$$

NPK is 35-0-0

(c) KH_2PO_4

$$?\% K_2O = \frac{94.2\ g\ K_2O}{mol\ K_2O} \times \frac{1\ mol\ K_2O}{2\ mol\ KH_2PO_4} \times \frac{mol\ KH_2PO_4}{136\ g\ KH_2PO_4} \times 100\% = 35\%$$

$$?\% P_2O_5 = \frac{142\ g\ P_2O_5}{mol\ P_2O_5} \times \frac{mol\ P_2O_5}{2\ mol\ KH_2PO_4} \times \frac{mol\ KH_2PO_4}{136\ g\ KH_2PO_4} \times 100\% = 52\%$$

K_2HPO_4

$$?\% K_2O = \frac{94.2\ g\ K_2O}{mol\ K_2O} \times \frac{mol\ K_2O}{mol\ K_2HPO_4} \times \frac{mol\ K_2HPO_4}{174\ g\ K_2HPO_4} \times 100\% = 54\%$$

$$?\% P_2O_5 = \frac{142\ g\ P_2O_5}{mol\ P_2O_5} \times \frac{mol\ P_2O_5}{2\ mol\ K_2HPO_4} \times \frac{mol\ K_2HPO_4}{174\ g\ K_2HPO_4} \times 100\% = 41\%$$

K_3PO_4

$$?\% K_2O = \frac{94.2\ g\ K_2O}{mol\ K_2O} \times \frac{3\ mol\ K_2O}{2\ mol\ K_3PO_4} \times \frac{mol\ K_3PO_4}{212\ g\ K_3PO_4} \times 100\% = 67\%$$

$$?\% P_2O_5 = \frac{142\ g\ P_2O_5}{mol\ P_2O_5} \times \frac{mol\ P_2O_5}{2\ mol\ K_3PO_4} \times \frac{mol\ K_3PO_4}{212\ g\ K_3PO_4} \times 100\% = 33\%$$

0.5 mol of KH_2PO_4 would provide ≈ 18% (0.5 × 35) of K_2O, so one compound is KH_2PO_4. It would also provide ≈ 26% of the P_2O_5.

$NH_4H_2PO_4$

$$?\% P_2O_5 = \frac{142\ g\ P_2O_5}{mol\ P_2O_5} \times \frac{mol\ P_2O_5}{2\ mol\ NH_4H_2PO_4} \times \frac{mol\ NH_4H_2PO_4}{115\ g\ NH_4H_2PO_4} \times 100\% = 62\%$$

$$?\% N = \frac{14.01\ g\ N}{mol\ N} \times \frac{mol\ N}{mol\ NH_4H_2PO_4} \times \frac{mol\ NH_4H_2PO_4}{115\ g\ NH_4H_2PO_4} \times 100\% = 12\%$$

That is too much P_2O_5 and not enough N.

$(NH_4)_2HPO_4$

$$?\% P_2O_5 = \frac{142\ g\ P_2O_5}{mol\ P_2O_5} \times \frac{mol\ P_2O_5}{2\ mol\ (NH_4)_2HPO_4} \times \frac{mol\ (NH_4)_2HPO_4}{132\ g\ (NH_4)_2HPO_4} \times 100 = 54\%\ P_2O_5$$

$$?\% N = \frac{14.01\ g\ N}{mol\ N} \times \frac{2\ mol\ N}{mol\ (NH_4)_2HPO_4} \times \frac{mol\ (NH_4)_2HPO_4}{132\ g\ (NH_4)_2HPO_4} \times 100\% = 21\%$$

A 1:1 mole ratio of $(NH_4)_2HPO_4$ and KH_2PO_4 would produce a 10-53-18 fertilizer. The 1:1 mole ratio would be a 132:136 $(NH_4)_2HPO_4$ to KH_2PO_4 mass ratio, so a 1:1 mass ratio would also provide about the same fertilizer.

(d) $$? g\ KNO_3 = 20.0\ g\ K_2O \times \frac{mol\ K_2O}{94.20\ g} \times \frac{2\ mol\ KNO_3}{mol\ K_2O} \times \frac{101.11\ g}{mol\ KNO_3}$$

$$= 42.9\text{ g KNO}_3$$

$$? \text{ g N} = 42.9\text{ g KNO}_3 \times \frac{\text{mol KNO}_3}{101.11\text{ g KNO}_3} \times \frac{\text{mol N}}{\text{mol KNO}_3} \times \frac{14.01\text{ g}}{\text{mol N}} = 5.94\text{ g N}$$

That is 5.94 % N. 14.06 % N is needed from NH_4NO_3.

$$? \text{ g NH}_4\text{NO}_3 = 14.06\text{ g N} \times \frac{\text{mol N}}{14.01\text{ g}} \times \frac{\text{mol NH}_4\text{NO}_3}{2\text{ mol N}} \times \frac{80.05\text{ g}}{\text{mol NH}_4\text{NO}_3}$$

$$= 40.2\text{ g NH}_4\text{NO}_3$$

$$? \text{ NaH}_2\text{PO}_4 = 20.0\text{ g P}_2\text{O}_5 \times \frac{\text{mol P}_2\text{O}_5}{141.94\text{ g}} \times \frac{2\text{ mol NaH}_2\text{PO}_4}{\text{mol P}_2\text{O}_5} \times \frac{119.98\text{ g}}{\text{mol NaH2PO4}}$$

$$= 33.8\text{ g NaH}_2\text{PO}_4$$

The 20-20-20 fertilizer cannot be made from these three compounds. To make 20 g of N, P_2O_5 and K_2O requires 116.9 g. The percentage of each is less then 20% (17.1%). A mixture could be made that would contain two of the components, but there would not be enough mass left to put in a sufficient amount of the third component.

130. $(CS_2 + 3\ Cl_2 \rightarrow CCl_4 + S_2Cl_2) \times 2$

$\underline{2\ S_2Cl_2 + CS_2 \rightarrow CCl_4 + 6\ S}$

$3\ CS_2 + 6\ Cl_2 \rightarrow 3\ CCl_4 + 6\ S$, which reduces to

$CS_2 + 2\ Cl_2 \rightarrow CCl_4 + 2\ S$

$$\%\text{ AE} = \frac{153.82\text{ g CCl}_4}{76.143\text{ g CS}_2 + 2 \times 70.906\text{ g Cl}_2} \times 100\ \% = 70.574\ \%$$

132. $2\ Na(s) + 2\ H_2O(l) \rightarrow 2\ NaOH(aq) + H_2(g)$

$$? \text{ g Na} = 250.0\text{ mL} \times \frac{10^{-3}\text{ L}}{\text{mL}} \times 0.315\text{ M} \times \frac{2\text{ mol Na}}{2\text{ mol NaOH}} \times \frac{22.99\text{ g Na}}{\text{mol Na}} = 1.81\text{ g Na}$$

133. $$? \text{ mol Cl}^- = 1.7272\text{ g AgCl} \frac{\text{mol Cl}^-}{143.321\text{ g AgCl}} = 0.012051\text{ mol Cl}^-$$

$$X\text{ mol NaCl} \times \frac{58.443\text{g}}{\text{mol NaCl}} + (0.012051 - X)\text{ mol Cl}^- \times \frac{\text{mol MgCl}_2}{2\text{ mol Cl}^-} \times \frac{95.211\text{ g}}{\text{mol MgCl}_2} = 0.6118\text{-g sample}$$

$58.443X\text{ g} + 0.5737\text{ g} - 47.606X\text{ g} = 0.6118\text{ g}$

$10.838X = 0.0381$

$$X = \frac{0.0381}{10.838} = 3.52 \times 10^{-3}\text{ mol NaCl}$$

$$3.52 \times 10^{-3}\text{ mol NaCl} \times \frac{58.443\text{ g}}{\text{mol NaCl}} = 0.206\text{ g NaCl}$$

$$(0.012051 - 0.00352)\text{ mol Cl}^- \times \frac{\text{mol MgCl}_2}{2\text{ mol Cl}^-} \times \frac{95.211\text{ g}}{\text{mol MgCl}_2} = 0.406\text{ g MgCl}_2$$

Chapter 4

Chemical Reactions in Aqueous Solutions

Exercises

4.1A $[Na^+] = 0.438\ M$

$[Mg^{2+}] = 0.0512\ M$

$[Cl^-] = 0.438\ M + (2 \times 0.0512\ M) = 0.540\ M$

4.1B $[\text{glucose}] = \dfrac{20.0\ g}{L} \times \dfrac{mol}{180.2\ g} = 0.111\ M$ glucose

$[C_6H_5O_7^{3-}] = = \dfrac{2.9\ g}{L} \times \dfrac{mol}{258.1\ g} \times \dfrac{1\ mol\ C_6H_5O_7^{3-}}{1\ mol\ Na_3C_6H_5O_7} = 0.011\ M$ citrate ion

$[K^+] = \dfrac{1.5\ g\ KCl}{L} \times \dfrac{mol\ KCl}{74.55\ g\ KCl} \times \dfrac{1\ mol\ K^+}{1\ mol\ KCl} = 0.020\ M\ K^+$

$[Na^+] = \dfrac{3.5\ g\ NaCl}{L} \times \dfrac{mol\ NaCl}{58.44\ g\ NaCl} \times \dfrac{1\ mol\ Na^+}{1\ mol\ NaCl} = 0.060\ M\ Na^+$

$[Cl^-] = 0.020\ M + 0.060\ M = 0.080\ M$

$[Na^+] = 0.060\ M + 3 \times 0.011\ M = 0.093\ M$

4.2A (a) $Ca(OH)_2(s) + 2\ HCl(aq) \rightarrow CaCl_2(aq) + 2H_2O(l)$

(b) $Ca^{2+}(aq) + 2\ OH^-(aq) + 2\ H^+(aq) + 2\ Cl^-(aq) \rightarrow Ca^{2+}(aq) + 2\ Cl^-(aq) + 2\ H_2O(l)$

(c) $OH^-(aq) + H^+(aq) \rightarrow H_2O(l)$

4.2B (a) $2\ KHSO_4(aq) + 2\ NaOH(aq) \rightarrow Na_2SO_4(aq) + K_2SO_4(aq) + 2\ H_2O(l)$

(b) $K^+(aq) + HSO_4^-(aq) + Na^+(aq) + OH^-(aq) \rightarrow Na^+(aq) + K^+(aq) + SO_4^{2-}(aq) + H_2O(l)$

(c) $HSO_4^-(aq) + OH^-(aq) \rightarrow SO_4^{2-}(aq) + H_2O(l)$

4.3A The CH_3NH_2 will cause a dimly lit bulb, as it is a weak base. HNO_3 is a strong acid and will cause a brightly lit bulb. The combination will produce a strong electrolyte and a brightly lit bulb.

$CH_3NH_2(aq) + H_2O(l) \rightleftharpoons CH_3NH_3^+(aq) + OH^-(aq)$

$HNO_3(aq) \rightarrow H^+(aq) + NO_3^-(aq)$

$\underline{H^+(aq) + OH^-(aq) \rightarrow H_2O(l)}$

$CH_3NH_2(aq) + HNO_3(aq) \rightarrow CH_3NH_3^+(aq) + NO_3^-(aq)$

4.3B The $Mg(OH)_2$ slurry will cause a very dim glow. The vinegar will dissolve the $Mg(OH)_2$ because the H^+ from the acetic acid will react with the OH^- ions to form water. This will leave Mg^{2+} ions and acetate ions in solution. The bulb should shine brightly.

4.4A (a) $MgSO_4(aq) + 2\ KOH(aq) \rightarrow Mg(OH)_2(s) + 2\ K^+(aq) + SO_4^{2-}(aq)$
$Mg^{2+}(aq) + 2\ OH^-(aq) \rightarrow Mg(OH)_2(s)$
(b) $2\ FeCl_3(aq) + 3\ Na_2S(aq) \rightarrow Fe_2S_3(s) + 6\ Na^+(aq) + 6\ Cl^-(aq)$
$2\ Fe^{3+}(aq) + 3\ S^{2-}(aq) \rightarrow Fe_2S_3(s)$
(c) $Sr(NO_3)_2(aq) + Na_2SO_4(aq) \rightarrow SrSO_4(s) + 2\ NaNO_3(aq)$
$Sr^{2+}(aq) + SO_4^{2-}(aq) \rightarrow SrSO_4(s)$

4.4B (a) $ZnSO_4(aq) + BaS(aq) \rightarrow ZnS(s) + BaSO_4(s)$
$Zn^{2+}(aq) + SO_4^{2-}(aq) + Ba^{2+}(aq) + S^{2-}(aq) \rightarrow ZnS(s) + BaSO_4(s)$
(b) $Mg(OH)_2(s) + NaOH(aq) \rightarrow$ No reaction
(c) $NaHCO_3(aq) + Ca(OH)_2(aq) \rightarrow CaCO_3(s) + H_2O(l) + NaOH(aq)$
$HCO_3^-(aq) + Ca^{2+}(aq) + OH^-(aq) \rightarrow CaCO_3(s) + H_2O(l)$

4.5A The acid will cause the $Fe(OH)_3$ to dissolve because the H^+ will react with the OH^- to form H_2O. $Fe(OH)_3(s) + 3\ H^+(aq) \rightarrow Fe^{3+}(aq) + 3\ H_2O(l)$

4.5B $2CH_3(CH_2)_{14}COOK(aq) + Ca(Cl)_2(aq) \rightarrow (CH_3(CH_2)_{14}COO)_2Ca(s) + 2KCl(aq)$
$2CH_3(CH_2)_{14}COO^-(aq) + Ca^{2+}(aq) \rightarrow (CH_3(CH_2)_{14}COO)_2Ca(s)$

4.6A $$?\ g\ NaCl = 0.9372\ g\ AgCl \times \frac{mol\ AgCl}{143.32\ g\ AgCl} \times \frac{mol\ NaCl}{mol\ AgCl} \times \frac{58.443\ g\ NaCl}{mol\ NaCl} = 0.3822\ g\ NaCl$$

$$?\%\ NaCl = \frac{0.3822\ g\ NaCl}{0.9056\ g\ sample} \times 100\% = 42.20\%\ NaCl$$

4.6B (a) $$?\ g\ AgCl = 225\ mL \times \frac{10^{-3}\ L}{mL} \times 0.540\ M\ Cl^- \times \frac{mol\ AgCl}{mol\ Cl^-} \times \frac{143.3\ g\ AgCl}{mol\ AgCl} = 17.4\ g\ AgCl$$

(b) $$?\ g\ Mg(OH)_2 = 0.0512\ M\ Mg^{2+} \times 5.00\ L \times \frac{mol\ Mg(OH)_2}{mol\ Mg^{2+}} \times \frac{58.32\ g}{mol\ Mg(OH)_2} = 14.9\ g\ Mg(OH)_2$$

AgCl and $Mg(OH)_2$ are the precipitates.

4.7A

(a) $\overset{+3}{Al_2}\overset{-2}{O_3}$ (b) $\overset{0}{P_4}$ (c) $\overset{-2}{C}\overset{+1}{H_3}\overset{-1}{F}$ (d) $\overset{+1}{H}\overset{+5}{As}\overset{-2}{O_4}{}^{2-}$

(e) $\overset{+1}{Na}\overset{+7}{Mn}\overset{-2}{O_4}$ (f) $\overset{+3}{Cl}\overset{-2}{O_2}{}^-$ (g) $\overset{+1}{Cs}\overset{-\frac{1}{2}}{O_2}$

4.7B

(a) $H\underline{Sb}F_6$: +5 (b) $\underline{C}HCl_3$: +2 (c) $\underline{P_3}O_{10}^{5-}$: +5 (d) $\underline{S_4}O_6^{2-}$: +2.5

(e) $\underline{C_3}O_2$: +4/3 (f) $\underline{N}O_2^{+}$: +5 (g) $\underline{C_2}O_4^{2-}$: +3

4.8A $Cr_2O_7^{2-}$(aq) is an oxidizing agent; it will react with a reducing agent. HNO_3(aq) is also an oxidizing agent. There is no reaction between $Cr_2O_7^{2-}$(aq) and HNO_3(aq). HCl is a reducing agent (with Cl in the oxidation state, –1). It is oxidized by $Cr_2O_7^{2-}$(aq), probably to Cl_2(g). $Cr_2O_7^{2-}$(aq) is reduced to Cr^{3+}(aq) which causes the green color.

4.8B The solution in (a) would be Zn^{2+} (aq) in HCl(aq) instead of unreacted HCl(aq). The solution in (b) would be Zn^{2+}(aq), and Cu^{2+}(aq) in HNO_3(aq) instead of just Cu^{2+}(aq) in HNO_3(aq).

4.9A $$? \text{ mL HBr} = 0.01580 \text{ M Ba(OH)}_2 \times 25.00 \text{ mL Ba(OH)}_2 \times \frac{10^{-3}\text{ L}}{\text{mL}} \times \frac{2 \text{ mol HBr}}{1 \text{ mol Ba(OH)}_2} \times \frac{\text{L HBr}}{0.01060 \text{ mol HBr}} \times \frac{\text{mL}}{10^{-3}\text{ L}} = 74.53 \text{ mL HBr}$$

Notice that the two mL – L conversions can be left out to simplify the equation. Then the unit is millimoles instead of moles.

$$? \text{ mL HBr} = 25.00 \text{ mL Ba(OH)}_2 \times 0.01580 \text{ M Ba(OH)}_2 \times \frac{2 \text{ mol HBr}}{1 \text{ mol Ba(OH)}_2} \times \frac{\text{L HBr}}{0.01060 \text{ mol HBr}} = 74.53 \text{ mL HBr}$$

4.9B $$? \text{ mL KOH} = 2.000 \text{ g solution} \times \frac{96.5 \text{ g H}_2\text{SO}_4}{100 \text{ g solution}} \times \frac{\text{mol H}_2\text{SO}_4}{98.08 \text{ g H}_2\text{SO}_4} \times \frac{2 \text{ mol KOH}}{\text{mol H}_2\text{SO}_4} \times \frac{1 \text{ L KOH}}{0.3580 \text{ mol KOH}} \times \frac{\text{mL}}{10^{-3}\text{ L}} = 110. \text{ mL KOH}$$

4.10A $NaOH(aq) + HC_2H_3O_2(aq) \rightarrow C_2H_3O_2^-(aq) + Na^+(aq) + H_2O(l)$

$$? \text{ mL NaOH} = 10.00 \text{ mL} \times \frac{1.01 \text{ g}}{\text{mL}} \times \frac{4.12 \text{ g HC}_2\text{H}_3\text{O}_2}{100 \text{ g solution}} \times \frac{\text{mol HC}_2\text{H}_3\text{O}_2}{60.05 \text{ g}} \times \frac{\text{mol OH}^-}{\text{mol HC}_2\text{H}_3\text{O}_2} \times \frac{1 \text{ L}}{0.550 \text{ mol OH}^-} \times \frac{\text{mL}}{10^{-3}\text{ L}} = 12.6 \text{ mL NaOH}$$

4.10B $NaOH(aq) + H_2SO_4(aq) \rightarrow SO_4^{2-}(aq) + Na^+(aq) + H_2O(l)$

$$? \text{ g } H_2SO_4 = 32.44 \text{ mL} \times \frac{10^{-3}\text{ L}}{\text{mL}} \times 0.00986 \text{ M NaOH} \times \frac{250.0 \text{ mL}}{10.00 \text{ mL}} \times \frac{\text{mol } H_2SO_4}{2 \text{ mol NaOH}}$$
$$\times \frac{98.09 \text{ g}}{\text{mol}} = 0.3922 \text{ g } H_2SO_4$$

$$? \% \; H_2SO_4 = \frac{0.3922 \text{ g } H_2SO_4}{1.239 \text{ g sample}} \times 100\% = 31.7\% \; H_2SO_4$$

4.11A $Th^{4+}(aq) + H_2C_2O_4(aq) \rightarrow Th(C_2O_4)_2(s)$

$$? \text{ M} = 19.63 \text{ mL} \times 0.02500 \text{ M } H_2C_2O_4 \times \frac{\text{mol } Th^{+4}}{2 \text{ mol } C_2O_4^{\;2-}} \times \frac{1}{25.00 \text{ mL}}$$
$$= 9.815 \times 10^{-3} \text{ M } Th^{4+}$$

4.11B $$? \text{ mol } Cl^- = 0.1015 \text{ g NaCl} \times \frac{\text{mol } Cl^-}{58.443 \text{ g}} \; \frac{\text{mol}}{58.443 \text{ g}} + 0.1324 \text{g KCl} \times \frac{\text{mol } Cl^-}{74.551 \text{ g}}$$
$$= 3.513 \times 10^{-3} \text{ mol } Cl^-$$

$AgNO_3(aq) + Cl^-(aq) \rightarrow AgCl(s) + NO_3^-(aq)$

$$? \text{ mL } AgNO_3 = 3.513 \times 10^{-3} \text{ mol } Cl^- \times \frac{1}{0.1500 \text{ M}} \times \frac{\text{mL}}{10^{-3}\text{ L}} = 23.42 \text{ mL } AgNO_3$$

4.12A $$0.2865 \text{ g sample} \times \frac{58.01 \text{ g Fe}}{100 \text{ g sample}} \times \frac{\text{mol } Fe^{2+}}{55.847 \text{ g } Fe^{2+}} \times \frac{1 \text{ mol } Cr_2O_7^{2-}}{6 \text{ mol } Fe^{2+}}$$
$$\times \frac{1}{0.02250 \text{ M } Cr_2O_7^{2-}} \times \frac{\text{mL}}{10^{-3}\text{ L}} = 22.04 \text{ mL}$$

Alternatively, because the $K_2Cr_2O_7$ solution in Exercise 4.12 has the same molarity as the $KMnO_4$ of Example 4.12 and the stoichiometric factors are 1 mol $Cr_2O_7^{2-}$/ 6 mol Fe^{2+} and 1 mol MnO_4^- / 5 mol Fe^{2+}, the volume of $K_2Cr_2O_7(aq)$ required is 5/6 that of $KMnO_4(aq)$.

5/6 × 26.45 mL = 22.04 mL

4.12B $$? \text{ M } MnO_4^- = 0.2378 \text{g} \times \frac{\text{mol}}{134.00 \text{ g}} \times \frac{2 \text{ mol } MnO^-}{5 \text{ mol } C_2O_4^{\;2-}} \times \frac{1}{20.00 \text{ mL}} \times \frac{\text{mL}}{10^{-3}\text{ L}}$$
$$= 3.549 \times 10^{-2} \text{ M } MnO_4^-$$

$$? \text{ mL} = 25.00 \text{ mL} \times 0.1010 \text{ M } Fe^{2+} \times \frac{\text{mol } MnO_4^{\;-}}{5 \text{ mol } Fe^{2+}} \times \frac{1}{3.549 \times 10^{-2} \text{ M}} = 14.23 \text{ mL}$$

Review Questions

1. 0.10 M NaCl has more ions in solution. It is the only strong electrolyte. (b) and (d) are nonelectrolytes and (c) is a weak electrolyte.

2. (a) weak acid
(b) strong acid
(c) strong base
(d) weak base
(e) salt
(f) weak acid

3. (c) is highest, with $[NO_3^-] = 3 \times 0.040 = 0.12$ M.
(a) and (d) both have $[NO_3^-] = 0.10$ M.
(b) is lowest, with $[NO_3^-] = 0.080$ M.

4. (a) ?[ions] = $0.0012 \times 5 = 0.0060$ M
(b) ?[ions] = $0.030 \times 2 = 0.060$ M
(c) ?[ions] = $0.040 \times 4 = 0.160$ M
(d) ?[ions] = $0.025 \times 3 = 0.075$ M
(c) has the highest .

5. (a) ? ions = $0.08 \text{ M} \times 2 = 0.16$ M
(b) very few ions
(c) ? ions = $0.10 \text{ M} \times <1 = <0.10$ M
(d) very few ions
(a) is best conductor, it has the most ions.

6. (a) $[H^+] = 0.10 \text{ M} \times 1 = 0.10$ M
(b) $[H^+] = 0.10 \text{ M} \times <1 = <0.10$ M
(c) strong base $[H^+]$ very small
(d) weak base $[H^+]$ very small
(a) is largest $[H^+]$

9. $BaSO_3 + 2HCl \rightarrow Ba^{2+} + 2Cl^- + H_2SO_3 \rightarrow H_2O + SO_2(g)$

11. Nitrates, chlorides (except of Pb^{2+}, Ag^+, Hg_2^{2+}), and sulfates (except Sr^{2+}, Ba^{2+}, Pb^{2+}, and Hg_2^{2+}) are soluble. $PbSO_4$ must be insoluble.

12. Na_2CO_3. Most carbonates are insoluble.
$CO_3^{2-} + Mg^{2+} \rightarrow MgCO_3(s)$

17. $\overset{+6}{SO_4^{2-}} \rightarrow \overset{+4}{SO_2}$
(d) SO_4^{2-} is reduced.

18. (a) $\overset{0}{Cr}$ (b) $\overset{+3}{ClO_2^-}$ (c) $\overset{-2}{K_2Se}$ (d) $\overset{+6}{TeF_6}$ (e) $\overset{-3}{PH_4^+}$

(f) $\overset{+4}{CaRuO_3}$ (g) $\overset{+4}{SrTiO_3}$ (h) $\overset{+5}{P_2O_7^{4-}}$ (i) $\overset{+2.5}{S_4O_6^{2-}}$ (j) $\overset{-1}{NH_2OH}$

19. (a) −3 C_2H_6 (b) +2 CH_2O_2 (c) +1 $C_2H_2O_2$
(d) −2 C_2H_6O (e) +3 $C_2H_2O_4$

24. (a) $H_2SO_4(aq) + 2\ OH^-(aq) \rightarrow SO_4^-(aq) + H_2O(l)$

$$? \text{ mL NaOH} = 10.00 \text{ mL} \times 0.100 \text{ M } H_2SO_4 \times \frac{2 \text{ mol NaOH}}{\text{mol } H_2SO} \times \frac{1}{0.100 \text{ M}} = 20.0 \text{ mL} \quad \textbf{NO}$$

(b) $$? \text{ mL NaOH} = 10.00 \text{ mL} \times 0.100 \text{ M } H_2SO_4 \times \frac{2 \text{ mol NaOH}}{\text{mol } H_2SO} \times \frac{1}{0.0200 \text{ M}} = 100 \text{ mL} \quad \textbf{YES}$$

(c) $$? \text{ mL KOH} = 10.00 \text{ mL} \times 0.100 \text{ M } H_2SO_4 \times \frac{2 \text{ mol KOH}}{\text{mol } H_2SO} \times \frac{1}{0.100 \text{ M}} = 20.0 \text{ mL} \quad \textbf{NO}$$

(d) $$? \text{ mL Ba(OH)}_2 = 10.00 \text{ mL} \times 0.100 \text{ M } H_2SO_4 \times \frac{\text{mol Ba(OH)}_2}{\text{mol } H_2SO_4} \times \frac{1}{0.100 \text{ M}} = 10.0 \text{ mL} \quad \textbf{NO}$$

(b) is the answer.

Problems

25. nonelectrolytes:
strong electrolytes: (a),(b), (c), (e), (f)
weak electrolytes: (d)

27. All potassium salts are soluble and will cause the bulb to glow brightly.

29. (a) $[Li^+] = 0.647$ M, $[NO_3^-] = 0.647$ M
(b) $[Ca^{2+}] = 0.035$ M, $[I^-] = 0.070$ M
(c) $[Al^{3+}] = 2.14$ M, $[SO_4^{2-}] = 3.21$ M

31. $[Na^+] = 0.0554 + (2 \times 0.0145) = 0.0844$ M,
$[Cl^-] = 0.0554$ M, $[SO_4^{2-}] = 0.0145$ M

33. $$? \text{ L} = 16.11 \text{ g} \times \frac{\text{mol } MgCl_2}{95.211 \text{ g}} \times \frac{2 \text{ mol } Cl^-}{\text{mol } MgCl_2} \times \frac{1}{0.1000 \text{ M}} = 3.384 \text{ L}$$

35. KNO_3 $\quad ?\ [NO_3^-] = 0.10 \text{ M} \times 1 = 0.10$ M
$Al(NO_3)_3$ $\quad ?\ [NO_3^-] = 0.040 \times 3 = 0.12$ M
$Ca(NO_2)_2$ $\quad ?\ [NO_3^-] = 0.047 \times 2 = 0.094$ M
$Al(NO_3)_3 > KNO_3 > Ca(NO_2)_2$

37. $? [SO_4^{2-}] = \dfrac{18.3 \text{ g}}{285 \text{ mL}} \times \dfrac{\text{mL}}{10^{-3} \text{ L}} \times \dfrac{\text{mol}}{246.5 \text{ g}} = 0.260 \text{ M}$

39. $[Cl^-] = 0.540$ M

$? \text{ mg/L} = \dfrac{0.540 \text{ mol}}{\text{L}} \times \dfrac{35.45 \text{ g } Cl^-}{\text{mol } Cl^-} \times \dfrac{\text{mg}}{10^{-3} \text{ g}} = 1.91 \times 10^4 \text{ mg/L}$

41. $V = 0.2500 \text{ L} \times \dfrac{0.0135 \text{ mol } Cl^-}{\text{L}} \times \dfrac{1 \text{ mol } MgCl_2}{2 \text{ mol } Cl^-} \times \dfrac{1 \text{ L}}{0.0250 \text{ M } MgCl_2} \times \dfrac{\text{mL}}{10^{-3} \text{ L}}$

$= 67.5 \text{ mL}$

43. (a) < (d) < (c) < (b)

(b) is 3 × 0.45 M = 1.35 M, (c) is $\dfrac{1.20 \text{ mol}}{2.00 \text{ L}} = 0.600 \times 2 = 1.20$ M, (d) is 0.15 + 0.35 = 0.50 M, and (a) is 0.21 M.

45. (a) $HBr(aq) \xrightarrow{H_2O} H^+(aq) + Br^-(aq)$

(b) $LiOH(s) \xrightarrow{H_2O} Li^+(aq) + OH^-(aq)$

(c) $HF(aq) \rightleftharpoons H^+(aq) + F^-(aq)$

(d) $HIO_3(aq) \rightleftharpoons H^+(aq) + IO_3^-(aq)$

(e) $(CH_3)_2NH(aq) + H_2O(l) \rightleftharpoons (CH_3)_2NH_2^+(aq) + OH^-(aq)$

(f) $HCOOH(aq) \rightleftharpoons HCOO^-(aq) + H^+(aq)$

47. (a) HCl is a strong electrolyte so $[H^+] = 0.10$ M.

(b) 0.10 M H_2SO_4 produces $[H^+] > 0.10$ M, because ionization is complete in the first ionization step and also occurs to some extent in the second.

(c) Acetic acid is a weak acid so $[H^+] < 0.10$ M.

(d) Ammonia is a base.

(d) < (c) < (a) < (b)

49. $H^+(aq) + OH^-(aq) \rightarrow H_2O(l)$

(a) no reaction

(b) $OH^-(aq) + H^+(aq) \rightarrow H_2O(l)$

(c) $NH_3(aq) + H^+(aq) \rightarrow NH_4^+(aq)$

(b) is the same.

51. Major reaction: $CaCO_3(s) + 2\, H^+(aq) \rightarrow Ca^{2+}(aq) + H_2CO_3(aq)$

$\rightarrow Ca^{2+}(aq) + H_2O(l) + CO_2(g)$

$CaCO_3(s) + 2\, H^+(aq) \rightarrow Ca^{2+}(aq) + H_2O(l) + CO_2(g)$

Minor reaction: $CuCO_3(s) + 2\, H^+(aq) \rightarrow Cu^{2+}(aq) + H_2CO_3(aq)$

$\rightarrow Cu^{2+}(aq) + H_2O(l) + CO_2(g)$

$CuCO_3(s) + 2\ H^+(aq) \rightarrow Cu^{2+}(aq) + H_2O(l) + CO_2(g)$

53. (a) $2\ I^-(aq) + Pb^{2+}(aq) \rightarrow PbI_2(s)$
(b) no reaction
(c) $Cr^{3+}(aq) + 3\ OH^-(aq) \rightarrow Cr(OH)_3(s)$
(d) no reaction
(e) $OH^-(aq) + H^+(aq) \rightarrow H_2O(l)$
(f) $HSO_4^-(aq) + OH^-(aq) \rightarrow H_2O(l) + SO_4^{2-}(aq)$

55. (a) $Mg(OH)_2(s) + 2\ H^+(aq) \rightarrow Mg^{2+}(aq) + 2\ H_2O(l)$
(b) $HCOOH(aq) + NH_3(aq) \rightarrow NH_4^+\ (aq) + HCOO^-(aq)$
(c) no reaction
(d) $Cu^{2+}(aq) + CO_3^{2+}(aq) \rightarrow CuCO_3(s)$
(e) no reaction

57. Nitrates, chlorides (except of Pb^{2+}, Ag^+, Hg_2^{2+}), and sulfates (except Sr^{2+}, Ba^{2+},Pb^{2+}, and Hg_2^{2+}) are soluble. PbS must be insoluble.

59. No, it does not indicate what the powder is because $MgSO_4$ dissolves in aqueous solution and $Mg(OH)_2$ dissolves as a result of an acid-base reaction. Testing the pH ($Mg\ (OH)_2$ is more basic) or adding Ba^{2+} to precipitate $BaSO_4$ are other tests that would obtain results. The simplest test is to add water to the powder. $MgSO_4$ is soluble; $Mg(OH)_2$ is not.

61. $Cu^{2+}(aq) + CO_3^{2-}(aq) \rightarrow CuCO_3(s)$

63. Add $BaCl_2(aq)$ to one sample. If a precipitate ($BaSO_4$) forms, the solution is $Na_2SO_4(aq)$. If there is no precipitate, add $Na_2SO_4(aq)$ to a second sample. Here, a precipitate ($BaSO_4$) indicates the solution is $Ba(NO_3)_2(aq)$, and no precipitate, that it is $NH_3(aq)$. An odor of ammonia would also confirm $NH_3(aq)$.

65. (a) oxidation. The oxidation state of chromium increases from +2 to +3.
(b) neither. The oxidation states do not change.
(c) neither. The oxidation state of nitrogen does not change.

67. (a) $4\ HCl + O_2 \rightarrow 2\ Cl_2 + 2\ H_2O$
check 0 = 0
(b) $2\ NO + 5\ H_2 \rightarrow 2\ NH_3 + 2\ H_2O$
(c) $CH_4 + 4\ NO \rightarrow 2\ N_2 + CO_2 + 2\ H_2O$
(d) $3\ Ag + 4\ H^+ + NO_3^- \rightarrow 3\ Ag^+ + 2\ H_2O + NO$
(e) $IO_4^- + 7\ I^- + 8\ H^+ \rightarrow 4\ I_2 + 4\ H_2O$

69.

	Oxidizing Agent	Reducing Agent
(a)	O_2	HCl
(b)	NO	H_2
(c)	NO	CH_4
(d)	NO_3^-	Ag
(e)	IO_4^-	I^-

71.

(a) $\overset{+2}{PbO} \rightarrow \overset{+4}{PbO_2}$

$\overset{+3}{V^{3+}} \rightarrow \overset{+5}{VO^{2+}}$ PbO and V^{3+} are both oxidized, so nothing is reduced.

(b) $S^{2-} \rightarrow S^0$ Sulfur is oxidized, but nothing is reduced.

73. (a) $Zn(s) + 2H^+(aq) \rightarrow Zn^{2+}(aq) + H_2(g)$

(b) $Cu(s) + Zn^{2+}(aq) \rightarrow$ no reaction

(c) $Fe(s) + 2\,Ag^+(aq) \rightarrow Fe^{2+}(aq) + 2\,Ag(s)$

(d) $Au(s) + H^+(aq) \rightarrow$ no reaction

75. (a) $? \text{ mL} = 25.00 \text{ mL} \times \frac{10^{-3}\text{ L}}{\text{mL}} \times 0.0365 \text{ M KOH} \times \frac{\text{mol HCl}}{\text{mol KOH}} \times \frac{1}{0.0195 \text{ M}} \times \frac{\text{mL}}{10^{-3}\text{ L}} = 46.8 \text{ mL}$

(b) $? \text{ mL} = 10.00 \text{ mL} \times \frac{10^{-3}\text{ L}}{\text{mL}} \times 0.0116 \text{ M Ca(OH)}_2 \times \frac{2 \text{ mol HCl}}{\text{mol Ca(OH)}_2} \times \frac{1}{0.0195 \text{ M}} \times \frac{\text{mL}}{10^{-3}\text{ L}} = 11.9 \text{ mL}$

(c) $? \text{ mL} = 20.00 \text{ mL} \times \frac{10^{-3}\text{ L}}{\text{mL}} \times 0.0225 \text{ M NH}_3 \times \frac{\text{mol HCl}}{\text{mol NH}_3} \times \frac{1}{0.0195 \text{ M}} \times \frac{\text{mL}}{10^{-3}\text{ L}} = 23.1 \text{ mL}$

77. $CH_3COOH + OH^- \rightarrow H_2O + CH_3COO^-$

$? \text{ M} = 31.45 \text{ mL} \times \frac{10^{-3}\text{ L}}{\text{mL}} \times 0.2560 \text{ M KOH} \times \frac{\text{mol CH}_3\text{COOH}}{\text{mol KOH}} \times \frac{1}{10.00 \text{ mL}} \times \frac{\text{mL}}{10^{-3}\text{ L}} = 0.8051 \text{ M}$

79. $CaCO_3(s) + HCl(aq) \rightarrow CaCl_2(aq) + H_2O(l) + CO_2(g)$

(a) $? \text{ mg CaCO}_3 = 38.8 \text{ mL} \times 0.251 \text{ M HCl} \times \frac{\text{mole CaCO}_3}{2 \text{ mole HCl}} \times \frac{100.08 \text{ g CaCO}_3}{\text{mole CaCO}_3} = 487 \text{ mg CaCO}_3$

(b) $? \text{ mg } Ca^{2+} = 38.8 \text{ mL} \times 0.251 \text{ M HCl} \times \frac{\text{mole } CaCO_3}{\text{2 mole HCl}} \times \frac{40.08 \text{ g } Ca^{2+}}{\text{mole } CaCO_3}$
$= 195 \text{ mg } Ca^{2+}$.

81. $? \text{ mol HCl} = 1 \text{ g } NaHCO_3 \times \frac{\text{mol } NaHCO_3}{84 \text{ g } NaHCO_3} \times \frac{\text{mol HCl}}{\text{mol } NaHCO_3}$

$? \text{ mol HCl} = 1 \text{ g } CaCO_3 \times \frac{\text{mol } CaCO_3}{100 \text{ g } CaCO_3} \times \frac{\text{2 mol HCl}}{\text{mol } CaCO_3}$

Tums ($CaCO_3$) will neutralize more acid. $\left(\frac{2}{100} > \frac{1}{84}\right)$

83. All of the CH_3COOH is neutralized, and there are some excess OH^- ions; thus it is past the equivalence point of the titration (at 20.00 mL) by a factor of 10% (1 OH^- for each 10 CH_3COO^-). So (d) 22.00 mL.

85. (a) $? \text{ mL} = 25.00 \text{ mL} \times 0.1235 \text{ M KI} \times \frac{\text{mol } I^-}{\text{mol KI}} \times \frac{\text{mol } Ag^+}{\text{mol } I^-} \times \frac{1}{0.02091 \text{ M}}$
$= 147.7 \text{ mL}$

(b) $? \text{ mL} = 40.00 \text{ mL} \times 0.01944 \text{ M } FeCl_3 \times \frac{\text{3 mol } Cl^-}{\text{mol } FeCl_3} \times \frac{\text{mol } Ag^+}{\text{mol } Cl^-} \times \frac{1}{0.02091 \text{ M}}$
$= 111.6 \text{ mL}$

(c) $? \text{ mL} = 0.0323 \text{g } Na_2CO_3 \times \frac{\text{mol}}{105.99 \text{g}} \times \frac{\text{mol } CO_3^{2-}}{\text{mol } Na_2CO_3} \times \frac{\text{2 mol } Ag^+}{\text{mol } CO_3^{2-}}$
$\times \frac{1}{0.02091 \text{ M}} \times \frac{\text{mL}}{10^{-3} \text{ L}} = 29.1 \text{ mL}$

87. $Na_2SO_4(aq) + Ba(ClO_4)_2(aq) \rightarrow BaSO_4(s) + 2NaClO_4(aq)$
$SO_4^{2-}(aq) + Ba^{2+}(aq) \rightarrow BaSO_4(s)$

$? \text{ M} = 0.2558 \text{ g } Na_2SO_4 \times \frac{\text{mol } Na_2SO_4}{142.05 \text{ g } Na_2SO_4} \times \frac{\text{1 mol } SO_4^{2-}}{\text{1 mol } Na_2SO_4} \times \frac{\text{mol } Ba^{2+}}{\text{mol } SO_4^{2-}}$
$\times \frac{1}{41.60 \text{ mL}} \times \frac{\text{mL}}{10^{-3} \text{ L}} = 0.04329 \text{ M}$

89. (a) $? \text{ mL } KMnO_4 = 20.00 \text{ mL} \times \frac{0.3252 \text{ mol } Fe^{2+}}{\text{L}} \times \frac{\text{mol } MnO_4^-}{5 \text{ mol } Fe^{2+}}$
$\times \frac{\text{L}}{0.1050 \text{ mol } KMnO_4} = 12.39 \text{ mL } KMnO_4$

(b) $? \text{ mL KMnO}_4 = 1.065 \text{ g KNO}_2 \times \frac{\text{mol KNO}_2}{85.103 \text{ g KNO}_2} \times \frac{2 \text{ mol MnO}_4^-}{5 \text{ mol NO}_2^-}$

$\times \frac{\text{L}}{0.1050 \text{ mol KMnO}_4} \times \frac{\text{mL}}{10^{-3} \text{ L}} = 47.67 \text{ mL KMnO}_4$

91. $3 \text{ Mn}^{2+} + 2 \text{ MnO}_4^- + 4 \text{ OH}^- \rightarrow 5 \text{ MnO}_2 + 2 \text{ H}_2\text{O}$

$[\text{Mn}^{2+}] = 0.03477 \text{ L} \times 0.05876 \text{ M MnO}_4^- \times \frac{3 \text{ mol Mn}^{2+}}{2 \text{ mol MnO}_4^-} \times \frac{1}{0.02500 \text{ L}}$

$= 0.1226 \text{ M Mn}^{2+}$

Additional Problems

95. $? \text{ mol} = 6.85 \text{ g} \times \frac{98.8 \text{ g NaCl}}{100 \text{ g}} \times \frac{\text{mol NaCl}}{58.44 \text{ g NaCl}} = 0.116 \text{ mol Cl}^-$

$? \text{ mol} = 6.85 \text{ g} \times \frac{1.2 \text{ g MgCl}_2}{100 \text{ g}} \times \frac{\text{mol MgCl}_2}{95.21 \text{ g MgCl}_2} \times \frac{2 \text{ mol Cl}^-}{\text{mol MgCl}_2} = 0.0017 \text{ mol Cl}^-$

$[\text{Cl}^-] = \frac{(0.116 \text{ mol} + 0.0017 \text{ mol})}{500.0 \text{ mL}} = \frac{0.118 \text{ mol}}{500.0 \text{ mL}} \times \frac{\text{mL}}{10^{-3} \text{ L}} = 0.236 \text{ M}$

96. $[\text{I}^-] = 0.0240 \text{ M} + 2 \times 0.0146 \text{ M} = 0.0532 \text{ M}$

$V = 0.1000 \text{ L} \times \frac{0.0532 \text{ mol I}^-}{\text{L}} \times \frac{1\text{L}}{0.0500 \text{ mol I}^-} \times \frac{1 \text{ mL}}{10^{-3} \text{ L}} = 106.4 \text{ mL}$

Add 6.4 mL water to the 100.0 mL of solution.

99. $\text{H}_2\text{SO}_4 + \text{Na}_2\text{CO}_3 \rightarrow \text{Na}_2\text{SO}_4 + \text{H}_2\text{CO}_3 \rightarrow \text{Na}_2\text{SO}_4 + \text{H}_2\text{O} + \text{CO}_2$

$? \text{ kg} = 1.5 \times 10^3 \text{ kg} \times \frac{10^3 \text{ g}}{1 \text{ kg}} \times \frac{93.2 \text{ g}}{100\text{g}} \times \frac{\text{mol H}_2\text{SO}_4}{98.08 \text{ g H}_2\text{SO}_4} \times \frac{\text{mol Na}_2\text{CO}_3}{\text{mol H}_2\text{SO}_4}$

$\times \frac{105.99 \text{ g Na}_2\text{CO}_3}{\text{mol Na}_2\text{CO}_3} \times \frac{1 \text{ kg}}{10^3 \text{ g}} = 1.5 \times 10^3 \text{ kg Na}_2\text{CO}_3$

101. $? \text{ mol} = 220 \text{ mL} \times \frac{1.16 \text{ g}}{\text{mL}} \times \frac{31.4 \text{ g HCl}}{100 \text{ g solution}} \times \frac{\text{mol HCl}}{36.46 \text{ g HCl}} \times \frac{1 \text{ mol H}^+}{1 \text{ mol HCl}}$

$\times \frac{\text{mol OH}^-}{\text{mol H}^+} = 2.20 \text{ mol OH}^-$

$[\text{OH}^-] = \frac{2.20 \text{ mol OH}^-}{0.50 \text{ gal}} \times \frac{\text{gal}}{3.785 \text{ L}} = 1.2 \text{ M OH}^-$

104. $2 \text{ MnO}_4^- + 5 \text{ C}_2\text{O}_4^{2-} + 16 \text{ H}^+ \rightarrow 10 \text{ CO}_2(g) + 2 \text{ Mn}^{2+} + 8 \text{ H}_2\text{O}(l)$

$$? \text{ g} = 0.02140 \text{ M KMnO}_4 \times 25.82 \text{ mL} \times \frac{10^{-3}\text{ L}}{\text{mL}} \times \frac{5 \text{ mol } C_2O_4^{2-}}{2 \text{ mol } MnO_4^-} \times \frac{250.0 \text{ mL}}{5.00 \text{ mL}}$$

$$\times \frac{134.0 \text{ g } Na_2C_2O_4}{\text{mol } Na_2C_2O_4} = 9.26 \text{ g } Na_2C_2O_4$$

107. $? \text{ e}^- = 48.97 \text{ mol} \times 0.3000 \text{ M } Ce^{4+} \times \frac{1 \text{ mmol e}^-}{\text{mmol } Ce^{4+}} = 14.69 \text{ mmol e}^-$

$$? \frac{\text{mmol e}^-}{\text{mmol } V^{2+}} = \frac{14.69 \text{ mmol e}^-}{25.00 \text{ mL} \times 0.1996 \text{ M}} = \frac{2.944 \text{ mmol e}^-}{\text{mmol } V^{2+}} \approx \frac{3 \text{ mmol e}^-}{\text{mmol } V^{2+}}$$

V^{2+} goes to V^{5+}

$V^{2+} + Ce^{4+} \rightarrow V^{5+} + Ce^{3+}$

110. $$\frac{0.206 \text{ M KCl} \times 235 \text{ mL} + 0.185 \text{ M MgCl}_2 \times X \times \frac{2 \text{ mol Cl}^-}{\text{mol MgCl}_2}}{(235 + X) \text{ mL}} = 0.250 \text{ M}$$

$48.4 \text{ mmol} + 0.370 \text{ M} \times X = 58.8 \text{ mmol} + 0.250 \text{ M} \times X$

$0.120 \text{ M} \times X = 10.4 \text{ mmol}$

$X = 86.2 \text{ mL}$

113. $$? \text{ g residue} = ? \text{ g KCl} = 25.00 \text{ mL} \times 1.840 \text{ M} \times \frac{10^{-3}\text{ L}}{\text{mL}} \times \frac{\text{mol KCl}}{\text{mol HCl}} \times \frac{74.551 \text{ g}}{\text{mol KCl}} = 3.429 \text{ g KCl}$$

All the residue is KCl. All of the HCl became KCl. The calculation can be made using Cl^- as the limiting factor.

$KOH + KCl \rightarrow KCl + H_2O$

$K_2CO_3 + 2\,HCl \rightarrow 2\,KCl + H_2CO_3 \rightarrow H_2O + CO_2(g)$

$HCl + KOH \rightarrow KCl + H_2O$

115. $$? \text{ g S} = (25.00 \text{ mL} \times 0.00923 \text{ M NaOH} - 13.33 \text{ mL} \times 0.01007 \text{ M HCl}) \times \frac{\text{mol } H_2SO_4}{2 \text{ mol NaOH}} \times \frac{10^{-3}\text{ mol}}{\text{mmol}} \times \frac{\text{mol S}}{\text{mol } H_2SO_4} \times \frac{32.07 \text{ g S}}{\text{mol S}} = 1.6 \times 10^{-3} \text{ g S}$$

$$? \,\% \text{ S} = \frac{1.6 \times 10^{-3} \text{ g S}}{4.476 \text{ g sample}} \times 100\% = 0.035 \,\% \text{ S}$$

Apply Your Knowledge

116. $$? \text{ M } H_2SO_4 = 22.65 \text{ mL} \times 0.5510 \text{ M NaOH} \times \frac{\text{mol } H_2SO_4}{2 \text{ mol NaOH}} \times \frac{1}{25.00 \text{ mL}} = 0.2496 \text{ M } H_2SO_4$$

$$? \text{ g N} = (50.00 \text{ mL} \times 0.2496 \text{ M } H_2SO_4 - 19.90 \text{ mL} \times 0.5510 \text{ M NaOH} \times \frac{\text{mol } H_2SO_4}{2 \text{ mol NaOH}}) \times \frac{10^{-3} \text{ mol}}{\text{mmol}} \times \frac{2 \text{ mol } NH_3}{\text{mol } H_2SO_4} \times \frac{\text{mol N}}{\text{mol } NH_3} \times \frac{14.007 \text{ g}}{\text{mol N}} = 0.196 \text{ g N}$$

$$? \text{ \% protein} = \frac{0.1960 \text{ g N}}{2.500 \text{ g sample}} \times 100\% \times \frac{6.25 \text{ \% protein}}{\text{\%N}} = 49.0 \text{ \% protein}$$

118. $Ag^+ + SCN^- \rightarrow AgSCN\ (s)$

$$? \text{ g } Ag^+ = 0.1005 \text{ M} \times 43.56 \text{ mL} \times \frac{10^{-3} \text{ L}}{\text{mL}} \times \frac{\text{mol } Ag^+}{\text{mol } SCN^-} \times \frac{107.87 \text{ g}}{\text{mol } Ag^+} = 0.4722 \text{ g } Ag^+$$

$$? \text{ \% Ag} = \frac{0.4722 \text{ g } Ag^+}{0.5039 \text{ g sample}} \times 100\% = 93.72 \text{ \% Ag}$$

122. $H_2O_2 + 3\ I^- + 2\ H^+ \rightarrow 2\ H_2O + I_3^-$

$I_3^- + 2\ S_2O_3^{2-} \rightarrow 3\ I^- + S_4O_6^{2-}$

$$? \text{ g } H_2O_2 = 28.91 \text{ mL} \times \frac{10^{-3} \text{ L}}{\text{mL}} \times 0.1522 \text{ M } Na_2S_2O_3 \times \frac{\text{mol } I_3^-}{2 \text{ mol } S_2O_3^{2-}} \times \frac{\text{mol } H_2O_2}{\text{mol } I_3^-} \times \frac{34.02 \text{ g } H_2O_2}{\text{mol } H_2O_2} = 7.48 \times 10^{-2} \text{ g } H_2O_2$$

$$? \text{ g} = 10.00 \text{ mL} \times \frac{1.00 \text{ g}}{\text{mL}} = 10.00 \text{ g}$$

$$?\% = \frac{7.48 \times 10^{-2} \text{ g}}{10.00 \text{ g}} \times 100\% = 0.748\%$$

The $H_2O_2(aq)$ is not up to full strength.

Chapter 5

Gases

Exercises

5.1A (a) $? \text{ mmHg} = 0.947 \text{ atm} \times \frac{760 \text{ mmHg}}{\text{atm}} = 7.20 \times 10^2 \text{ mmHg}$

(b) $? \text{ Torr} = 98.2 \text{ kPa} \times \frac{760 \text{ Torr}}{101.325 \text{ kPa}} = 737 \text{ Torr}$

(c) $? \text{ Torr} = 29.95 \text{ in.Hg} \times \frac{2.54 \text{ cm}}{1.000 \text{ in}} \times \frac{10^{-2} \text{ m}}{\text{cm}} \times \frac{\text{mm}}{10^{-3} \text{ m}} \times \frac{\text{Torr}}{\text{mmHg}} = 760.7 \text{ Torr}$

(d) $? \text{ atm} = 768 \text{ Torr} \times \frac{\text{atm}}{760 \text{ Torr}} = 1.01 \text{ atm}$

5.1B $? \text{ kPa} = \frac{1.00 \times 10^2 \text{ N}}{5.00 \text{ cm}^2} \times \left(\frac{\text{cm}}{10^{-2} \text{ m}}\right)^2 \times \frac{\text{Pa}}{\frac{\text{N}}{\text{m}^2}} \times \frac{\text{kPa}}{10^3 \text{ Pa}} = 2.00 \times 10^2 \text{ kPa}$

$? \text{ atm} = 2.00 \times 10^2 \text{ kPa} \times \frac{10^3 \text{ Pa}}{\text{kPa}} \times \frac{\text{atm}}{101325 \text{ Pa}} = 1.97 \text{ atm}$

5.2A $h_{CCl_4} d_{CCl_4} = d_{Hg} h_{Hg}$

$$h_{CCl_4} = \frac{d_{Hg} h_{Hg}}{d_{CCl_4}} = \frac{\frac{13.6 \text{ g}}{\text{cm}^3}}{\frac{1.59 \text{ g}}{\text{cm}^3}} \times 7.60 \times 10^2 \text{ mm} = 6.50 \times 10^3 \text{ mm} = 6.50 \text{ m}$$

5.2B $h_{Hg} = \frac{d_{H_2O} \times h_{H_2O}}{d_{Hg}} = 30.0 \text{ m} \times \frac{\text{mm}}{10^{-3} \text{ m}} \times \frac{1.00 \text{ g/cm}^3}{13.6 \text{ g/cm}^3} = 2.21 \times 10^3 \text{ mmHg}$

$P_{H_2O} = 2.21 \times 10^3 \text{ mmHg} \times \frac{\text{atm}}{760 \text{ mmHg}} = 2.90 \text{ atm}$

The total pressure is 3.90 atm because the atmospheric pressure is 1.00 atm on top of the water.

5.3A (e) $? \text{ mmHg} = 101 \text{ kPa} \times \frac{760 \text{ mmHg}}{101 \text{ kPa}} = 7.60 \times 10^2 \text{ mmHg}$

(f) $? \text{ mmHg} = 103 \text{ kPa} \times \frac{760 \text{ mmHg}}{101 \text{ kPa}} = 775 \text{ mmHg}$

(d) < (a) < (c) <(e) < (b) or (f)

mmHg 735 <745 <750. <760. <(>762) 775

(f) cannot be placed in relation to (b) as it is not known how much (b) is above 762. (f) is greater than all of the other values.

5.3B (a) When one sucks on a straw, it lowers the pressure above the liquid inside the straw. The atmospheric pressure on the liquid outside the straw pushes the liquid up the straw.

(b) Thirty feet of water is equivalent to 760 mmHg or 1 atm. A hand-operated pump can cause a vacuum that will push water to the height that is equivalent to the difference between the vacuum and the atmospheric pressure. Since atmospheric pressure is about 1 atm, the pump can only raise water about 30 ft. In earlier times, deep mines had to have a series of pumps to remove water from the mines.

5.4A $P_1V_1 = P_2V_2 \qquad P_2 = \frac{P_1V_1}{V_2} = \frac{535\text{ mL x }988\text{ Torr}}{1.05\text{ L x }\frac{\text{mL}}{10^{-3}\text{ L}}} = 503\text{ Torr}$

5.4B $P_1V_1 = P_2V_2 \qquad V_2 = \frac{P_1V_1}{P_2} = \frac{98.7\text{ kPa x }73.3\text{ mL}}{4.02\text{ atm}} \times \frac{\text{atm}}{101.325\text{ kPa}} = 17.8\text{ mL}$

5.5A $P_1V_1 = P_2V_2$

$$P_2 = \frac{P_1V_1}{V_2} = \frac{10.2\text{ L x }1208\text{ Torr}}{30.0\text{ L}}$$

10.2 L is about $\frac{1}{3}$ of 30, so $\frac{1}{3}$ of 1208 Torr is about 400 Torr, or 400 mmHg.

5.5B ? $PV = P_1V_1 = 10.2\text{ L} \times 1208\text{ Torr} = 1.23 \times 10^4\text{ L Torr}$

$1.23 \times 10^4\text{ L Torr} = 1.23 \times 10^4\text{ L mmHg}$

$$1.23 \times 10^4\text{ L Torr} \times \frac{\text{mL}}{10^{-3}\,L} \times \frac{\text{cm}^3}{\text{mL}} \times \left(\frac{10^{-2}\text{ m}}{\text{cm}}\right)^3 \times \frac{\text{atm}}{760\text{ Torr}} \times \frac{101.325\text{ kPa}}{\text{atm}}$$

$$= 1.64\text{ kPa m}^3$$

kg/m^2 is not a pressure volume unit.

(d) 1.6 kPa m^3

5.6A $\frac{V_1}{T_1} = \frac{V_2}{T_2} \qquad V_2 = \frac{V_1T_2}{T_1} = \frac{692\text{ L x }(273 + 23)\text{K}}{(273 + 602\text{)K}} = 234\text{ L}$

5.6B $T_f = T_i \times \frac{V_f}{V_i} = 300\text{ K} \times \frac{2.25\text{ L}}{2.00\text{ L}} = 338\text{ K}$

T(°C) = T(K) – 273.15 = 338 – 273.15 = 65 °C

5.7A $V_f = V_i \times \frac{T_f}{T_i} = 2.50\text{ L} \times \frac{180\,°\text{C} + 273}{-120\,°\text{C} + 273} = 2.50\text{ L} \times \frac{453\text{ K}}{153\text{ K}}$

T_f is about 3 times the T_i, so the V_f is about 7.5 L. (actual 7.40 L)

5.7B $\frac{P_1}{T_1} = \frac{P_2}{T_2}$

$$P_2 = \frac{P_1 T_2}{T_1} = \frac{0.60\text{ atm} \times 423\text{ K}}{323\text{ K}}$$

$$P_2 \approx \frac{0.60\text{ atm} \times 4\text{ K}}{3\text{ K}} = 0.8\text{ atm},$$ actual 0.79 atm

5.8A ? g C_3H_8 = 50.0 L $C_3H_8 \times \frac{\text{mol } C_3H_8}{22.4\text{ L } C_3H_8} \times \frac{44.09\text{ g } C_3H_8}{\text{mol } C_3H_8} = 98.4$ g C_3H_8

5.8B ? L CO_2 = 12.0 in. × 12.0 in. × 2.00 in. × $\left(\frac{2.54\text{ cm}}{1.00\text{ in.}}\right)^3 \times \frac{1.56\text{ g}}{\text{cm}^3} \times \frac{\text{mol } CO_2}{44.01\text{ g } CO_2}$

$\times \frac{22.4\text{ L } CO_2}{\text{mol } CO_2} = 3.75 \times 10^3$ L CO_2

5.9A $\frac{P_1 V_1}{T_1} = \frac{P_2 V_2}{T_2}$ $V_1 = V_2$

$$T_2 = \frac{P_2 T_1}{P_1} = \frac{8.0\text{ atm x } (273 + 22)\text{K}}{2.5\text{ atm}}$$

$T_2 = 944\text{ K} - 273 = 671\,°\text{C}$

5.9B $\frac{P_1 V_1}{n_1 T_1} = \frac{P_2 V_2}{n_2 T_2}$, $T_1 = T_2$ and $V_1 = V_2$

$\frac{P_1}{n_1} = \frac{P_2}{n_2}$ or $\frac{P}{n}$ = constant

If the number of molecules increases, the pressure must increase as more molecular collisions with the wall occur. See the graph below.

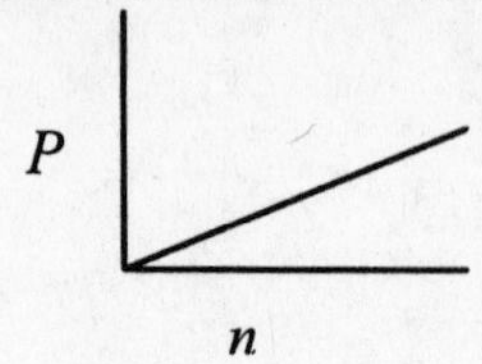

5.10A $n = \frac{PV}{RT} = \frac{3.15\text{ atm} \times 35.0\text{ L}}{\frac{0.08206\text{ L atm}}{\text{K mol}} \times 852\text{ K}} = 1.58$ moles

5.10B $? \text{ mol} = \frac{PV}{RT} = \frac{5.00 \text{ atm} \times 35.0 \text{ L}}{0.08206 \frac{\text{L atm}}{\text{K mol}} \times 852 \text{ K}} = 2.50 \text{ mol}$

$? \text{ molecules} = 2.50 \text{ mol} - 1.58 \text{ mol} = 0.92 \text{ mol added}$

5.11A $T = \frac{PVM}{mR} = \frac{785 \text{ Torr} \times \frac{\text{atm}}{760 \text{ Torr}} \times 5.00 \text{ L} \times \frac{32.00 \text{ g}}{\text{mol}}}{15.0 \text{ g} \times \frac{0.08206 \text{ L atm}}{\text{K mol}}} = 134 \text{ K}$

$134 \text{ K} - 273 = -139 \text{ °C}$ where M is the molar mass and $n = \frac{m}{M}$

5.11B $? \text{ mol } O_2 = 25.0 \text{ g} \times \frac{\text{mol } O_2}{32.00 \text{ g } O_2} = 0.781 \text{ mol } O_2$

$? \text{ mol } N_2 = \frac{n_{O_2} R T_{O_2}}{P_{O_2}} \times \frac{P_{N_2}}{R T_{O_2}} = \frac{0.781 \text{ mol } O_2 \times (30.0 + 273.2) \text{K}}{755 \text{ Torr}} \times \frac{734 \text{ Torr}}{(25.0 + 273.2) \text{K}}$

$= 0.772 \text{ mol } N_2$

$? \text{ g } N_2 = 0.772 \text{ mol } N_2 \times \frac{28.02 \text{ g } N_2}{\text{mol } N_2} = 21.6 \text{ g } N_2$

5.12A $M = \frac{mRT}{PV} = \frac{0.440 \text{ g} \times \frac{0.08206 \text{ L atm}}{\text{K mol}} \times (86 + 273) \text{ K}}{741 \text{ mmHg} \times \frac{\text{atm}}{760 \text{ mmHg}} \times 79 \text{ mL} \times \frac{10^{-3} \text{ L}}{\text{mL}}}$ $M = 74.3 \text{ g/mol}$

5.12B $PV = nRT = \frac{mRT}{M}$

$$\frac{m_1 R T_1}{P_1 V_1 M_1} = \frac{m_2 R T_2}{P_2 V_2 M_2}$$

$$P_2 = \frac{P_1 V_1 M_1}{m_1 R T_1} \frac{m_2 R T_2}{V_2 M_2}$$

$P_2 = \frac{P_1 M_1}{M_2}$ R cancels, $T_1 = T_2$, $V_1 = V_2$, $m_1 = m_2$

$$P_2 = \frac{1.00 \text{ atm} \times 32.00 \frac{\text{g}}{\text{mol}}}{16.04 \frac{\text{g}}{\text{mol}}} = 2.00 \text{ atm}$$

5.13A $M = \frac{mRT}{PV} = \frac{0.471\ \text{g} \times \frac{0.08206\ \text{L atm}}{\text{K mol}} \times (98+273)\ \text{K}}{715\ \text{mmHg} \times 121\ \text{mL} \times \frac{10^{-3}\ \text{L}}{\text{mL}} \times \frac{\text{atm}}{760\ \text{mmHg}}}$ $M = 126$ g/mol

5.13B $?\ \text{mol C} = 55.80\ \text{g C} \times \frac{\text{mol C}}{12.011\ \text{g C}} = 4.646\ \text{mol C} \times \frac{1}{2.323\ \text{mol O}} = 2.00\ \frac{\text{mol C}}{\text{mol O}}$

$?\ \text{mol H} = 7.03\ \text{g H} \times \frac{\text{mol H}}{1.008\ \text{g H}} = 6.97\ \text{mol H} \times \frac{1}{2.323\ \text{mol O}} = 3.00\ \frac{\text{mol H}}{\text{mol O}}$

$?\ \text{mol O} = 37.17\ \text{g O} \times \frac{\text{mol O}}{16.00\ \text{g O}} = 2.323\ \text{mol O} \times \frac{1}{2.323\ \text{mol O}} = 1.00\ \frac{\text{mol O}}{\text{mol O}}$

Empirical formula is C_2H_3O, at 43.04 g/formula unit.

$M = \frac{mRT}{PV} = \frac{0.3060\ \text{g} \times \frac{0.08206\ \text{L atm}}{\text{K mol}} \times (100+273)\ \text{K}}{747\ \text{mmHg} \times \frac{\text{atm}}{760\ \text{nnHg}} \times 111\ \text{mL} \times \frac{10^{-3}\ \text{L}}{\text{mL}}}$

$M = 85.8$ g/mol or 85.8 u/molecule

$\frac{85.8\ \text{u}}{\text{molecule}} \times \frac{\text{formula}}{43.04\ \text{u}} = 2$ formula/mol Molecular formula is $C_4H_6O_2$.

5.14A $PV = nRT$

$P = \frac{mRT}{VM} = \frac{dRT}{M}$

$d = \frac{PM}{RT} = \frac{748\ \text{Torr} \times \frac{\text{atm}}{760\ \text{Torr}} \times \frac{30.07\ \text{g}}{\text{mol}}}{\frac{0.08206\ \text{L atm}}{\text{K mol}} \times (15+273)\ \text{K}} = 1.25\ \text{g/L}$

5.14B $M = \frac{dRT}{P}$

$M = \frac{2.42\ \frac{\text{g}}{\text{L}} \times \frac{62.36\ \text{L Torr}}{\text{K mol}} \times 293.2\ \text{K}}{762\ \text{Torr}} = 58.1\ \text{g/mol}$

C_4H_{10} is the only alkane that has the molar mass of 58.1 g/mol.

5.15A $T = \frac{PM}{Rd} = \frac{785\ \text{Torr} \times \frac{\text{atm}}{760\text{Torr}} \times \frac{44.09\ \text{g}}{\text{mol}}}{\frac{0.08206\ \text{L atm}}{\text{K mol}} \times 1.51\ \text{g/L}} = 367\ \text{K}$

$?\ \text{T} = 367\ \text{K} - 273 = 94\ °\text{C}$

5.15B $? \, d = \dfrac{P_{NH_3} M_{NH_3}}{RT_{NH_3}} = \dfrac{1.45 \text{ atm} \times \dfrac{17.03 \text{ g}}{\text{mol}}}{\dfrac{0.08206 \text{ L atm}}{\text{mol K}} \times 295.7 \text{ K}} = \dfrac{1.02 \text{ g}}{\text{L}}$

$? \, T = \dfrac{P_{O_2} M_{O_2}}{Rd_{O_2}} = \dfrac{725 \text{ Torr} \times \dfrac{32.00 \text{ g}}{\text{mol}}}{\dfrac{62.36 \text{ L Torr}}{\text{K mol}} \times \dfrac{1.02 \text{ g}}{\text{L}}} = 365 \text{ K} - 273 = 92 \text{ °C}$

5.16A $C_3H_8 + 5\, O_2 \rightarrow 3\, CO_2 + 4\, H_2O$

$? \text{ L } O_2 = 0.556 \text{ L } C_3H_8 \times \dfrac{5 \text{ L } O_2}{1 \text{ L } C_3H_8} = 2.78 \text{ L } O_2$

5.16B $CH_3OCH_3 + 3\, O_2 \rightarrow 2\, CO_2 + 3\, H_2O$

$? \text{ L } O_2 = 125 \text{ g } C_2H_6O \times \dfrac{\text{L } C_2H_6O}{1.81 \text{ g } C_2H_6O} \times \dfrac{3 \text{ L } O_2}{1 \text{ L } C_2H_6O} = 207 \text{ L } O_2$

5.17A $V = \dfrac{nRT}{P}$

$$45.8 \text{ kg } CaCO_3 \times \frac{10^3 \text{ g}}{\text{kg}} \times \frac{1 \text{ mol } CaCO_3}{100.09 \text{ g } CaCO_3} \times \frac{\text{mol } CO_2}{\text{mol } CaCO_3} \times \frac{\dfrac{0.08206 \text{ L atm}}{\text{K mol}} \text{ x } (825 + 273) \text{ K}}{754 \text{ Torr x } \dfrac{\text{atm}}{760 \text{ Torr}}} = V = 4.16 \times 10^4 \text{ L}$$

5.17B $2\, C_5H_{10} + 15\, O_2 \rightarrow 10\, CO_2 + 10\, H_2O$

$$? \text{ L } C_5H_{10} = \frac{736 \text{ Torr} \times 1.00 \times 10^6 \text{ L}}{\dfrac{62.36 \text{ L Torr}}{\text{K mol}} \times 298.2 \text{ K}} \times \frac{2 \text{ mol } C_5H_{10}}{10 \text{ mol } CO_2} \times \frac{70.13 \text{ g } C_5H_{10}}{\text{mol } C_5H_{10}} \times \frac{\text{mL}}{0.7445 \text{ g}} \times \frac{10^{-3} \text{ L}}{\text{mL}} = 746 \text{ L } C_5H_{10}$$

5.18A $P_{O_2} = \dfrac{n_{O_2} RT}{V} = \dfrac{0.00856 \text{ mol x } \dfrac{0.08206 \text{ L atm}}{\text{mol K}} \text{ x } 298 \text{ K}}{1.00 \text{ L}}$

$P_{O_2} = 0.209 \text{ atm}$

$P_{Ar} = \dfrac{n_{Ar} RT}{V} = \dfrac{0.000381 \text{ mol} \times \dfrac{0.08206 \text{ L atm}}{\text{k mol}} \times 298 \text{ K}}{1.00 \text{ L}} = 0.00932 \text{ atm}$

$$P_{CO_2} = \frac{n_{CO_2}RT}{V} = \frac{0.00002\,\text{mol} \times \frac{0.08206\,\text{L atm}}{\text{mol K}} \times 298\text{K}}{1.00\,\text{L}} = 0.0005\text{ atm}$$

$$P_{total} = P_{N_2} + P_{O_2} + P_{Ar} + P_{CO_2} = (0.780 + 0.209 + 0.00932 + 0.0005)\text{ atm}$$

$$P_{total} = 0.999\text{ atm}$$

5.18B $$P_{N_2} = \frac{m_{N_2}RT}{MV} = \frac{4.05\,\text{g} \times \frac{0.08206\,\text{L atm}}{\text{K mol}} \times (25+273)\,\text{K}}{28.01\,\text{g/mol} \times 6.10\,\text{L}}$$

$$P_{N_2} = 0.580\text{ atm}$$

$$P_{H_2} = \frac{m_{H_2}RT}{MV} = \frac{3.15\,\text{g} \times \frac{0.08206\,\text{L atm}}{\text{K mol}} \times (25+273)\,\text{K}}{2.016\,\text{g/mol} \times 6.10\,\text{L}}$$

$$P_{H_2} = 6.26\text{ atm}$$

$$P_{He} = \frac{m_{He}RT}{MV} = \frac{6.05\,\text{g} \times \frac{0.08206\,\text{L atm}}{\text{K mol}} \times (25+273)\,\text{K}}{4.003\,\text{g/mol} \times 6.10\,\text{L}}$$

$$P_{He} = 6.06\text{ atm}$$

$$P_{total} = P_{N_2} + P_{H_2} + P_{He} = 12.90\text{ atm}$$

5.19A $P_{N_2} = 0.741 \times 1.000\text{ atm} = 0.741\text{ atm}$

$P_{O_2} = 0.150 \times 1.000\text{ atm} = 0.150\text{ atm}$

$P_{H_2O} = 0.060 \times 1.000\text{ atm} = 0.060\text{ atm}$

$P_{Ar} = 0.009 \times 1.000\text{ atm} = 0.009\text{ atm}$

$P_{CO_2} = 0.040 \times 1.000\text{ atm} = 0.040\text{ atm}$

5.19B $$P = \frac{mRT}{MV} = \frac{10.5\,\text{g C}_4\text{H}_{10} \times \frac{62.36\,\text{L Torr}}{\text{K mol}} \times 296.7\,\text{K}}{\frac{58.12\,\text{g}}{\text{mol}} \times 75.0\,\text{L}} = 44.6\text{ Torr (butane)}$$

$$P_{total} = P_{CH_4} + P_{C_2H_6} + P_{C_3H_8} + P_{C_4H_{10}}$$

$$P_{total} = (505 + 201 + 43 + 44.6)\text{ Torr} = 794\text{ Torr}$$

$$\chi_{CH_4} = \frac{505\text{ Torr}}{794\text{ Torr}} = 0.636$$

$$\chi_{C_2H_6} = \frac{201 \text{ Torr}}{794 \text{ Torr}} = 0.253$$

$$\chi_{C_3H_8} = \frac{43 \text{ Torr}}{794 \text{ Torr}} = 0.054$$

$$\chi_{C_4H_{10}} = \frac{44.6 \text{ Torr}}{794 \text{ Torr}} = 0.0562$$

5.20A The increase to 3.00 atm could be achieved by adding only hydrogen, but the P_{H_2} would be 2.50 atm, not 2.00 atm. To achieve a 3.00 atm pressure with P_{H_2} = 2.00 atm, other gases must supply 1.00 atm pressure. The original He supplies only 0.50 atm. 0.5 mol of any inert gas may be added to supply the partial pressure of 1.00 atm. The partial pressure of helium is not listed only total and hydrogen. Adding helium only could raise the total pressure to 3.00 atm but would not raise the partial pressure of hydrogen to 2.00 atm.
It is necessary to raise the partial pressure of hydrogen as it is listed. The partial pressure of helium is not listed so adding helium is not necessary.

5.20B $$n = \frac{PV}{RT} = \frac{1.00 \text{ atm} \times 2.2 \text{ L}}{0.08206 \dfrac{\text{L atm}}{\text{K mol}} \times 273 \text{ K}} = 0.098 \text{ mol}$$

$$n = \frac{PV}{RT} = \frac{2.0 \text{ atm} \times 2.2 \text{ L}}{0.08206 \dfrac{\text{L atm}}{\text{K mol}} \times 673 \text{ K}} = 0.080 \text{ mol}$$

0.098 mol − 0.080 mol = 0.018 mol to be removed.

$$V = \frac{nRT}{P} = \frac{0.018 \text{ mol} \times 0.08206 \dfrac{\text{L atm}}{\text{K mol}} \times 273 \text{ K}}{1.00 \text{ atm}} = 0.40 \text{ L}$$

(d) Releasing 0.46 L is closer to the answer than any of the other answers.

5.21A $P_{H_2} = P_{total} - P_{H_2O} = 738 \text{ Torr} - 16 \text{ Torr} = 722 \text{ Torr}$

$$m_{H_2} = \frac{MPV}{RT} = \frac{2.016 \text{ g/mol} \times 722 \text{ Torr} \times \dfrac{\text{atm}}{760 \text{ Torr}} \times 246 \text{ mL} \times \dfrac{10^{-3} \text{ L}}{\text{mL}}}{\dfrac{0.08206 \text{ L atm}}{\text{K mol}} \times (18 + 273) \text{ K}}$$

$$= 0.0197 \text{ g } H_2$$

$$m_{H_2O} = \frac{18.02\text{ g/mol x }15.5\text{ Torr x }246\text{ mL x }\dfrac{10^{-3}\text{ L}}{\text{mL}}\text{ x }\dfrac{\text{atm}}{760\text{ Torr}}}{\dfrac{0.08206\text{ L atm}}{\text{K mol}}\text{ x }(18+273)\text{K}} = 0.00379\text{ g }H_2O$$

$m_{total} = 0.0235$ g

5.21B $2\ KClO_3 \rightarrow 2\ KCl + 3O_2$

$P_{H_2} = P_{total} - P_{H_2O} = 746\text{ mmHg} - 19\text{ mmHg} = 727\text{ mmHg}$

$$?\text{ g }KClO_3 = \frac{727\text{ mmHg}\times\dfrac{\text{atm}}{760\text{ mmHg}}\times 155\text{mL}\times\dfrac{10^{-3}\text{L}}{\text{mL}}}{\dfrac{0.08206\text{Latm}}{\text{Kmol}}\times(21+273)\text{K}}\times\frac{2\text{ mol }KClO_3}{3\text{ mol }O_2}$$

$$\times\frac{122.55\text{ g}}{\text{mol}} = 0.502\text{ g }KClO_3$$

5.22A $u_{rms} = \sqrt{\dfrac{3RT}{M}}$

The greatest u_{rms} will have the largest $\dfrac{T}{M}$.

(a) $\dfrac{273}{16}$ (b) $\dfrac{523}{32}$ (c) $\dfrac{1023}{64}$ (d) $\dfrac{308}{17}$

273 is more than $\frac{1}{2}$ of 523 and 523 is more than $\frac{1}{2}$ of 1023, so (a) is larger than (b) or (c).

$\dfrac{273}{16} = 17.1$

$\dfrac{308}{17} = 18.1$ so (d) is the greatest.

$$\text{actual } u_{rms} = \sqrt{\frac{3\times\dfrac{8.3145\text{ J}}{\text{K mol}}\times 273\text{ K}}{\dfrac{0.01604\text{ kg}}{\text{mol}}}}$$

(a) 652 m/s (b) 638 m/s (c) 631 m/s (d) 672 m/s

5.22B The O_2-to-H_2 molar mass ratio is 32/2 = 16. For u_{rms} of O_2 to be greater than 1838 m/s (u_{rms} of H_2 at 0 °C), T/16 must be greater than 273 K.
T must be greater than 16 × 273, which simplifies to 15 × 300 = 4500 K. Answer: 5000 K; Actual: 16 × 273 = 4368 K.

5.23A $\frac{r_{N2}}{r_{Ar}} = \sqrt{\frac{M_{Ar}}{M_{N_2}}} = \sqrt{\frac{39.95}{28.01}} = 1.194$

N_2 is 1.194 times faster.

5.23B $\frac{r_{NO}}{r_{(CH_3)_2O}} = \sqrt{\frac{M_{(CH_3)_2O}}{M_{NO}}} = \sqrt{\frac{46.07}{30.01}} = 1.239$

NO effuses 1.239 times faster.

$?\% = \frac{0.239}{1.00} \times 100\% = 23.9\%$ faster

5.24A $\frac{r_{N2}}{r_{unk}} = \frac{t_{unk}}{t_{N2}} = \frac{83\ s}{57\ s} = \sqrt{\frac{M_{unk}}{M_{N_2}}} = 1.46$

$\mathcal{M}_{unk} = (1.46)^2 \times 28.01$ g/mol = 59 g/mol

5.24B $\frac{t_{C4H10}}{t_{O2}} = \sqrt{\frac{M_{C_4H_{10}}}{M_{O_2}}} = \frac{t_{C4H10}}{123\ s} = \sqrt{\frac{58.12}{32.00}}$

$t_{CH4} = 166$ s

Self-Assessment Questions

3. (a) Charles's law states that the volume of a fixed amount of gas at a constant pressure is directly proportional to its Kelvin temperature: $V = bT$.
   (b) Boyle's law states that for a given amount of gas at a constant temperature, the volume of a gas varies inversely with its pressure: $PV = a$ or $V = a/P$.
   (c) Avogadro's law states that at a fixed temperature and pressure, the volume of gas is directly proportional to the amount of gas: $V = cn$.
   (b) is the answer because $V \approx P$.

5. (a) $? \text{ atm} = 21.92 \text{ inHg} \times \frac{2.540 \text{ cm}}{1.000 \text{ in}} \times \frac{10^{-2} \text{ m}}{\text{cm}} \times \frac{\text{mm}}{10^{-3} \text{ m}} \times \frac{\text{atm}}{760 \text{ mmHg}} = 0.7326 \text{ atm}$

   (b) $? \text{ atm} = 77.2 \text{ kPa} \times \frac{\text{atm}}{101.325 \text{ kPa}} = 0.762 \text{ atm}$

   (c) $? \text{ atm} = 525 \text{ Torr} \times \frac{\text{atm}}{760 \text{ Torr}} = 0.691 \text{ atm}$

   (d) $? \text{ atm} = 892 \text{ mmHg} \times \frac{\text{atm}}{760 \text{ mmHg}} = 1.17 \text{ atm}$

6. (a) The volume will decrease.
(b) The volume will decrease.
(c) The effect is uncertain, as the increase in pressure will cause a volume decrease, and the increase in temperature will cause a volume increase.
(d) The volume will increase.

7. (a) The pressure will increase.
(b) The pressure will increase.
(c) The pressure will increase.
(d) The effect is uncertain, as the decrease in temperature will cause a pressure decrease while the decrease in volume will cause a pressure increase.

9. $V = \dfrac{nRT}{P}$
(a) increase V
(b) decrease V
(c) decrease V
(d) decrease V

10. $P = \dfrac{dRT}{M}$ $\qquad d = \dfrac{PM}{RT}$
The greatest density comes from the largest molar mass, (d) PF_3.

(a) $d = \dfrac{1.00 \text{ atm} \times 70.90 \text{ g/mol}}{0.08206 \dfrac{\text{L atm}}{\text{K mol}} \times 273 \text{ K}} = 3.16 \text{ g/L}$

(b) $d = \dfrac{1.00 \text{ atm} \times 80.07 \text{ g/mol}}{0.08206 \dfrac{\text{L atm}}{\text{K mol}} \times 273 \text{ K}} = 3.57 \text{ g/L}$

(c) $d = \dfrac{1.00 \text{ atm} \times 44.01 \text{ g/mol}}{0.08206 \dfrac{\text{L atm}}{\text{K mol}} \times 273 \text{ K}} = 1.96 \text{ g/L}$

(d) $d = \dfrac{1.00 \text{ atm} \times 87.97 \text{ g/mol}}{0.08206 \dfrac{\text{L atm}}{\text{K mol}} \times 273 \text{ K}} = 3.93 \text{ g/L}$

11. $V = \dfrac{nRT}{P}$ $\qquad 100°\text{C} = 373 \text{ K}$
A T decrease decreases volume so (a).

12. $n = \dfrac{PV}{RT} = \dfrac{0.500\text{ atm} \times 4.48\text{ L}}{0.08206\dfrac{\text{L atm}}{\text{K mol}} \times 273\text{ K}} = 0.100$ mol, not (a) or (d)

$0.100\text{ mol} \times \dfrac{17.04\text{g}}{\text{mol}} = 1.7$ g, not (b)

$0.100\text{ mol} \times \dfrac{6.02 \times 10^{23}\text{ molecules}}{\text{mol}} = 6.02 \times 10^{22}$, (c)

(c) is the answer

13. (a) A; a higher pressure in container A means more molecules and thus more mass and a higher density.
(b) They are equal. Equal pressures and temperatures mean the same number of molecules per volume.
(c) B; a higher temperature means that container A has fewer molecules and thus less mass and a lower density.

14. ? mol = 0.250 mol + 0.250 mol + 0.250 mol = 0.750 mol

$P = \dfrac{nRT}{V} = \dfrac{0.750\text{ mol} \times 0.08206\dfrac{\text{L atm}}{\text{K mol}} \times 298\text{ K}}{10.0\text{ L}}$

$P = 1.83$ atm

(a) is answer

15. $\chi_{O_2} = \dfrac{0.25\text{ mol}}{0.25\text{ mol} + 0.25\text{ mol} + 0.25\text{ mol}} = 0.33$

(b) is answer

16. $n = \dfrac{PV}{RT} = \dfrac{2.00\text{ atm} \times 2.24\text{ L}}{0.08206\dfrac{\text{L atm}}{\text{K mol}} \times 273\text{ K}} = 0.200$ mol gas

? mol = $1.60\text{g} \times \dfrac{\text{mol}}{32.00\text{g}} = 0.0500$ mol gas so 0.150 mol of gas must be added.

(a) 1.60g is only 0.0500 mol
(b) add not release
(c) ? mol = $2.00\text{g He} \times \dfrac{\text{mol}}{4.003\text{g}} = 0.500$ mol
(d) ? mol = $0.60\text{g He} \times \dfrac{\text{mol}}{4.003\text{ g}} = 0.150$ mol

(d) is answer.

17. $P_{total} = P_{O_2} + P_{H_2O}$

$$P_{O_2} = 751 \text{ Torr} - 21 \text{ Torr} = 730 \text{ Torr}$$

$$P_{O_2} = 730 \text{ Torr} \times \frac{\text{atm}}{760 \text{ Torr}} = 0.961 \text{ atm} \quad \text{(c) is the answer.}$$

18. (a) Kinetic energy is dependent on temperature.
(b) Molecular speed depends on molar mass.
(c) Volume depends on number of moles.
(d) Effusion rates depend on molar mass.
(a) is the answer.

19. H_2 effuses faster than SO_2 so (b).

20. At high pressure and low temperatures, gas molecules are close enough together and moving slowly enough that the intermolecular forces are no longer negligible. Also, the volume of the gas molecules becomes a significant part of the total volume of the gas. Low pressure and high temperature is where a gas will act like an ideal gas. (b) is the best choice.

Problems

21. (a) $? \text{ mmHg} = 0.934 \text{ atm} \times \frac{760 \text{ Torr}}{\text{atm}} = 710 \text{ Torr}$

(b) $? \text{ atm} = 767 \text{ Torr} \times \frac{\text{atm}}{760 \text{ Torr}} = 1.01 \text{ atm}$

(c) $? \text{ kPa} = 698 \text{ mmHg} \times \frac{101.3 \text{ kPa}}{760 \text{ mmHg}} = 93.0 \text{ kPa}$

(d) $? \text{ mmHg} = 33.5 \text{ lb/in}^2 \times \frac{760 \text{ mmHg}}{14.70 \text{ lb/in}^2} = 1.73 \times 10^3 \text{ mmHg}$

23. $$P_{Hg} = P_{H_2O} \times \frac{d_{H_2O}}{d_{Hg}} = 1250 \text{ m } H_2O \times \frac{\frac{1.00 \text{ g}}{\text{cm}^3}}{\frac{13.6 \text{ g}}{\text{cm}^3}} = 91.9 \text{ m Hg} \times \frac{\text{mm}}{10^{-3} \text{ m}}$$

$$= 9.19 \times 10^4 \text{ mm Hg} \times \frac{\text{atm}}{760 \text{ mm}} = 121 \text{ atm}$$

25. $$h_{CCl_4} = \frac{d_{Hg}h_{Hg}}{d_{CCl_4}} = \frac{\frac{13.6 \text{ g}}{\text{cm}^3}}{\frac{1.59 \text{ g}}{\text{cm}^3}} \times 11.2 \text{ cm} \times \frac{10^{-2} \text{ m}}{\text{cm}} = 0.958 \text{ m}$$

27. $$h_{oil} \times \frac{d_{oil}}{d_{Hg}} = h_{Hg}$$

$$h_{Hg} = 83 \text{ mm oil} \times \frac{0.901 \text{ g/mL}}{13.6 \text{ g/mL}} = 5.5 \text{ mmHg} = 5.5 \text{ Torr}$$

? Torr = 751 Torr + 6 Torr = 757 Torr

29. The height for the closed end of the manometer must be 760 mm for each atmosphere of pressure to be measured. One atmosphere of gas pressure would cause a difference of 760 mm in the heights of the two columns. Two atmospheres would cause a 1520-mm difference.

31. $P_1V_1 = P_2V_2$

(a) $V_2 = \frac{P_1V_1}{P_2} = 882 \text{ mL} \times \frac{752 \text{ Torr}}{719 \text{ Torr}} = 922 \text{ mL}$

(b) $V_2 = \frac{P_1V_1}{P_2} = 882 \text{ mL} \times \frac{752 \text{ Torr}}{1.38 \text{ atm} \times \frac{760 \text{ Torr}}{\text{atm}}} = 632 \text{ mL}$

(c) $V_2 = \frac{P_1V_1}{P_2} = 882 \text{ mL} \times \frac{752 \text{ Torr}}{125 \text{ kPa} \times \frac{760 \text{ Torr}}{101.3 \text{ kPa}}} = 707 \text{ mL}$

33. $V_2 = \frac{P_1V_1}{P_2} = 10.3 \text{ m}^3 \times \frac{4.50 \text{ atm}}{1.00 \text{ atm}} = 46.4 \text{ m}^3$

35. (a) $V_2 = \frac{P_1V_1}{P_2} = 43.0 \text{ L} \times \frac{150. \text{ atm}}{750.0 \text{ Torr}} \times \frac{760 \text{ Torr}}{\text{atm}} = 6.54 \times 10^3 \text{ L}$

(b) $? \text{ h} = 6.54 \times 10^3 \text{ L} \times \frac{\text{min}}{8.00 \text{ L}} = 817 \text{ min}$

37. $\frac{V_1}{T_1} = \frac{V_2}{T_2} \qquad V_2 = \frac{V_1T_2}{T_1} = 641 \text{ mL} \times \frac{(5.0 + 273.2)\text{K}}{(99.8 + 273.2)\text{K}} = 478 \text{ mL}$

39. ? mL = 832 mL × 1.10 = 915 mL

$\frac{V_1}{T_1} = \frac{V_2}{T_2} \qquad \frac{832 \text{ mL}}{T} = \frac{915 \text{ mL}}{T + 15.0}$

$832\,T + 1.25 \times 10^4 = 915\,T$

$T = 1.5 \times 10^2 \text{ K}$

$\frac{V_1}{V_2} = \frac{T_1}{T_2}$ Rearranging the equation to this form makes it obvious that for a 10% increase in volume, the temperature increase is 15 degrees, and T must be about 150 K.

41. $? \text{ kg} = 4.55 \text{ L} \times 10^3 \times \frac{\text{mol}}{22.4 \text{ L}} \times \frac{16.04 \text{ g}}{\text{mol}} \times \frac{\text{kg}}{10^3 \text{ g}} = 3.26 \text{ kg } CH_4$

43. $? \text{ mL} = 125 \text{ mg Ar} \times \frac{\text{mol Ar}}{39.95 \text{ g Ar}} \times \frac{22.4 \text{ L}}{\text{mol}} = 70.1 \text{ mL}$

$? \text{ mL} = 505 \text{ mL} + 70.1 \text{ mL} = 575 \text{ mL}$

45. The answer can be determined by estimation.

(a) 5.0 g $H_2 \approx 2.5$ mol

(b) 50 L SF_6 at STP is a little more than 2.0 mol but less than 2.5 mol.

(c) 1.0×10^{24} is less than 2 mol.

(d) $67 \text{ L} \times \frac{\text{mol}}{22.4 \text{ L}} \times \approx 3 \text{ mol}$

(d) is the largest number of molecules.

<u>Actual calculations:</u>

(a) $? \text{ molecules} = 5.0 \text{ g } H_2 \times \frac{\text{mol } H_2}{2.016 \text{ g } H_2} \times \frac{6.022 \text{ x } 10^{23} \text{ molecules}}{\text{mol}}$

$= 1.5 \times 10^{24}$ molecules

(b) $? \text{ molecules} = 50 \text{ L } SF_6 \times \frac{\text{mol } SF_6}{22.4 \text{ L } SP_6} \times \frac{6.022 \text{ x } 10^{23} \text{ molecules}}{\text{mol}}$

$= 1.3 \times 10^{24}$ molecules

(c) $? \text{ molecules} = 1.0 \times 10^{24}$ molecules

(d) $? \text{ molecules} = 67 \text{ L} \times \frac{\text{mol}}{22.4 \text{ L}} \times \frac{6.022 \text{ x } 10^{23} \text{ molecules}}{\text{mol}}$

$= 1.8 \times 10^{24}$ molecules

(d) is the largest number of molecules.

47. $\frac{P_1}{T_1} = \frac{P_2}{T_2} \; P_2 = P_1 \times \frac{T_2}{T_1} = 721 \text{ Torr} \times \frac{(755 + 273)\text{K}}{(25 + 273)\text{K}} = 2.49 \times 10^3 \text{ Torr}$

49. 1 mol at 273 K and 1 atm is 22.4 L.

$\frac{P_1V_1}{T_1} = \frac{P_2V_2}{T_2}, \; P_1 = P_2, V_2 = V_1 \times \frac{T_2}{T_1} = \frac{22.4 \text{ L x } (25 + 273)\text{K}}{273\text{K}} = 24.5 \text{ L}$

51. $V_2 = V_1 \times \frac{P_1T_2}{P_2T_1} = 2.53 \text{ m}^3 \times \frac{191 \text{ Torr}}{1142 \text{ Torr}} \times \frac{(25 + 273)\text{K}}{(-15 + 273)\text{K}} = 0.489 \text{ m}^3$

53. (a) $V = \frac{nRT}{P} = \frac{1.88 \text{ mol} \times \frac{0.08206 \text{ L atm}}{\text{K mol}} \times (55 + 273)\text{K}}{1.08 \text{ atm}} = 46.9 \text{ L}$

(b) $P = \frac{nRT}{V} = \frac{137 \text{ g CO} \times \frac{\text{mol}}{28.01 \text{ g}} \times \frac{0.08206 \text{ L atm}}{\text{K mol}} \times (28 + 273)\text{K}}{4.49 \text{ L}} = 26.9 \text{ atm}$

(c) $? \text{ mg} = \dfrac{PVM}{RT} = \dfrac{744 \text{ Torr} \times \dfrac{\text{atm}}{760 \text{ Torr}} \times 97.4 \text{ mL} \times \dfrac{2.016 \text{ g H}_2}{\text{mol H}_2}}{\dfrac{0.08206 \text{ L atm}}{\text{K mol}} \times (273 - 2) \text{ K}} = 8.64 \text{ mg H}_2$

(d) $P = \dfrac{mRT}{M \quad V} = \dfrac{19.6 \text{ g N}_2 \times \dfrac{\text{mol N}_2}{28.01 \text{ g N}_2} \times \dfrac{0.08206 \text{ L atm}}{\text{K mol}} \times (273 - 0) \text{ K}}{6.41 \text{ L}}$

$\times \dfrac{101.3 \text{ kPa}}{\text{atm}} = 248 \text{ kPa}$

55. $\dfrac{P_1V_1}{T_1} = \dfrac{P_2V_2}{T_2}$

$V_2 = \dfrac{V_1P_1T_2}{P_2T_1}$

$V_2 = \dfrac{760 \text{ Torr} \times 7.30 \text{ L}}{273 \text{ K}} \times \dfrac{(17 + 273) \text{ K}}{788 \text{ Torr}} = 7.48 \text{ L}$

57. $\dfrac{P_1V_1}{T_1} = \dfrac{P_2V_2}{T_2}$

$V_2 = \dfrac{V_1P_1T_2}{P_2T_1} = 4.20 \times 10^3 \text{ L} \times \dfrac{2.50 \text{ atm}}{151 \text{ atm}} \times \dfrac{(25 + 273)\text{K}}{(17 + 273)\text{K}}$

$V_2 = 71.5 \text{ L}$

59. $M = \dfrac{mRT}{PV} = 0.808 \text{ g} \times \dfrac{\dfrac{0.08206 \text{ L atm}}{\text{K mol}} \times (98 + 273) \text{ K}}{756 \text{ Torr} \times \dfrac{\text{atm}}{760 \text{ Torr}} \times 139 \text{ mL} \times \dfrac{10^{-3} \text{ L}}{\text{mL}}} = 178 \text{ g/mol}$

Molecular mass = 178 u

61. $? \text{ mol H} = 8.75 \text{ g H} \times \dfrac{\text{mol H}}{1.008 \text{ g H}} = 8.68 \text{ mol H} \times \dfrac{1}{7.597 \text{ mol C}} = \dfrac{1.14 \text{ mol H}}{\text{mol C}}$

$? \text{ mol C} = 91.25 \text{ g C} \times \dfrac{\text{mol C}}{12.011 \text{ g C}} = 7.597 \text{ mol C} \times \dfrac{1}{7.597 \text{ mol C}} = \dfrac{1 \text{ mol C}}{\text{mol C}}$

$C_{(1\times7)}H_{(1.14\times7)} \Rightarrow C_7H_8$ formula mass = 92.1 u/formula

$n = \dfrac{PV}{RT} = \dfrac{761 \text{ Torr x } 435 \text{ mL x } \dfrac{10^{-3} \text{ L}}{\text{mL}}}{\dfrac{62.36 \text{ L Torr}}{\text{K mol}} \text{ x } (273 + 115) \text{ K}} = 0.0137 \text{ mol}$

$M = \dfrac{1.261 \text{ g}}{0.0137 \text{ mol}} = 92.2 \text{ g/mol or } 92.2 \text{ u/molecule}$

$$? = \frac{\text{molar mass}}{\text{formula mass}} = \frac{92.2\ \text{u/molecule}}{92.1\ \text{u/formula unit}} = \frac{1\ \text{formula unit}}{\text{molecule}}$$

Molecular formula is C_7H_8.

63. (a) $d = \frac{m}{V} = \frac{PM}{RT} = \frac{1.00\ \text{atm} \times \frac{20.18\ \text{g}}{\text{mol}}}{\frac{0.08206\ \text{L atm}}{\text{K mol}} \times 273\ \text{K}} = 0.901\ \text{g/L}$ or $\frac{20.18\ \text{g/mol}}{22.4\ \text{L/mol}} = d$

(b) $d = \frac{PM}{RT} = \frac{1.19\ \text{atm} \times \frac{28.05\ \text{g}}{\text{mol}}}{\frac{0.08206\ \text{L atm}}{\text{K mol}} \times (147 + 273)\ \text{K}} = 0.969\ \text{g/L}$

65. $P = \frac{dRT}{M} = \frac{\frac{1.01\ \text{g}}{\text{L}} \times \frac{0.08206\ \text{L atm}}{\text{K mol}} \times (37 + 273)\ \text{K}}{32.00\ \text{g/mol}} = 0.803\ \text{atm}$

67. $M = \frac{dRT}{P} = \frac{\frac{4.33\ \text{g}}{\text{L}}\ \text{x}\ \frac{0.08206\ \text{L atm}}{\text{K mol}}\ \text{x}\ (445 + 273)\text{K}}{755\ \text{mmHg x}\ \frac{\text{atm}}{760\ \text{mmHg}}} = 256.8\ \text{g/mol}$

$\frac{256.8\ \text{u}}{\text{molecule}} \times \frac{\text{formula unit}}{32.06\ \text{u}} = 8.01\ \frac{\text{formula unit}}{\text{molecule}}$ S_8 is molecular formula.

69. $d = \frac{PM}{RT}$ actual

(a) $\frac{745\ \text{Torr x 2 g/mol}}{\frac{62.4\ \text{L Torr}}{\text{K mol}}\ \text{x}\ (273 - 15)\ \text{K}}$ $\frac{0.0926\ \text{g}}{\text{L}}$

(b) $\frac{760\ \text{Torr x 4 g/mol}}{\frac{62.4\ \text{L Torr}}{\text{K mol}}\ \text{x}\ 273\ \text{K}}$ $\frac{0.178\ \text{g}}{\text{L}}$

(c) $\frac{1.15\ \text{atm x}\ \frac{760\ \text{Torr}}{\text{atm}}\ \text{x 16 g/mol}}{\frac{62.4\ \text{L Torr}}{\text{K mol}}\ \text{x}\ (273 - 10)\ \text{K}}$ $\frac{0.852\ \text{g}}{\text{L}}$

(d) $\frac{435\ \text{Torr x 30 g/mol}}{\frac{62.4\ \text{L Torr}}{\text{K mol}}\ \text{x}\ (273 + 50)\ \text{K}}$ $\frac{0.647\ \text{g}}{\text{L}}$

By comparing equations, it can be seen that (c) is larger than (b) or (a), because the molar mass in (c) is four times the molar mass in (b) and eight times the molar mass in (a), while temperature and pressure are only slightly changed. Comparing (c) and (d), the molar mass of (c) is more than half that of (d), the pressure is more than twice that of (d), and the temperature is much less than (d). So, (c) has the greatest density.

71. $? \text{ L} = 37.6 \text{ L NH}_3 \times \dfrac{\text{L N}_2}{2 \text{ L NH}_3} = 18.8 \text{ L N}_2$

73. $? \text{ L SO}_3 = 6.06 \text{ L SO}_2 \times \dfrac{2 \text{ L SO}_3}{2 \text{ L SO}_2} = 6.06 \text{ L SO}_3$

$? \text{ L SO}_3 = 2.25 \text{ L O}_2 \times \dfrac{2 \text{ L SO}_3}{\text{L O}_2} = 4.50 \text{ L SO}_3$

O_2 is limiting. 4.50 L SO_3 is produced.

75. $\dfrac{PV}{RT} = n$

$$? \text{ mg Mg} = \frac{758 \text{ Torr x } \dfrac{\text{atm}}{760 \text{ Torr}} \text{ x } 28.50 \text{ mL x } \dfrac{10^{-3} \text{ L}}{\text{mL}}}{\dfrac{0.08206 \text{ L atm}}{\text{K mol}} \text{ x } (26 + 273)\text{K}} \times \frac{\text{mol Mg}}{\text{mol H}_2} \times \frac{24.31 \text{ g Mg}}{\text{mol Mg}}$$

$$\times \frac{\text{mg}}{10^{-3} \text{ g}} = 28.2 \text{ mg Mg}$$

77. $2 \text{ CH}_3\text{CH}_2\text{CH}_2\text{OH} + 9 \text{ O}_2 \rightarrow 6 \text{ CO}_2 + 8 \text{ H}_2\text{O}$

$$? \text{ L CO}_2 = 125 \text{ mL C}_3\text{H}_8\text{O} \times \frac{0.804 \text{ g}}{\text{mL}} \times \frac{\text{mol C}_3\text{H}_8\text{O}}{60.09 \text{ g C}_3\text{H}_8\text{O}} \times \frac{6 \text{ mol CO}_2}{2 \text{ mol C}_3\text{H}_8\text{O}}$$

$$\times \frac{\dfrac{62.36 \text{ L Torr}}{\text{mol K}} \text{ x } (273 + 26) \text{ K}}{767 \text{ Torr}} = 122 \text{ L CO}_2$$

79. $\text{CaCO}_3 \text{ (s)} + 2\text{HCl (aq)} \rightarrow \text{CaCl}_2\text{(aq)} + \text{H}_2\text{CO}_3\text{(aq)} \rightarrow \text{H}_2\text{O(l)} + \text{CO}_2 \text{ (g)}$

$$? \text{ mL} = \frac{748 \text{ Torr} \times 565 \text{ mL}}{62.36 \dfrac{\text{L Torr}}{\text{K mol}} \times 294 \text{ K}} \times \frac{10^{-3} \text{ L}}{\text{mL}} \times \frac{2 \text{ mol HCl}}{\text{mol CO}_2} \times \frac{1}{0.375 \text{ M HCl}} \times \frac{\text{mL}}{10^{-3} \text{ L}}$$

$$= 123 \text{ mL HCl}$$

81. $P_{N_2} = \chi_{N_2} P_T \qquad \chi_{N_2} = \dfrac{\text{mol \%}}{100\%}$

$P_{N_2} = 762 \text{ mmHg} \times 0.768 = 585 \text{ mmHg}$

$P_{O_2} = 762 \text{ mmHg} \times 0.201 = 153 \text{ mmHg}$

$P_{CO_2} = 762 \text{ mmHg} \times 0.031 = 24 \text{ mmHg}$

83. (a) $? \text{ mol He} = \frac{1.96 \text{ g He}}{4.003 \text{ g He}} = 0.490 \text{ mol}$

$? \text{ mol } O_2 = \frac{60.8 \text{ g } O_2}{32.00 \text{ g } O_2} = 1.90 \text{ mol}$

$? \text{ mol total} = 0.490 \text{ mol} + 1.90 \text{ mol} = 2.39 \text{ mol}$

$\chi_{He} = \frac{0.490 \text{ mol}}{2.39 \text{ mol}} = 0.205$

$\chi_{O_2} = \frac{1.90 \text{ mol}}{2.39 \text{ mol}} = 0.795$

(b) $$P_{He} = \frac{mRT}{VM} = \frac{1.96 \text{ g He x } \frac{0.08206 \text{ L atm}}{\text{K mol}} \text{ x } (25.0 + 273.2)\text{K}}{5.00 \text{ L x } 4.003 \text{ g/mol}} = 2.40 \text{ atm}$$

$$P_{O_2} = \frac{60.8 \text{ g } O_2 \text{ x } \frac{0.08206 \text{ L atm}}{\text{K mol}} \text{ x } (25.0 + 273.2)\text{K}}{5.00 \text{ L x } 32.00 \text{ g/mol}} = 9.30 \text{ atm}$$

(c) $P_{total} = P_{He} + P_{O_2} = 2.40 \text{ atm} + 9.30 \text{ atm} = 11.70 \text{ atm}$

85. $P_{total} = P_{O_2} + P_{H_2O}$

$P_{O_2} = 756 \text{ Torr} - 19.8 \text{ Torr} = 736 \text{ Torr}$

$$\chi_{O_2} = \frac{P_{O_2}}{P_{total}} = \frac{736 \text{ Torr}}{756 \text{ Torr}} = 0.974$$

87. $P_{O_2} = 743 \text{ Torr - } 19 \text{ Torr} = 724 \text{ Torr}$

$$n = \frac{PV}{RT} = \frac{724 \text{ Torr x } \frac{\text{atm}}{760 \text{ Torr}} \text{ x } 122 \text{ mL x } \frac{10^{-3} \text{ L}}{\text{mL}}}{\frac{0.08206 \text{ L atm}}{\text{K mol}} \text{ x } (21 + 273)\text{K}} = 4.82 \times 10^{-3} \text{ mol } O_2$$

$$? \text{ g } O_2 = 4.82 \times 10^{-3} \text{ mol } O_2 \times \frac{32.00 \text{ g } O_2}{\text{mol } O_2} = 0.154 \text{ g } O_2$$

$$? \text{ g} = 4.82 \times 10^{-3} \text{ mol } O_2 \times \frac{\text{mol } C_6H_{12}O_6}{6 \text{ mol } O_2} \times \frac{180.16 \text{ g}}{\text{mol}} = 0.145 \text{ g } C_6H_{12}O_6$$

89. $$\frac{r_{H_2}}{r_{He}} = \sqrt{\frac{4.003}{2.016}} = 1.41$$

H_2 will diffuse somewhat faster than He, so b—with a ratio of H_2 to He molecules of 6:4 = 1.5—is the answer.

91. The partial pressures depend on the number of moles of each gas; they can easily be different. However, there can be only one average e_k of the gas molecules in the mixture. Because the average e_k depends on temperature, the two gases must be at the same average e_k.

93. (a) $r \propto \frac{1}{t}$

$$\frac{r_{N_2}}{r_X} = \frac{\frac{1}{44\text{ s}}}{\frac{1}{75\text{ s}}} = \sqrt{\frac{M_X}{M_{N_2}}} = \sqrt{\frac{M_X}{28.01\text{ g/mol}}}$$

$$1.70 = \sqrt{\frac{M_X}{28.01\text{ g/mol}}}$$

$$2.91 = \frac{M_X}{28.01\text{ g/mol}}$$

$M_X = 81.5$ g/mol. Molecular mass = 81 u.

(b) $$\frac{r_{N_2}}{r_X} = \frac{\frac{1}{44\text{ s}}}{\frac{1}{42\text{ s}}} = \sqrt{\frac{M_X}{28.01\text{ g/mol}}}$$

$$0.955 = \sqrt{\frac{M_X}{28.01\text{ g/mol}}}$$

$M_X = 0.911 \times 28.01$ g/mol $= 26$ g/mol. Molecular mass = 26 u.

Additional Problems

95. $$?\ \frac{\text{lb}}{\text{in.}^2} = \frac{32\text{ lb}}{\text{in.}^2} + \frac{15\text{ lb}}{\text{in.}^2} = \frac{47\text{ lb}}{\text{in.}^2}$$

$$?\text{ atm} = \frac{47\text{ lb}}{\text{in.}^2} \times \frac{\text{atm}}{14.7\text{ lb/in}^2} = 3.2\text{ atm}$$

$$PV = nRT = \frac{m}{M}RT$$

$$?\text{ g} = m = \frac{PVM}{RT} = \frac{3.2\text{ atm} \times 21.1\text{ atm} \times \frac{28.96\text{g}}{\text{mol}}}{0.08206\frac{\text{L atm}}{\text{R mol}} \times 294\text{ K}} = 81\text{ g}$$

$$\frac{P_1}{T_1} = \frac{P_2}{T_2} \qquad P_2 = \frac{P_1 T_2}{T_1} = \frac{\frac{47\text{lb}}{\text{in}^2} \times 317\text{K}}{294\text{K}} = \frac{51\text{ lb}}{\text{in.}^2}$$

$$?\ \frac{\text{lb}}{\text{in}^2} = \frac{51\text{ lb}}{\text{in.}^2} - \frac{15\text{ lb}}{\text{in.}^2} = \frac{36\text{ lb}}{\text{in.}^2}$$

97. $V = 5.00\ \text{in.}^3 \times \left(\dfrac{2.54\ \text{cm}}{\text{in.}}\right)^3 \times \dfrac{\text{mL}}{\text{cm}^3} \times \dfrac{10^{-3}\ \text{L}}{\text{mL}} = 0.08194\ \text{L}$

$P = 195\ \text{lb/in.}^2 \times \dfrac{\text{atm}}{14.696\ \text{lb/in.}^2} = 13.27\ \text{atm}$

$$m = \frac{PVM}{RT} = \frac{13.27\ \text{atm} \times 0.08194\ \text{L} \times 4.003\ \text{g/mol}}{\dfrac{0.08206\ \text{L atm}}{\text{K mol}} \times (20 + 273)\text{K}} = 0.181\ \text{g}$$

100. $h_{Hg} = \dfrac{h_{H_2O} d_{H_2O}}{d_{Hg}} = \dfrac{3.8\ \text{cm} \times 1.00\ \text{g/cm}^3}{13.6\ \text{g/cm}^3} = 0.28\ \text{cm}$

$?\ \text{mmHg} = 0.28\ \text{cm} \times \dfrac{10\ \text{mm}}{\text{cm}} = 2.8\ \text{mmHg}$

$P_{\text{in bottle}} = 753.5\ \text{mmHg} - 2.8\ \text{mmHg} = 750.7\ \text{mmHg}$

$P_{O_2} = P_{\text{total}} - P_{H_2O} = 753.5\ \text{mmHg} - 18.7\ \text{mmHg} = 734.8\ \text{mmHg}$

$$?\ \text{mL}\ O_2 = 3.275\ \text{g} \times \frac{65.82\ \text{g}\ KClO_3}{100\ \text{g sample}} \times \frac{\text{mol}\ KClO_3}{122.55\ \text{g}\ KClO_3} \times \frac{3\ \text{mol}\ O_2}{2\ \text{mol}\ KClO_3}$$

$$\times \frac{\dfrac{62.36\ \text{L Torr}}{\text{K mol}} \times (273 + 21)\ \text{K}}{734.8\ \text{mmHg}} \times \frac{\text{mL}}{10^{-3}\ \text{L}} = 658\ \text{mL}\ O_2$$

102. Use the freon data to calculate the glass volume.

192.8273 g freon + glass	45.2217 g B + glass
– 45.0143 g glass	–45.0143 g glass
147.8130 g freon	0.2074 g B

$147.8130\ \text{g} \times \dfrac{\text{mL}}{1.576\ \text{g}} = 93.79\ \text{mL volume glass}$

$$M = \frac{mRT}{PV} = \frac{0.2074\ \text{g} \times \dfrac{0.082057\ \text{L atm}}{\text{K mol}} \times (21.48 + 273.15)\ \text{K}}{751.2\ \text{mmHg} \times \dfrac{\text{atm}}{760\ \text{mmHg}} \times 93.79\ \text{mL} \times \dfrac{10^{-3}\ \text{L}}{\text{mL}}}$$

$M = 54.09$ g/mol. Molecular mass = 54.09 u

105. $?\ \text{mol}\ C_3H_6 = \dfrac{PV}{RT} = \dfrac{25.0\ \text{atm} \times 1.50\ \text{L}}{\dfrac{0.08206\ \text{L atm}}{\text{K mol}} \times 298\ \text{K}} = 1.53\ \text{mol}$

$$?\ \text{L}\ C_3H_6 = \frac{nRT}{P} = \frac{1.53\ \text{mol} \times \dfrac{0.08206\ \text{L atm}}{\text{K mol}} \times 298\ \text{K}}{755\ \text{mm Hg} \times \dfrac{1\ \text{atm}}{760\ \text{mm Hg}}} = 37.7\ \text{L}$$

$$? \text{ ft}^3\ C_3H_6 = 37.7\text{ L} \times \frac{\text{mL}}{10^{-3}\text{ L}} \times \frac{\text{cm}^3}{\text{mL}} \times \left(\frac{\text{in.}}{2.54\text{ cm}}\right)^3 \times \left(\frac{\text{ft}}{12\text{ in.}}\right)^3 = 1.33\text{ ft}^3$$

$$\%\ C_3H_6 = \frac{1.33\text{ ft}^3}{72\text{ ft}^3} \times 100\ \% = 1.9\ \%\ C_3H_6$$

No, it is not an explosive mixture.

107. $$? \text{ tons} = \frac{14.696\text{ lb}}{\text{in.}^2} \times \frac{\text{ton}}{2000\text{ lb}} \times \left(\frac{12\text{ in}}{\text{ft}}\right)^2 \times \left(\frac{5280\text{ ft}}{\text{mi}}\right)^2 \times 1.95 \times 10^8\text{ mi}^2$$

$$= 5.75 \times 10^{15}\text{ tons}$$

110. $$? \text{ mol } H_2 = n = \frac{PV}{RT} = \frac{(746\text{-}17)\text{Torr x } 40.71\text{ mL}\ \frac{10^{-3}}{\text{mL}}}{\frac{62.364\text{ L Torr}}{\text{Kmol}}(273 + 19)\text{ K}} = 1.63 \times 10^{-3}\text{ mol } H_2$$

Let X = mass of Lithium

$$X\frac{\text{mol Li}}{6.941\text{ g}} \times \frac{\text{mol } H_2}{2\text{ mol Li}} + (0.0297\text{ g} - X) \times \frac{\text{mol Mg}}{24.305\text{ g}} \times \frac{\text{mol } H_2}{\text{mol Mg}} =$$

$$1.63 \times 10^{-3}\text{ mol } H_2$$

$0.07204\ X + 0.00122\text{ mol } H_2 - 0.04114\ X = 1.63 \times 10^{-3}\text{ mol}$

X = 0.013 g Li

$$? \%\text{ Li} = \frac{0.013\text{ g}}{0.0297\text{ g}} \times 100\% = 44\%\text{ Li}$$

56% Mg

113. $CH_4 + 2\ O_2 \rightarrow CO_2 + 2\ H_2O$

$$? \text{ L } O_2 = 1.00 \times 10^3\text{ L gas} \times \frac{77.3\text{ L } CH_4}{100\text{ L gas}} \times \frac{2\text{ L } O_2}{\text{L } CH_4} = 1.55 \times 10^3\text{ L } O_2$$

$2\ C_2H_6 + 7\ O_2 \rightarrow 4\ CO_2 + 6\ H_2O$

$$? \text{ L } O_2 = 1.00 \times 10^3\text{ L} \times \frac{11.2\text{ L } C_2H_6}{100\text{ L gas}} \times \frac{7\text{ L } O_2}{2\text{ L } C_2H_6} = 392\text{ L } O_2$$

$C_3H_8 + 5\ O_2 \rightarrow 3\ CO_2 + 4\ H_2O$

$$? \text{ L } O_2 = 1.00 \times 10^3\text{ L} \times \frac{5.8\text{ L } C_3H_8}{100\text{ L gas}} \times \frac{5\text{ L } O_2}{\text{L } C_3H_8} = 290\text{ L } O_2$$

$2\ C_4H_{10} + 13\ O_2 \rightarrow 8\ CO_2 + 10\ H_2O$

$$? \text{ L } O_2 = 1.00 \times 10^3\text{ L} \times \frac{2.3\text{ L } C_4H_{10}}{100\text{ L gas}} \times \frac{13\text{ L } O_2}{2\text{ L } C_4H_{10}} = 150\text{ L } O_2$$

$? \text{ L } O_2\text{ total} = 1.55 \times 10^3\text{ L } O_2 + 392\text{ L } O_2 + 290\text{ L } O_2 + 150\text{ L } O_2$

$= 2.38 \times 10^3\text{ L } O_2$ at STP

$$V_{O_2} = \frac{P_1V_1T_2}{T_1P_2} = \frac{760\ \text{mmHg} \times 2.38 \times 10^3\ \text{L} \times (273.2 + 23.0)\ \text{K}}{273\ \text{K} \times 741\ \text{mmHg}}$$

$V_{O_2} = 2.65 \times 10^3$ L O_2

$$?\ \text{L air} = 2.65 \times 10^3\ \text{L O}_2 \times \frac{100\ \text{L air}}{20.95\ \text{L O}_2} = 1.26 \times 10^4\ \text{L air}$$

$CH_4 + 2\ O_2 \rightarrow CO_2 + 2\ H_2O$

$$?\ \text{L CO}_2 = 1.00 \times 10^3\ \text{L gas} \times \frac{77.3\ \text{L CH}_4}{100\ \text{L gas}} \times \frac{\text{L CO}_2}{\text{L CH}_4} = 7.73 \times 10^2\ \text{L CO}_2$$

$2\ C_2H_6 + 7\ O_2 \rightarrow 4\ CO_2 + 6\ H_2O$

$$?\ \text{L CO}_2 = 1.00 \times 10^3\ \text{L} \times \frac{11.2\ \text{L C}_2\text{H}_6}{100\ \text{L gas}} \times \frac{4\ \text{L CO}_2}{2\ \text{L C}_2\text{H}_6} = 224\ \text{L CO}_2$$

$C_3H_8 + 5\ O_2 \rightarrow 3\ CO_2 + 4\ H_2O$

$$?\ \text{L CO}_2 = 1.00 \times 10^3\ \text{L} \times \frac{5.8\ \text{L C}_3\text{H}_8}{100\ \text{L gas}} \times \frac{3\ \text{L CO}_2}{\text{L C}_3\text{H}_8} = 174\ \text{L CO}_2$$

$2\ C_4H_{10} + 13\ O_2 \rightarrow 8\ CO_2 + 10\ H_2O$

$$?\ \text{L CO}_2 = 1.00 \times 10^3\ \text{L} \times \frac{2.3\ \text{L C}_4\text{H}_{10}}{100\ \text{L gas}} \times \frac{8\ \text{L CO}_2}{2\ \text{L C}_4\text{H}_{10}} = 92\ \text{L CO}_2$$

? L CO_2 total = 773 L CO_2 + 224 L CO_2 + 174 L CO_2 + 92 L CO_2

= 1.263×10^3 L CO_2 at STP

$$V_{CO_2} = \frac{P_1V_1T_2}{T_1P_2} = \frac{1013\ \text{mb} \times 1.263 \times 10^3\ \text{L} \times (273.2 + 35.0)\ \text{K}}{273.2\ \text{K} \times 985\ \text{mb}}$$

$V_{CO_2} = 1.47 \times 10^3$ L CO_2

117. (a) $P = \dfrac{nRT}{V} = \dfrac{1.00\ \text{mol CO}_2 \times \dfrac{0.08206\ \text{L atm}}{\text{K mol}} \times 298\ \text{K}}{2.50\ \text{L}}$

$P = 9.78$ atm

(b) $P = \dfrac{nRT}{V\text{-}nb} - \dfrac{n^2a}{V^2}$

$$P = \left(\frac{1.00\ \text{mol CO}_2 \times 0.08206\ \text{L atm} \times 298\ \text{K}}{2.50\ \text{L} - 1.00\ \text{mol} \times 0.0427\ \text{L/mol}}\right) - \left(\frac{(1.00\ \text{mol})^2\ \dfrac{3.59\ \text{L}^2\ \text{atm}}{\text{mol}^2}}{(2.50\ \text{L})^2}\right)$$

$P = 9.95$ atm $- 0.57$ atm $= 9.38$ atm

(c) At high pressures, the molecules are forced closer together. The increased intermolecular interaction of the molecules prevents them from colliding with the wall as often as at lower pressures. Because the molecules exert less force on the wall, the pressure is decreased.

119. (a) $? \text{ mol C} = 4.305 \text{ g } CO_2 \times \frac{\text{mol C}}{44.009 \text{ g } CO_2} = 0.09782 \text{ mol C} \times \frac{1}{0.09782 \text{ mol C}}$

$$= 1.000 \frac{\text{mol C}}{\text{mol C}}$$

$? \text{ mol H} = 2.056 \text{ g } H_2O \times \frac{2 \text{ mol H}}{18.015 \text{ g } H_2O} = 0.2283 \text{ mol H} \times \frac{1}{0.09782 \text{ mol C}}$

$$= 2.333 \frac{\text{mol H}}{\text{mol C}}$$

$C_{(1\times3)}H_{(2.333\times3)} = C_3H_7$ formula mass = 43.09

$$PV = \frac{mRT}{M}$$

$$M = \frac{mRT}{PV} = \frac{0.403 \text{ g x } \frac{62.36 \text{ L Torr}}{\text{K mol}} \text{ x } (273.2 + 99.8) \text{ K}}{749 \text{ Torr x } 145 \text{ mL x } \frac{10^{-3} \text{ L}}{\text{mL}}} = 86.3 \text{ g/mol}$$

$$? \frac{\text{formula units}}{\text{molecule}} = \frac{86.3 \text{ u/molecule}}{43.09 \text{ u/formula units}} = \frac{2 \text{ formula units}}{\text{molecule}}$$

$C_3H_7 \times 2 = C_6H_{14}$

(b) $CH_3CHCH_2CH_2CH_3$ with CH_3 on C-2 — 2-methyl pentane

$CH_3CH_2CHCH_2CH_3$ with CH_3 on C-3 — 3-methyl pentane

Two isomers for a 6 C alkane with one methyl group.

Apply Your Knowledge

121. ? g sample = 104.772g − 103.868g = 0.904g

$$? V_{\text{water}} = (300.623\text{g} - 104.772\text{g}) \times \frac{\text{mol}}{1.00\text{g}} = 196 \text{ mL}$$

$$? \text{ mol} = \frac{PV}{RT} = \frac{739 \text{ Torr} \times 196 \text{ mL} \times \frac{10^{-3} \text{ L}}{\text{mL}}}{\frac{62.36 \text{ L Torr}}{\text{K mol}} \times 372.4 \text{ K}} = 6.23 \times 10^{-3} \text{ mol}$$

$$? \text{ g/mol} = \frac{0.904\text{g}}{6.23 \times 10^{-3} \text{ mol}} = 145 \text{ g/mol}$$

123. $$m_{H_2} = \frac{PVM}{RT} = \frac{1.00 \text{ atm} \times 120 \text{ ft}^3 \times \left(\frac{12 \text{ in}}{\text{ft}}\right)^3 \times \left(\frac{2.54 \text{ cm}}{\text{in}}\right)^3 \times \frac{\text{mL}}{\text{cm}^3} \times \frac{10^{-3} \text{ L}}{\text{mL}}}{\frac{0.08206 \text{ L atm}}{\text{K mol}} \times 273 \text{ K}} \times \frac{2.016 \text{ g}}{\text{mol}} = 306 \text{ g } H_2$$

$m_{\text{total}} = 1200 \text{ g} + 1700 \text{ g} + 306 \text{ g} = 3206 \text{ g} = 3.21 \times 10^3 \text{ g}$

When the mass of the air displaced by the bag is more than the total mass of the bag, payload, etc., the bag will rise.

$$V=\frac{4}{3}\pi r^3=\frac{4}{3}\pi(12.5\text{ ft})^3\times\left(\frac{12\text{ in.}}{\text{ft}}\right)^3\times\left(\frac{2.54\text{ cm}}{1.00\text{ in.}}\right)^3\times\frac{\text{mL}}{\text{cm}^3}\times\frac{10^{-3}\text{ L}}{\text{mL}}=2.32\times10^5\text{ L}$$

$$0\text{ km}\quad V_{H_2}=\frac{306\text{ g }H_2\times\frac{\text{mol}}{2.016\text{ g}}\times\frac{0.08206\text{ L atm}}{\text{K mol}}\times290\text{ K}}{1.0\times10^3\text{ mb}\times\frac{\text{atm}}{1013\text{ mb}}}=3.66\times10^3\text{ L}$$

$$m_{air}=\frac{PVM}{RT}=\frac{1.0\times10^3\text{ mb}\times\frac{\text{atm}}{1013\text{ mb}}}{\frac{0.08206\text{ L atm}}{\text{K mol}}\times290\text{ K}}\times3.66\times10^3\text{ L}\times\frac{28.96\text{ g}}{\text{mol}}=4.4\times10^3\text{ g}$$

will rise: $m_{total}<m_{air}$

$$5\text{ km}\quad V_{H_2}=\frac{306\text{ g }H_2\times\frac{\text{mol}}{2.016\text{ g}}\times\frac{0.08206\text{ L atm}}{\text{K mol}}\times266\text{ K}}{5.4\times10^2\text{ mb}\times\frac{\text{atm}}{1013\text{ mb}}}=6.21\times10^3\text{ L}$$

$$m_{air}=\frac{5.4\times10^2\text{ mb}\times\frac{\text{atm}}{1013\text{ mb}}\times6.21\times10^3\text{ L}\times\frac{28.96\text{ g}}{\text{mol}}}{\frac{0.08206\text{ L atm}}{\text{K mol}}\times266\text{ K}}=4.4\times10^3\text{ g}$$

will rise: $m_{total}<m_{air}$

$$10\text{ km}\quad V_{H_2}=\frac{306\text{ g }H_2\times\frac{\text{mol }H_2}{2.016\text{ g }H_2}\times\frac{0.08206\text{ L atm}}{\text{K mol}}\times235\text{ K}}{2.7\times10^2\text{ mb}\times\frac{\text{atm}}{1013\text{ mb}}}=1.10\times10^4\text{ L}$$

$$m_{air}=\frac{2.7\times10^2\text{ mb}\times\frac{\text{atm}}{1013\text{ mb}}\times1.10\times10^4\text{ L}\times\frac{28.96\text{ g}}{\text{mol}}}{\frac{0.08206\text{ L atm}}{\text{K mol}}\times235\text{ K}}=4.4\times10^3\text{ g}$$

will rise: $m_{total}<m_{air}$

$$20\text{ km}\quad V_{H_2}=\frac{306\text{ g }H_2\times\frac{\text{mol }H_2}{2.016\text{ g }H_2}\times\frac{0.08206\text{ L atm}}{\text{K mol}}\times217\text{ K}}{55\text{ mb}\times\frac{\text{atm}}{1013\text{ mb}}}=4.98\times10^4\text{ L}$$

$$m_{air}=\frac{55\text{ mb}\times\frac{\text{atm}}{1013\text{ mb}}\times4.98\times10^4\text{ L}\times\frac{28.96\text{ g}}{\text{mol}}}{\frac{0.08206\text{ L atm}}{\text{K mol}}\times217\text{ K}}=4.4\times10^3\text{ g}$$

will rise: $m_{total} < m_{air}$

$$30\text{ km}\quad V_{H_2} = \frac{306\text{ g }H_2 \times \frac{\text{mol }H_2}{2.016\text{ g }H_2} \times \frac{0.08206\text{ L atm}}{\text{K mol}} \times 239\text{ K}}{12\text{mb} \times \frac{\text{atm}}{1013\text{ mb}}} = 2.51 \times 10^5\text{ L}$$

This is larger than maximum volume of 2.32×10^5 L.

$$m_{air} = \frac{12\text{ mb} \times \frac{\text{atm}}{1013\text{mb}} \times 2.32 \times 10^5\text{ L} \times \frac{28.96\text{g}}{\text{mol}}}{\frac{0.08206\text{L atm}}{\text{K mol}} \times 239\text{K}} = 4.1 \times 10^3\text{ g}$$

will rise: $m_{total} < m_{air}$

$$40\text{ km}\quad m_{air} = \frac{2.9\text{ mb} \times \frac{\text{atm}}{1013\text{mb}} \times 2.32 \times 10^5\text{ L} \times \frac{28.96\text{g}}{\text{mol}}}{\frac{0.08206\text{L atm}}{\text{K mol}} \times 267\text{K}} = 8.8 \times 10^2\text{ g}$$

will not rise: $m_{total} > m_{air}$

The balloon will rise to between 30 km and 40 km.

124. $65.01 \times 0.494 = 32.1$
$49.01 \times 0.327 = 16.0$
$86.07 \times 0.186 = 16.0$
$102.07 \times 0.314 = 32.0$
The atomic mass must be 16.0. The atom is oxygen. There must be two atoms of oxygen in the first and last molecules and one in the others.

Chapter 6

Thermochemistry

Exercises

6.1A $\Delta U = q + w = (-567\text{ J}) + (+89\text{ J}) = -478\text{ J}$

6.1B $\Delta U = q + w = 41.4\text{ J} = q + (-81.2\text{ J})$

$q = 122.6\text{ J}$

Heat is absorbed by the system.

6.2A (a) Work is done by the gas.

(b) $\Delta U = w$ Since work is done by the system, the internal energy decreases.

(c) Since the internal energy is related to the temperature, the temperature must go down.

6.2B $\Delta w = -152\text{ kJ} + 209\text{ J} \times \dfrac{\text{kJ}}{10^3\text{ J}} = -152\text{ kJ}$

The process can occur. The values of w are different because it is path dependent and the paths for compression and expansion are different. Since the initial and final conditions are the same, $\Delta u = 0$.

6.3A $\Delta H_{rxn} = \dfrac{1}{2}(285.4\text{ kJ}) = 142.7\text{ kJ}$

6.3B $\Delta H_{rxn} = -\dfrac{1}{4}(595.5\text{ kJ}) = -148.9\text{ kJ}$

6.4A $H_2O\text{ (s)} \rightarrow H_2O\text{ (l)}$ $\Delta H = 334\text{ J/g}$

6.4B $N_2O_3(g) \xrightarrow{25\,°\text{C}} NO(g) + NO_2(g)$ $\Delta H = 40.5\text{ kJ}$

$$?\text{ kJ/mol} = \frac{0.533\text{ kJ}}{\text{g } N_2O_3} \times \frac{76.02\text{ g}}{\text{mol}} = \frac{40.5\text{ kJ}}{\text{mol}}$$

6.5A $\Delta H_{rxn} = 12.8\text{ g } H_2 \times \dfrac{\text{mol } H_2}{2.016\text{ g } H_2} \times \dfrac{-184.6\text{ kJ}}{\text{mol}} = -1.17 \times 10^3\text{ kJ}$

6.5B $?\text{ L} = -1.00 \times 10^6\text{ kJ} \times \dfrac{\text{mol}}{-890.3\text{ kJ}} \times \dfrac{0.08206\text{ L atm}}{\text{K mol}} \times \dfrac{(273 + 25)\text{K}}{745\text{ Torr x } \dfrac{\text{atm}}{760\text{ Torr}}}$

$= 2.80 \times 10^4\text{ L } CH_4$

6.6A $C = \dfrac{q}{\Delta T} = \dfrac{911\text{ J}}{100°\text{C} - 15°\text{C}} = 11\text{ J/°C}$

6.6B $q = \dfrac{345\text{ J}}{\text{K}} \times (23 - 467)\text{ K} \times \dfrac{\text{kJ}}{10^3\text{ J}} = -153\text{ kJ}$

6.7A $q = m \times \text{specific heat} \times \Delta T$

$$q = 814\text{ g} \times \frac{4.18\text{ J}}{\text{g °C}} \times (100.0\text{ °C} - 18.0\text{ °C}) \times \frac{1.00\text{ cal}}{4.184\text{ J}} = 6.67 \times 10^4\text{ cal}$$

$$q = 6.67 \times 10^4\text{ cal} \times \frac{\text{kcal}}{10^3\text{ cal}} = 66.7\text{ kcal}$$

6.7B $\text{Mass } H_2O = \dfrac{q}{\text{specific heat} \times \Delta T} = \dfrac{9.09 \times 10^{10}\text{ J}}{\dfrac{4.18\text{ J}}{\text{g °C}} \times (55.0 - 5.5)\text{°C}}$

$$= 4.39 \times 10^8\text{ g} = 4.39 \times 10^5\text{ kg}$$

6.8A $T_f - T_i = \dfrac{q}{m \times \text{specific heat}}$

$$T_f = 22.5\text{ °C} + \frac{4.22\text{ kJ} \times \dfrac{10^3\text{ J}}{\text{kJ}}}{454\text{ g} \times 0.128\text{ J/g °C}}$$

$T_f = 95.1\text{ °C}$

6.8B $q_{Cu} = q_{H_2O}$

$$m_{Cu} \times \text{specific heat}_{Cu} \times \Delta T = m_{H_2O} \times \text{specific heat}_{H_2O} \times \Delta T$$

$$m_{Cu} \times \frac{0.385\text{ J}}{\text{g °C}} \times \Delta T = 145\text{ g } H_2O \times \frac{4.18\text{ J}}{\text{g °C}} \times \Delta T \quad \Delta T \text{ cancels.}$$

$m_{Cu} = 1.57 \times 10^3\text{ g Cu}$

6.9A $q_{Ir} = -q_{H_2O}$

$$m_{Ir} \times \text{specific heat}_{Ir} \times \Delta T_{Ir} = -m_{H_2O} \times \text{specific heat}_{H_2O} \times \Delta T_{H_2O}$$

$$23.9\text{ g Ir} \times \text{specific heat}_{Ir} \times (22.6\text{ °C} - 89.7\text{ °C}) = -20.0\text{ g } H_2O \times \frac{4.182\text{ J}}{\text{g °C}} \times (22.6\text{ °C} - 20.1\text{ °C})$$

$\text{specific heat}_{Ir} = 0.13\text{ J/g °C}$

6.9B $q_{gly} = -q_{Fe}$

$m_{gly}\text{sp.ht.}_{gly}\ \Delta T_{gly} = -m_{Fe}\text{sp.ht.}_{Fe}\ \Delta T_{Fe}$

$$250.0\ \text{mL} \times \frac{1.261\ \text{g}}{\text{mL}} \times \text{sp.ht.}_{gly} \times (44.7\ °\text{C} - 23.5\ °\text{C}) = -135\ \text{g} \times \frac{0.449\ \text{J}}{\text{g}\ °\text{C}} \times (44.7\ °\text{C} - 225\ °\text{C})$$

$6683.3\ \text{sp.ht.}_{gly}\ \text{g}\ °\text{C} = 10929\ \text{J}$

$$\text{sp.ht.}_{gly} = \frac{1.64\ \text{J}}{\text{g}\ °\text{C}}$$

6.10A The heat exchange produces twice the temperature increase in the 100-mL sample, $2X$ as the temperature decreases, X, in the 200-mL sample. The temperature changes must be 80 °C – 20 °C = 60 °C

$X + 2X = 60°$ $\quad X = 20°$

The final temperature is 20 °C + (2 × 20°) = 80 °C – 20 °C = 60 °C

6.10B $q_{Fe} = -q_{water}$

$m_{Fe} \times \text{specific heat}_{Fe} \times \Delta T_{Fe} = -m_{water} \times \text{specific heat}_{water} \times \Delta T_{water}$

$$100\ \text{g} \times \frac{0.45\ \text{J}}{\text{g}\ °\text{C}} \times (T_f - 100\ °\text{C}) = -100\ \text{g} \times \frac{4.18\ \text{J}}{\text{g}\ °\text{C}} \times (T_f - 20\ °\text{C})$$

$45\ \text{J/°C}\ T_f - 4500\ \text{J} = -418\ \text{J/°C}\ T_f + 8360\ \text{J}$

$463\ \text{J/°C}\ T_f = 12860\ \text{J}$

$T_f = 27.8\ °\text{C}$

6.11A $$?\ \text{mol} = 100.0\ \text{mL} \times \frac{10^{-3}\ \text{L}}{\text{mL}} \times 0.500\ \text{M} = 0.0500\ \text{mol product}$$

$q_{rxn} = -q_{\text{calorim}} = -m \times \text{specific heat} \times \Delta T$

$$q = -200.0\ \text{mL} \times \frac{1.00\ \text{g}}{\text{mL}} \times \frac{4.182\ \text{J}}{\text{g}\ °\text{C}} \times (23.65\ °\text{C} - 20.29\ °\text{C})$$

$q = -2.81 \times 10^3\ \text{J}$

6.11B The only difference will be the difference between the heats of dilution of chloride ion and iodide ion. The reaction in both cases is the acid-base neutralization reaction.

6.12A $$\Delta H = -q_{\text{calorim}} = \frac{2.81 \times 10^3\ \text{J}}{0.0500\ \text{mol}} \times \frac{\text{kJ}}{10^3\ \text{J}} = -56.2\ \text{kJ/mol}$$

6.12B $$?\ \text{mol}\ H_2O = 125\ \text{mL} \times \frac{10^{-3}\ \text{L}}{\text{mL}} \times 1.33\ \text{M HCl} \times \frac{\text{mol}\ H_2O}{\text{mol HCl}} = 0.166\ \text{mol}\ H_2O$$

$$?\ \text{mol}\ H_2O = 225\ \text{mL} \times \frac{10^{-3}\ \text{L}}{\text{mL}} \times 0.625\ \text{M NaOH} \times \frac{\text{mol}\ H_2O}{\text{mol NaOH}} = 0.141\ \text{mol}\ H_2O$$

NaOH is limiting.

$q_{cal} = -q_{rxn}$

$$q_{cal} = \frac{+57.2\text{ kJ}}{\text{mol}} \times 0.141\text{ mol} = 350\text{ mL} \times \frac{1.00\text{ g}}{\text{mL}} \times \frac{4.18\text{ J}}{\text{g °C}} \times \frac{\text{kJ}}{10^3\text{ J}} \times (T - 24.4\text{ °C})$$

$5.51\text{ °C} = T - 24.4$

$T = 24.4\text{ °C} + 5.51\text{ °C} = 29.9\text{ °C}$

6.13A $q_{\text{ calorim}} = 6.52\text{ kJ} \cdot \text{°C}^{-1} \times (21.26 - 20.00)\text{ °C} = 8.22\text{ kJ}$

$q_{\text{ rxn}} = -q_{\text{ calorim}} = -8.22\text{ kJ}$

$$\Delta H = \frac{-8.22\text{ kJ}}{0.250\text{ g}} \times \frac{12.01\text{ g}}{\text{mol}} = \frac{-395\text{ kJ}}{\text{mol}}$$

$C(\text{diamond}) + O_2(g) \rightarrow CO_2(g) \qquad \Delta H = -395\text{ kJ}$

6.13B $? \text{ kJ} = 0.8082\text{ g} \times \frac{\text{mol}}{180.16\text{ g}} \times \frac{-2803\text{ kJ}}{\text{mol}} = -12.57\text{ kJ}$

$q_{\text{ calorim}} = -q_{\text{ rxn}} = -(-12.57\text{ kJ})$

$$\text{heat capacity} = \frac{q_{\text{calorim}}}{\Delta T} = \frac{12.57\text{ kJ}}{(27.21 - 25.11)\text{°C}} = \frac{5.99\text{ kJ}}{\text{°C}}$$

6.14A

	ΔH
$2\ CH_4(g) + 4\ O_2(g) \rightarrow 2\ CO_2(g) + 4\ H_2O(l)$	$2 \times (-890.3\text{ kJ})$
$2\ CO_2(g) \rightarrow 2\ CO(g) + O_2(g)$	$-2 \times (-283.0\text{ kJ})$
$2\ CH_4(g) + 3\ O_2(g) \rightarrow 2\ CO(g) + 4\ H_2O(l)$	$\Delta H = -1215\text{ kJ}$

6.14B

$C_2H_4(g) + 3\ O_2(g) \rightarrow 2\ CO_2(g) + 2\ H_2O(l)$	-1410.9 kJ
$3\ H_2O(l) + 2\ CO_2(g) \rightarrow \frac{7}{2}\ O_2(g) + C_2H_6(g)$	$-\frac{1}{2} \times (-3119.4\text{ kJ})$
$H_2(g) + \frac{1}{2}\ O_2(g) \rightarrow H_2O(l)$	$\frac{1}{2} \times (-571.6\text{ kJ})$
$C_2H_4(g) + H_2(g) \rightarrow C_2H_6(g)$	$\Delta H = -137.0\text{ kJ}$

6.15A $\Delta H° = \Sigma\nu_p\Delta H°_f\text{ (products)} - \Sigma\nu_r\Delta H°_f\text{ (reactants)}$

$\Delta H° = (1\text{ mol } CH_3CH_2OH \times -277.7\text{ kJ/mol})$

$\quad - (1\text{ mol } H_2O \times -285.8\text{ kJ/mol} + 1\text{ mol } C_2H_4 \times 52.26\text{ kJ/mol}) = -44.2\text{ kJ}$

6.15B $2\ C_4H_{10}(g) + 13\ O_2(g) \rightarrow 8\ CO_2(g) + 10\ H_2O(l)$

$$\Delta H° = 8\text{ mol } CO_2 \times \frac{-393.5\text{ kJ}}{\text{mol}} + 10\text{ mol } H_2O \times \frac{-285.8\text{ kJ}}{\text{mol}}$$

$$-2\text{ mol } C_4H_{10} \times \frac{-125.7\text{ kJ}}{\text{mol}} - 13\text{ mol} \times 0$$

$\Delta H° = -5754.6\text{ kJ}$

6.16A $? \text{ kJ} = \left(1 \text{ mol } C_2Cl_4 \times \Delta H^\circ_f\right) + \left(4 \text{ mol } H_2O(l) \text{ x } \frac{-285.8 \text{ kJ}}{\text{mol}}\right)$

$- \left(1 \text{mol} C_2H_4 \times \frac{52.26 \text{ kJ}}{\text{mol}}\right) - \left(4 \text{ mol HCl x } \frac{-92.31 \text{ kJ}}{\text{mol}}\right) - (2 \text{ mol } O_2 \text{ x } 0)$

$= -878.5 \text{ kJ}$

$? \text{ kJ} = \left(1 \text{ mol } C_2Cl_4 \times \Delta H^\circ_f\right) + -826.2 \text{ kJ} = -878.5 \text{ kJ}$

$\Delta H^\circ_f = \frac{-878.5 \text{ kJ} + 826.2 \text{ kJ}}{1 \text{ mol } C_2Cl_4}$

$\Delta H^\circ_f = \frac{-52.3 \text{ kJ}}{\text{mol } C_2Cl_4}$

6.16B $C_4H_4S(l) + 6\ O_2(g) \rightarrow 4\ CO_2(g) + SO_2(g) + 2\ H_2O(l)$

$\Delta H^\circ = \Sigma \nu_p \Delta H^\circ_f \text{ (products)} - \Sigma \nu_r \Delta H^\circ_f \text{ (reactants)}$

$-2523 \text{ kJ} = (4 \text{ mol } CO_2 \times -393.5 \text{ kJ/mol}) + (1 \text{ mol } SO_2 \times -296.8 \text{ kJ/mol})$

$+ (2 \text{ mol } H_2O \times -285.8 \text{ kJ /mol}) - \left(1 \text{ mol } C_4H_4S \times \Delta H^\circ_f\right) - (6 \text{ mol } O_2 \times 0)$

$-2523 \text{ kJ} = -2442.4 \text{ kJ} - \left(1 \text{ mol } C_4H_4S \times \Delta H^\circ_f\right)$

$\Delta H^\circ_f = 81 \text{ kJ /mol}$

6.17A The enthalpy of formation of CH_3OH is 39 kJ/mol greater than that of CH_3CH_2OH, but the enthalpies of the products of the combustion of CH_3CH_2OH are much lower than those of the combustion of CH_3OH [by one mole each of $H_2O(l)$ and $CO_2(g)$]. CH_3CH_2OH has the greater negative heat of combustion.

6.17B The heat of formation of acetic acid is more negative (– 432.3 kJ) than either ethane (–84.68 kJ) or ethanol (–277.7 kJ) and the products from the combustion of acetic acid are one mole of H_2O (–285.8 kJ) less than either. Since the difference for acetic acid is less, acetic acid will liberate less heat then ethanol or ethane.

6.18A $\Delta H^\circ = \left(1 \text{ mol} \times \Delta H^\circ_f BaSO_4\right) - 1 \text{mol } Ba^{2+} \times \frac{-537.6 \text{ kJ}}{\text{mol } Ba^{2+}}$

$- \left(1 \text{ mol } SO_4^{2-} \text{ x } \frac{-909.3 \text{ kJ}}{\text{mol } SO_4^{2-}}\right) = 26 \text{ kJ}$

$\Delta H^\circ_f = \frac{-1446.9 \text{ kJ} + 26 \text{ kJ}}{\text{mol } BaSO_4} = \frac{-1421 \text{ kJ}}{\text{mol}}$

6.18B $MgCl_2(aq) + 2\ KOH(aq) \rightarrow Mg(OH)_2(s) + 2\ K^+ + 2\ Cl^-$

$Mg^{2+}(aq) + 2\ Cl^- + 2\ K^+ + 2\ OH^-(aq) \rightarrow Mg(OH)_2(s) + 2\ K^+ + 2\ Cl^-$

The Cl^- and K^+ will cancel to produce a net ionic equation.

$Mg^{2+}(aq) + 2\ OH^-(aq) \rightarrow Mg(OH)_2(s)$

$$\Delta H° = 1 \text{ mol Mg(OH)2} \times \frac{-924.5 \text{ kJ}}{\text{mol}} - 2 \text{ mol OH}^- \times \frac{-230.0 \text{ kJ}}{\text{mol}}$$

$$-1 \text{ mol Mg}^{2+} \times \frac{-466.9 \text{ kJ}}{\text{mol}} = 2.4 \text{ kJ}$$

Review Questions

3. $q > 0$ heat is absorbed by the system
   $w > 0$ work done on the system by the surroundings
   All are correct

6. $\Delta H < 0$ is an exothermic reaction
   (a) is incorrect

8. The system becomes cold $\Delta H < 0$, the process is endothermic, (c) is correct.

9. $$\Delta T = \frac{q}{m \times sp.ht.} = \frac{236 \text{ J}}{14.5 \text{ g} \times 0.444 \text{ J/g °C}} = 36.7\ °\text{C (a)}$$

10. Aluminum has the highest specific heat and will absorb the most heat. (b)

12. $$? \text{ kJ} = 4.301 \text{ g} \times \frac{\text{mol}}{2.0158 \text{ g}} \times \frac{-483.6 \text{ kJ}}{2 \text{ mol}} = -515.9 \text{ kJ (d)}$$

13. The equation for the standard enthalpy of formation is from the elements in their standard state. The standard state for oxygen is O_2 so (a) is not correct.

14. $2\ Fe(s) + \frac{3}{2}\ O_2(g) \rightarrow Fe_2O_3(s)$

18 carbohydrate $9.0 \text{ g} \times \frac{4.0 \text{ kcal}}{\text{g}} = 36 \text{ kcal}$

protein $2.0 \text{ g} \times \frac{4.0 \text{ kcal}}{\text{g}} = 8.0 \text{ kcal}$

fat $1.0 \text{ g} \times \frac{9.0 \text{ kcal}}{\text{g}} = 9.0 \text{ kcal}$

53 kcal

$$? \% \text{ calories from fat} = \frac{9.0 \text{ kcal}}{53 \text{ kcal}} \times 100\% = 17\% \text{ calories from fat}$$

Problems

19. $\Delta U = -89 \text{ J} + 531 \text{ J} = 442 \text{ J}$

21. $\Delta U = w + q = 0$

$w = -123$ cal of work is done by the system. $q = 123 \text{ cal} \times \frac{4.184 \text{ J}}{\text{cal}} = 515 \text{ J}$

23. Yes, it can do both.
Yes, if the system absorbs the same amount of heat as the work that it does.

25. Endothermic $q = \frac{902\text{ kJ}}{67.3\text{ g}} \times \frac{18.02\text{ g}}{\text{mol}} = 242\text{ kJ/mol}$

$q_p = \Delta H$ At constant pressure there is pressure–volume work that is included in q_p.

27. $CaO(s) + H_2O(l) \rightarrow Ca(OH)_2(s)$ $\Delta H = \frac{-9.78\text{ kJ}}{0.1500\text{ mol}} = \frac{-65.2\text{ kJ}}{\text{mol}}$

29. (a) $\Delta H°_{rxn} = \frac{2}{3}(-46\text{ kJ}) = -31\text{ kJ}$

(b) $\Delta H°_{rxn} = -3 \times (-46\text{ kJ}) = 138\text{ kJ}$

31. $\frac{1}{4}P_4(s) + \frac{3}{2}Cl_2(g) \rightarrow PCl_3(g)$ $\Delta H° = -287\text{ kJ}$

$? \text{ kJ} = \frac{-9.27\text{ kJ}}{1.00\text{ g } P_4} \times \frac{123.88\text{ g}}{\text{mol}} \times 0.25\text{ mol} = -287\text{ kJ}$

33. $? \text{ kJ} = 0.500\text{ kg CaO} \times \frac{10^3\text{ g}}{\text{kg}} \times \frac{1\text{ mol CaO}}{56.08\text{ g CaO}} \times \frac{-178.4\text{ kJ}}{\text{mol}} = -1.59 \times 10^3\text{ kJ}$

35. $? \text{ mol } O_2 = 226\text{ g } Na_2O_2 \times \frac{\text{mol } Na_2O_2}{77.98\text{ g } Na_2O_2} \times \frac{\text{mol } O_2}{2\text{ mol } Na_2O_2} = 1.45\text{ mol } O_2$

$? \text{ mol } O_2 = 98.5\text{ g } H_2O \times \frac{\text{mol } H_2O}{18.02\text{ g}} \times \frac{\text{mol } O_2}{2\text{ mol } H_2O} = 2.73\text{ mol } O_2$

Na_2O_2 is limiting.

$\Delta H = \frac{-287\text{ kJ}}{2\text{ mol } Na_2O_2} \times 226\text{ g } Na_2O_2 \times \frac{\text{mol } Na_2O_2}{77.98\text{ g } Na_2O_2} = -416\text{ kJ}$

37. $? \text{ L } C_2H_6 = -2.75 \times 10^4\text{ kJ} \times \frac{2\text{ mol } C_2H_6}{-3.12 \times 10^3\text{ kJ}} \times \frac{62.36\text{ L Torr}}{\text{K mol}} \times \frac{(273 + 17)\text{ K}}{714\text{ Torr}}$

$= 4.46 \times 10^2\text{ L}$

39. $? \text{ g CaO} = \frac{24.4\text{ L } CH_4 \times 753\text{ Torr}}{\frac{62.36\text{ L Torr}}{\text{K mol}} \times (273.2 + 24.7)\text{ K}} \times \frac{-890.3\text{ kJ}}{\text{mol } CH_4} \times \frac{\text{mol CaO}}{-65.2\text{ kJ}}$

$\times \frac{56.08\text{ g CaO}}{\text{mol CaO}} = 757\text{ g CaO}$

41. $\Delta H = \frac{3.12\times10^3\text{ kJ}}{2\text{ mol }C_2H_6} \times \frac{\text{mol }C_2H_6}{30.07\text{ g}}$

$\Delta H = \frac{5.76\times10^3\text{ kJ}}{2\text{ mol }C_4H_{10}} \times \frac{\text{mol }C_4H_{10}}{58.12\text{ g}}$

$\frac{3.12}{30} > \frac{5.76}{58}$, so ethane gives off more heat per gram.

43. $q = C\Delta T$

$C = \frac{672\text{ J}}{69\text{ °C} - 18\text{ °C}} = 13\text{ J/°C}$

45. (a) $q = 320.0\text{ g} \times \frac{4.18\text{ J}}{\text{g °C}} \times (96.0\text{ °C} - 15.0\text{ °C}) \times \frac{\text{kJ}}{10^3\text{ J}} = 108\text{ kJ}$

(b) $q = 74.3\text{ g} \times \frac{2.46\text{ J}}{\text{g °C}} \times (44.5\text{ °C} - (-36.4\text{ °C})) \times \frac{\text{kJ}}{10^3\text{ J}} = 14.8\text{ kJ}$

47. $\Delta T = \frac{q}{m\text{ x specific heat}} = \frac{2044\text{ J}}{638\text{ g} \times \frac{0.128\text{ J}}{\text{g °C}}} = 25.0\text{ °C}$

$t_f = \Delta T + t_i = 25.0\text{ °C} + 27.0\text{ °C} = 52.0\text{ °C}$

49. $-q_V = q_{\text{water}}$

$q_V = 9.13\text{ g} \times \text{specific heat} \times (24.46\text{ °C} - 99.10\text{ °C})$

$q_{\text{water}} = 20.0\text{ g} \times \frac{4.18\text{ J}}{\text{g °C}} \times (24.46\text{ °C} - 20.51\text{ °C})$

$q_{\text{water}} = 330.2\text{ J} = -q_V = -(-681.5\text{ g°C} \times \text{specific heat})$

specific heat = 0.485 J/g°C

51. $q_{Fe} = -q_{H_2O}$

$m_{Fe}\text{ sp. ht.}_{(Fe)}\,\Delta T = -m_{H_2O}\text{ sp. ht.}_{(H_2O)}\,\Delta T$

$t_i = \frac{5}{9} \times (375 - 32) = 191\text{ °C}$

$2.29\text{ kg} \times \frac{10^3\text{ g}}{\text{kg}} \times \frac{0.449\text{ J}}{\text{g °C}} \times (t_f - 191)\text{ °C} = -453\text{ g} \times \frac{4.18\text{ J}}{\text{g °C}} \times (t_f - 11.5)\text{ °C}$

$\frac{1.03\times10^3\text{ J}}{\text{°C}}\,t_f - 1.96\times10^5\text{ J} = \frac{-1894\text{ J}}{\text{°C}}\,t_f + 2.18\times10^4\text{ J}$

$t_f = \frac{2.18\times10^5\text{ J}}{2.92\times10^3\text{ J/°C}} = 74.7\text{ °C}$, no, the water will not boil.

53. $q_{Cu} = 2.25 \text{ g Cu} \times \frac{0.385 \text{ J}}{\text{g °C}} \times (100 - 25) \text{ °C}$

$q_{Pb} = 4.50 \text{ g Pb} \times \frac{0.128 \text{ J}}{\text{g °C}} \times (100 - 25) \text{ °C}$

$q_{Ag} = 1.87 \text{ g Ag} \times \frac{0.235 \text{ J}}{\text{g °C}} \times (100 - 25) \text{ °C}$

By estimating	actual
$q_{Cu} = 2 \times 0.4 \times 75 = 0.8 \times 75$	65 J
$q_{Pb} = 5 \times 0.1 \times 75 = 0.5 \times 75$	43 J
$q_{Ag} = 2 \times 0.2 \times 75 = 0.4 \times 75$	33 J

Thus, Cu must absorb the most heat, with Pb next.

55. $q_{\text{ calorim}} = \text{mass} \times \text{specific heat} \times \Delta T$

$q_{\text{ calorim}} = 1000. \text{ mL} \times \frac{1.00 \text{ g}}{\text{mL}} \times \frac{4.18 \text{ J}}{\text{g °C}} \times (23.21 - 20.00)\text{°C}$

$q_{\text{ calorim}} = 1.342 \times 10^4 \text{ J} \times \frac{\text{kJ}}{10^3 \text{ J}} = 13.42 \text{ kJ}$

$? \text{ mol} = 500.0 \text{ mL soln} \times \frac{10^{-3} \text{ L}}{\text{mL}} \times 0.500 \text{ M NaOH} \times \frac{\text{mol } H_2O}{1 \text{ mol NaOH}} = 0.2500 \text{ mol}$

$\Delta H = q_{rxn} = -q_{\text{ calorim}} = \frac{-13.42 \text{ kJ}}{0.2500 \text{ mol}} = -\frac{53.7 \text{ kJ}}{\text{mol}}$

57. $q_{rxn} = -q_{calorimeter} = -n\text{sp.ht.}\Delta T$

$q_{rxn} = -35.0 \text{ g } H_2O \times \frac{4.182 \text{ J}}{\text{g °C}} \times (19.4 \text{ °C} - 22.7 \text{ °C})$

$q_{rxn} = 48 \text{ J}$

$\Delta H = \frac{48 \text{ J}}{1.50 \text{ g}} \times \frac{\text{kJ}}{10^3 \text{ J}} \times \frac{80.05 \text{ g}}{\text{mol } NH_4NO_3} = \frac{26 \text{ kJ}}{\text{mol } NH_4NO_3}$

59. $q_{calorim} = \text{heat capacity} \times \Delta T$

$q_{calorim} = \frac{4.62 \text{ kJ}}{\text{°C}} \times (25.05 - 20.11)\text{°C} = 22.82 \text{ kJ}$

$q_{rxn} = -q_{calorim} = -\frac{22.82 \text{ kJ}}{0.7971 \text{ g}} = -28.6 \text{ kJ/g coal}$

61. $q_{\text{ calorim}} = \text{heat capacity} \times \Delta T = -q_{rxn}$

$\text{heat capacity} = \frac{\frac{(16.5 \text{ kJ})}{\text{g}} \times 1.83 \text{ g}}{(21.88 - 18.64) \text{ °C}} = \frac{9.32 \text{ kJ}}{\text{°C}}$

63. $q_{naphthalene} = -q_{calorimeter} = -\text{heat capacity} \times \Delta T$

$$\text{heat capacity} = \frac{\frac{5153.5\,\text{kJ}}{\text{mol}} \times \frac{\text{mol}}{128.16\,\text{g}} \times 1.108\,\text{g}}{5.92\,°\text{C}}$$

heat capacity = 7.53 kJ /°C

$$q_{\text{thymol}} = -q_{\text{calorimeter}} = -\text{heat capacity} \times \Delta T = -\frac{7.53\,\text{kJ}}{°\text{C}} \times 6.74\,°\text{C}$$

$q_{\text{thymol}} = -50.7$ kJ

$$\Delta H_{\text{comb}}(\text{thymol}) = \frac{q}{\text{mol}} = \frac{-50.7\,\text{kJ}}{1.351\,\text{g} \times \frac{\text{mol}}{150.2\,\text{g}}} = -5.64 \times \frac{10^3\,\text{kJ}}{\text{mol}}$$

65. $2\,C\,(\text{graphite}) + 2\,O_2 \rightarrow 2\,CO_2 \qquad 2 \times (-393.5\,\text{kJ})$
$2\,CO_2 \rightarrow O_2 + 2\,CO \qquad -(-566.0\,\text{kJ})$
$2\,C\,(\text{graphite}) + O_2 \rightarrow 2\,CO \qquad \Delta H° = -221.0\,\text{kJ}$

67. $N_2H_4 + O_2 \rightarrow N_2 + 2\,H_2O \qquad -622.2\,\text{kJ}$
$2\,H_2O_2 \rightarrow 2\,O_2 + 2\,H_2 \qquad -2 \times (-187.8\,\text{kJ})$
$2\,H_2 + O_2 \rightarrow 2\,H_2O \qquad 2 \times (-285.8\,\text{kJ})$
$N_2H_4 + 2\,H_2O_2 \rightarrow N_2 + 4\,H_2O \qquad \Delta H° = -818.2\,\text{kJ}$

69.

	ΔH (kJ)
$4\,NH_3(g) \rightarrow 6\,H_2(g) + 2\,N_2(g)$	-2 x -99.22
$2\,N_2(g) + 2\,O_2(g) \rightarrow 4NO(g)$	2 x 180.5
$6\,H_2(g) + 3\,O_2(g) \rightarrow 6\,H_2O\,(l)$	3 x -571.6
$4\,NH_3(g) + 5\,O_2(g) \rightarrow 4\,NO(g) + 6\,H_2O\,(l)$	ΔH = -1155.4 kJ

71. $\Delta H°_{rxn} = \Sigma\nu_p \Delta H°_f(\text{products}) - \Sigma\nu_r \Delta H°_f(\text{reactants})$

(a) $\Delta H°_{rxn} = 2\Delta H°_f[Al(OH)_3(s)] - \Delta H°_f[Al_2O_3(s)] - 3\Delta H°_f[H_2O(l)]$
$\Delta H°_{rxn} = 2\,\text{mol} \times -1276\,\text{kJ/mol} - (-1676\,\text{kJ} + 3\,\text{mol} \times -285.8\,\text{kJ/mol})$
$= -19\,\text{kJ}$

(b) $\Delta H°_{rxn} = 2\Delta H°_f[CO_2(g)] + \Delta H°_f[N_2(g)] - \Delta H°_f[C_2N_2(g)] - 2\Delta H°_f[O_2(g)]$
$\Delta H°_{rxn} = 2\,\text{mol} \times -393.5\,\text{kJ/mol} + 0 - (308.9\,\text{kJ} + 0\,\text{kJ}) = -1095.9\,\text{kJ}$

(c) $\Delta H°_{rxn} = \Delta H°_f[NH_3(g)] + \Delta H°_f[H_2O(g)] + \Delta H°_f[CO_2(g)] - \Delta H°_f[NH_4HCO_3(s)]$
$\Delta H°_{rxn} = -46.11\,\text{kJ} + -241.8\,\text{kJ} + -393.5\,\text{kJ} - (-849.4\,\text{kJ}) = -168.0\,\text{kJ}$

(d) $\Delta H°_{rxn} = 3\Delta H°_f[SiO_2(s)] + 4\Delta H°_f[Fe(s)] - 2\Delta H°_f[Fe_2O_3(s)] - 3\Delta H°_f[Si(s)]$
$\Delta H°_{rxn} = 3\,\text{mol} \times -910.9\,\text{kJ/mol} + 0\,\text{kJ} - (2\,\text{mol} \times -824.2\,\text{kJ/mol} + 0)$
$= -1084.3\,\text{kJ}$

73. $\Delta H^\circ_{rxn} = \Sigma \nu_p \Delta H^\circ_f(\text{products}) - \Sigma \nu_r \Delta H^\circ_f(\text{reactants})$

$\Delta H^\circ_{rxn} = 2\Delta H^\circ_f[SO_2(g)] + 2\Delta H^\circ_f[ZnO(s)] - 2\Delta H^\circ_f[ZnS(s)] - 3\Delta H^\circ_f[O_2(g)]$

$- 2\Delta H^\circ_f[ZnS(s)] = -878.2\text{ kJ} - 2\text{ mol} \times -296.8\text{ kJ/mol} - 2\text{ mol} \times -348.3\text{ kJ/mol}$

$+ 3\text{ mol} \times 0\text{ kJ/mol} = 412.0\text{ kJ}$

$$\Delta H^\circ_f[ZnS\ (s)] = \frac{412.0\text{ kJ}}{-2\text{ mol}} = -206.0\text{ kJ/mol}$$

75. $$\frac{-23.50\text{ kJ}}{1.050\text{ g}} \times \frac{106.12\text{ g}}{\text{mol}} = \frac{-2.3751 \times 10^3\text{ kJ}}{\text{mol}}$$

$C_4H_{10}O_3(l) + 5\ O_2(g) \rightarrow 4\ CO_2(g) + 5\ H_2O(l)$

$\Delta H^\circ_{rxn} = \Sigma \nu_p \Delta H^\circ_f(\text{products}) - \Sigma \nu_r \Delta H^\circ_f(\text{reactants})$

$-2375.1\text{ kJ} = 4\text{ mol } CO_2 \times -393.5\text{ kJ/mol} + 5\text{ mol } H_2O \times -285.8\text{ kJ/mol}$

$-5\text{ mol } O_2 \times 0 - 1\text{ mol } C_4H_{10}O_6\ \Delta H^\circ_f$

$\Delta H^\circ_f = -628\text{ kJ/mol}$

77. $\Delta H^\circ = \Delta H^\circ_f(NH_3) + \Delta H^\circ_f(H_2O(l)) - \Delta H^\circ_f(NH_4) - \Delta H^\circ_f(OH^-)$

$\Delta H^\circ = -46.11\text{ kJ} + (-285.8\text{ kJ}) - (-132.5\text{ kJ}) - (-230.0\text{ kJ})$

$\Delta H^\circ = 30.6\text{ kJ}$

79.

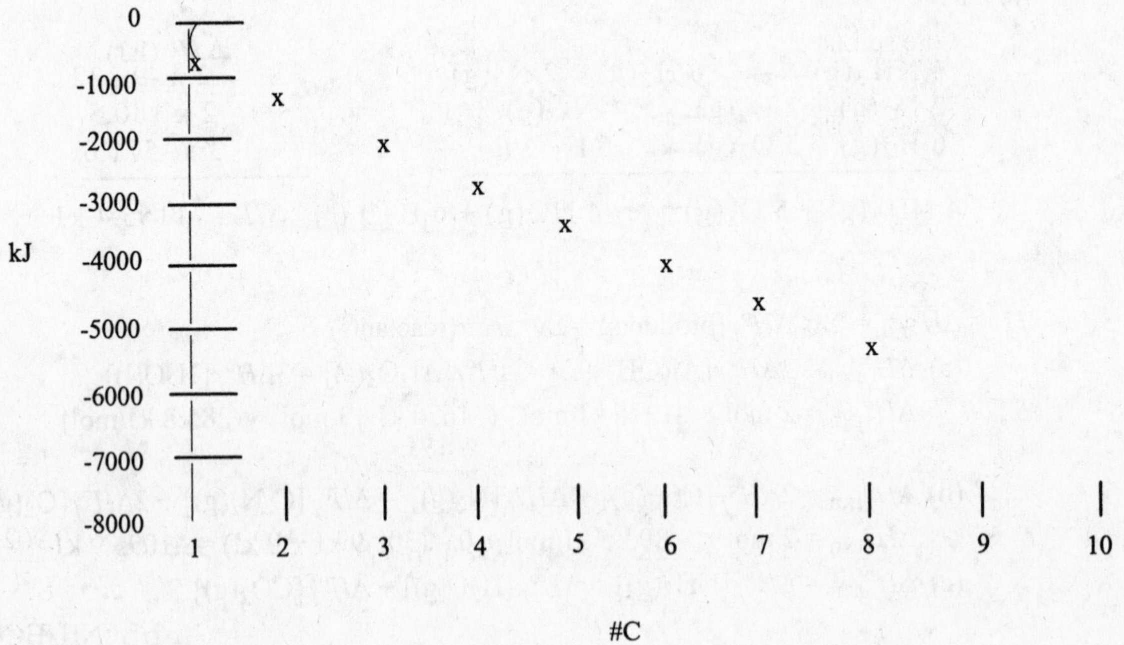

Extending that line produces

$\Delta H\ (C_{10}H_{22}) \approx -6800\text{ kJ}$

$\Delta H\ (C_{12}H_{26}) \approx -8100\text{ kJ}$

81. $? \% = 15.0 \text{ g} \times \frac{50.1 \text{ g fat}}{100 \text{ g}} \times \frac{9.0 \text{ Cal}}{\text{g}} \times \frac{100\%}{0.20 \text{ x } 2500 \text{ Cal}} = 14\%$

Additional Problems

83. (a) $\text{KE} = \frac{1}{2} m\overline{u^2} = \frac{1}{2}\frac{M}{N_A}\left(\frac{3RT}{M}\right) = \frac{3}{2}\frac{RT}{N_A}$

$$\text{KE} = \frac{3}{2} \times \frac{8.3145 \frac{\text{J}}{\text{mol K}} \text{ x } 273 \text{ K}}{6.022 \text{ x } 10^{23} \text{ atoms/mol}} = 5.65 \times 10^{-21} \text{ J}$$

The KE of a single molecule, no matter how fast it is moving, is not likely to match the KE of the much more massive "BB," thus the calculation in (a) could be skipped.

(b) $\text{KE} = \frac{1}{2} m\overline{u^2} = \frac{1}{2} \times 1.0 \text{ g} \times \frac{\text{kg}}{10^3 \text{ g}} \times \left(100 \frac{\text{m}}{\text{s}}\right)^2$

KE = 5.00 J

(c) $q = m \times \text{specific heat} \times \Delta T$

$$q = 10 \text{ mL} \times \frac{1.0 \text{ g}}{\text{mL}} \times \frac{4.18 \text{ J}}{\text{g °C}} \times (21 \text{ °C} - 20 \text{ °C})$$

$q = 42$ J

(c) is the greatest

85. $q_{\text{Fe}} = -q_{\text{H}_2\text{O}}$

$m\text{Fe sp.ht.(Fe)} \Delta T = -m_{\text{H}_2\text{O}} \text{ sp.ht.(H}_2\text{O)}\Delta T$

$$100 \text{ g } \frac{0.449 \text{ J}}{\text{g °C}}(t_f - 100) \text{ °C} = -100 \text{ g } \frac{4.18 \text{ J}}{\text{g °C}}(t_f - 20) \text{ °C}$$

$$-4.49 \times 10^3 \text{ J} + \frac{44.9 \text{ J}}{\text{°C}} t_f = +8.364 \times 10^3 \text{ J} - \frac{418 \text{ J}}{\text{°C}} t_f$$

$$\frac{463.1 \text{ J}}{\text{°C}} t_f = 12.85 \times 10^3 \text{ J}$$

$t_f = 27.8$ °C

88. (a) $nC_p = \frac{4}{3}\pi (1.10 \text{ cm})^3 \times \frac{7.83 \text{ g}}{\text{cm}^3} \times \frac{0.45 \text{ J}}{\text{g °C}} = 19.6 \text{ J/°C}$

(b) $nC_p = 2.00 \text{ m} \times \frac{\text{cm}}{10^{-2} \text{m}} \times 5.0 \text{ cm} \times 0.10 \text{ mm} \times \frac{\text{cm}}{10 \text{ mm}} \times \frac{2.70 \text{ g}}{\text{cm}^3} \times \frac{0.902 \text{ J}}{\text{g °C}}$

$= 24.4$ J/°C

(c) $nC_p = 5.0 \text{ mL} \times \frac{1.0 \text{ g}}{\text{mL}} \times \frac{4.18 \text{ J}}{\text{g °C}} = 20.9 \text{ J/°C}$

The greatest lowering of temperature will be caused by the aluminum, then the water, and least of all the steel ball. Instead of having to calculate a final temperature, the heat capacities (mass times the specific heat) were compared.

90. $? \text{ mol} = 15.5 \text{ g } H_2 \times \dfrac{\text{mol}}{2.016 \text{ g}} \times \dfrac{\text{mol } H_2O}{\text{mol } H_2} = 7.688 \text{ mol}$

$? \text{ mol} = 84.5 \text{ g } O_2 \times \dfrac{\text{mol}}{32.00 \text{ g}} \times \dfrac{\text{mol } H_2O}{0.500 \text{ mol } O_2} = 5.281 \text{ mol}$

Oxygen is limiting.

$? \text{ kJ} = 84.5 \text{ g } O_2 \times \dfrac{\text{mol}}{32.00 \text{ g}} \times \dfrac{241.8 \text{ kJ}}{0.500 \text{ mol } O_2} = 1.28 \times 10^3 \text{ kJ}$

93. heat generated $= 500.0 \text{ mL} \times \dfrac{10^{-3} \text{ L}}{\text{mL}} \times 6.0 \text{ M} \times \dfrac{-42 \text{ kJ}}{\text{mol}} \times \dfrac{10^3 \text{ J}}{\text{kJ}} = -1.26 \times 10^5 \text{ J}$

$q = -m \text{ sp.ht. } \Delta T$

$-1.26 \times 10^5 \text{ J} = -500 \text{ g} \times \dfrac{4.18 \text{ J}}{\text{g °C}} \times (t_f - 20 \text{ °C})$

$t_f = 80°C$

Assuming specific heat of solution is the same as water, 500 mL equals 500 g, and the initial temperature is 20°C.

95. $NaOH + HCl \rightarrow H_2O + Na^+ + Cl^-$ $\qquad \Delta H_{rxn} = -55.90 \text{ kJ/mol}$

$? \text{ mol} = 50.00 \text{ mL} \times \dfrac{10^{-3} \text{ L}}{\text{mL}} \times 1.16 \text{ M} \times \dfrac{\text{mol } H_2O}{\text{mol NaOH}} = 0.05800 \text{ mol } H_2O$

$? \text{ mol} = 25.00 \text{ mL} \times \dfrac{10^{-3} \text{ L}}{\text{mL}} \times 1.79 \text{ M} \times \dfrac{\text{mol } H_2O}{\text{mol HCl}} = 0.04475 \text{ mol } H_2O$

HCl is limiting.

$\Delta T = \dfrac{44.75 \text{ mmol}}{75.00 \text{ mL}} \times \dfrac{10^{-3} \text{ mol}}{\text{mmol}} \times \dfrac{55.90 \text{ kJ}}{\text{mol}} \times \dfrac{10^3 \text{ J}}{\text{kJ}} \times \dfrac{\text{g °C}}{4.18 \text{ J}} \times \dfrac{\text{mL}}{1.00 \text{ g}} = 7.98 \text{ °C}$

T before neutralization reaction is due to mixing of two solutions.

$q_{gain} = -q_{lost}$

$m\text{sp.ht.}\Delta T = -m\text{sp.ht.}\Delta T$

$50.00 \text{ mL sp.ht.}(T_f - 25.15 \text{ °C}) = -25.00 \text{ mL sp.ht.}(T_f - 26.34 \text{ °C})$

$50.00\ T_f - 1258 = -25.00\ T_f + 659$

$75.00 T_f = 1917 \text{ °C}$

$T_f = 25.56 \text{ °C}$

From neutralization reaction $7.98 \text{ °C} = \Delta T = t_f - t_v = t_f - 25.56 \text{ °C}$

$t_f = 33.54 \text{ °C}$

98. (a) Because ΔH°_f is a state function, ΔH°_f values of H_2SO_4(aq) depend on concentration.

(b) When concentrated solutions are diluted, there is a temperature increase.

$\Delta H^\circ_{rxn} = \Sigma \Delta H^\circ_f(\text{prod}) - \Sigma \Delta H^\circ_f(\text{react})$

$\Delta H^\circ_{rxn} = \Delta H^\circ_f$ dilute $- \Delta H^\circ_f$ concentrated

The dilute solution has a more negative value of ΔH, making the ΔH°_{rxn} negative and the dilution an exothermic process; heat is given off.

99. $PV = \frac{m}{M}RT \qquad M = \frac{mRT}{PV}$

$$M = \frac{1.103\text{ g} \times \frac{62.364\text{ L Torr}}{\text{K mol}} \times 298.15\text{ K}}{765.5\text{ Torr} \times 582\text{ mL} \times \frac{10^{-3}\text{ L}}{\text{mL}}} = 46.0\frac{\text{g}}{\text{mol}}$$

$$?\text{ g C} = 2.108\text{ g CO}_2 \times \frac{12.011\text{ g C}}{44.011\text{ g CO}_2} = 0.5753\text{ g C}$$

$$?\text{ g H} = 1.294\text{ g H}_2\text{O} \times \frac{2.0159\text{ g H}}{18.015\text{ g H}_2\text{O}} = 0.1448\text{ g H}$$

$$?\text{ g O} = 1.103\text{ g} - 0.5753\text{ g C} - 0.1448\text{ g H} = 0.383\text{ g O}$$

$$0.5753\text{ g C} \times \frac{\text{mol C}}{12.011\text{ g C}} = 0.04790\text{ mol C} \times \frac{1}{0.0239\text{ mol O}} = \frac{2.00\text{ mol C}}{\text{mol O}}$$

$$0.1448\text{ g H} \times \frac{\text{mol H}}{1.0079\text{ g H}} = 0.1437\text{ mol H} \times \frac{1}{0.0239\text{ mol O}} = \frac{6.01\text{ mol H}}{\text{mol O}}$$

$$0.383\text{ g O} \times \frac{\text{mol O}}{16.00\text{ g O}} = 0.0239\text{ mol O} \times \frac{1}{0.0239\text{ mol O}} = 1\frac{\text{mol O}}{\text{mol O}}$$

C_2H_6O

$C_2H_6O + 3\,O_2 \rightarrow 2\,CO_2 + 3\,H_2O$

$$\Delta H^\circ = \frac{5.015\text{ kJ}}{^\circ\text{C}} \times (31.94 - 25.00)^\circ\text{C} \times \frac{1}{1.103\text{ g}} \times \frac{46.07\text{ g}}{\text{mol}} = 1.45 \times 10^3\text{ kJ}$$

102. $q_{\text{calorim}} = 9.96\text{ kJ} \cdot {}^\circ\text{C}^{-1} \times (23.10 - 19.44)\ {}^\circ\text{C} = 36.5\text{ kJ}$

$q_{\text{rxn}} = -q_{\text{calorim}} = -36.5\text{ kJ}$

$2\,C_7H_{14} + 21\,O_2(g) \rightarrow 14\,CO_2(g) + 14\,H_2O(g)$

$$\Delta H = \frac{98.18\text{ g}}{\text{mol}} \times \frac{-36.5\text{ kJ}}{0.7782\text{ g}} = \frac{-4.61\times 10^3\text{ kJ}}{\text{mol}}$$

Apply Your Knowledge

104. $$3.294\text{ g CO}_2 \times \frac{\text{mol C}}{44.010\text{ g CO}_2} = 0.07485\text{ mol C} \times \frac{1}{0.07485\text{ mol C}} = \frac{1\text{ mol C}}{\text{mol C}}$$

$$1.573\text{ g H}_2\text{O} \times \frac{2\text{ mol H}}{18.015\text{ g H}_2\text{O}} = 0.1746\text{ mol H} \times \frac{1}{0.07485\text{ mol C}} = \frac{2.333\text{ mol H}}{\text{mol C}}$$

$C_{(1\times 3)}H_{(2.333\times 3)} \Rightarrow C_3H_7$

Molecular formula is twice the empirical formula.

Molecular formula = C_6H_{14}

(a) $2\ C_6H_{14} + 19\ O_2 \rightarrow 12\ CO_2\ (g) + 14\ H_2O\ (l)$

$\Delta H_{rxn} = \Sigma\Delta H_f°\ (prod) - \Sigma\Delta H_f°\ (react)$

$$\Delta H_{rxn} = 12\ mol \times -\frac{393.5\ kJ}{mol} + 14\ mol \times -\frac{285.8\ kJ}{mol} - 2\ mol \times -\frac{204.6\ kJ}{mol}$$

$$= -8314.0\ kJ$$

(b) $CH_3CH(CH_3)CH_2CH_2CH_3$ or $CH_3CH_2CH(CH_3)CH_2CH_3$

(c) $$?\ mL\ H_2O = 25.0\ m^3 \times \left(\frac{cm}{10^{-2}\ m}\right)^3 \times \frac{mL}{cm^3} \times\ = 2.50 \times 10^7\ mL$$

$$?\ gal\ C_6H_{14} = 2.50 \times 10^7\ mL \times \frac{1.00\ g}{mL} \times (33.0 - 19.2)°C \times \frac{4.18\ J}{g°C} \times \frac{kJ}{10^3\ J}$$

$$\times \frac{mol}{4157\ kJ} \times \frac{86.178\ g}{mol} \times \frac{mL}{0.6532\ g} \times \frac{10^{-3}\ L}{mL} \times \frac{1.057\ qt}{1.000\ L} \times \frac{gal}{4\ qt} = 12.1\ gal$$

107. (a) Al 0.902 J/g °C × 26.98 g/mol = 24.3 J/°C mol

Cu 0.385 J/g °C × 63.55 g/mol = 24.5 J/°C mol

Fe 0.449 J/g °C × 55.85 g/mol = 25.1 J/°C mol

Pb 0.128 J/g °C × 207.2 g/mol = 26.5 J/°C mol

Hg 0.139 J/g °C × 200.6 g/mol = 27.9 J/°C mol

Ag 0.235 J/g °C × 107.9 g/mol = 25.4 J/°C mol

S 0.706 J/g °C × 32.07 g/mol = 22.6 J/°C mol

Average 25.2 J/°C mol

(b) y = mx + b is the straight line equation. To make Dulong and Petit's law fit a straight line, plot either $\frac{1}{M}$ versus specific heat or $\frac{1}{\text{specific heat}}$ versus molar mass.

$C = M \times$ specific heat

specific heat = $\frac{C}{M}$ where $X = \frac{1}{M}$

(c) $$?\ g/mol = \frac{\frac{25.2\ J}{°C\ mol}}{\frac{0.421\ J}{g\ °C}} = 59.9\ g/mol$$ actual 58.9 g/mol

(d) $$?\ J/g\ °C = \frac{\frac{25.2\ J}{°C\ mol}}{\frac{47.88\ g}{mol}} = \frac{0.526\ J}{g\ °C}$$

$$-q_{Ti} = q_{H_2O}$$

$$-25.5\ g \times \frac{0.526\ J}{g\ °C} \times (27.4 - 99.7)\ °C = V_{H_2O} \times \frac{1.00\ g}{mL} \times \frac{4.18\ J}{g\ °C}$$

$\times (27.4 - 24.6)\ ^\circ C$

$$970.\ J = V_{H_2O} \times \frac{11.7\ J}{mL}$$

$$V_{H_2O} = 83\ mL$$

109. Experiment 1:

$$q_{water} = 106.6\ g \times \frac{4.18\ J}{g\ ^\circ C} \times (24.97\ ^\circ C - 23.28\ ^\circ C)$$

$$q_{water} = 753\ J$$

$$q_{hyd} = -q_{water} = -753\ J$$

$$\Delta H_{hyd} = \frac{-753\ J}{8.56\ g} \times \frac{142.04\ g}{mol} \times \frac{kJ}{10^3\ J} = -12.5\ \frac{kJ}{mol}$$

Experiment 2:

$$q_{water} = 105.4\ g \times \frac{4.18\ J}{g\ ^\circ C} \times (17.34\ ^\circ C - 24.40\ ^\circ C)$$

$$q_{water} = -3.11\ kJ$$

$$q_{dehyd} = -q_{water} = 3.11\ kJ$$

$$\Delta H_{dehyd} = \frac{3.11\ kJ}{19.3\ g} \times \frac{322.21\ g}{mL} = 51.9\ \frac{kJ}{mol}$$

$$\Delta H_{hyd} = -\Delta H_{dehyd} = -51.9\ \frac{kJ}{mol}$$

$$\Delta H_{hyd} = \sum \Delta H_f(\text{prod}) - \sum \Delta H_f(\text{react})$$

$$\Delta H_{hyd} = \Delta H_f Na_2SO_4 \cdot 10H_2O - \Delta H_f Na_2SO_4 - 10\Delta H_f H_2O$$

$$\Delta H_{hyd} = 1\ mol \times -4327\ \frac{kJ}{mol} - 1\ mol \times (-1387\ \frac{kJ}{mol}) - 10 \times (-285.8\ \frac{kJ}{mol})$$

$$\Delta H_{hyd} = -82\ kJ$$

If the molecules did not gain all of their water in the first experiment or lose all of their water in the second, the value will be less negative than expected. Other error causes include assumption of specific heat of solution, heat absorbed by calorimeter, and solutions not at 1M.

Chapter 7

Atomic Structure

Exercises

7.1A $\nu = \frac{c}{\lambda} = \frac{3.00 \text{ x } 10^8 \text{ m/s}}{1.07 \text{ mm}} \times \frac{\text{mm}}{10^{-3} \text{ m}} = 2.80 \times 10^{11} \text{ Hz}$

7.1B $\lambda = \frac{c}{\nu} = \frac{3.00 \text{ x } 10^8 \text{ m/s}}{9.76 \text{ x } 10^{13} \text{ s}^{-1}} \times \frac{\text{nm}}{10^{-9} \text{ m}} = 3.07 \times 10^3 \text{ nm}$

7.2A The microwave has a longer wavelength than the visible light of the television set.

7.2B (a) green light is about 550 nm

$$v = \frac{c}{\lambda} = \frac{3.00 \times 10^8 \text{ m/s}}{550 \text{ nm}} \times \frac{\text{nm}}{10^{-9} \text{ m}} = 5.45 \times 10^{14} \text{ s}^{-1}$$

(b) $v = \frac{c}{\lambda} = \frac{3.00 \times 10^8 \text{ m/s}}{4610 \text{ Å}} \times \frac{\text{Å}}{10^{-10} \text{ m}} = 6.51 \times 10^{14} \text{ s}^{-1}$

(c) $v = 91.9 \text{ MHz} \times \frac{10^6 \text{ Hz}}{\text{MHz}} \times \frac{\text{s}^{-1}}{\text{Hz}} = 91.9 \times 10^6 \text{ s}^{-1}$

(d) $v = \frac{c}{\lambda} = \frac{3.00 \times 10^8 \text{ m/s}}{622 \text{ nm}} \times \frac{\text{nm}}{10^{-9} \text{ m}} = 4.82 \times 10^{14} \text{ s}^{-1}$

(b) is the highest frequency

7.3A $E = hv = \frac{6.626 \times 10^{-34} \text{ J} \cdot \text{s}}{\text{photon}} \times \frac{2.89 \times 10^{10}}{\text{s}} = \frac{1.91 \times 10^{-23} \text{ J}}{\text{photon}}$

7.3B $E = \frac{hc}{\lambda} = \frac{\frac{6.626 \times 10^{-34} \text{ J} \cdot \text{s}}{\text{photon}} \times \frac{3.00 \times 10^8 \text{ m}}{\text{s}}}{235 \text{ nm} \times \frac{10^{-9} \text{ m}}{\text{nm}}} = \frac{8.46 \times 10^{-19} \text{ J}}{\text{photon}}$

7.4A $E = \frac{N_A hc}{\lambda} = \frac{6.022\times10^{23}\,\text{photon}}{\text{mole}} \times \frac{\frac{6.626\times10^{-34}\,\text{J}\cdot\text{s}}{\text{photon}} \times \frac{3.00\times10^{8}\,\text{m}}{\text{s}}}{400\,\text{nm}\times\frac{10^{9}\,\text{m}}{\text{nm}}} \times \frac{\text{kJ}}{10^{3}\,\text{J}}$

$= \frac{2.99\times10^{2}\,\text{kJ}}{\text{mol}}$

7.4B $\lambda = \frac{N_A hc}{E} = \frac{\frac{6.022\times10^{23}\,\text{photon}}{\text{mol}}\times6.626\times10^{-34}\,\text{J}\cdot\text{s}\times3.00\times10^{8}\,\text{m/s}}{\frac{100\,\text{kJ}}{\text{mol}}\times\frac{10^{3}\,\text{J}}{\text{kJ}}} \times \frac{\text{nm}}{10^{9}\,\text{m}}$

$= 1.20 \times 10^3$ nm

7.5A $E_6 = \frac{-B}{n^2} = \frac{-2.179 \times 10^{-18}\ \text{J}}{6^2} = -6.053 \times 10^{-20}$ J

7.5B $E_n = \frac{-B}{n^2} = \frac{-2.179\times10^{-18}\ \text{J}}{n^2} = -2.179 \times 10^{-19}$ J

$n^2 = \frac{-2.179\times10^{-18}\ \text{J}}{-2.179\times10^{-19}\ \text{J}} = 10$

$n = 3.16$

Values of n must be integer numbers. There is no level at -2.179×10^{-19} J.

7.6A $\Delta E = B \times \left(\frac{1}{n_i^2} - \frac{1}{n_f^2}\right) = 2.179 \times 10^{-18} \times \left(\frac{1}{2^2} - \frac{1}{4^2}\right)$

$= 2.179 \times 10^{-18} \times (0.1875) = 4.086 \times 10^{-19}$ J

7.6B $\Delta E = B \times \left(\frac{1}{n_i^2} - \frac{1}{n_f^2}\right) = 2.179 \times 10^{-18} \times \left(\frac{1}{n_i^2} - \frac{1}{n_f^2}\right) = 4.269 \times 10^{-20}$ J

$\left(\frac{1}{n_i^2} - \frac{1}{n_f^2}\right) = 0.0196$

$\frac{1}{2^2} = \frac{1}{4} = 0.250$

$\frac{1}{3^2} = \frac{1}{9} = 0.111$

$\frac{1}{4^2} = \frac{1}{16} = 0.0625$

$$\frac{1}{5^2} = \frac{1}{25} = 0.0400$$

$$\frac{1}{6^2} = \frac{1}{36} = 0.0278$$

$$\frac{1}{7^2} = \frac{1}{49} = 0.0204$$

Yes. The difference between n=7 and n = 5 is 4.269×10^{-20} J.

7.7A $E_i = \frac{-B}{n^2} = \frac{-2.179 \text{ x } 10^{-18} \text{ J}}{4^2} = -1.362 \times 10^{-19} \text{ J}$

$$E_f = \frac{-B}{n^2} = \frac{-2.179 \text{ x } 10^{-18} \text{ J}}{1^2} = -2.179 \times 10^{-18} \text{ J}$$

$$\Delta E = E_f - E_i = -2.043 \times 10^{-18} \text{ J}$$

$$\nu = \frac{?E}{h} = \frac{2.043 \times 10^{-18} \text{ J}}{6.626 \times 10^{-34} \text{ J} \cdot \text{s}} = 3.083 \times 10^{15} \text{ s}^{-1}$$

7.7B $\Delta E = B \times \left(\frac{1}{n_i^2} - \frac{1}{n_f^2}\right) = 2.179 \times 10^{-18} \times \left(\frac{1}{5^2} - \frac{1}{2^2}\right)$

$$\Delta E = 2.179 \times 10^{-18} \times (-0.21) = -4.576 \times 10^{-19} \text{ J}$$

$$\nu = \frac{\Delta E}{h} = \frac{4.576 \times 10^{-19} \text{ J}}{6.626 \times 10^{-34} \text{ J} \cdot \text{s}} = 6.906 \times 10^{14} \text{ s}^{-1}$$

$$\lambda = \frac{c}{\nu} = \frac{2.998 \text{ x } 10^8 \text{ m/s}}{6.906 \text{ x } 10^{14} \text{ s}^{-1}} = 4.341 \times 10^{-7} \text{ m} \times \frac{\text{nm}}{10^{-9} \text{ m}} = 434.1 \text{ nm} \quad \text{visible region}$$

7.8A (c) requires energy to be emitted. Since the energy difference between two successive energy levels is smaller at higher energy levels, (a) is larger than (d). Comparing (a) and (b) produces $\left(\frac{1}{1^2} - \frac{1}{2^2}\right)$ versus $\left(\frac{1}{3^2} - \frac{1}{8^2}\right)$ or 0.75 versus 0.095, so (a) represents the greatest amount of energy absorbed.

7.8B $\Delta E = B \times \left(\frac{1}{n_i^2} - \frac{1}{n_f^2}\right) = 2.179 \times 10^{-18} \times \left(\frac{1}{n_i^2} - \frac{1}{n_f^2}\right) = 4.90 \times 10^{-20} \text{ J}$

$$\left(\frac{1}{n_i^2} - \frac{1}{n_f^2}\right) = 0.0225$$

$$\frac{1}{n_f^2} = 0.0225 + \frac{1}{4^2} = 0.0225 + 0.0625 = 0.0850$$

$$n_f^2 = 11.8$$

$$n_i^2 = 3.43$$

No, n_f has to be an integer number.

$$7.9A\ \lambda = \frac{h}{mv} = \frac{\frac{6.626 \times 10^{-34}\ \text{kg m}^2\ \text{s}}{\text{s}^2}}{1.67 \times 10^{-27}\ \text{kg} \times 3.79 \times 10^3\ \text{m/s}}$$

$$\lambda = 1.05 \times 10^{-10}\ \text{m} \times \frac{\text{nm}}{10^{-9}\ \text{m}} = 0.105\ \text{nm}$$

$$7.9B\ \lambda_e = \frac{h}{m_e v_e} = \frac{6.626\times10^{-34}\ \text{kg m}^2\ \text{s}^{-2}\ \text{s}}{9.11\times10^{-31}\ \text{kg}\times2.74\times10^{6}\ \text{m/s}} = 2.65 \times 10^{-10}\ \text{m}$$

$$\lambda_e = \lambda_p = \frac{h}{m_p v_p} = \frac{6.626\times10^{-34}\ \text{kg m}^2\ \text{s}^{-2}\ \text{s}}{1.07\times10^{-27}\ \text{kg}\ v} = 2.65 \times 10^{-12}\ \text{m}$$

$$v = 6.69 \times 10^{-6}\ \text{m/s}$$

7.10A (a) not possible. For $l = 1$, m_l must be between +1 and –1.
(b) All values are possible.
(c) All values are possible.
(d) not possible. For $l = 2$, m_l must be between +2 and –2.

7.10B (a) $m_l = -1, 0,$ or 1
(b) $l = 3, 2, 1, 0$
(c) $n = 4, 5, 6, \ldots$ $m_l = -3, -2, -1, 0, 1, 2,$ or 3

7.11A (a) $5p$ $l = 1$ $m_l = -1, 0, 1$ 3 orbitals
(b) $l = 3$ $m_l = -3, -2, -1, 0, 1, 2, 3$ *f* subshells consist of seven orbitals

7.11B (a)

$l = 0$	$m_l = 0$	1 orbital
$l = 1$	$m_l = -1, 0, 1$	3 orbitals
$l = 2$	$m_l = -2, -1, 0, 1, 2$	5 orbitals
$l = 3$	$m_l = -3, -2, -1, 0, 1, 2, 3$	7 orbitals
		16 orbitals

(b) $n^2 = 4^2 = 16$ orbitals

Review Questions

1. Cathode rays are beams of negatively charged particles that are very tiny. They are electrons. Cathode rays travel in straight lines unless deflected by electric or magnetic fields. Because of the direction the rays were deflected, it was determined that they were negatively charged. Cathode rays hitting a fluorescent screen cause light to be emitted by the screen.
(b) is not a property.

5. $v = \dfrac{c}{\lambda}$

Wavelength, speed, and frequency are mathematically related. Amplitude is not.

6. The highest frequency is the shortest wavelength.

(a) $735\text{ nm} \times \dfrac{10^{-9}\text{ m}}{\text{nm}} = 7.35 \times 10^{-7}\text{ m}$

(b) $6.3 \times 10^{-5}\text{ cm} \times \dfrac{10^{-2}\text{ m}}{\text{cm}} = 6.3 \times 10^{-7}\text{ m}$

(c) $1.05\text{ mm} \times \dfrac{10^{-3}\text{ m}}{\text{mm}} = 1.05 \times 10^{-3}\text{ m}$

(d) $3.5 \times 10^{-6}\text{ m} = 3.5 \times 10^{-6}\text{ m}$

(e) 6.3×10^{-7} cm is the shortest.

7. (a) Frequency is inverse to wavelength.
(b) The range of visible wavelengths is 390 − 760 nm
(c) All radiation has the same speed in vacuum.
(d) Energy is inverse to wavelength.
(a) is true.

9. (b) Changing the intensity of the light has no effect on the photoelectric effect.

10. At the microscopic level, energy changes are quantized—that is, occurring only in discrete steps or intervals. At the macroscopic level, those available energy intervals are so close together that they appear to be continuous.

12. The negative value is an indication that this energy is due to a force of attraction. The closer to the nucleus, the more attraction, so the more negative the energy. If the equation were $E_n = -B \times n^2$, the value would become more negative the farther from the nucleus. The equation $E_n = -B/n^2$ gives a smaller negative value the farther from the nucleus, becoming zero for an infinite value of n.

13. The difference between (a) $\dfrac{1}{3^2} - \dfrac{1}{1^2} = \left(\dfrac{1}{9} - 1\right) = \dfrac{-8}{9}$ is greater than the difference

between (b) $\dfrac{1}{3^2} - \dfrac{1}{2^2} = \left(\dfrac{1}{9} - \dfrac{1}{4}\right) = \left(\dfrac{4}{36} - \dfrac{9}{36}\right) = \dfrac{-5}{36}$

or (c) $\dfrac{1}{4^2} - \dfrac{1}{2^2} = \left(\dfrac{1}{16} - \dfrac{1}{4}\right) = \left(\dfrac{1}{16} - \dfrac{4}{16}\right) = \dfrac{-3}{16}$

or (d) $\dfrac{1}{4^2} - \dfrac{1}{3^2} = \left(\dfrac{1}{16} - \dfrac{1}{9}\right) = \left(\dfrac{9}{144} - \dfrac{16}{144}\right) = \dfrac{-7}{144}$.

17. (a) $3d$ (b) $2s$ (c) $4p$ (d) $4f$

18. $n = 2$, $l = 2$, $m_l = 0$ is not allowed. l cannot be equal to n.

19. (a) $n = 3$ $l = 0$, $l = 1$, $l = 2$ 3 subshells

(b) $n = 2$ $l = 0,\ l = 1$ 2 subshells
(c) $n = 4$ $l = 0,\ l = 1,\ l = 2,\ l = 3$ 4 subshells

21. A *p* orbital is like a solid three-dimensional figure eight. A 3*p* orbital will be larger than a 2*p* and have areas of zero probability within the lobes.

Problems

23. Charge = 1.602×10^{-19} C
1.044×10^{-8} kg/C $\times 1.602 \times 10^{-19}$ C $= 1.672 \times 10^{-27}$ kg

25. 6.4×10^{-19} C $= 4 \times 1.6 \times 10^{-19}$ C
3.2×10^{-19} C $= 2 \times 1.6 \times 10^{-19}$ C
4.8×10^{-19} C $= 3 \times 1.6 \times 10^{-19}$ C
8.0×10^{-19} C $= 5 \times 1.6 \times 10^{-19}$ C
All values are integral multiples of 1.6×10^{-19} C, the charge on an electron or on a proton.

27. Values are atomic units of mass and charge followed by SI units.

(a) $^{80}Br^-$	–80:1	-8.29×10^{-7} kg/C
(b) $^{18}O^{2-}$	–9:1	-9.33×10^{-8} kg/C
(c) $^{40}Ar^+$	+40:1	4.15×10^{-7} kg/C

Values are approximate because mass numbers were used instead of actual atomic masses.

29. $69.9243 \times 0.205 = 14.33$
$71.9217 \times 0.274 = 19.71$
$72.9234 \times 0.078 = 5.69$
$73.9219 \times 0.365 = 26.98$
$75.9214 \times 0.078 = \underline{5.92}$
72.63 => 72.6

31. (a) $\lambda = \frac{c}{\nu} = \frac{2.998 \times 10^8 \text{ m/s}}{725 \text{ kHz} \times \frac{10^3 \text{ Hz}}{\text{kHz}}} = 414 \text{ m}$ radio

(b) $\lambda = \frac{c}{\nu} = \frac{2.998 \times 10^8 \text{ m/s}}{555 \text{ MHz} \times \frac{10^6 \text{ Hz}}{\text{MHz}}} = 0.540 \text{ m}$ microwave

(c) $\lambda = \frac{c}{\nu} = \frac{2.998 \times 10^8 \text{ m/s}}{1.05 \times 10^{19} \text{ s}^{-1}} \times \frac{\text{nm}}{10^{-9} \text{ m}} = 2.86 \times 10^{-2} \text{ nm}$ X-ray

33. $\frac{1}{4}\lambda = 175$ nm

$\lambda = 700 \text{ nm} \qquad \text{red}$

$$\nu = \frac{2.998\times10^{8}\text{ m/s}}{700.\text{ nm}\times\frac{10^{-9}\text{ m}}{\text{nm}}} = 4.28 \times 10^{14}\text{ s}^{-1}$$

35. $$?\text{ s} = 4.4 \times 10^{9}\text{ km} \times \frac{10^{3}\text{ m}}{\text{km}} \times \frac{\text{s}}{3.00 \times 10^{8}\text{ m}} = 1.5 \times 10^{4}\text{ s}$$

37. $$E = h\nu = 6.626 \times 10^{-34}\text{ J}\cdot\text{s} \times 7.42 \times 10^{14}\text{ s}^{-1} = 4.92 \times 10^{-19}\text{ J}$$

$$E = \frac{hc}{\nu} = \frac{6.626 \times 10^{-34}\text{ J s} \times \frac{3.00 \times 10^{8}\text{ m}}{\text{s}}}{655\text{ nm} \times \frac{10^{-9}\text{ m}}{\text{nm}}} = 3.03 \times 10^{-19}\text{ J}$$

The violet light has greater energy than the red 655-nm light. Energy increases as wavelength decreases and frequency increases.

39. $$\nu = \frac{c}{\lambda} = \frac{2.998 \times 10^{8}\text{ m/s}}{780\text{ nm} \times \frac{10^{-9}\text{ m}}{\text{nm}}} = 3.84 \times 10^{14}\text{ s}^{-1}$$

$$E = h\nu = \frac{6.626 \times 10^{-34}\text{ J s}}{\text{photon}} \times 3.84 \times 10^{14}\text{ s}^{-1} = \frac{2.55 \times 10^{-19}\text{ J}}{\text{photon}}$$

$$E = \frac{2.55 \times 10^{-10}\text{ J}}{\text{photon}} \times \frac{6.022 \times 10^{23}\text{ photons}}{\text{mol}} \times \frac{1\text{ kJ}}{1000\text{ J}} = 154\text{ kJ/mol}$$

41. $$\nu = \frac{E}{h} = \frac{4.65\times10^{-19}\text{ J}}{6.626\times10^{-34}\text{ J}\cdot\text{s}} = 7.018 \times 10^{14}\text{ s}^{-1}$$

$$\lambda = \frac{c}{\nu} = \frac{2.998\times10^{8}\text{ m/s}}{7.018\times10^{14}\text{ s}^{-1}} = 4.27 \times 10^{-7}\text{ m} \times \frac{\text{nm}}{10^{-9}\text{ m}} = 427\text{ nm}$$

43. $$\frac{?\text{ J}}{\text{photon}} = \frac{487\text{ kJ}}{\text{mol}} \times \frac{\text{mol}}{6.022 \times 10^{23}\text{ photons}} \times \frac{10^{3}\text{ J}}{\text{kJ}} = 8.087 \times 10^{-19}\text{ J/photon}$$

$$\nu = \frac{E}{h} = \frac{8.087 \times 10^{-19}\text{ J/photon}}{6.626 \times 10^{-34}\text{ J s/photon}} = 1.220 \times 10^{15}\text{ s}^{-1}$$

$$\lambda = \frac{c}{\nu} = \frac{2.998 \times 10^{8}\text{ m/s}}{1.220 \times 10^{15}\text{ s}^{-1}} \times \frac{\text{nm}}{10^{-9}\text{ m}} = 246\text{ nm}; \qquad \text{ultraviolet region}$$

45. (a) $$E_{\text{i}} = \frac{-B}{n^2} = \frac{-2.179\times10^{-18}\text{ J}}{6^2} = -6.053 \times 10^{-20}\text{ J}$$

$$E_{\text{f}} = \frac{-B}{n^2} = \frac{-2.179\times10^{-18}\text{ J}}{1^2} = -2.179 \times 10^{-18}\text{ J}$$

$$\Delta E_{\text{atom}} = E_{\text{f}} - E_{\text{i}} = -2.179 \times 10^{-18}\text{ J} - (-6.053 \times 10^{-20}\text{ J})$$

$\Delta E_{\text{atom}} = -2.119 \times 10^{-18}$ J

$\Delta E_{\text{photon}} = 2.119 \times 10^{-18}$ J

(b) $E_i = \frac{-B}{n^2} = \frac{-2.179 \times 10^{-18}\ \text{J}}{6^2} = -6.053 \times 10^{-20}$ J

$E_f = \frac{-B}{n^2} = \frac{-2.179 \times 10^{-18}\ \text{J}}{4^2} = -1.362 \times 10^{-19}$ J

$\Delta E_{\text{atom}} = E_f - E_i = -1.362 \times 10^{-19}\ \text{J} - (-6.053 \times 10^{-20}\ \text{J}) = -7.57 \times 10^{-20}$ J

$\Delta E_{\text{photon}} = 7.57 \times 10^{-20}$ J

47. (a) $E_i = \frac{-B}{n^2} = \frac{-2.179 \times 10^{-18}\ \text{J}}{5^2} = -8.716 \times 10^{-19}$ J

$E_f = \frac{-B}{n^2} = \frac{-2.179 \times 10^{-18}\ \text{J}}{2^2} = -5.448 \times 10^{-19}$ J

$\Delta E_{\text{atom}} = E_f - E_i = -3.268 \times 10^{-19}$ J

$\Delta E_{\text{photon}} = 3.268 \times 10^{-19}$ J

$\nu = \frac{\Delta E}{h} = \frac{3.268 \times 10^{-19}\ \text{J}}{6.626 \times 10^{-34}\ \text{J s}} = 4.932 \times 10^{14}\ \text{s}^{-1}$

(b) $E_i = \frac{-B}{n^2} = \frac{-2.179 \times 10^{-18}\ \text{J}}{3^2} = -2.421 \times 10^{-19}$ J

$E_f = \frac{-B}{n^2} = \frac{-2.179 \times 10^{-18}\ \text{J}}{1^2} = -2.179 \times 10^{-18}$ J

$\Delta E_{\text{atom}} = E_f - E_i = -1.937 \times 10^{-18}$ J

$\Delta E_{\text{photon}} = 1.937 \times 10^{-18}$ J

$\nu = \frac{\Delta E}{h} = \frac{1.937 \times 10^{-18}\ \text{J}}{6.626 \times 10^{-34}\ \text{J s}} = 2.923 \times 10^{15}\ \text{s}^{-1}$

49. The lowest possible energy level is $E_1 = -2.179 \times 10^{-18}$ J. The energy $E_n = -1.00 \times 10^{-17}$ J would be below E_1 and would thus be impossible.

$E_n = -1.00 \times 10^{-17}\ \text{J} = -B \times \frac{1}{n^2} = -2.179 \times 10^{-18} \times \frac{1}{n^2}$

That would make $\frac{1}{n^2} = 4.59$; $n^2 = 0.218$ or $n = 0.467$, which is not an integer number.

51. $\Delta E = B \times \left(\frac{1}{1^2} - \frac{1}{\infty^2}\right)$

$= 2.179 \times 10^{-18}\ \text{J} \times (1 - 0)$

$= 2.179 \times 10^{-18}\ \text{J} \times \frac{6.022 \times 10^{23}\ \text{atoms}}{\text{mol atoms}} \times \frac{\text{kJ}}{10^3\ \text{J}}$

$$\Delta E = 1.312 \times \frac{10^3\ \text{kJ}}{\text{mol}}$$

53. $\lambda = \dfrac{h}{mu} = \dfrac{6.626\times10^{-34}\ \text{kg m}^2\ \text{s}^{-1}}{1.67\times10^{-27}\ \text{kg}\times1.96\times10^{6}\ \text{m/s}} = 2.02 \times 10^{-13}\ \text{m} \times \dfrac{\text{nm}}{10^{-9}\ \text{m}}$

$= 2.02 \times 10^{-4}$ nm.

55. $u = \dfrac{h}{m\lambda} = \dfrac{6.626\times10^{-34}\ \text{kg m}^2\ \text{s}^{-1}}{9.109\times10^{-31}\ \text{kg}\times0.00510\ \text{nm}\times\dfrac{10^{-9}\ \text{m}}{\text{nm}}} = 1.43 \times 10^{8}$ m/s

57. (a) For p orbitals, $l = 1$; n must be at least 2 for $l = 1$.
(b) For f orbitals, $l = 3$; n must be at least 4 for $l = 3$.

59. If $n = 4$, $l = 0, 1, 2, 3$.
If $l = 1$, $m_l = -1, 0, 1$.

61. (a) m_l cannot be greater than l.
(b) permissible
(c) permissible
(d) $n = 0$ is not permissible.

63. (a) For 4p, $n = 4$, $l = 1$, $m_l = -1, 0, 1$
(b) For 4f, $n = 4$, $l = 3$, $m_l = -3, -2, -1, 0, 1, 2, 3$
(c) 3d subshell

65. (a) $n = 2, 3, 4, 5$, etc. $m_s = \pm\frac{1}{2}$ posssible orbitals: 2p, 3p, 4p, 5p, etc.
(b) $l = 1$ or 2 3p or 3d orbital
(c) $l = 4, 3$, or 2 $m_s = \pm\frac{1}{2}$ possible orbitals 5g, 5f, or 5d
(d) $n = 1, 2, 3, \ldots$ $m_l = 0$ possible orbitals: 1s, 2s, 3s, …

67. $n = 3, l = 2, m_l = 0, m_s = -\dfrac{1}{2}$

$n = 3, l = 2, m_l = \pm1, \pm2, m_s = \dfrac{1}{2}$

69. $196 \times 0.00146 = 0.286$
$198 \times 0.1002 = 19.8$
$199 \times 0.1684 = 33.5$
$200 \times 0.2313 = 46.3$
$201 \times 0.1322 = 26.6$
$202 \times 0.2980 = 60.2$

$$204 \times 0.0685 = \frac{14.0}{200.7}$$

It is approximate because mass numbers are used instead of the isotopic masses.

71. $T = \frac{1}{\nu}$ Period and frequency are inversely related.

$$\frac{1\text{s}}{60 \text{ cycles}} = 1.7 \times 10^{-2} \text{ s/cycle}$$

74. $? \frac{\text{J}}{\text{molecule}} = \frac{946 \text{ kJ}}{\text{mol}} \times \frac{\text{mol}}{6.022 \times 10^{23} \text{ molecules}} \times \frac{10^3}{\text{kJ}} = 1.57 \times 10^{-18} \frac{\text{J}}{\text{molecule}}$

$E = h\nu$

$$\nu = \frac{E}{h} = \frac{1.57 \times 10^{-18} \text{ J}}{6.626 \times 10^{-34} \text{ J s}} = 2.37 \times 10^{15} \text{ s}^{-1}$$

A photon emitted from a level 2 → level 1 change in a hydrogen atom has enough energy to break the N to N bond.

$$\Delta E_{\text{atom}} = E_1 - E_2 = \frac{-2.179 \times 10^{-18} \text{ J}}{1^2} - \frac{-2.179 \times 10^{-18} \text{ J}}{2^2} = -1.634 \times 10^{-18} \text{ J}$$

$\Delta E_{\text{photon}} = 1.634 \times 10^{-18}$ J

77. $\Delta E_{\text{atom}} = E_1 - E_\infty = \frac{-2.179 \times 10^{-18} \text{ J}}{1^2} - 0 = -2.179 \times 10^{-18}$ J

$\Delta E_{\text{photon}} = 2.179 \times 10^{-18}$ J

$$\lambda = \frac{hc}{E} = \frac{6.626 \times 10^{-34} \text{ J s} \times 2.998 \times 10^8 \text{ m/s}}{2.179 \times 10^{-18} \text{ J}} \times \frac{\text{nm}}{10^{-9} \text{ m}} = 91.16 \text{ nm} \quad \text{minimum}$$

$$\Delta E_{\text{atom}} = E_1 - E_2 = \frac{-2.179 \times 10^{-18} \text{ J}}{1^2} - \frac{-2.179 \times 10^{-18} \text{ J}}{2^2} = -1.634 \times 10^{-18} \text{ J}$$

$\Delta E_{\text{photon}} = 1.634 \times 10^{-18}$ J

$$\lambda = \frac{hc}{E} = \frac{6.626 \times 10^{-34} \text{ J s} \times 2.998 \times 10^8 \text{ m/s}}{1.634 \times 10^{-18} \text{ J}} \times \frac{\text{nm}}{10^{-9} \text{ m}} = 121.6 \text{ nm} \quad \text{maximum}$$

79. $E_{\text{photon}} = \frac{hc}{\lambda} = \frac{6.626 \times 10^{-34} \text{ J s} \times 2.998 \times 10^8 \text{ m/s}}{486.1 \text{ nm} \times 10^{-9} \text{ m/nm}} = 4.087 \times 10^{-19}$ J

$\Delta E_{\text{atom}} = -4.087 \times 10^{-19}$ J

$$\Delta E = \frac{-B}{n^2} - \frac{-B}{m^2} = -B\left(\frac{1}{n^2} - \frac{1}{m^2}\right)$$

$$-4.087 \times 10^{-19} \text{ J} = -2.179 \times 10^{-18} \text{ J} \times \left(\frac{1}{n^2} - \frac{1}{m^2}\right)$$

$$\frac{1}{n^2} - \frac{1}{m^2} = 0.1876$$

$$\frac{1}{1^2} = 1 \quad \frac{1}{2^2} = 0.2500 \quad \frac{1}{3^2} = 0.1111 \quad \frac{1}{4^2} = 0.0625 \quad \frac{1}{5^2} = 0.0400$$

$\frac{1}{2^2} - \frac{1}{4^2} = 0.1875$ Thus, the transition is $n = 4 \rightarrow n = 2$

82. $$\frac{hc}{\lambda \mathbf{B}} = \frac{6.626\times10^{-34}\ \text{J s}\times\dfrac{2.9979\times10^{8}\ \text{m}}{\text{s}}}{7400\ \text{nm}\times\dfrac{10^{-9}\ \text{m}}{\text{nm}}\times2.179\times10^{-18}\ \text{J}} = -\left(\frac{1}{n_i^2} - \frac{1}{n_f^2}\right)$$

$$\left(\frac{1}{n_i^2} - \frac{1}{n_f^2}\right) = -0.01232$$

By simple trial and error, $n_f = 5$ and $n_i = 6$. The Pfund series consists of electron transitions down to level 5.

84. $$\lambda = \frac{h}{mu} = \frac{6.626\ \text{x}\ 10^{-34}\ \text{J s}}{25.0\ \text{g x}\ \dfrac{\text{kg}}{10^3\ \text{g}}\ \text{x}\ 110\ \text{m/s}} = 2.41\times10^{-34}\ \text{m}$$

87. $$-\Delta E_{\text{atom}} = \Delta E_{\text{photon}} = \text{B}\times\left(\frac{1}{1^2} - \frac{1}{5^2}\right) = 2.179\times10^{-18}\ \text{J}\times\left(1 - \frac{1}{25}\right)$$

$$= 2.092\times10^{-18}\ \text{J}$$

$$\lambda = \frac{hc}{E} = \frac{6.626\ \text{x}\ 10^{-34}\ \text{J s x}\ \dfrac{2.998\ \text{x}\ 10^{8}\ \text{m}}{\text{s}}}{2.092\ \text{x}\ 10^{-18}\ \text{J}} = 9.496\times10^{-8}\ \text{m}$$

$$u = \frac{h}{m\lambda} = \frac{\dfrac{6.626\times10^{-34}\ \text{kg m}^2\ \text{s}}{\text{s}^2}}{9.109\times10^{-31}\ \text{kg}\times9.496\times10^{-8}\ \text{m}} = 7.66\times10^{3}\ \text{m/s}$$

89. $E = h\nu = 6.626\times10^{-34}\ \text{J s}\times8.4\times10^{9}\ \text{s}^{-1} = 5.566\times10^{-24}\ \text{J/photon}$

$$\frac{4\ \text{x}\ 10^{-21}\ \text{watt}}{5.566\ \text{x}\ 10^{-24}\ \text{J /photon}}\times\frac{\text{J}}{\text{watt s}} = 7\times10^{2}\ \text{photon/s}$$

93. $$?\ \frac{\text{kg}}{\text{molecule}} = \frac{720\ \text{g}}{\text{mol}}\times\frac{\text{kg}}{10^3\ \text{g}}\times\frac{\text{mole}}{6.022\times10^{23}\ \text{molecules}} = 1.20\times10^{-24}\ \frac{\text{kg}}{\text{molecule}}$$

$$\lambda = \frac{h}{mv} = \frac{6.626\times10^{-34}\ \dfrac{\text{kg m}^2}{\text{s}}}{1.20\times10^{-24}\ \text{kg}\times220\dfrac{\text{m}}{\text{s}}} = 2.51\times10^{-12}\ \text{m}$$

$$?\ \text{nm} = 2.51\times10^{-12}\ \text{m}\times\frac{\text{nm}}{10^{-9}\ \text{m}} = 2.51\times10^{-3} = 0.00251\ \text{nm}$$

The calculated value is a factor of 10 less than the observed value.

97. (a) $\nu = \frac{E}{h} = \frac{3.42 \times 10^{-19}\ \text{J}}{6.626 \times 10^{-34}\ \text{Js}} = 5.16 \times 10^{14}\ \text{s}^{-1}$

(b) $\lambda = \frac{c}{\nu} = \frac{2.998 \times 10^{8}\ \text{m/s}}{5.16 \times 10^{14}\ \text{s}^{-1}} = 5.81 \times 10^{-7}\ \text{m} \times \frac{\text{nm}}{10^{-9}\ \text{m}} = 581\ \text{nm}$

The longer wavelength (1000 nm) has less energy and would not have enough energy to expel an electron.

$$E = \frac{hc}{\lambda} = \frac{6.626 \times 10^{-34}\ \text{J s} \times \frac{2.998 \times 10^{8}\ \text{m}}{\text{s}}}{1000\ \text{nm} \times \frac{10^{-9}\ \text{m}}{\text{nm}}} = 1.99 \times 10^{-19}\ \text{J}$$

which is less than the work function for cesium.

(c) $$E = \frac{hc}{\lambda} = \frac{6.626 \times 10^{-34}\ \text{J s} \times \frac{2.998 \times 10^{8}\ \text{m}}{\text{s}}}{425\ \text{nm} \times \frac{10^{-9}\ \text{m}}{\text{nm}}} = 4.67 \times 10^{-19}\ \text{J}$$

$E_{\text{excess}} = 4.67 \times 10^{-19}\ \text{J} - 3.42 \times 10^{-19}\ \text{J} = 1.25 \times 10^{-19}\ \text{J}$

$E = \frac{1}{2} mu^2$

$$u = \sqrt{\frac{2E}{m}} = \sqrt{\frac{2 \times 1.25 \times 10^{-19}\ \text{J}}{9.109 \times 10^{-31}\ \text{kg}}} = 5.24 \times 10^{5}\ \text{m/s}$$

Apply Your Knowledge

101. $r_n = n^2 a_o = 3^2 \times 53\ \text{pm} = 477\ \text{pm}$

(a) $\frac{nh}{2\pi} = \text{mvr}$

$\nu = \frac{nh}{2\pi mr}$

$$\nu = \frac{3 \times 6.626 \times 10^{-34}\ \text{kg m}^2\ \text{s}^{-1}}{2 \times 3.1416 \times 9.109 \times 10^{-31}\ \text{kg} \times 477\ \text{pm} \times \frac{10^{-12}\ \text{m}}{\text{pm}}} = 7.3 \times 10^{5}\ \frac{\text{m}}{\text{s}}$$

(b) $C = \pi d = 3.142 \times 2 \times 477\ \text{pm} \times \frac{10^{-12}\ \text{m}}{\text{pm}} = 3.00 \times 10^{-9}\ \text{m}$

$$?\ \text{revolutions/s} = 7.3 \times 10^{5}\ \frac{\text{m}}{\text{s}} \times \frac{\text{revolution}}{3.00 \times 10^{-9}\ \text{m}} = 2.43 \times 10^{14}\ \text{revolutions/s}$$

103. (a) $\frac{N^*}{N^\circ} = \left(\frac{g_m}{g_n}\right) e^{\frac{-\Delta E}{kT}}$

$$\Delta E = 2.10\ \text{eV} \times \frac{1.602 \times 10^{-19}\ \text{J}}{\text{eV}} = 3.36 \times 10^{-19}\ \text{J}$$

$$\frac{N^*}{N^\circ} = 2 \times e^{\frac{-3.36\times10^{-19}\ \text{J}}{1.38\times10^{-23}\ \text{J/K}\cdot 298\ \text{K}}} = 2 \times e^{-81.7} = 6.57 \times 10^{-36}$$

$N^* = N^\circ \times 6.57 \times 10^{-36}$

$N^* + N^\circ = N_{total}$

$N^\circ + N^\circ \times 6.57 \times 10^{-36} = N_{total}$

$N^\circ \times (1 + 6.57 \times 10^{-36}) = N_{total}$

$$\frac{N^*}{N_{total}} \times 100\% = \frac{N^\circ \times 6.57\times10^{-36}}{N^\circ \times (1+6.57\times10^{-36})} \times 100\%$$

$= 6.57 \times 10^{-34}\%$ **atoms in excited state**

$$\frac{N^\circ}{N_{total}} \times 100\% = \frac{N^\circ}{N^\circ \times (1+6.57\times10^{-36})} \times 100\% = 100\%$$ **atoms in ground state**

(b) $$\frac{N^*}{N^\circ} = 2 \times e^{\frac{-3.36\times10^{-19}\ \text{J}}{1.38\times10^{-23}\ \text{J/K}\times 2700\ \text{K}}} = 2 \times e^{-9.02}$$

$$\frac{N^*}{N^\circ} = 2.42 \times 10^{-4}$$

$N^\circ = 99.98\%$

$N^* = 2.42 \times 10^{-2}\%$

(c) $$\frac{N^*}{N^\circ} = 2 \times e^{\frac{-3.36\times10^{-19}\ \text{J}}{1.38\times10^{-23}\ \text{J/K}\times 2200\ \text{K}}} = 2 \times e^{-11.1}$$

$$\frac{N^*}{N^\circ} = 3.02 \times 10^{-5}$$

$N^\circ = 99.997\%$

$N^* = 3.02 \times 10^{-3}\%$

(d) The signal from atomic emission sectroscopy will change with temperature because the number of excited atoms changes with temperature. An increase of 500 K makes an increase of eight times the number of atoms in the excited state and increases the emission intensity.

(e) The number of ground state-atoms changes insignificantly with temperature. There is little difference between $1 - 10^{-4}$ and $1 - 10^{-5}$. Both values are about 1.

Chapter 8

Electron Configurations, Atomic Properties, and the Periodic Table

Exercises

8.1A (a) Cl $1s^22s^22p^63s^23p^5$

(b) [Ne] $3s^23p^5$

(c)

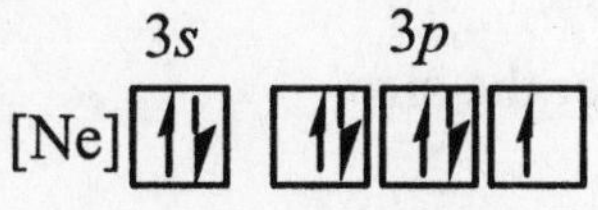

8.1B (b) and (c) are valid. (a) can not have two electrons in degenerate orbitals until all orbitals have an electron (Hund's rule). In (d), single electrons in degenerate orbitals have the same spin.

8.2A (a) Mo $1s^22s^22p^63s^23p^63d^{10}4s^24p^64d^45s^2$
is the answer from the periodic table. $4d^55s^1$ is actual.
[Kr]$4d^55s^1$

(b) Bi $1s^22s^22p^63s^23p^63d^{10}4s^24p^64d^{10}4f^{14}5s^25p^65d^{10}6s^26p^3$
[Xe]$4f^{14}5d^{10}6s^26p^3$

8.2B (a) Sn [Kr]$4d^{10}5s^25p^2$

(b)

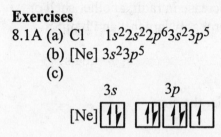

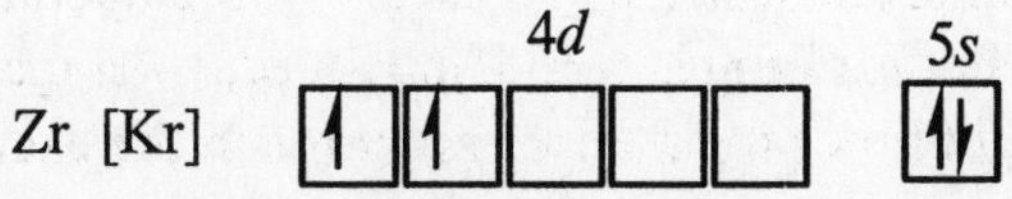

8.3A Se^{2-} [Ar]$3d^{10}4s^24p^6$
Pb^{2+} [Xe]$4f^{14}5d^{10}6s^26p^0$

8.3B I^- [Kr]$4d^{10}5s^25p^6$
Cr^{3+} [Ar]$3d^34s^0$

8.4A (a) 2 (b) 4 (c) 0 (d) 0 (e) 0 (f) 4

8.4B

(a) K	[Ar]$4s^1$	paramagnetic	one odd e^-
(b) Hg	[Xe]$4f^{14}5d^{10}6s^2$	diamagnetic	all paired
(c) Ba^{2+}	[Xe]	diamagnetic	all paired
(d) N	[He]$2s^22p^3$	paramagnetic	3 unpaired e^-
(e) F^-	[Ne]	diamagnetic	all paired

(f) Ti^{2+} $[Ar]3d^2$ paramagnetic 2 unpaired e^-
(g) $Cu^{2+}[Ar]3d^9$ paramagnetic one unpaired e^-

8.5A (a) F < N < Be
(b) Be < Ca < Ba
(c) F < Cl < S
(d) Mg < Ca < K

8.5B P < Ge < Sn < Pb < Cs is obvious from trends. Where to place Ca is less obvious. It is difficult to determine which will cause more increase in radii, another shell or being more to the left on the periodic table. Ca is probably larger than Pb or Sn even though it is in a lower shell.
P < Ge < Sn < Pb < Ca < Cs probable order.

8.6A All are isoelectronic, so the more protons, the smaller the size.
$Y^{3+} < Sr^{2+} < Rb^+ < Br^- < Se^{2-}$

8.6B	Ca^{2+}	Cr^{2+}	Cs^+	Cr^{3+}	K^+
p^+	20	24	55	24	19
e^-	18	22	54	21	18

smallest $Cr^{3+} < Cr^{2+} < Ca^{2+} < K^+ < Cs^+$ largest
Cr^{3+} has one less electron than Cr^{2+}.
Cr^{2+} has more protons than Ca^{2+}.
Ca^{2+} has more protons than K^+.

8.7A (a) Be < N < F
(b) Ba < Ca < Be
(c) S < P< F
(d) K < Ca < Mg

8.7B H < Al < S < Mg < Ar < K

8.8A Because the Se atom is larger than the S atom, the value must be less than 450 kJ/mol but a positive number. The value must be 400 kJ/mol.

8.8B (a) Adding one electron to silicon would produce a half-filled 3*p* subshell. Adding one electron to Al results in only two electrons in the 3*p* subshell.
(b) Adding one elecron to phosphorus would be adding to a half-filled 3*p* subshell, but adding one electron to silicon would produce a half-filled 3*p* subshell.

8.9A (a) O is more nonmetallic than P.
(b) S is more nonmetallic than As.
(c) F is more nonmetallic than P.

8.9B Ba ~ Na > Fe ~ Co > Sb > Se > S

1A elements are usually more metallic than 2A but Ba is so much farther down the periodic table that the metallic chatacter is about the same as Na. Fe and Co are both transitions elements of about the same size and ionization energies; so they are similar in matallic character. The others can be placed by their position on the periodic table.

8.10A

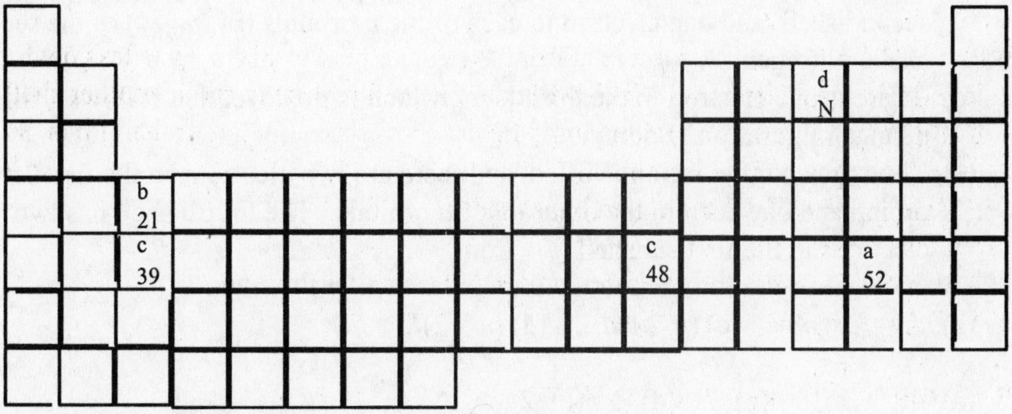

(c) Cadmium is the highest; yttrium is the lowest.

8.10B

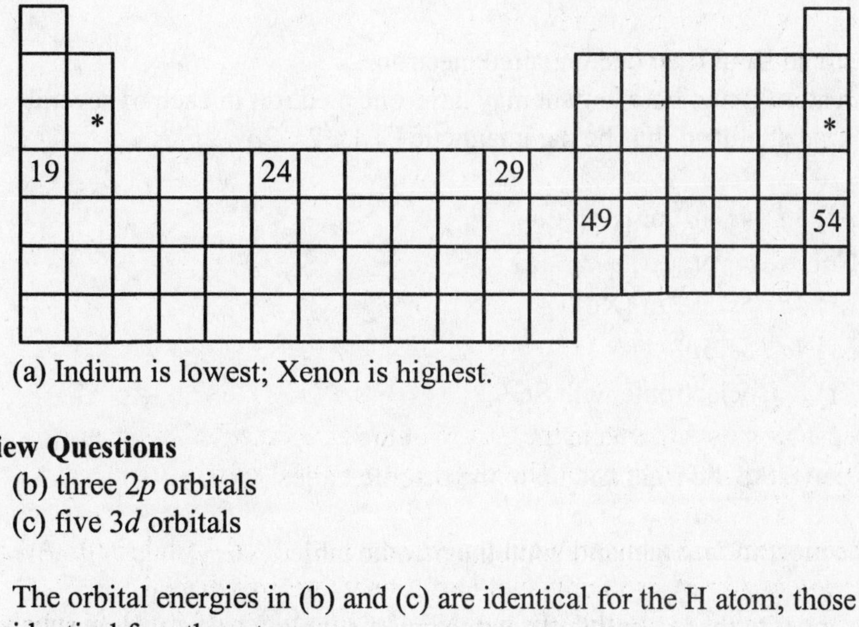

(a) Indium is lowest; Xenon is highest.

Review Questions

1. (b) three 2*p* orbitals
   (c) five 3*d* orbitals

2. The orbital energies in (b) and (c) are identical for the H atom; those in (c) are also identical for other atoms.

3. Hund's rule states that electrons filling degenerate orbitals will fill all orbitals with one electron with spins aligned before making pairs.

   Thus carbon is 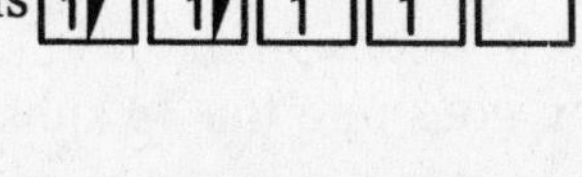

.

4. (a) There are two electrons in the first shell, two electrons in the first subshell (*s*), and five electrons in the second subshell (*p*) of the second shell. Fluorine.
   (b) There are two electrons in the first shell, two electrons in the first subshell (*s*), and six electrons in the second subshell (*p*) of the second shell. The third shell has two electrons in the first subshell (*s*), six in the second (*p*), and ten in the third (*d*). The fourth shell has one electron in the first subshell (*s*). Copper.
   (c) There are two electrons in the first shell, two electrons in the *s* subshell of the second shell, and one electron in each of the *p* orbitals (p_x, p_y, p_z) of the second shell. Nitrogen.
   (d) There is one electron in the third level, which is the level that follows a filled neon configuration. Sodium.
   (e) The argon configuration is filled, and there are two electrons in the first 3*d* orbital and one each in the other four 3*d* orbitals. The fourth shell has two electrons in the first subshell (*s*). Iron.

5. (a) *ns*; (b) *np* (c) $(n-1)d$ (d) $(n-2)f$.

7. (a) 4 (b) 8 (c) 7 (d) 3 (e) 2

11. (c) Fe and (e) Pt are d-block elements. (a) Ca is s-block and (b) Al and (d) Pb are p-block elements.

12. Yes, there has to be at least one unpaired electron.
No, an even atomic-number element may have one electron in each of several orbitals in a subshell and thus be paramagnetic C: $1s^2 2s^2 2p_x^1 2p_y^1$.

15. Se^{2-} [Ar] $3d^{10}4s^24p^6$ or [Kr]
   (a) S^{2-} [Ne] $3s^23p^6$ or [Ar]
   (b) I^- [Kr] $4d^{10}5s^25p^6$ or [Xe]
   (c) Xe [Kr] $4d^{10}5s^25p^6$
   (d) Sr^{2+} [Kr] isoelectronic with Se^{2-}

16. (b) Cl is the most to the right and up in the peiodic table.

17. (a) Br is the most to the right and up in the peiodic table.

19. (c) K is the most to the left and down in the peiodic table.

Problems

21. (a) One pair of electrons in a 2*p* orbital has parallel spins.
   (b) There are three electrons in one 2*p* orbital.
   (c) The spins of the unpaired electrons in the 2*p* orbital should all be the same.

23. (a) The 3*s* subshell fills before any electron goes into the 3*p* subshell.

(b) The 4*s* subshell fills before the 3*d* subshell begins to fill.
(c) The 2*d* subshell does not exist.

25. (a) The 2*s* orbital can have only two electrons. Also, the 2*p* subshell is lower in energy than the 3*s*. Pauli exclusion
(b) The 2*p* subshell should contain only six electrons. Pauli exclusion and Aufbau
(c) There is no 2*d* subshell. rules for quantum numbers *n* and *l*

27. (a) Mg $1s^22s^22p^63s^2$
(b) F $1s^22s^22p^5$
(c) O $1s^22s^22p^4$
(d) Be $1s^22s^2$
(e) P $1s^22s^22p^63s^23p^3$
(f) Si $1s^22s^22p^63s^23p^2$
(g) C $1s^22s^22p^2$
(h) K $1s^22s^22p^63s^23p^64s^1$
(i) Sc $1s^22s^22p^63s^23p^63d^14s^2$

29. (a) Cs [Xe] $6s^1$
(b) Sr [Kr] $5s^2$
(c) Ti [Ar] $3d^24s^2$
(d) Sb [Kr] $4d^{10}5s^25p^3$
(e) Br [Ar] $3d^{10}4s^24p^5$
(f) Pb [Xe] $4f^{14}5d^{10}6s^26p^2$

31.

(a) N 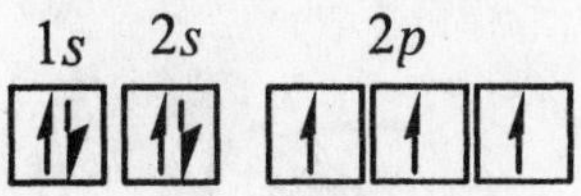

(b) B 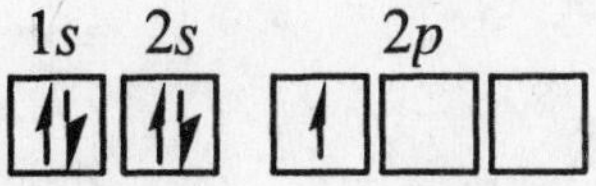

(c) Si 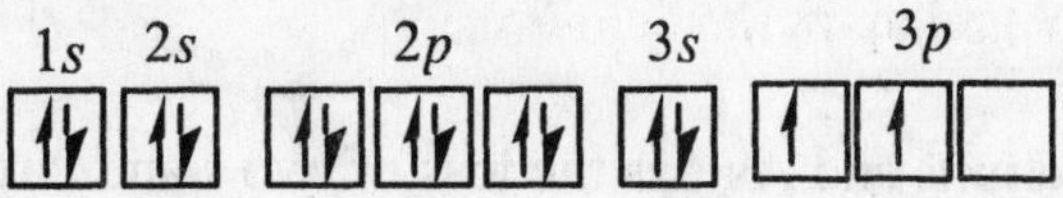

(d) Ca 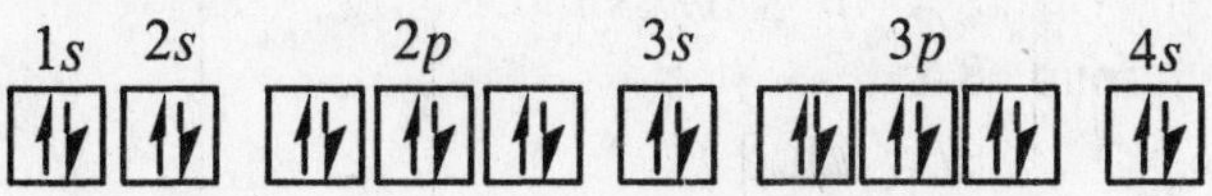

(e) Cl 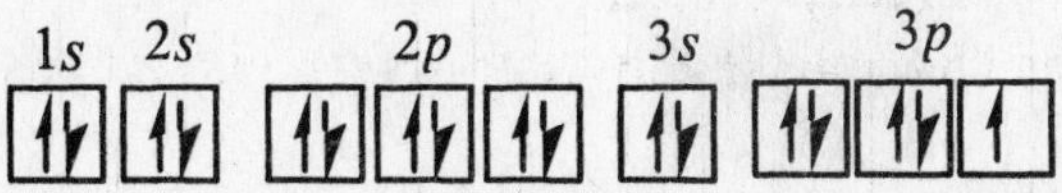

(f) Sc 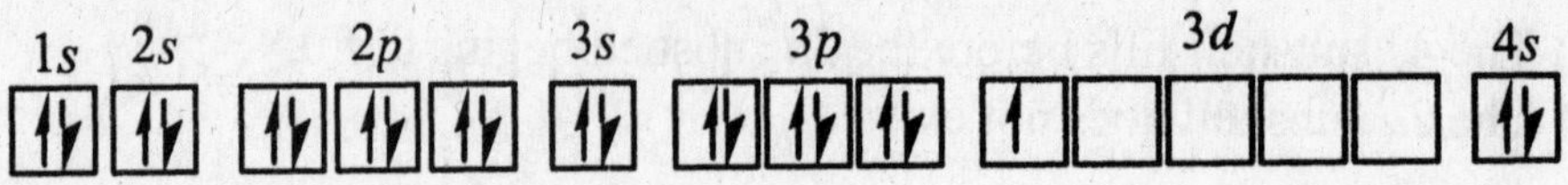

33. Re [Xe] $4f^{14}5d^56s^2$ Re is in the 6th period and Group 7B, and it is the fifth 5*d* transition element.

[Xe] 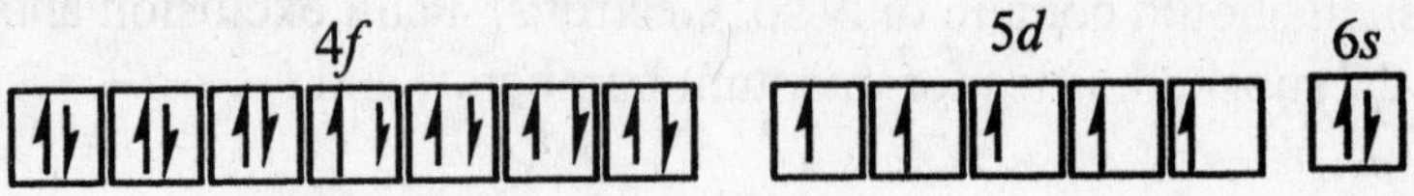

35.

(a) Cl^- [Ne] 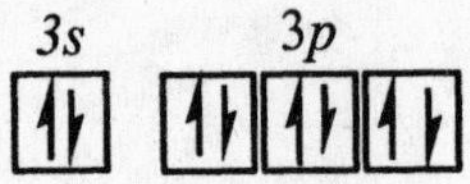

or [Ar]

(b) Zn^{2+} [Ar] 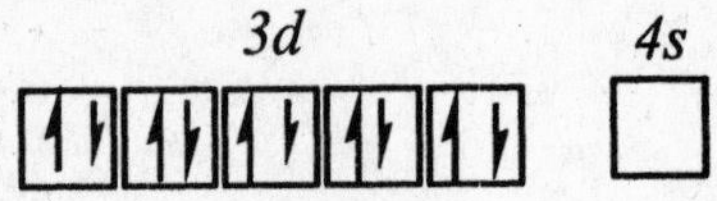

(c) Pb^{2+} [Xe] 

(d) Mg^{2+} [He] 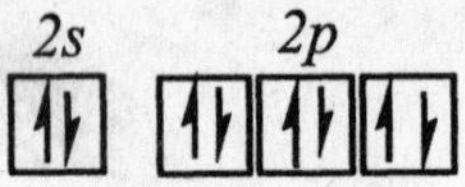

or [Ne]

37. $1s^22s^22p^63s^23p^63d^7$
(a) Mn - $1s^22s^22p^63s^23p^64s^23d^5$
(b) Co^{3+} - $1s^22s^22p^63s^23p^63d^6$
(c) Ni^{2+} - $1s^22s^22p^63s^23p^63d^8$
(d) Cu^{2+} - $1s^22s^22p^63s^23p^63d^9$
None are correct.

39. As [Ar] $3d^{10}4s^24p^3$
As^{3+} is the loss of three *p* electrons and As^{5+} is the loss of two *s* and three *p* electrons.

41.

	period	group
(a)	2nd	8A
(b)	3rd	4A
(c)	3rd	1A
(d)	2nd	2A
(e)	2nd	5A
(f)	3rd	3A

43. (a) 5 ($6s^2 6p^3$)
(b) 32 ($4s^2 4p^6 4d^{10} 4f^{14}$)
(c) 5 (nitrogen, phosphorus, arsenic, antimony, bismuth)
(d) 2
(e) 10 (yttrium, zirconium, niobium, molybdenum, technetium, ruthenium, rhodium, palladium, silver, cadmium)

45. (a) Sr $[Kr]\ 5s^2$ all paired, diamagnetic
(b) F $[He]\ 2s^2 2p_x^2 2p_y^2 2p_z^1$ odd atomic #, 1 unpaired e^-, para
(c) V^{5+} [Ar] all paired, dia
(d) Al $[Ne]\ 3s^2 3p_x^1$ odd atomic #, 1 unpaired e^-, para
(e) Zn^{2+} $[Ar]\ 3d^{10}$ all paired, dia

47. K, Sc, Ti, V, Cr, Mn, Fe, Co, Ni, Cu, Ga, Ge, As, Se, Br

49. (a) Al is to the left of S and is larger.
(b)

	Ca^{2+}	Cl^-	$Ca^{2+} < Cl^-$
e^-	18	18	
p^+	20	17	

(c) Sn < Ba Ba has one more principal level of electrons.
(d) Na^+ < K K has one more principal level of electrons than Na and Na is larger than Na^+.

51. (a) B < Al < Mg < K
B has one less shell of electrons than Al and Mg. K has one more shell. Mg is to the left of Al.
(b) Cl < P < Br < Br^-
Cl has one less shell of electrons than Br and is to the right of phosphorus. The negative ion is larger than the atoms.

53. The lower left-hand corner of the periodic table will have the largest atoms. Size increases to the left and down the periodic table.

55. (a) Ca is larger than Mg, which in turn is larger than Cl, so Ca is larger than Cl.
(b) K^+ is smaller than Cl^- because it has more protons. F^- is also smaller than Cl^-, so it is impossible to predict which is larger.

57. (a) Ionization energies decrease with increasing size or down the periodic table, so Ba < Ca < Mg.
(b) Ionization energies decrease with increasing size or to the left on the periodic table. Al < P < Cl

(c) Ne is a noble gas with a high *I* (ionization energy). Cl is below (larger than) F, so $I_{Cl} < I_F$. Na is to the left of Cl, so $I_{Na} < I_{Cl}$. The transition elements have *I* slightly larger than Ca, which has *I* larger than Na but less than Cl. Na < Fe < Cl < F < Ne.

59. The ionization energy increases for each electron removed, as positive charge builds up on the ions formed. Removing the 3 outer-shell electrons leaves a stable noble-gas configuration. The 4th electron is very difficult to remove because it must come from the $n = 2$ shell, which is at a much lower energy than the $n = 3$ shell.

61. The 7A atoms have the most negative electron affinities in their periods. An electron to be gained can get closer to the positively charged nucleus of a 7A atom than to other atoms in the period, and so it is gained most readily. The process also produces a very stable noble gas electron configuration.

63. An extra electron is more readily accepted by a Si atom, where it can enter an empty 3*p* orbital producing a stable half-filled subshell. For a P atom, the electron would have to pair up with an existing 3*p* electron because each 3*p* orbital in P is already half-filled.

65. The atomic mass divided by density gives a molar volume, which should be related to the atomic size. Na - 24, Mg - 14.0, Al - 9.99, Si - 12.1, P - 14.1, S - 15.5, Cl - 17.5, Ar - 24.1,K - 45, Ca - 25.9, Sc - 15.0, Cr - 7.23, Co - 6.62, Zn - 9.17, Ga - 11.8, As - 15.9, Br - 19.7, Kr - 29.7, Rb - 55.9, Sr - 34.5. (See next page.)

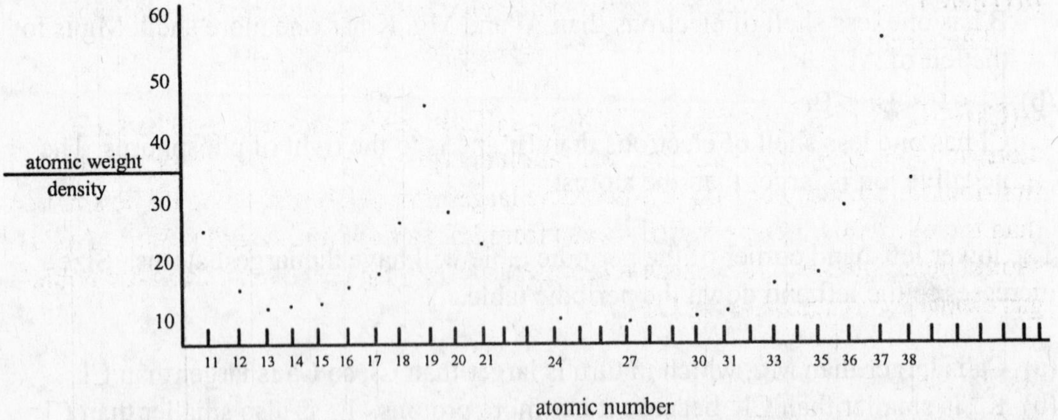

The overall shape of the curve is similar to that for atomic radius versus atomic number. The alkali metals (Group 1A) are at the peaks and there is less variation for the elements in between. The values for the elements 13–17 and 30–36 increase with atomic number rather than decrease.

67. Metallic character increases as electron removal becomes easier—that is, larger radius, lower ionization potential, and location in the periodic table (including designation as a metal, nonmetal, or metalloid). K is to the left of Ca, which is to

the left and below Al. Rb is below K. $I_{Al} < I_{Ca}$. Bi is to the right of Al. Ge is a metalloid. P is a nonmetal but is also to the right and above the other elements. Least P < Ge < Bi < Al < Ca < K < Rb most metallic.

69. (a) Ge (b) S (c) Tl (d) Ne (e) Se
Tl < Ge < Se < S < Ne greatest I_1

71. (a) $Cl_2(g) + 2\ Br^-(aq) \rightarrow 2\ Cl^-(aq) + Br_2(l)$
(b) $I_2(s) + F^-(aq) \rightarrow$ no reaction
(c) $Br_2(l) + 2\ I^-(aq) \rightarrow 2\ Br^-(aq) + I_2(s)$

73. (a) $N_2O_5(s) + H_2O(l) \rightarrow 2\ HNO_3(aq)$
(b) $MgO(s) + 2\ CH_3COOH(aq) \rightarrow Mg(CH_3COO)_2(aq) + H_2O(l)$
(c) $Li_2O(s) + H_2O(l) \rightarrow 2\ LiOH(aq)$

75.

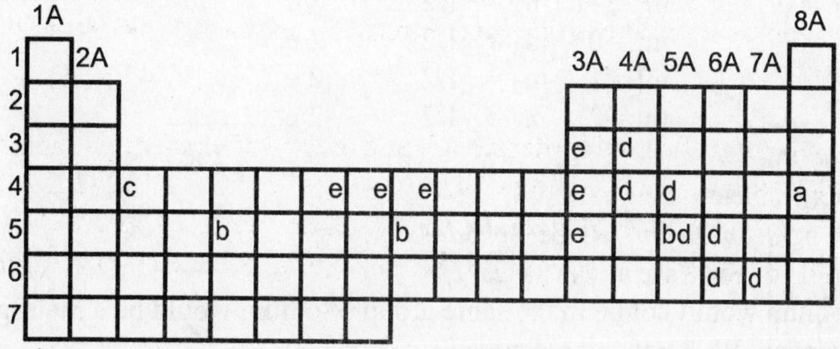

(d)There is some disagreement about which elements are considered metalloids. The ones noted here are often those designated.

78. $I_1(Cs) < I_1(B) < I_2(Sr) < I_2(In) < I_2(Xe) < I_3(Ca)$. A Cs atom is larger than a B atom and has a lower I_1. Second-ionization energies of atoms are larger than their first-ionization energies. $I_2(Sr)$ should be larger than $I_1(B)$, but it should be smaller than the other I_2 values, which increase from left to right in the fifth period. $I_3(Ca)$ should be greatest of all because it represents removing an electron from the noble-gas electron configuration of Ca^{2+}.

81. (a) 37 electrons

n = 1	l = 0	$m_l = 0$	$m_s = -1/2, 0, +1/2$	3 e⁻	
n = 2	l = 0	$m_l = 0$	$m_s = -1/2, 0, +1/2$	3 e⁻	3 e⁻
	l = 1	$m_l = -1$	$m_s = -1/2, 0, +1/2$	3 e⁻	
		$m_l = 0$	$m_s = -1/2, 0, +1/2$	3 e⁻	
		$m_l = 1$	$m_s = -1/2, 0, +1/2$	3 e⁻	
n = 3	l = 0	$m_l = 0$	$m_s = -1/2, 0, +1/2$	3 e⁻	15 e⁻
	l = 1	$m_l = -1$	$m_s = -1/2, 0, +1/2$	3 e⁻	
		$m_l = 0$	$m_s = -1/2, 0, +1/2$	3 e⁻	
		$m_l = 1$	$m_s = -1/2, 0, +1/2$	3 e⁻	

	$l = 2$	$m_l = -2$	$m_s = -1/2,\ 0,\ +1/2$	3 e⁻	
		$m_l = -1$	$m_s = -1/2,\ 0,\ +1/2$	3 e⁻	
		$m_l = 0$	$m_s = -1/2,\ 0,\ +1/2$	3 e⁻___	
		$m_l = 1$	$m_s = -1/2,\ 0,\ +1/2$	3 e⁻	36 e⁻
		$m_l = 2$	$m_s = -1/2,\ 0,\ +1/2$	3 e⁻	37–39 e⁻
$n = 4$	$l = 0$	$m_l = 0$	$m_s = -1/2,\ 0,\ +1/2$	3 e⁻	

Rb = $1s^3 2s^3 2p^9 3s^3 3p^9 3d^7 4s^3$

(b) $n = 1$	$l = 0$	$m_l = 0$	$m_s = \pm 1/2$	2 e⁻ ______	
	$l = 1$	$m_l = -1$	$m_s = \pm 1/2$	2 e⁻	2e⁻
		$m_l = 0$	$m_s = \pm 1/2$	2 e⁻	
		$m_l = 1$	$m_s = \pm 1/2$	2 e⁻ ______	
$n = 2$	$l = 0$	$m_l = 0$	$m_s = \pm 1/2$	2 e⁻ ________	8e⁻
	$l = 1$	$m_l = -1$	$m_s = \pm 1/2$	2 e⁻	10e⁻
		$m_l = 0$	$m_s = \pm 1/2$	2 e⁻	
		$m_l = 1$	$m_s = \pm 1/2$	2 e⁻ _____	
	$l = 2$	$m_l = 2$	$m_s = \pm 1/2$	2 e⁻	16 e⁻
		$m_l = -1$	$m_s = \pm 1/2$	2 e⁻	
		$m_l = 0$	$m_s = \pm 1/2$	2 e⁻	
		$m_l = 1$	$m_s = \pm 1/2$	2 e⁻	
		$m_l = 2$	$m_s = \pm 1/2$	2 e⁻ _____	
					26e⁻

etc., etc., etc.
Rb = $1s^2 1p^6 2s^2 2p^6 2d^{10} 3s^2 3p^6 3d^1 4s^2$
Na 11e⁻ for Case a, Na $1s^3 2s^3 2p^5$
Sodium would not be in the same group. Sodium would be a main-group element, Rb a transition element.
For Case b, Na $1s^2 1p^6 2s^2 2p^1$
Na and Rb would not be in the same group. Na would be a main-group element, Rb a transition element.

82. The level is not the same because of the greater attraction of two protons instead of one, and the repulsion of the two electrons.

$$E_1 = \frac{-2.179\times10^{-18}\ \text{J}\,(2)^2}{1^2} \times \frac{6.022\times10^{23}\ \text{e}^-}{\text{mol e}^-} \times \frac{\text{kJ}}{10^3\ \text{J}} = \frac{-5.25\times10^3\ \text{kJ}}{\text{mol}}$$

This calculation is for the level of 1*s* in a He^+ ion. The -2.37×10^3 kJ/mol value is for the two electrons in a He atom.

85. (a) $E_1 = \dfrac{-2.179\times10^{-18}\ \text{J}}{1^2} \times \dfrac{6.022\times10^{23}\ \text{e}^-}{\text{mol e}^-} \times \dfrac{\text{kJ}}{10^3\ \text{J}} = -1.312\times10^3\ \text{kJ/mol}$

$I_1 = E_\infty - E_1 = 1.312\times10^3$ kJ/mol

(b) The shortest wavelength is the largest *E*.

The shortest wavelength in the Balmer series is that associated with the $\lambda_2 - \lambda_\infty$ emisiion. The longest wavelength in the Lyman series is that associated with the $\lambda_1 - \lambda_2$ emmision.

In energy terms, $I_1 = (E_\infty - E_2) + (E_2 - E_1)$. $I_1 = E_\infty - E_1$, which is the same value calculated above, because $E_\infty = 0$.

$$I_1 = -B \times \left(\frac{1}{\infty} - \frac{1}{2^2}\right) + -B \times \left(\frac{1}{2^2} - \frac{1}{1^2}\right)$$

$$I_1 = -2.179 \times 10^{-18}\ \text{J} \times \left(0 - \frac{1}{4}\right) - 2.179 \times 10^{-18}\ \text{J}\left(\frac{1}{4} - 1\right)$$

$$I_1 = 0.545 \times 10^{-18}\ \text{J} + 1.634 \times 10^{-18}\ \text{J}$$

$$I_1 = 2.179 \times 10^{-18}\ \text{J} \times \frac{6.022 \times 10^{23}\ \text{e}^-}{\text{mol e}^-} \times \frac{\text{kJ}}{10^3\ \text{J}}$$

$$I_1 = 1.312 \times 10^3\ \text{kJ/mol}$$

87. $Cl(g) + e^- \rightarrow Cl^-(g)$ ΔH = electron affinity = –349 kJ is the second step.

For the overall process, $\Delta H = \left(\frac{1}{2} \times 242.8\ \text{kJ}\right) - 349\ \text{kJ} = -228$ kJ./mol Cl^-. The overall process is exothermic.

Apply Your Knowledge

92. (a) In would have been put where Sr is. In is 82.5%.

$0.175 \times X = 16.00$

$X = 91.4$ g/mol for InO

0.825×91.4 g/mol $= 75.4$ g/mol for In

With the atomic mass of 75.4 g/mol, In would have been between As and Se, but the 2+ charge is not consistent with that placement.

$$\text{Atomic weight} = \frac{6.4}{0.055} = 116$$

$$17.5\ \text{g O} \times \frac{\text{mole}}{16.00\ \text{g O}} = 1.09\ \text{mol O} \times \frac{\text{mol O}}{0.711\ \text{mol In}} = \frac{1.54\ \text{mol O}}{\text{mol In}}$$

$$82.5\ \text{g In} \times \frac{\text{mol In}}{116\ \text{g In}} = 0.711\ \text{mol} \times \frac{\text{mol In}}{0.711\ \text{mol In}} = \frac{1\ \text{mol In}}{\text{mol In}}$$

$In_{1\times2}O_{1.5\times2} \Rightarrow In_2O_3$

A 3+ ion of atomic mass 116 goes where In is now.

Mendelev would have put it where it is now.

(b) $\text{Atomic mass} = \frac{6.4}{0.0276} = 232$ g/mol

$$62.66\ \text{g U} \times \frac{\text{mol}}{240\ \text{g}} = \frac{0.261\ \text{mol U}}{} \qquad \frac{0.261\ \text{mol U}}{0.261\ \text{mol U}} = \frac{1\ \text{mol U}}{1\ \text{mol U}}$$

$$37.34\ \text{g Cl} \times \frac{\text{mol}}{35.45\ \text{g}} = \frac{1.05\ \text{mol Cl}}{} \qquad \frac{1.05\ \text{mol Cl}}{0.261\ \text{mol U}} = \frac{4.04\ \text{mol Cl}}{1\ \text{mol U}}$$

UCl_4
The Dulong and Petit molar mass is 232 g/mol, and the number of Cl atoms per U atom is very close to an integer. U^{4+} is consistent with losing the *d* and *f* electrons, that is, back to the *n*+2 electron configuration.

94. (a)

Element	Z	ν s^{-1}	$\sqrt{\nu}$
Ca	20	8.901×10^{17}	9.435×10^{8}
Ti	22	1.087×10^{18}	1.043×10^{9}
V	23	1.190×10^{18}	1.091×10^{9}
Cr	24	1.303×10^{18}	1.141×10^{9}
Mn	25	1.420×10^{18}	1.192×10^{9}
Fe	26	1.541×10^{18}	1.241×10^{9}
Co	27	1.667×10^{18}	1.291×10^{9}
Ni	28	1.804×10^{18}	1.343×10^{9}
Cu	29	1.935×10^{18}	1.391×10^{9}
Zn	30	2.075×10^{18}	1.440×10^{9}

$$\nu = \frac{C}{\lambda} = \frac{2.9979\times10^{8}\ \text{m/s}}{3.368\times10^{-8}\ \text{cm}} \times \frac{\text{cm}}{10^{-2}\ \text{m}} = 8.901 \times 10^{-17}\ \text{s}^{-1}$$

$$\sqrt{\nu} = \sqrt{A} \times Z - \sqrt{A}\ b$$

plot Z on the horizontal axis and $\sqrt{\nu}$ on the vertical axis.

slope $= 5.0 \times 10^{7} = \sqrt{A}$

$A = 2.5 \times 10^{15}$

Intercept $= -5.3 \times 10^{7}$

$b = 1.1$

(b) $$\sqrt{A} = \frac{1.241\times10^{9} - 1.141\times10^{9}}{26-24}$$

$\sqrt{A} = 5.00 \times 10^{7}$

$A = 2.50 \times 10^{15}$

$\sqrt{\nu} = \sqrt{A} \times Z - \sqrt{A}\ b$ thus $\sqrt{A}\ b = \sqrt{A} \times Z - \sqrt{\nu}$

$$b = \frac{1.20\times10^{9} - 1.142\times10^{9}}{5.00\times10^{7}} = 1.2$$

The A values are the same, but the b values differ slightly. The long distance on the horizontal axis probably limited the accuracy of the graphical value.

(c) $\nu = A(Z-b)^2 = 2.50 \times 10^{15} \times (43.00 - 1.16)^2 = 4.38 \times 10^{18}\ s^{-1}$

$$\lambda = \frac{C}{\nu} = \frac{3.00\times10^{8}\ \text{m/s}}{4.38\times10^{18}\ \text{s}^{-1}} \times \frac{\text{cm}}{10^{-2}\ \text{m}} = 6.85 \times 10^{-9}\ \text{cm}$$

Chapter 9

Chemical Bonds

Exercises

9.1A

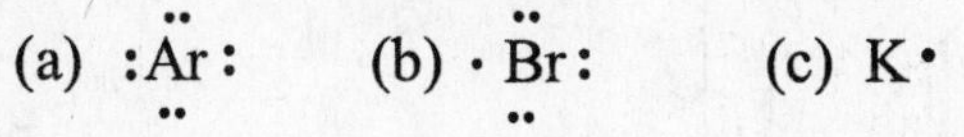

(a) $:\overset{\cdot\cdot}{\underset{\cdot\cdot}{Ar}}:$ (b) $\cdot\overset{\cdot\cdot}{\underset{\cdot\cdot}{Br}}:$ (c) $K\cdot$

9.1B

(a) $:\overset{\cdot}{\underset{\cdot}{As}}\cdot$ (b) $Rb\cdot$ (c) $\cdot\overset{\cdot\cdot}{\underset{\cdot\cdot}{Te}}\cdot$

9.2A

$\cdot Ba\cdot + 2:\overset{\cdot}{\underset{\cdot\cdot}{I}}: \longrightarrow :\overset{\cdot\cdot}{\underset{\cdot\cdot}{I}}: Ba^{2+} :\overset{\cdot\cdot}{\underset{\cdot\cdot}{I}}:^-$

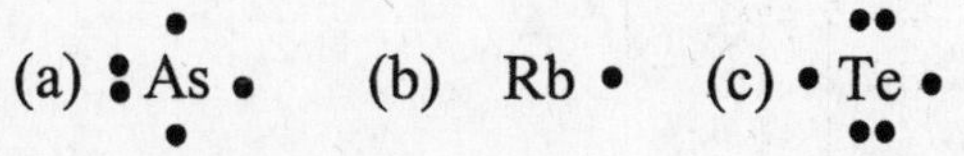

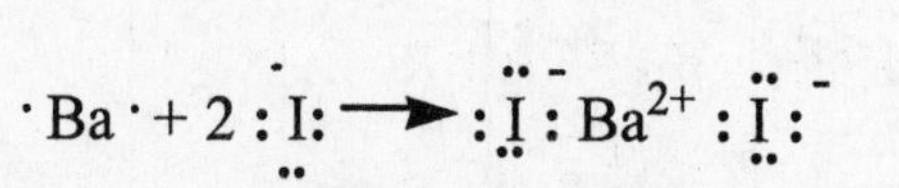

barium iodide BaI_2

9.2B

$\cdot Al\cdot$ $\cdot\overset{\cdot\cdot}{O}\cdot$

        $\cdot\overset{\cdot\cdot}{O}\cdot \longrightarrow$ $[Al]^{3+}$ $\left[:\overset{\cdot\cdot}{\underset{\cdot\cdot}{O}}:\right]^{2-}$

$\cdot Al\cdot$ $\cdot\overset{\cdot\cdot}{O}\cdot$ $[Al]^{3+}$ $\left[:\overset{\cdot\cdot}{\underset{\cdot\cdot}{O}}:\right]^{2-}$

                                     $\left[:\overset{\cdot\cdot}{\underset{\cdot\cdot}{O}}:\right]^{2-}$

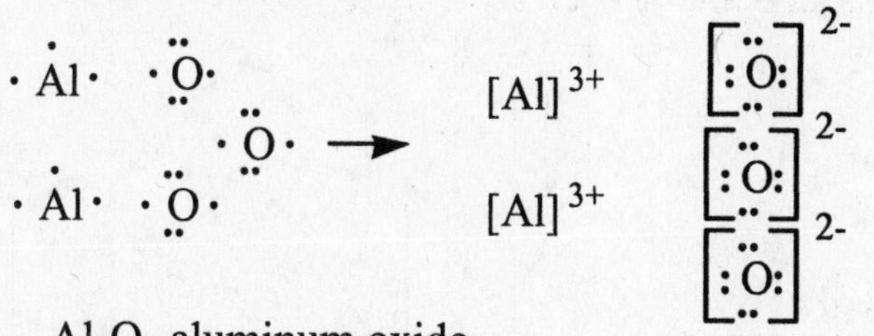

Al_2O_3 aluminum oxide

9.3A $Li(s) + \frac{1}{2} F_2(g) \rightarrow LiF(s)$ $\Delta H^\circ_f = ?$

This is the sum of the steps 1–5 in the Born-Haber cycle.

$Li(s) \rightarrow Li(g)$	$\Delta H_1 = 159 \text{ kJ}$
$\frac{1}{2} F_2(g) \rightarrow F(g)$	$\Delta H_2 = \frac{1}{2}(159 \text{ kJ})$
$Li(g) \rightarrow Li^+(g) + e^-$	$\Delta H_3 = 520 \text{ kJ}$
$F(g) + e^- \rightarrow F^-(g)$	$\Delta H_4 = -328 \text{ kJ}$
$\underline{Li^+(g) + F^-(g) \rightarrow LiF(s)}$	$\underline{\Delta H_5 = -1047}$
$Li(s) + \frac{1}{2} F_2(g) \rightarrow LiF(s)$	$\Delta H^\circ_f =$

$\Delta H^\circ_f = (159 + 80 + 520 + (-328) - 1047) \text{ kJ} = -616 \text{ kJ}$
$\Delta H^\circ_f = -616 \text{ kJ/mol } LiF(s)$

9.3B $Li^+(g) + Cl^-(g) \rightarrow LiCl(s)$ $\Delta H_{LE} = ?$

       $Li(s) \rightarrow Li(g)$ $\Delta H_1 = 159 \text{ kJ}$

$$\frac{1}{2} \, Cl_2(g) \rightarrow Cl(g) \qquad\qquad \Delta H_2 = \frac{1}{2}(243 \text{ kJ})$$

$$Li(g) \rightarrow Li^+(g) + e^- \qquad\qquad \Delta H_3 = 520 \text{ kJ}$$

$$Cl(g) + e^- \rightarrow Cl^-(g) \qquad\qquad \Delta H_4 = -349 \text{ kJ}$$

$$\underline{Li^+(g) + Cl^-(g) \rightarrow LiCl(s)} \qquad\qquad \underline{\Delta H_5 = \text{lattice energy}}$$

$$Li(s) + \frac{1}{2} \, Cl_2(g) \rightarrow LiCl(s) \qquad\qquad \Delta H^\circ_f = -409 \text{ kJ}$$

$\Delta H^\circ_f = -409 \text{ kJ} = (159 + 122 + 520 + (-349)) \text{ kJ} + \text{lattice energy}$

lattice energy $= (-409 - 161 - 122 - 520 + 349) \text{ kJ} = -861 \text{ kJ}$

The lattice energy is -861 kJ/mol LiCl (s)

9.4A (a) Ba < Ca < Be   (b) Ga < Ge < Se   (c) Te < S < Cl

9.4B Cl > S > B > Fe > Sc > Na > Rb

9.5A
C - Cl	C - H	C - Mg	C - O	C - S
0.5	0.4	1.3	1.0	-

C - S < C - H < C - Cl < C - O < C – Mg

9.5B (a) Si and F form the bond with the most ionic character.

(b) S and C form the bond with the least ionic character. Both elements have an eletronegativity of 2.5.

9.6A
```
   ..  ..
  H-N-N-H
    | |
    H H
```

9.6B
```
    H  H
    |  |   ..
H - C - C - Cl :
    |  |   ..
    H  H
```

9.7A
```
 ..        ..
 S = C = O
 ..        ..
```

9.7B
```
 ••   ••   ••
 O==N—Cl :
 ••   ••   ••
```

9.8A
```
        ┌           ┐ -
        |    ••      |
        |    O       |
        |  •• •• ••   |
        |  O Cl O    |
        |  •• •• ••   |
        |    O       |
        |    ••      |
        └           ┘
```

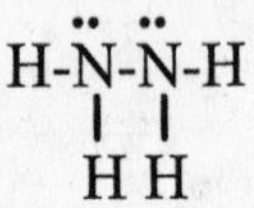

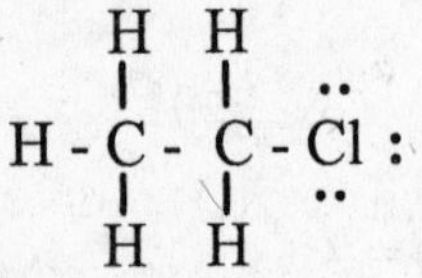

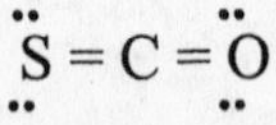

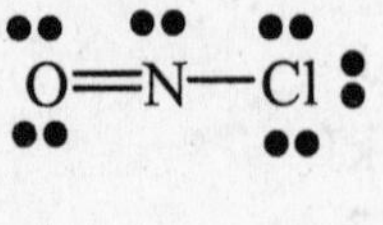

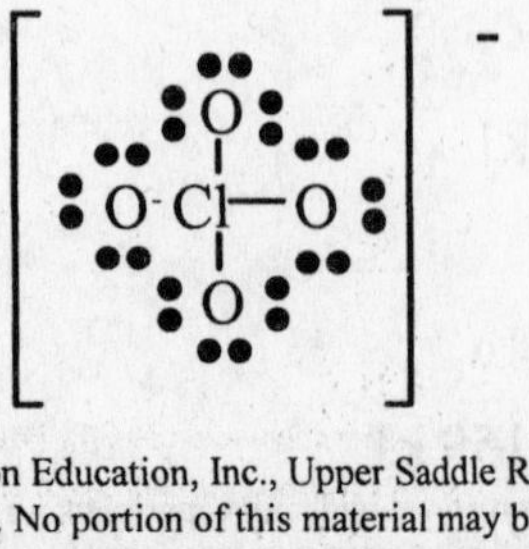

9.8B

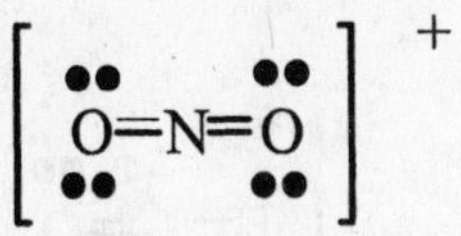

9.9A

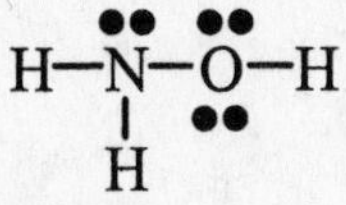

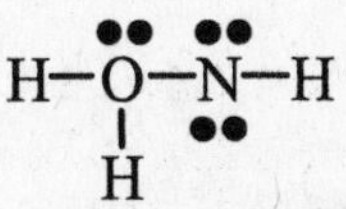

	FC	|FC|
H	1 − 1 = 0	0
H	1 − 1 = 0	0
H	1 − 1 = 0	0
N	5 − 5 = 0	0
O	6 − 6 = 0	0
		0

	FC	|FC|
H	1 − 1 = 0	0
H	1 − 1 = 0	0
H	1 − 1 = 0	0
N	5 − 6 = −1	1
O	6 − 5 = 1	1
		2

The structure on the left is the more plausible.

9.9B 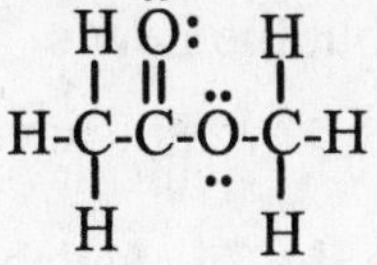

9.10A

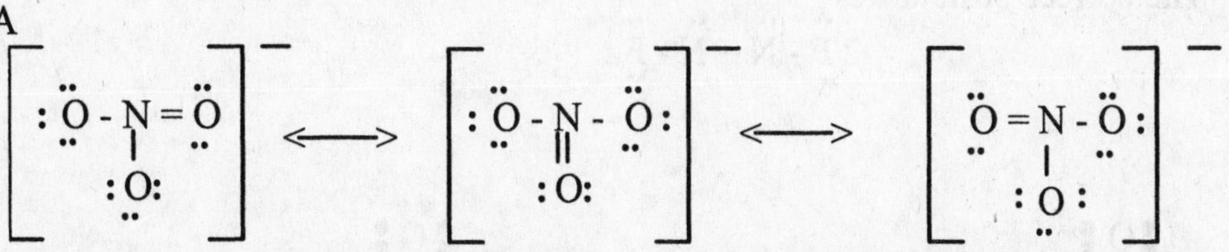

This resonance hybrid, which involves equal contributions from these three equivalent resonance structures, has N-to-O bonds with bond lengths and bond energies intermediate between single and double bonds.

9.10B (a)

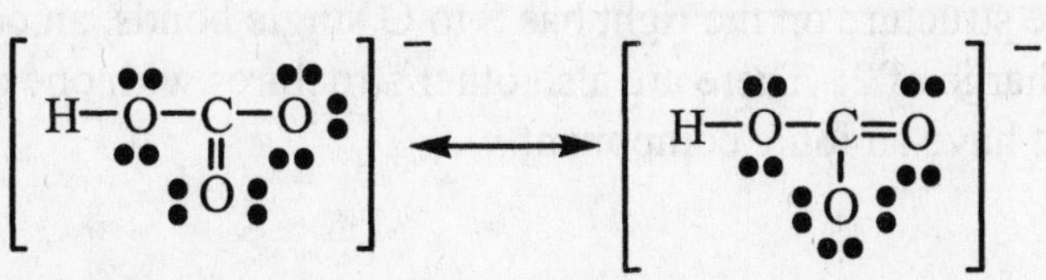

(b)

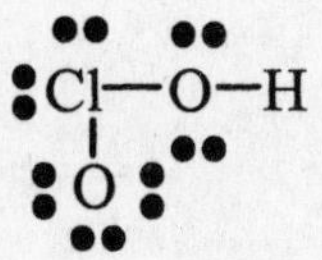

(c) [: C ≡ N:]⁻

(d)   (e)

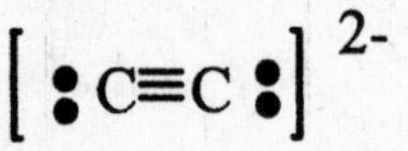

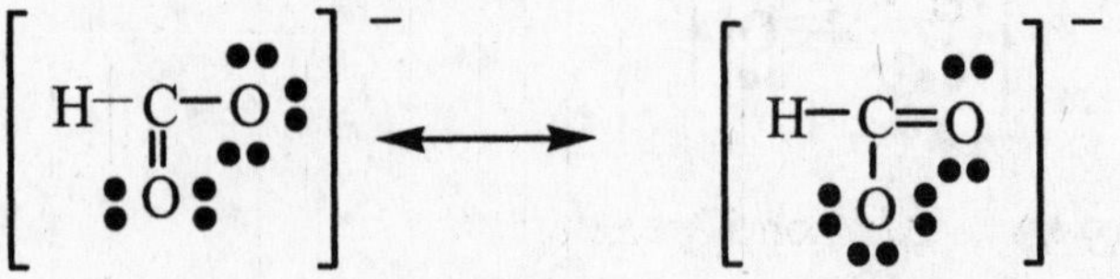

$$\left[\; :C \equiv C: \; \right]^{2-} \qquad \left[H - \overset{..}{\underset{..}{O}}{-}\overset{..}{\underset{..}{O}}{:} \right]^{-} \longleftrightarrow \left[H - C = O \atop \overset{..}{\underset{..}{O}} \right]^{-}$$

9.11A $\;:\overset{..}{\underset{..}{Cl}} - P - \overset{..}{\underset{..}{Cl}}:$
                $\;\;\;\;\;\;\;|$
                $\;\;\;\;\;:\overset{..}{\underset{..}{Cl}}:$

9.11B (a)   (b)

                                   $:\overset{..}{F}:$
                                   $\;\;|$
$\;:\overset{..}{\underset{..}{F}} - \overset{..}{\underset{..}{Cl}} - \overset{..}{\underset{..}{F}}:$ $:\overset{..}{\underset{..}{F}} - S - \overset{..}{\underset{..}{F}}:$
$\;\;\;\;\;\;|$ $\;\;\;\;\;|$
$\;\;\;\;:\overset{..}{F}:$ $\;\;\;:\overset{..}{F}:$

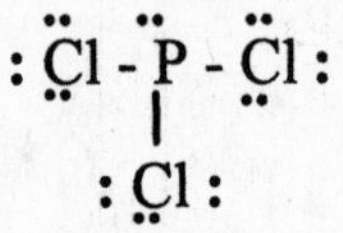

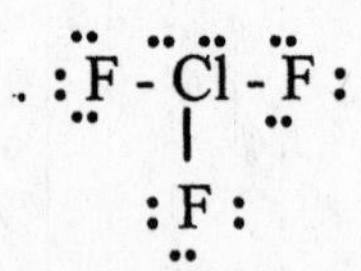

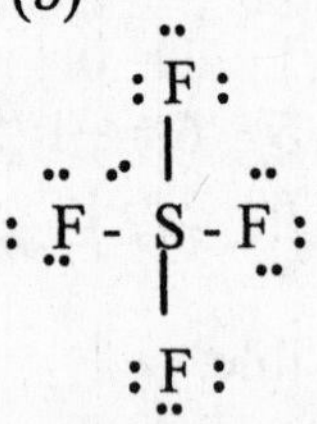

9.12A (a) Incorrect. The molecule ClO_2 has 19 valence electrons, but the Lewis structure shows 20, one too many.

(b) Correct.

(c) Incorrect. The structure shown has 26 electrons rather than the 24 available. The correct structure is

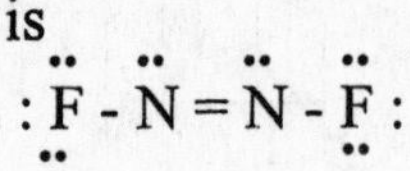

$$:\overset{..}{F} - \overset{..}{N} = \overset{..}{N} - \overset{..}{F}:$$

9.12B

$$H - O - \overset{O}{\underset{O}{S}} - O - H \qquad H - O - \overset{O}{\underset{O}{S}} - O - H$$

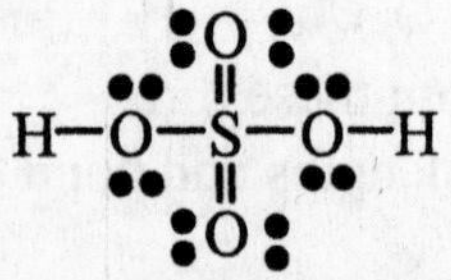

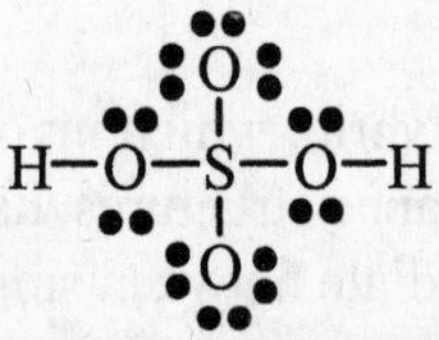

The structure on the left has S to O double bonds; S has an expanded shell and a formal charge of 0. The structure on the right has S to O single bonds, an octet on S, and S has a formal charge of 2. There are also other structures with one double bond and structures that have an ionic component.

9.13A

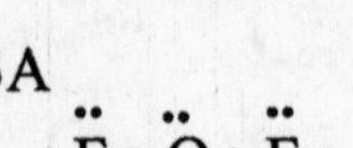

$:\overset{..}{\underset{..}{F}} : \overset{..}{\underset{..}{O}} : \overset{..}{\underset{..}{F}} :$ B.L. $= \dfrac{1}{2}(145 \text{ pm}) + \dfrac{1}{2}(143 \text{ pm}) = 144 \text{ pm}$

9.13B

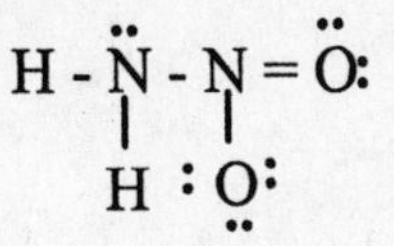

H - N̈ - N = Ö: N-to-N bond length = 145
| |
H   :Ö:

9.14A

ΔH bond breakage		ΔH bond formation	
1 mol CH	414 kJ	1 mol C - Cl	-339 kJ
1 mol Cl_2	243 kJ	1 mol H - Cl	-431 kJ
sum	657 kJ	sum	-770 kJ

$\Delta H = \Delta H_{breakage} + \Delta H_{formation}$
$\Delta H = 657$ kJ $+ (-770$ kJ$) = -113$ kJ

9.14B

O=N—F: N=O—F:

FC = 0   FC = 2

more plausible

½ N_2 + ½ O_2 + ½ F_2 → ONF

ΔH = ½ BE(O-O) + ½ BE(N-N) + ½ BE(F-F) − (Be(N-O) + BE(N-F))

−66.5 kJ = 249 kJ + 473 kJ + 80 kJ − (590 kJ + BE(N-F))

BE(N-F) = 279 kJ

Review Questions

3.

    (a) Na ·   (b) :Ö: (c) · Si · (d) :B̈r : (e) · **Ca** · (f) · Äs ·

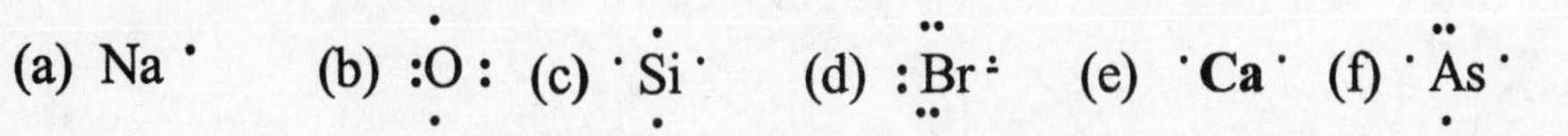

5.

(a) · Ca · + 2   · Br :   ⟶   [:B̈r:]⁻ Ca²⁺ [:B̈r:]⁻

(b)· Ba · + · Ö ·   ⟶ Ba²⁺ [:Ö:]²⁻

(c) 2 · Äl · + 3 : S̈ :   ⟶   2 [Al]³⁺ + 3 [:S̈:]²⁻

6.

or : Ï— Ï : where the dash line

represents the bonding pair

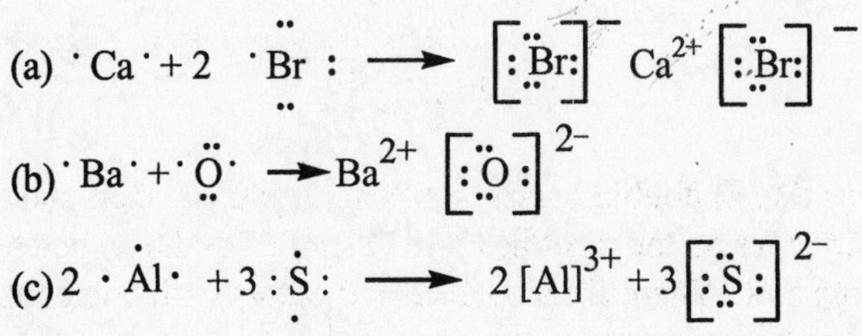

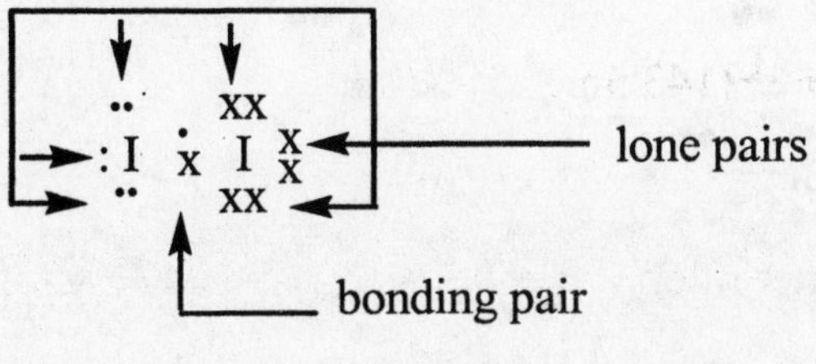

7. δ+ H $\overset{\times}{\cdot}$F : δ- or δ+ H−F : δ−

9. (a) N is more electronegative than S.
(b) Cl is more electronegative than B.
(c) F is more electronegative than As.
(d) O is more electronegative than S.

10. (a) F is more electronegative than Br.
(b) Br is more electronegative than Se.
(c) Cl is more electronegative than As.
(d) N is more electronegative than H.

11. (a) ionic (b) polar covalent
(c) ionic (d) polar covalent
(e) ionic (f) ionic
(g) nonpolar covalent (h) nonpolar covalent
(i) polar covalent

12. (a) 1 (b) 4 (c) 2 (d) 1 (e) 3 (f) 1

13. (b) cannot have three bonds.
(c) C needs four bonds.
(a) is best.

14.

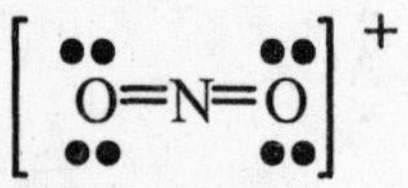

O 6 − 6 = 0
O 6 − 6 = 0
N 5 − 4 = 1
(c) is the answer.

15. (a)

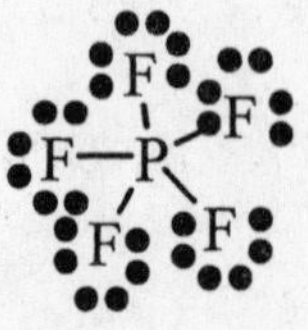

(b)

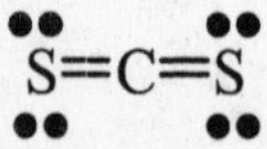

(c)

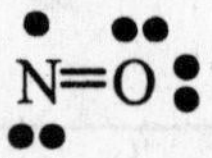

(d)

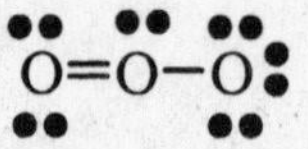

(e)

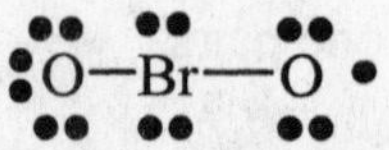

(a), (c), and (e) are exceptions to the octet rule.

16. (a)

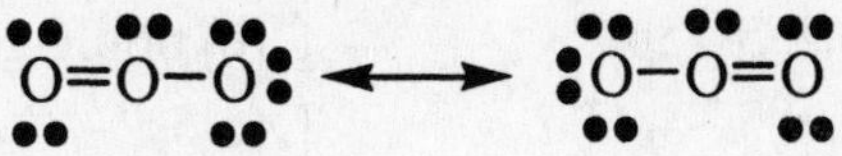

(b)

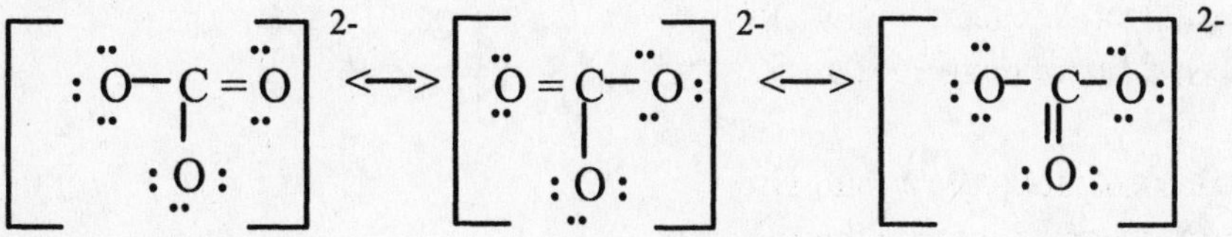

(c)

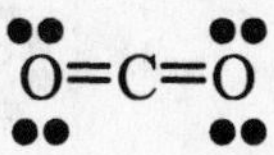

(d)

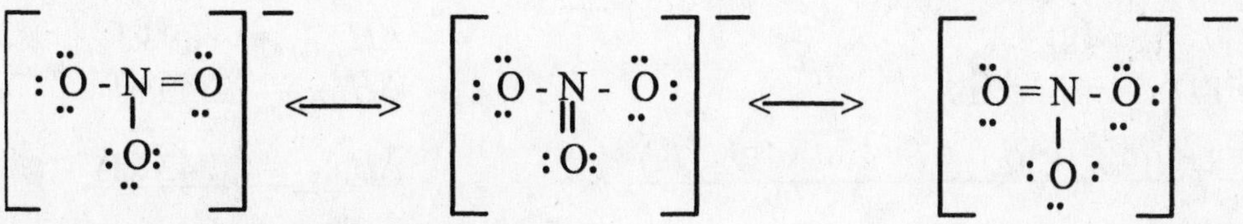

(a), (b), and (d) exhibit resonance.

17. (c) $[:C\equiv N:]^-$

18. (b) N_2 < (a) O_2 < (d) ICl < (c) I_2 N_2 has a triple bond, O_2 has a double bond, and I is larger than Cl.

19. (b) N_2

20. $\Delta H = \text{BE(products)} - \text{BE(reactants)}$

$$\Delta H = 2\text{ mol} \times \frac{3\text{ bonds}}{\text{mol}} \times \frac{389\text{ kJ}}{\text{bond}} + 2\text{ mol} \times \frac{2\text{ bonds}}{\text{mol}} \times \frac{464\text{ kJ}}{\text{bond}}$$

$$- 2\text{ mol} \times \frac{3\text{ bonds}}{\text{mol}} \times \frac{628\text{ kJ}}{\text{bond}} - 5\text{ mol} \times \frac{1\text{ bonds}}{\text{mol}} \times \frac{435\text{ kJ}}{\text{bond}} = 759\text{ kJ}$$

21. Unsaturated hydrocarbons are carbon-hydrogen compounds, which have double or triple bonds.
(a), (b), and (c) are unsaturated.

Problems

23. (a) Mo^{6+} - $1s^22s^22p^63s^23p^63d^{10}4s^24p^6$ [Kr] noble gas
(b) Te^{2-} - $1s^22s^22p^63s^23p^63d^{10}4s^24p^64d^{10}5s^25p^6$ [Xe] noble gas
(c) Ti^{4+} - $1s^22s^22p^63s^23p^6$ [Ar] noble gas

(d) Zn^{2+} - $1s^22s^22p^63s^23p^63d^{10}$ not noble gas
(e) La^{3+} - $1s^22s^22p^63s^23p^63d^{10}4s^24p^64d^{10}5s^25p^6$ [Xe] noble gas
(f) Hg^{2+} - $1s^22s^22p^63s^23p^63d^{10}4s^24p^64d^{10}5s^25p^65d^{10}$ [Xe]$5d^{10}$

25. (a)

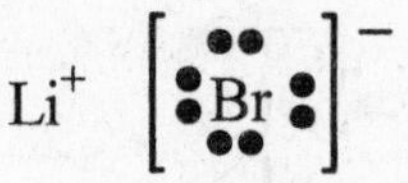

(b)

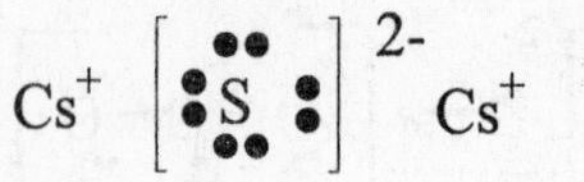

(c)

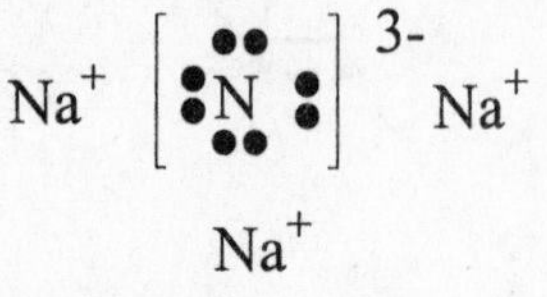

(d)

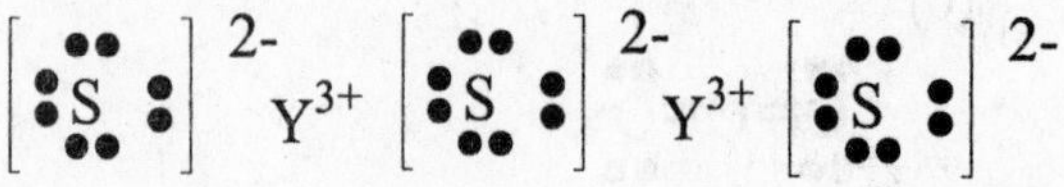

27. Equations must add together to produce the ΔH_f.

$K^+(g) + Cl^-(g) \rightarrow KCl(s)$	$\Delta H_{LE} =$ -701 kJ
$K(g) \rightarrow K^+(g) + e^-$	$\Delta H_{IE} =$ 419 kJ
$K(s) \rightarrow K(g)$	$\Delta H_{sub} =$ 89 kJ
$Cl(g) + e^- \rightarrow Cl^-(g)$	$\Delta H_{EA} =$ -349 kJ
$\frac{1}{2}Cl_2(g) \rightarrow Cl(g)$	$\Delta H_{dis} = \frac{1}{2}(243\ kJ)$
$K(s) + \frac{1}{2}Cl_2(g) \rightarrow KCl(s)$	ΔH_f = - 420 kJ

$[\Delta H_{dis} = \Delta H_f Cl(g) - \Delta H_f \frac{1}{2} Cl_2(g)]$

Appendix C ΔH_f = - 436.7 kJ

29. Equations must add together to produce the lattice energy equation.

$Na^+(g) + I^-(g) \rightarrow NaI(s)$	
$Na^+(g) + e^- \rightarrow Na(g)$	$-\Delta H_{IE}$ = - (496 kJ/mol)
$Na(g) \rightarrow Na(s)$	$-\Delta H_{sub}$ = - (107 kJ/mol)
$Na(s) + \frac{1}{2} I_2(s) \rightarrow NaI(s)$	$-\Delta H_f$ = - 288 kJ/mol
$\frac{1}{2} I_2(g) \rightarrow \frac{1}{2} I_2(s)$	$-\frac{1}{2}\Delta H_{sub} = -\frac{1}{2}$(62 kJ/mol)
$I(g) \rightarrow \frac{1}{2} I_2(g)$	$-\frac{1}{2}\Delta H_{dis} = -\frac{1}{2}$(151 kJ/mol I_2)
$I^-(g) \rightarrow I(g) + e^-$ (electron affinity)	$-\Delta H_{EA}$ = - (- 295 kJ/mol)
$Na^+(g) + I^-(g) \rightarrow NaI(s)$	$\Delta H°$ = lattice energy

$\Delta H°$ = - 496 kJ - 107 kJ - 288 kJ - 31 kJ - 76 kJ + 295

$\Delta H°$ = - 703 kJ

$[\Delta H_{dis} = \Delta H_f\ I(g) - \Delta H_f \frac{1}{2} I_2(g)]$

31. Equations must add together to produce the lattice energy equation.

$Ca^{2+}(g) + 2\ Br^-(g) \rightarrow CaBr_2(s)$

$Ca^{2+}(g) + e^- \rightarrow Ca^+(g)$	$-\Delta H_{IE2}$ =	- (1145 kJ/mol)
$Ca^+(g) + e^- \rightarrow Ca(g)$	$-\Delta H_{IE1}$ =	- (-590 kJ/mol)
$Ca(g) \rightarrow Ca(s)$	$-\Delta H_{sub}$ =	- (178.2 kJ/mol)
$Ca(s) + Br_2(l) \rightarrow CaBr_2(s)$	ΔH_f =	-682.8 kJ/mol
$Br_2(g) \rightarrow Br_2(l)$	ΔH_{vap} =	- (30.9 kJ/mol)
$2\ Br(g) \rightarrow Br_2(g)$	$-2\Delta H_{diss}$ =	-2(96.4 kJ/mol Br)
$2\ Br^-(g) \rightarrow 2\ Br(g) + 2\ e^-$	$-2\Delta H_{EA}$ =	- (325 kJ/mol
$Ca^{2+}(g) + 2\ Br^-(g) \rightarrow CaBr_2(s)$	$-\Delta H_{LE}$ =	-1965 kJ

33. (a)

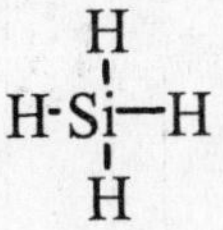

(b)

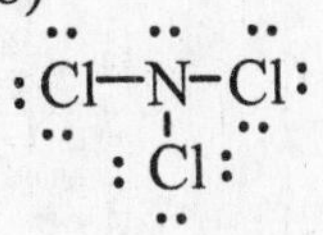

35.

(a)

(b) $H-C\equiv C-H$

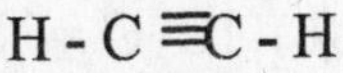

(c)

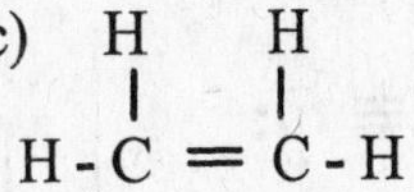

(d)

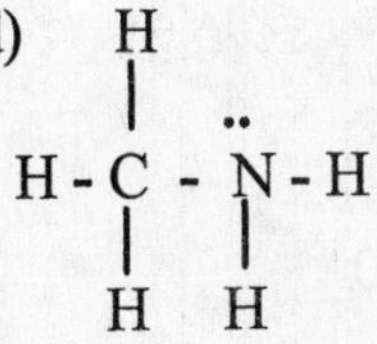

(e)

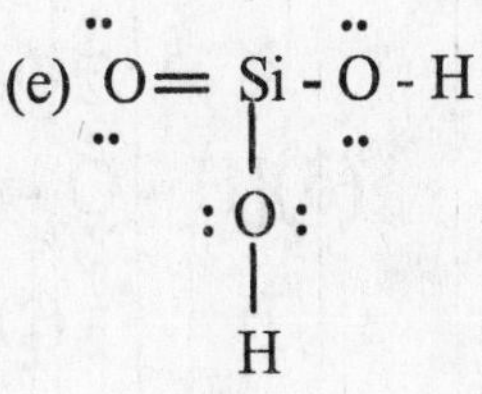

(f) $H-C\equiv N:$

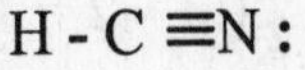

37. (a) MgO should be ionic.

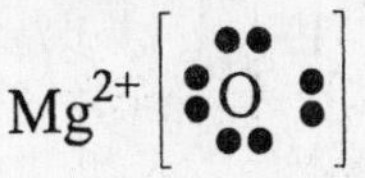

(b) Cl_2O is covalent.

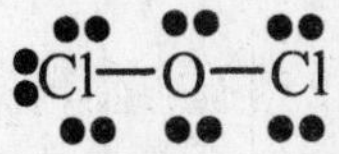

(c) This structure should have only 16 electrons, not 17, with two N-to-O double bonds.

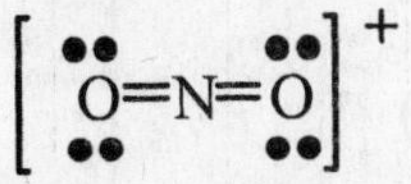

(d) This structure should have 16 electrons, not 14, and a C-to-S double bond.

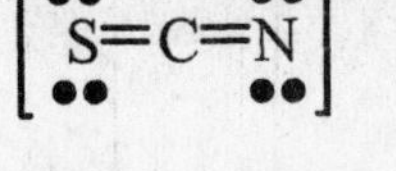

39. (a) K < As < Br
    (b) Cs < Ca < Be
    (c) Pb < Sb < Cl

41.   δ- δ+ δ+ δ- δ+ δ-   δ+ δ-
    (a) H - H < H - C < H - N < H - O < H - F
         2.1   2.1 2.5 2.1 3.0 2.1 3.5 2.1 4.0
                    δ+ δ-   δ+ δ-   δ+ δ-   δ+ δ-
    (b) C - C ≈ C - I < C - Br < C - Cl < C - F
         2.5   2.5 2.5 2.5 2.8 2.5 3.0 2.5 4.0

43. (a) As → P periodic table
    (b) Al → P periodic table
    (c) I ← P This set requires a table.
    (d) Cl ← P periodic table

45.
    (a) : C ≡ O :

    C $4 - \frac{1}{2}(6) - 2 = -1$

    O $6 - \frac{1}{2}(6) - 2 = 1$

    (c) [: N = N = N̈ :]⁻

    N $5 - \frac{1}{2}(4) - 4 = -1$

    N $5 - \frac{1}{2}(8) - 0 = 1$

    N $5 - \frac{1}{2}(4) - 4 = -1$

    (b) H - C = N:

    H $1 - \frac{1}{2}(2) - 0 = 0$

    C $4 - \frac{1}{2}(8) - 0 = 0$

    N $5 - \frac{1}{2}(6) - 2 = 0$

    (d) [H - Ö - Ö:]⁻

    H $1 - \frac{1}{2}(2) - 0 = 0$

    O $6 - \frac{1}{2}(4) - 4 = 0$

    O $6 - \frac{1}{2}(2) - 6 = -1$

47. (b)

    : N≡N-Ö :
      1 2

    $N_1 \; 5 - \frac{1}{2}(6) - 2 = 0$

    $N_2 \; 5 - \frac{1}{2}(8) \quad = 1$

    $O \; 6 - \frac{1}{2}(2) - 6 = -1$

    (a)

    : N̈=N=Ö :
       1 2

    $N_1 \; 5 - \frac{1}{2}(4) - 4 = -1$

    $N_2 \; 5 - \frac{1}{2}(8) - 0 = 1$

    $O \; 6 - \frac{1}{2}(4) - 4 = 0$

    (b) is better, since the negative charge is on the more electronegative atom.

49.

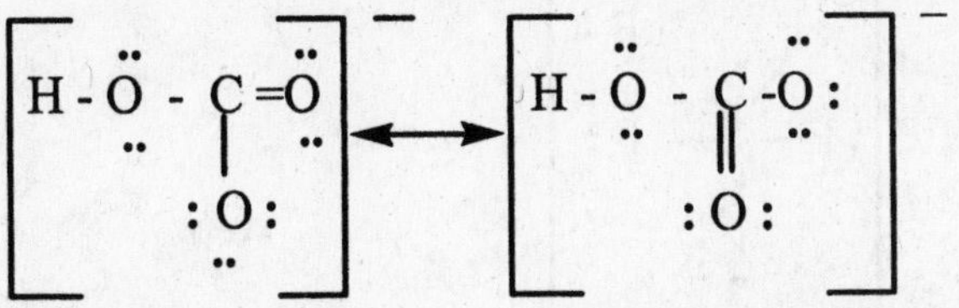

The resonance hybrid is a combination of these two structures, where the bonds of the C atom to the terminal O atoms are intermediate between single and double bonds, and the C atom and the terminal O atoms share a pair of delocalized electrons.

51.

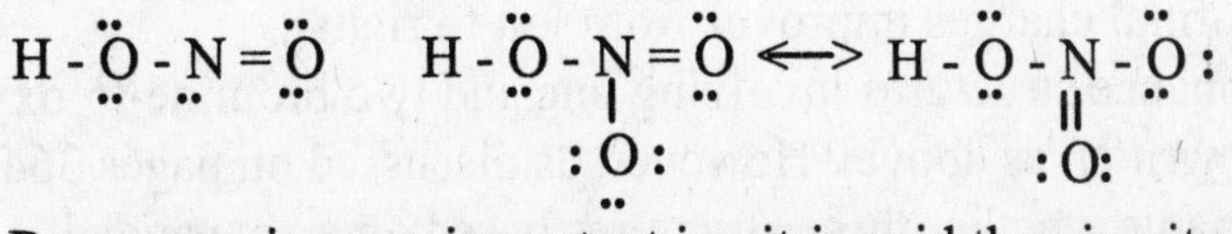

Resonance is more important in nitric acid than in nitrous acid. In nitric acid two oxygens are equivalent. The O-N bonds are between a single and a double bond. A pair of electrons is shared by both oxygens and the nitrogen.

53.

(a) 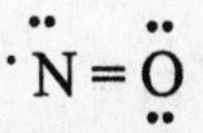 (b) 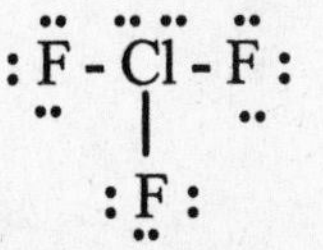 (c) 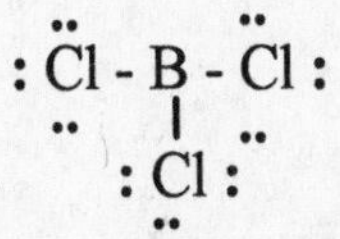 (d) 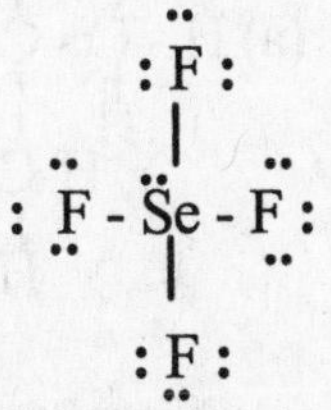

NO is an odd-electron molecule; BCl_3 has an incomplete octet. ClF_3 and SeF_4 have expanded shells.

55.

(a) 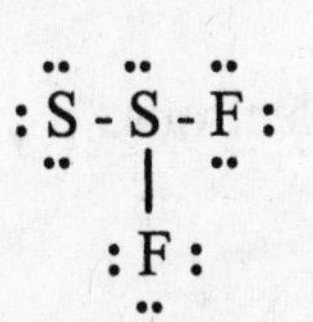 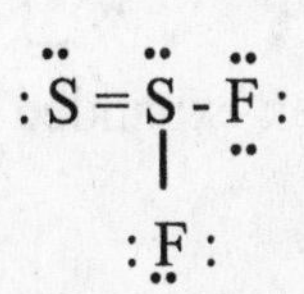

F 7 - 7 = 0	S 6 - 6 = 0
F 7 - 7 = 0	S 6 - 6 = 0
S 6 - 7 = -1	F 7 - 7 = 0
S 6 - 5 = 1	F 7 - 7 = 0

The first structure does not have an expanded valence shell. The second has an expanded valence shell but no formal charges. Both structures are likely to contribute to a resonance hybrid. A molecule bonded F-S-S-F would be an isomer, not a resonance structure.

(b) 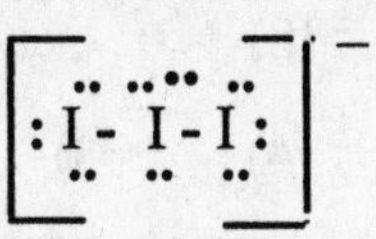 (c) 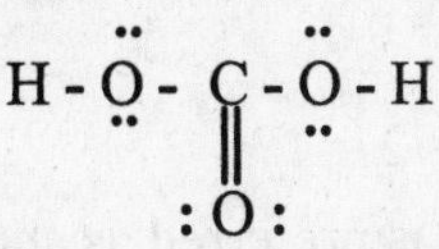

$$\begin{bmatrix} \begin{array}{cc} :\ddot{F}: & \ddot{F}: \\ \quad \diagdown \quad \diagup \\ :\ddot{F}-S-\ddot{F}: \\ \quad | \\ :\ddot{F}: \end{array} \end{bmatrix}$$

(d) $[:C\equiv N:]^-$ (e)

(f)

$$\begin{bmatrix} :\ddot{O}: \\ | \\ :\ddot{O}-Br-\ddot{O}: \\ \ddots \quad \ddots \end{bmatrix}^- \longleftrightarrow \begin{bmatrix} :\ddot{O}: \\ | \\ :\ddot{O}=Br-\ddot{O}: \\ \ddots \quad \ddots \end{bmatrix}^- \longleftrightarrow \begin{bmatrix} :\ddot{O}: \\ | \\ :\ddot{O}=Br=\ddot{O}: \\ \ddots \end{bmatrix}^-$$

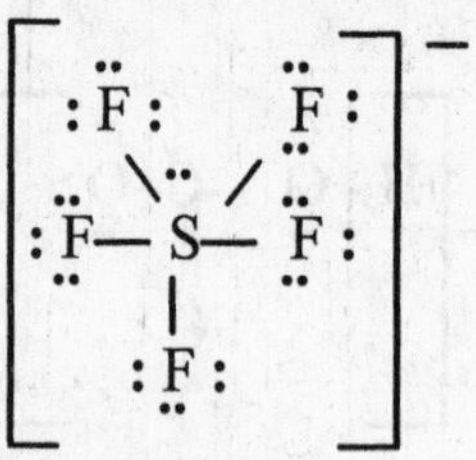

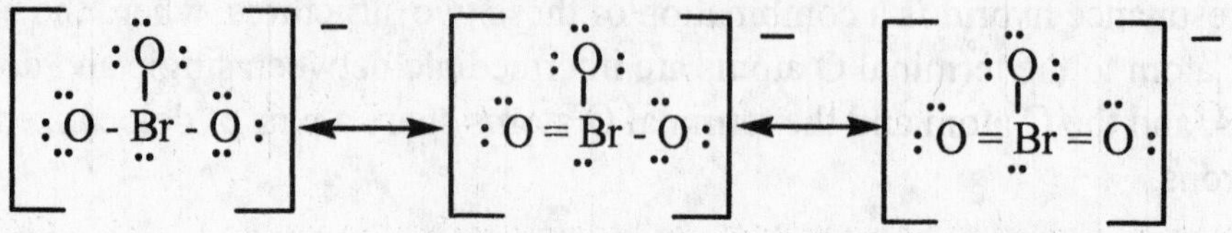

(3 equivalent structures) (3 equivalent structures)

The distribution of formal charges improves from left to right.

Several possible resonance structures involving one and two bromine-to-oxygen double bonds can be written as above. However, as discussed on pages 364 of the text, it is questionable whether these structures based on an expanded valence shell are important contributors to the resonance hybrid.

57.

(a)
$$\begin{array}{c} H \quad H \\ | \quad | \\ H-C-C-\ddot{O}-H \\ | \quad | \\ H \quad | \\ \quad H-C-H \\ \quad | \\ \quad H \end{array}$$

(b)
$$\begin{array}{c} :O: \\ \| \\ H-C-\ddot{O}-H \end{array}$$

(c)
$$\begin{array}{c} H \quad H \\ | \quad | \\ H-C-\ddot{O}-C-H \\ | \quad | \\ H \quad H \end{array}$$

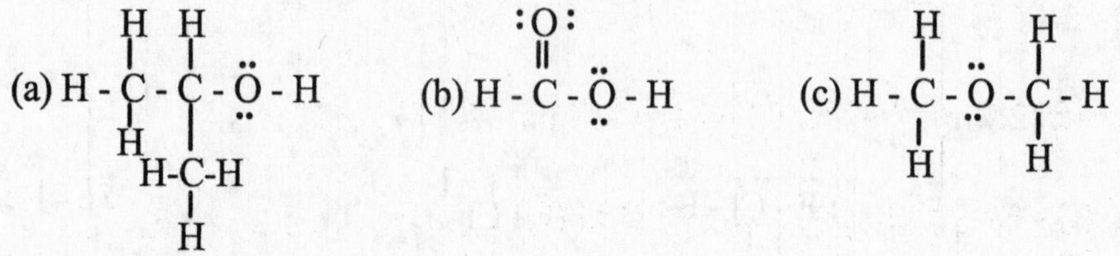

59. (a) Nitrogen only has six electrons.

$$\begin{bmatrix} :\ddot{O}: \\ | \\ O=N-\ddot{O}: \\ \ddots \end{bmatrix} \longleftrightarrow \begin{bmatrix} :O: \\ \| \\ :\ddot{O}-N-\ddot{O}: \\ \ddots \end{bmatrix} \longleftrightarrow \begin{bmatrix} :\ddot{O}: \\ | \\ :\ddot{O}-N=O: \\ \ddots \end{bmatrix}$$

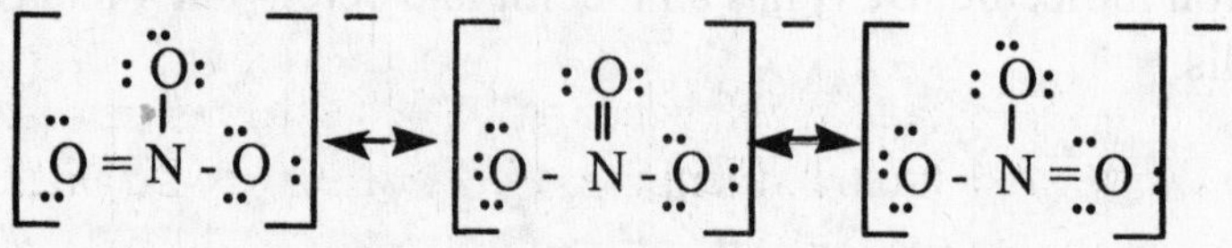

(b) Nitrogen cannot have an expanded shell and needs 8 electrons.

$$\begin{array}{c} H \\ | \\ H-N-C \equiv N: \\ \ddots \end{array}$$

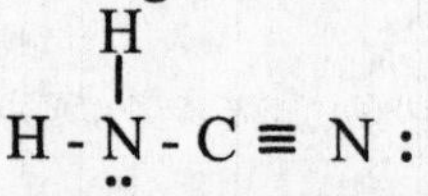

(c) The structure causes formal charges that can be reduced by placing the bonding electrons elsewhere.

$$H-\ddot{N}-C\equiv O:$$

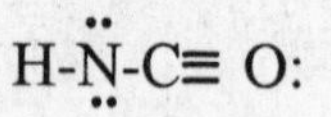

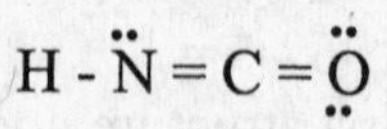

N $5-\frac{1}{2}(4)-4=-1$

H $1-\frac{1}{2}(2)=0$

C $4-\frac{1}{2}(2)=0$

O $6-\frac{1}{2}(6)-2=1$

$$H-\ddot{N}=C=\ddot{O}$$

H $1-\frac{1}{2}(2)=0$

N $5-\frac{1}{2}(6)-2=0$

C $4-\frac{1}{2}(8)=0$

O $6-\frac{1}{2}(4)-4=0$

The structure on the right is preferred as shown by lack of formal charges.

61. $? \text{ mol C} = 53.31 \text{ g C} \times \dfrac{\text{mol C}}{12.011 \text{ g C}} = 4.438 \text{ mol C} \times \dfrac{1}{2.220 \text{ mol O}} = \dfrac{1.999 \text{ mol C}}{\text{mol O}}$

$? \text{ mol H} = 11.18 \text{ g H} \times \dfrac{\text{mol H}}{1.0079 \text{ g H}} = 11.09 \text{ mol H} \times \dfrac{1}{2.220 \text{ mol O}} = \dfrac{4.995 \text{ mol H}}{\text{mol O}}$

$? \text{ mol O} = 35.51 \text{ g O} = \dfrac{\text{mol O}}{15.999 \text{ g O}} = 2.220 \text{ mol O} \times \dfrac{1}{2.220 \text{ mol O}} = \dfrac{1.000 \text{ mol O}}{\text{mol O}}$

$$C_2H_5O \qquad \text{H-}\underset{\overset{|}{H}}{\overset{\overset{H}{|}}{C}}\text{=C-O-H} \qquad \text{H-}\underset{\overset{|}{H}}{\overset{\overset{H}{|}}{C}}\text{-C=O-H} \qquad \text{H-}\ddot{\text{O}}\text{-C-C}$$

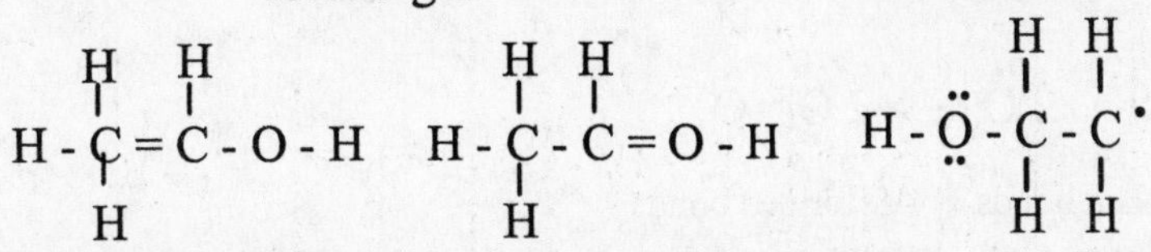

These are not good structures; there are too many bonds on the C or the O, or the structure has an odd electron. $C_4H_{10}O_2$ is better.

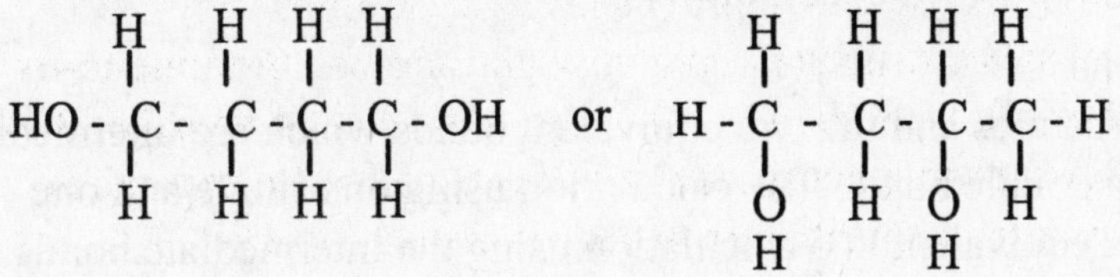

$$\text{HO-C-C-C-C-OH} \quad \text{or} \quad \text{H-C-C-C-C-H}$$

63. (a) (B)L. (I - Cl) $= \dfrac{1}{2}$ B.L. (I - I) $+ \dfrac{1}{2}$ B.L. (Cl - Cl)

B.L. $= \dfrac{1}{2}$ (266 pm) $+ \dfrac{1}{2}$ (199 pm) $= 233$ pm

(b) B.L. (C - F) $= \dfrac{1}{2}$ B.L. (C - C) $+ \dfrac{1}{2}$ B.L. (F - F)

B.L. $= \dfrac{1}{2}$ (154 pm) $+ \dfrac{1}{2}$ (143 pm)

B.L. $= 149$ pm

Estimates are likely to be too high because the bonds are polar. In general, the more polar the bond, the more the bond length is shortened from the calculated value.

65. The smallest percent difference should be between those atoms that are closest in electronegativity. H 2.1 F 4.0 Cl 3.5 Br 2.8 I 2.5
The smallest percent difference should be in HI.

67. NN double bond and NF single bonds as F forms only single bonds.
$: \ddot{\text{F}} - \ddot{\text{N}} = \ddot{\text{N}} - \ddot{\text{F}} :$

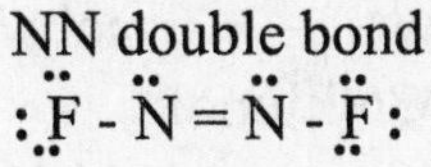

69. $\Delta H = \text{BE } (H_2) + \text{BE } (F_2) - 2\text{BE } (HF)$
$\Delta H = 436 \text{ kJ} + 159 \text{ kJ} - 2 \text{ mol} \times 565 \text{ kJ/mol} = -535 \text{ kJ}$

71. It is convenient to consider only the bonds that change instead of all of the bonds.

broken bonds		formed bonds	
C - H	414 kJ	C - Cl	− 339 kJ
Cl-Cl	243 kJ	H - Cl	− 431 kJ
	657 kJ		−770 kJ

exothermic $\Delta H = 657\text{ kJ} - 770\text{ kJ} = -113\text{ kJ}$

73.

:Ö=Ö-Ö: + :Ö: → 2 Ö=Ö

$\Delta H = \Delta H_{\text{broken bonds}} + \Delta H_{\text{formed bonds}}$

$-391.9\text{ kJ} = 2\Delta H_{O_3} - 2\,\Delta H_{O_2}$

$-391.9\text{ kJ} = 2\Delta H_{O_3} + 2\text{ mol }(-498\text{ kJ/mol})$

$\Delta H_{O_3} = 302\text{ kJ}/O_3\text{ bond}$

O_3 has 2 resonance structures and the two equivalent bonds which are intermediate between a single and a double bond. The calculation using one single and one double bond should be equivalent to a calculation using the intermediate bonds.

75. $\Delta H_{rxn} = \Delta H_{\text{broken bonds}} + \Delta H_{\text{formed bonds}} = \text{BE(reactants)} - \text{BE(products)}$

C(graphite) + 2 H_2(g) → CH_4(g)

$-74.81\text{ kJ} = -4\text{ mol }\Delta H_{CH} + 2\text{ mol }\Delta H_{H\text{-}H} + 1\text{ mol }\Delta H_{C\text{-}C}$

$-74.81\text{ kJ} = -4\text{ mol} \times \Delta H_{CH} + 2\text{ mol} \times 436\text{ kJ/mol} + 1\text{ mol} \times 717\text{ kJ}$

$-1664\text{ kJ} = -4\text{ mol} \times \Delta H_{CH}$

$$\Delta H_{CH} = \frac{-1664\text{ kJ}}{-4\text{ mol}} = 416\text{ kJ/mol}$$

The tabulated value of 414 kJ/mol compares very well.

77. (a) $H_2C = CHCH_3$ (b) $HC \equiv CCH_2CH_3$
(c) $H_2C = CHCH_2CH_2CH_3$ (d) $CH_3CH_2C \equiv CCH_2CH_3$

79. (a) $H_2C = CH_2 + H_2 \rightarrow H_3C - CH_3$
(b) $HC \equiv CH + 2H_2 \rightarrow H_3C - CH_3$

81. (a) $\Delta H = \Delta H_{\text{broken bonds}} + \Delta H_{\text{formed bonds}} = \text{BE(reactants)} - \text{BE(products)}$

$\Delta H = 1\text{ mol} \times \Delta H_{C=C} + 4\text{ mol} \times \Delta H_{C\text{-}H} + 1\text{ mol} \times \Delta H_{H\text{-}H} - 1\text{ mol }\Delta H_{C\text{-}C} - 6\text{ mol}\Delta H_{C\text{-}H}$

$\Delta H = 1\text{ mol} \times 611\text{ kJ/mol} + 4\text{ mol} \times 414\text{ kJ/mol} + 1\text{ mol} \times 436\text{ kJ/mol} - 1\text{ mol} \times (347\text{ kJ/mol}) - 6\text{ mol} \times (414\text{ kJ/mol})$

$\Delta H = -128\text{ kJ}$

(b) $\Delta H = \Delta H_f(C_2H_6) - \Delta H_f(C_2H_4) - \Delta H_{f_{H_2}}$

$\Delta H = 1\text{ mol }C_2H_6 \times (-84.68\text{ kJ/mol}) - 1\text{ mol }C_2H_4 \times 52.26\text{ kJ/mol} - 1\text{ mol }H_2 \times 0$

$\Delta H = -136.94\text{ kJ}$

The ΔH_f value is more negative.

83. (a) $-[-CH_2CH_2CH_2CH_2CH_2CH_2CH_2CH_2-]-$

(b) $\overset{}{+}\underset{|}{CH}-\underset{|}{CH}-\underset{|}{CH}-\underset{|}{CH}-\underset{|}{CH}-\underset{|}{CH}-\underset{|}{CH}-\underset{|}{CH}\overset{}{+}$

        Cl H   Cl H   Cl H   Cl H

(c) $\overset{}{+}\underset{|}{CH}-\underset{|}{CH}-\underset{|}{CH}-\underset{|}{CH}-\underset{|}{CH}-\underset{|}{CH}-\underset{|}{CH}-\underset{|}{CH}\overset{}{+}$

      CH_3 H CH_3 H CH_3 H CH_3 H

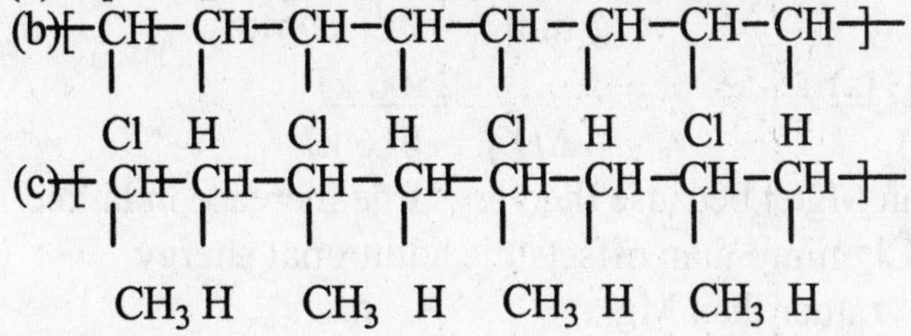

Additional Problems

85. In alkanes, the Lewis structures and structural formulas are identical because all electron pairs are bonding pairs (represented by a single dash line). There are no lone-pair electrons. In other organic compounds there may be lone-pair electrons to be shown as dots. For example, in organic compounds containing N or O atoms, there will be lone-pair electrons to be shown as dots on the O or N atoms.

87.

    dimethyl ether   ethanol

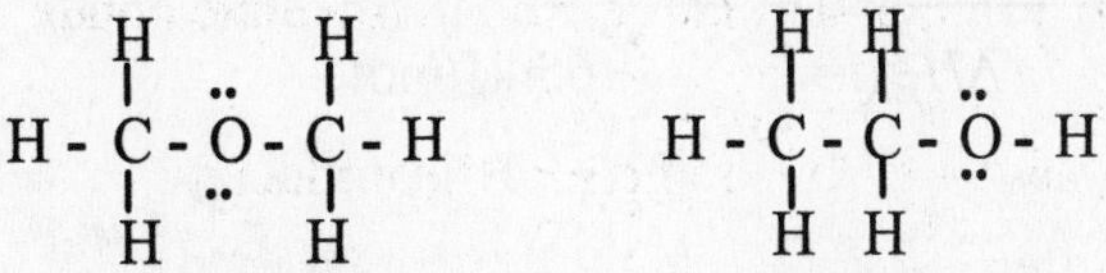

91.

$$H-\ddot{N}-N\equiv N: \longleftrightarrow H-\ddot{N}=N=\ddot{N}:$$

The single bond is about 145 pm, the triple bond is about 110 pm, and double bonds are about 123 pm. A resonance structure would be somewhere between the two resonance structures.

93. $2\ C\ (graphite) + 3\ H_2(g) \rightarrow C_2H_6(g)$ $\Delta H_f = -84.68$ kJ/mol

$\underline{C_2H_6(g) \rightarrow \cdot C_2H_5(g) + \frac{1}{2}\ H_2(g)\ \ \Delta H_{RXN} = ?}$

$2\ C\ (graphite) + \frac{5}{2}\ H_2(g) \rightarrow \cdot C_2H_5(g)\ \ \Delta H_f = ?$

$\Delta H_{RXN} = \Delta H_{broken\ bonds} + \Delta H_{formed\ bonds} = BE(reactants) - BE(products)$

$\Delta H_{RXN} = 1\ mol\ \Delta H_{CH} - \frac{1}{2}\ mol\ \Delta H_{H_2}$

$\Delta H_{RXN} = 1\ mol \times 414\ kJ/mol - \frac{1}{2}\ mol \times (436\ kJ/mol) = 196\ kJ/mol$

$\Delta H_f\ (\cdot C_2H_5) = \Delta H_f\ (C_2H_6) + \Delta H_{RXN} = -84.68\ kJ/mol + 196\ kJ/mol$

$\Delta H_f\ (\cdot C_2H_5) = 111\ kJ/mol$

95. $Mg\ (s) + Cl_2\ (g) \rightarrow MgCl_2\ (s)$ $\Delta H°_f = ?$

    $Mg\ (s) \rightarrow Mg\ (g)$ 150 kJ

    $Mg\ (g) \rightarrow Mg^+\ (g) + e^-$ 738 kJ

    $Mg^+\ (g) \rightarrow Mg^{2+}\ (g) + e^-$ 1451 kJ

$Cl_2(g) \rightarrow 2\,Cl(g)$ 243 kJ
$2\,Cl(g) + e^- \rightarrow 2\,Cl^-$ 2 mol × -349 kJ/mol
$Mg^{2+}(g) + 2\,Cl^-(g) \rightarrow MgCl_2(s)$ - 2500 kJ
$Mg(s) + Cl_2(g) \rightarrow MgCl_2(s)$ $\Delta H^\circ_f = -616$ kJ

$MgCl_2$ is much more stable than MgCl because the very large increase in lattice energy between MgCl and $MgCl_2$ more than offsets the additional energy requirement in producing Mg^{2+} rather than Mg^+.

96. $H(g) + e^- \rightarrow H^-(g)$ $\Delta H_{EA} = ?$
The equations have to be added together to generate the above equation.

$Na(s) + \frac{1}{2}H_2(g) \rightarrow NaH(s)$ $\Delta H_f = -56.27$ kJ/mol
$Na(g) \rightarrow Na(s)$ $-\Delta H_{sub} = -(107)$ kJ/mol
$Na^+(g) + e^- \rightarrow Na(g)$ $-\Delta H_{IE} = -(496)$ kJ/mol
$NaH(s) \rightarrow Na^+(g) + H^-(g)$ $-\Delta H_{LE} = -(-812)$ kJ/mol
$H(g) \rightarrow \frac{1}{2}H_2(g)$ $-\Delta H_{diss} = -(218.0)$ kJ/mol
$H(g) + e^- \rightarrow H^-(g)$ $\Delta H_{EA} = -65$ kJ/mol

98.

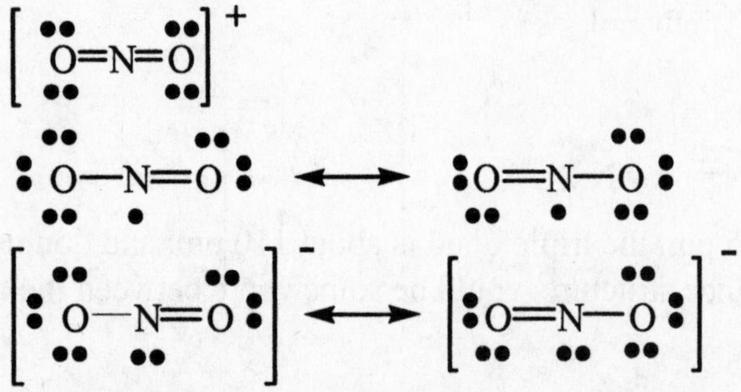

Both NO_2 and NO_2^- have two resonance structures with each structure having one single bond and one double bond. The bonds lengths will be about the same length, intermediate between the N-O single bond length and the double bond length. The bond energies for the two bonds are about the same, intermediate between the N-O single bond energy and the double bond energy.

102. $N_2(g) + O_2(g) \rightarrow 2\,NO(g)$ $\Delta H_f = 90.25$ kJ/mol

$\Delta H = \Delta H_{broken\ bonds} + \Delta H_{formed\ bonds} = BE(reactants) - BE(products)$

$2\text{ mol }\Delta H_f = \Delta H_{N_2} + \Delta H_{O_2} - 2\text{ mol }\Delta H_{NO}$

$$2\text{ mol} \times 90.25\text{ kJ/mol} = \frac{946\text{ kJ}}{\text{mol}} + \frac{498\text{ kJ}}{\text{mol}} - 2\text{ mol }\Delta H_{NO}$$

$$\Delta H_{NO} = \frac{-1264\text{ kJ/mol}}{2\text{ mol}} = -\Delta H_{diss}$$

$\Delta H_{diss} = 632$ kJ/mol
Table value 590 kJ/mol

Apply Your Knowledge

104. (a) $?\text{m} = 75000\text{ u} \times \dfrac{\text{unit}}{28.05\text{ u}} \times \dfrac{1.3\text{ cm}}{\text{unit}} \times \dfrac{10^{-2}\text{ m}}{\text{cm}} = 35\text{ m}$

(b) $?\text{pages} = 75000\text{ u} \times \dfrac{\text{unit}}{28.05\text{ u}} \times \dfrac{1.3\text{ cm}}{\text{unit}} \times \dfrac{\text{page}}{12.2\text{ cm}} = 2.8 \times 10^{2}\text{ pages}$

105.
$\frac{1}{2}\,Cl_2(g) + O_2(g) \rightarrow ClO_2(g)$ $\qquad \Delta H^\circ_f = \quad 102.5\text{ kJ/mol}$

$O(g) \rightarrow \frac{1}{2}\,O_2(g)$ $\qquad \Delta H_{BE} = \frac{1}{2}(-498\text{ kJ/mol})$

$\underline{ClO_2 \rightarrow {\bullet}ClO(g) + O(g)}$ $\qquad \underline{\Delta H_{BE} = \quad (243\text{ kJ/mol})}$

$\frac{1}{2}\,Cl_2(g) + \frac{1}{2}\,O_2(g) \rightarrow {\bullet}ClO(g)$ $\qquad \Delta H_f = \quad 97\text{ kJ/mol}$

The tabulated value is 101.84 kJ/mol. This is a pretty good agreement.

106. (a) $\Delta = 431\text{ kJ} - \sqrt{436\text{ kJ} \times 243\text{ kJ}} = 431 - 325\text{ kJ} = 106\text{ kJ}$

(b) $x = 0.18 \times \sqrt{\Delta} = 0.18 \times \sqrt{\dfrac{106\text{ kJ}}{\text{mol}} \times \dfrac{1\text{ kcal}}{4.184\text{ kJ}}} = 0.91$

(c) *EN* value = 0.91 + 2.05 = 2.96, which rounds to 3.0

Chapter 10

Bonding Theory and Molecular Structure

Exercises

10.1A

(a) $SiCl_4$ 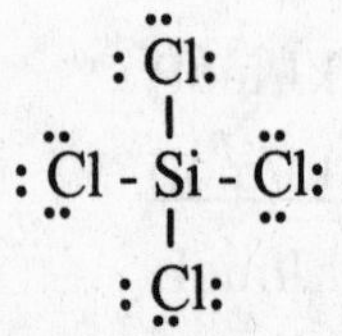

The four bonding groups of electrons (AX_4) produce a tetrahedral electron-group geometry and molecular geometry.

(b) $SbCl_5$ 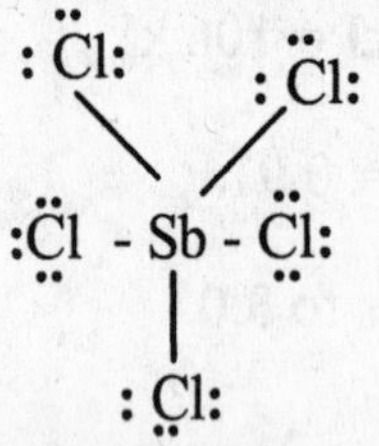

The five bonding groups of electrons (AX_5) produce both a trigonal bipyramidal electron-group geometry and molecular geometry.

10.1B

(a) 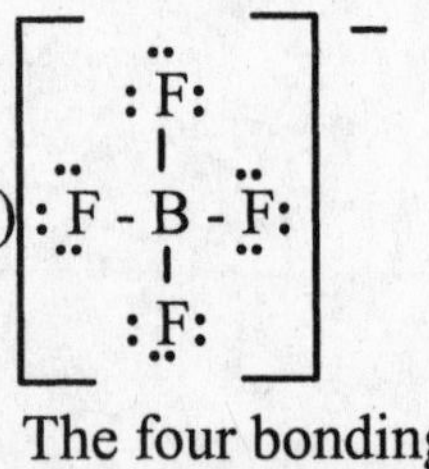

The four bonding groups of electrons (AX_4) produce a tetrahedral electron-group geometry and molecular geometry.

(b) 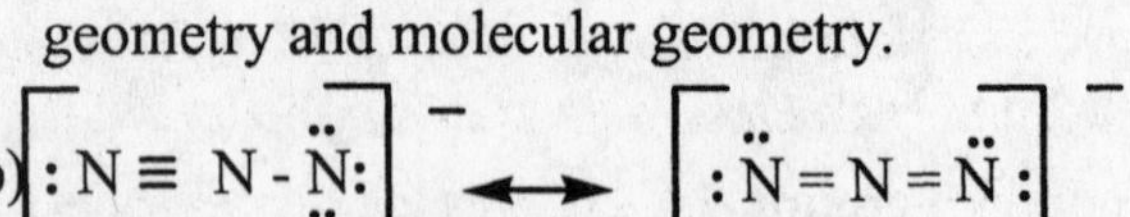

The two bonding groups of electrons (AX_2) produce a linear electron-group geometry and molecular geometry.

10.2A

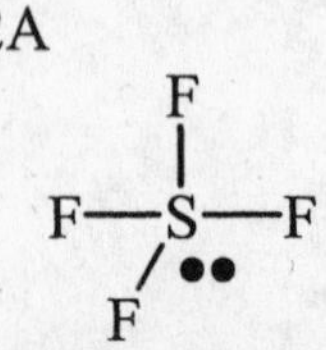

SF_4: There are five groups of electrons, but one pair is a lone pair of electrons. The VSEPR notation is AX_4E. Electron-group geometry is trigonal bipyramidal. In the seesaw structure (Table 10.1), there are two LP-BP repulsions at 90° and two at 120°. Because 90° LP-BP interactions are especially unfavorable, the seesaw structure is adopted so that the lone pair is on the equatorial plane of the structure instead of in the axial position which would have three of the unfavorable 90° LP-BP interactions. A tetrahedral structure would have only four groups of electrons.

10.2B

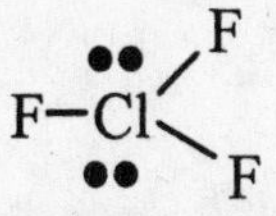

There are five groups of electrons in ClF_3, but two pairs are lone pairs of electrons. The VSEPR notation is (AX_3E_2). Those two pairs will be on the equatorial plane of the structure, producing a T-shape molecular shape from the trigonal bipyramidal electron-group geometry. In the T-shaped structure (Table 10.1), there are four unfavorable LP-BP repulsions at 90°. If the lone pairs were in the axial positions, the trigonal planar structure, there would be six LP-BP repulsions at 90°. The T-shaped structure is observed.

10.3A

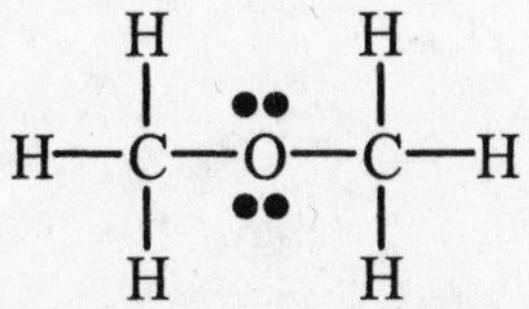

For both carbon atoms there are four bonding groups and no lone pairs of electrons (AX_4). So, the molecular shape around each carbon atom is tetrahedral. Around the O atom are four electron-groups but two of the electron groups are lone pairs (AX_2E_2), so the C-O-C molecular shape is bent.

10.3B

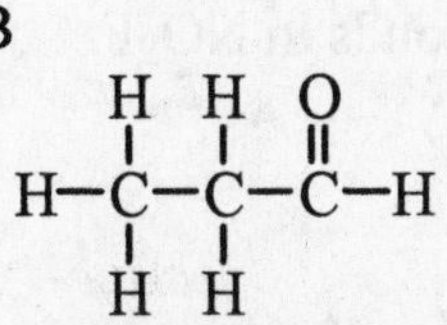

For the first two carbon atoms, there are four bonding groups and no lone pairs of electrons, (AX_4), thus the shape around each carbon atom is tetrahedral. The carbonyl carbon atom has three bonding groups, (AX_3), and the geometry is trigonal planar.

10.4A

(a)

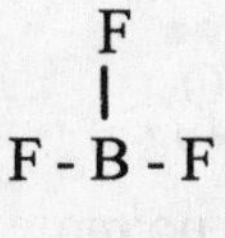

trigonal planar
symmetric
nonpolar

(b)

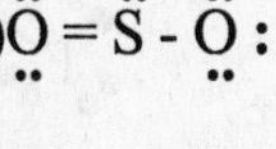

bent
nonsymmetric
polar

(c) : Br - Cl :

linear
different electronegativities
polar

(d) : N ≡ N:

linear
same electronegativity
nonpolar

10.4B

(a)

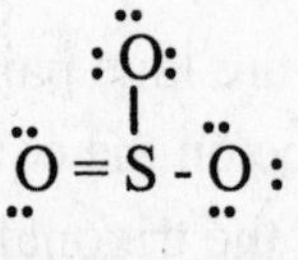

3 resonance structures
trigonal planar
symmetric
nonpolar

(b)

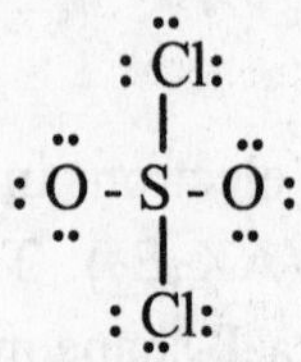

tetrahedral
symmetric
different electronegativities
polar

(c)

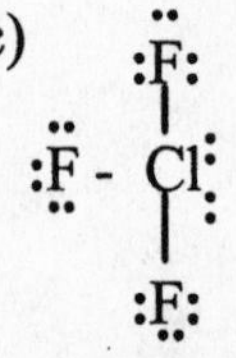

T-shape
nonsymmetric
polar

(d)

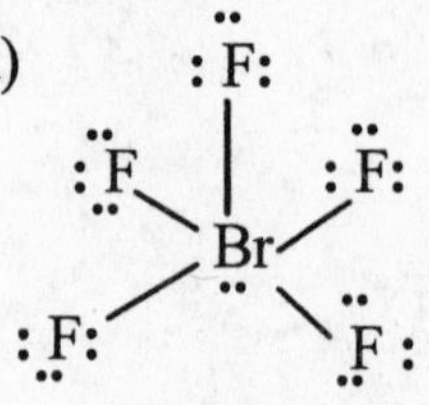

square pyramidal
nonsymmetric
polar

10.5A NOF is probably 110°, and NO_2F is 118°. The lone pair of electrons on N in NOF is more repulsive than are the bonding pairs in the N-to-O bonds in NO_2F.

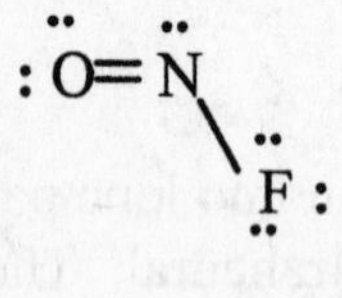

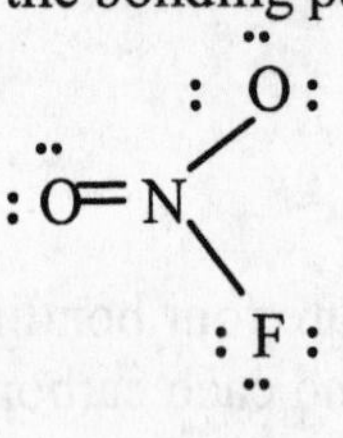

10.5B

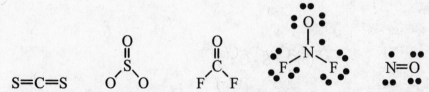

CS_2 and SO_3 are symmetric, the bond dipoles cancel, and there is no molecular dipole. Although COF_2 and NOF_2 (tetrahedral) are symmetric, the bond dipoles do not exactly cancel, and there is a small dipole for each molecule. In NO there is no canceling effect and a resultant molecular dipole is present. NO has the strongest dipole moment.

10.6A

$SiCl_4$ 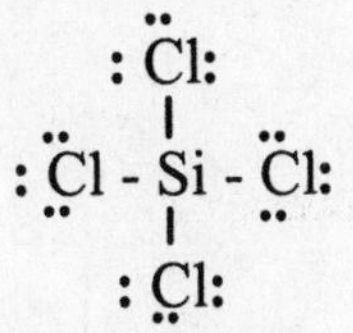

VSEPR notation AX_4
electron group-geometry: tetrahedral
molecular geometry: tetrahedral
hybridization scheme sp^3

Si [Ne] 

sp^3

10.6B

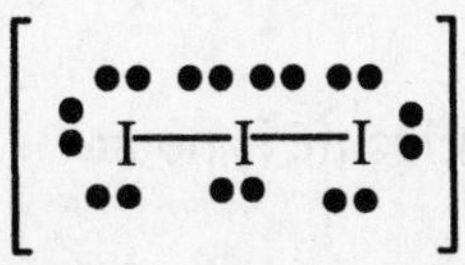

VSEPR notation: AX_2E_3
electron-group geometry: trigonal bipyramidal
molecular geometry: linear
hybridization scheme: sp^3d

I - $[Kr]4d^{10}$ 

sp^3d

10.7A

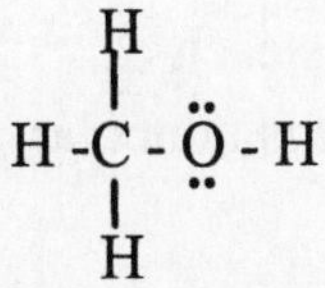

(a) Around the C is tetrahedral. AX_4
Around the O is bent. AX_2E_2
(b) C is sp^3, O is sp^3
(c)

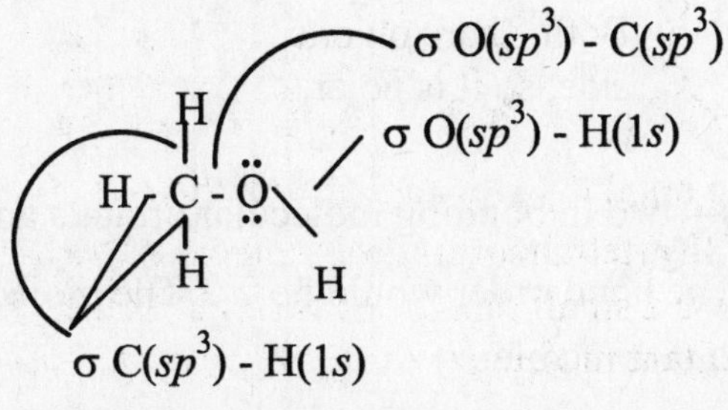

10.7B :N ≡ C - C ≡ N:

(a) Around each C is linear. AX_2

(b) C is *sp*, N is 2*p*

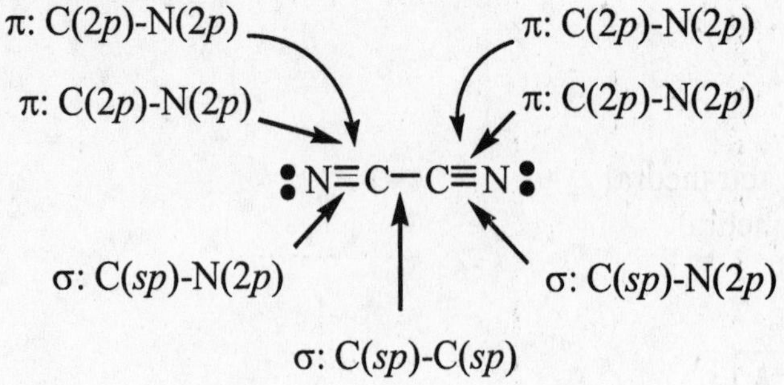

10.8A

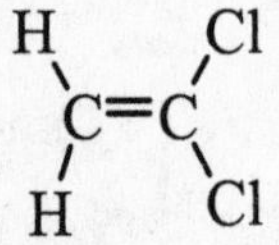

Because one end of the molecule has both chlorines, the molecule is polar.

10.8B

(a) 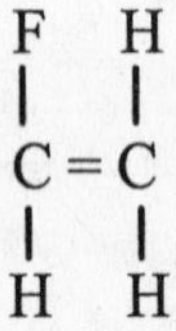

Fluorine has so much greater electronegativity that the molecule is polar.

(b) 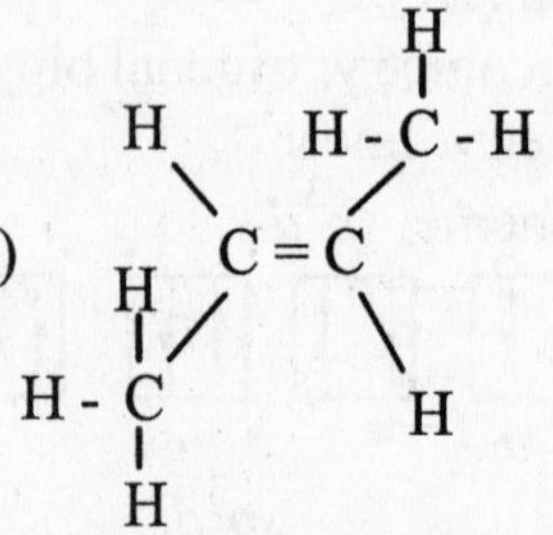

There is little electronegativity difference between H and C, and the CH_3 groups balance each other. So the molecule is nonpolar.

(c) H - C ≡ C - H

It is all symmetric, so it is nonpolar.

(d) 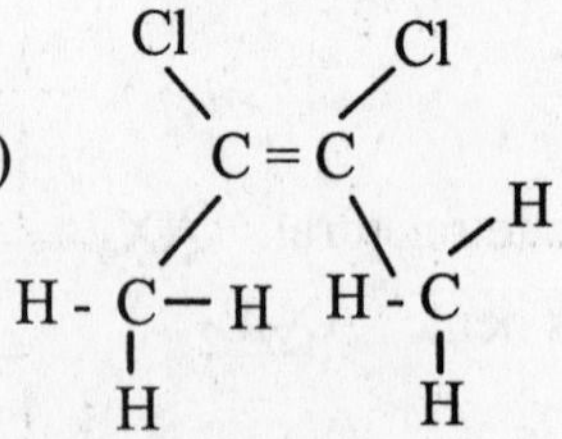

Both Cl are on one side, so it is polar.

10.9A H_2^- would have three electrons—two in bonding molecular orbitals and one in an antibonding molecular orbital. The bond order would be $\frac{1}{2}$. The molecular ion would exist but would be only slightly stable.

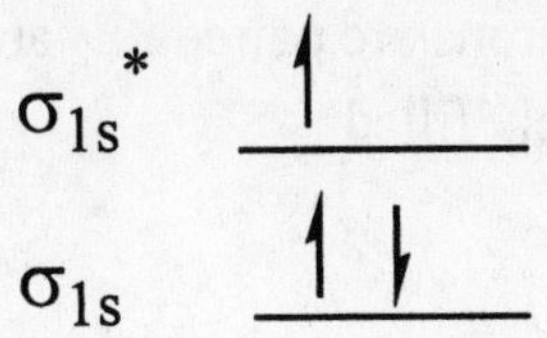

10.9B

He_2 He_2^+

He_2 is not stable, the bond order is zero. There is as much repulsion as attraction and nothing to hold the molecule together.
He_2^+ has two bonding and only one antibonding electron, a bond order of ½, and should exist. He_2^{2+} and He_2^{3+} should also exist.

10.10A Li_2 and Be_2 would have the same bond order and magnetic properties because only σ_{2s} and σ^*_{2s} molecular orbitals are involved.

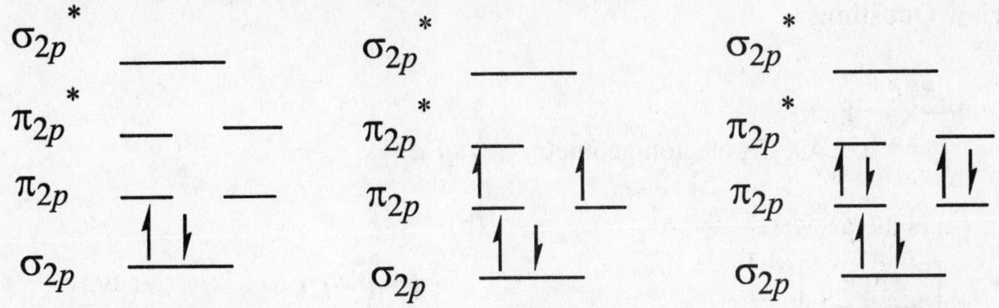

O_2 diagram B_2	C_2	N_2
BO = 1	BO = 2	BO = 3
diamagnetic	paramagnetic	diamagnetic

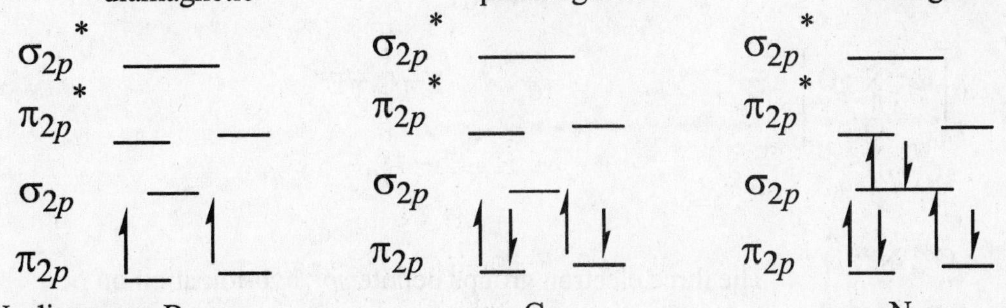

N_2 diagram B_2	C_2	N_2
BO = 1	BO = 2	BO = 3
paramagnetic	diamagnetic	diamagnetic

The bond order would not change because electrons are going into bonding orbitals in both cases. The magnetic properties of B_2 and C_2 switch from one diagram to the other. N_2 is diamagnetic in both cases because all the electrons are paired. F_2 and O_2 are not changed as the orbitals that change are completely filled.

10.10B $BO = \frac{8-3}{2} = \frac{5}{2}$

Either the N2 or O2 molecular orbital diagram is useful, and both produce the same result because the σ_{2p} and π_{2p} are completely filled in both diagrams, with one electron in the π_{2p}^* orbital in each case.

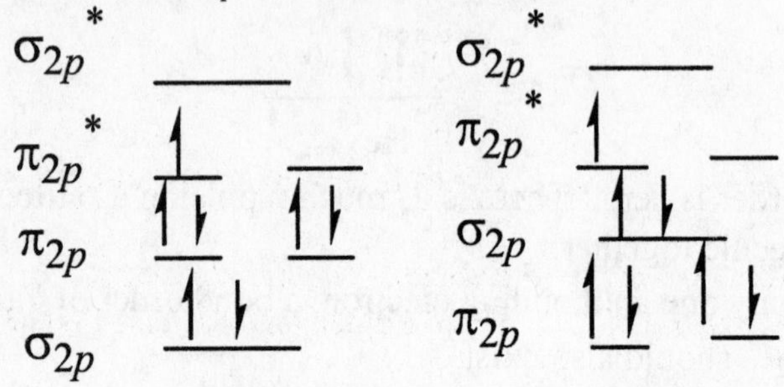

NO O_2 diagram   NO N_2 diagram

Review Questions

3.

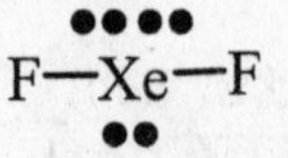

AX_2E_3 electon geometry (c) sp^3d

4. (a) is the answer.

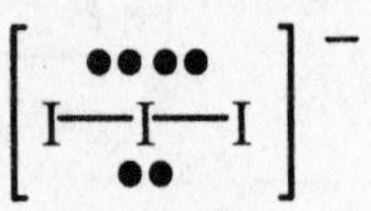

sp^3d

(b)

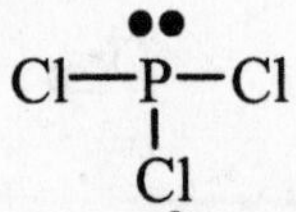

sp^3

(c)

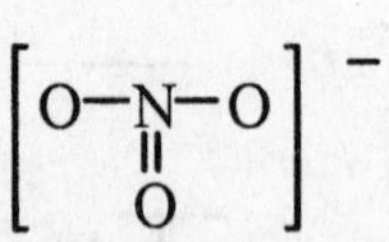

sp^2

(d)

sp^3

5.

:Ö :N::Ö: The three electron groups denote sp^2 hybridization on N.

:C̈l :N::Ö: Three electron groups denote sp^2 hybridization on N.

H:Ö :N̈:H
H Four electron groups denote sp^3 hybridization on N.

6.

(a) 180° $: \ddot{C}l - Be - \ddot{C}l :$

(b) 120° $\ddot{O} = \ddot{S} - \ddot{O} :$

(c) 109.5° $H - \ddot{O} - H$

7.   (a)

H—P—H
    |
    H
electron geometry: tetrahedral bipyramidal
molecular shape: pyramidal
Bond angles are just less than 109.5°.

(b)

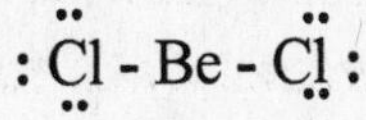

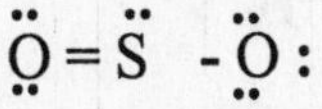

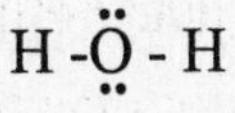

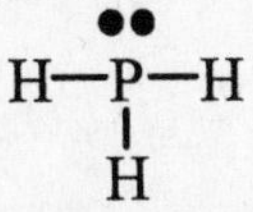

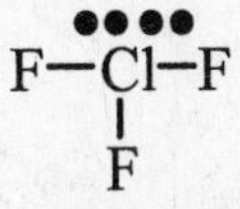

F—Cl—F
    |
    F
electron geometry: trigonal
molecular shape: T-shape
Bond angles are just less than 90°.

(c)

Cl—O—Cl
electron geometry: tetrahedral
molecular shape: bent
Bond angles are slightly less than 109.5°

(b) is the answer.

(d)

        O
        ‖
H—C—O—H
trigonal planar around C
electron geometry around O: tetrahedral
molecular shape around O: bent
H-C-O bond angle is slightly less than 120°.
C-O-H bond angle is slightly less than 109.5°.

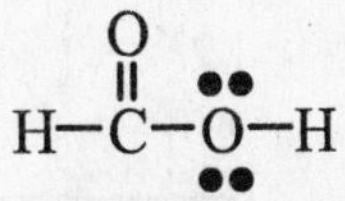

10.   (a)

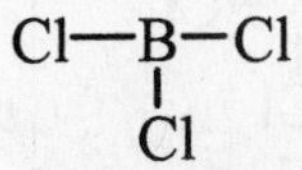

Cl—B—Cl
    |
    Cl
trigonal planar
like bonds
nonpolar

(b)

    H
    |
H—C—Cl
    |
    Cl
tetrahedral
not like bonds
polar

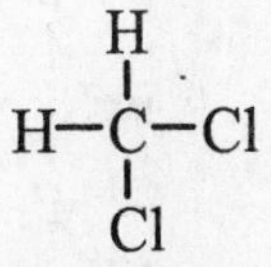

(c)

H—N—H
    |
    H
pyramidal
polar
(a) is the answer.

(d)

F—N=O
bent
polar

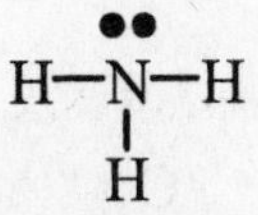

14. (a)

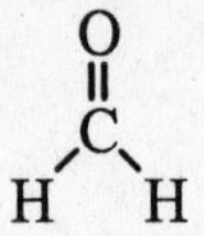

(b)

H−C≡C−H

(c)

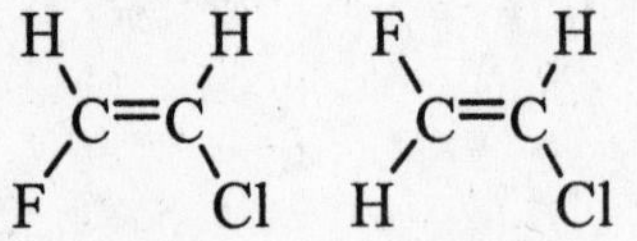

cis trans

(d)

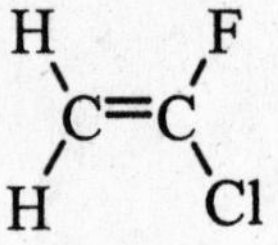

(c) is answer.

15. (a) 5 (b) 1

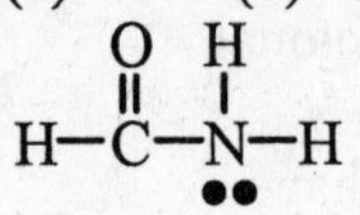

17.

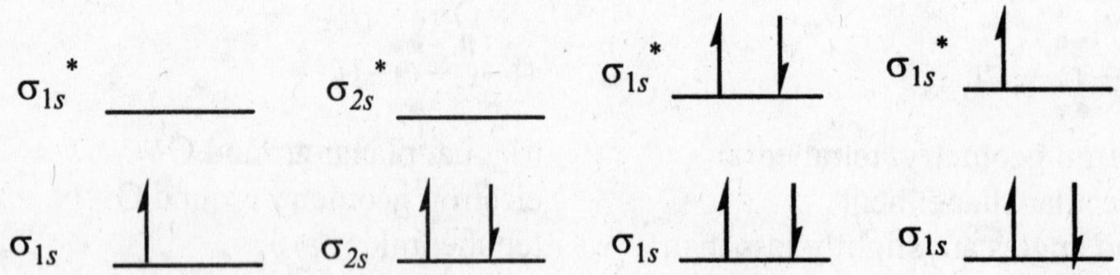

H_2^+	Li_2	He_2	H_2^-
BO = ½	BO = 1	BO = 0	BO = ½

(b) Li_2

Problems

21. (a)

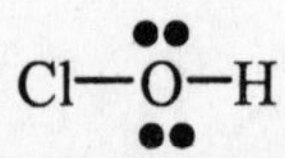

AX_2E_2

(b)

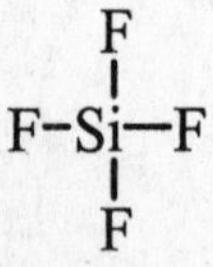

AX_4

(c)

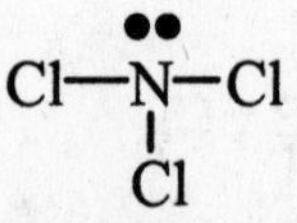

AX_3E

23. (a) bent (b) tetrahedral (c) pyramidal

25. (a)

S=C=S

2 electron groups
2 bond pairs
linear

(b)

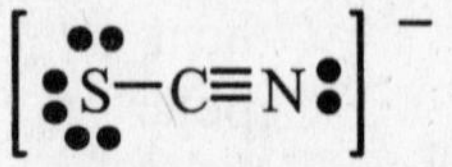

2 electron groups
2 bond pairs
linear

(c)

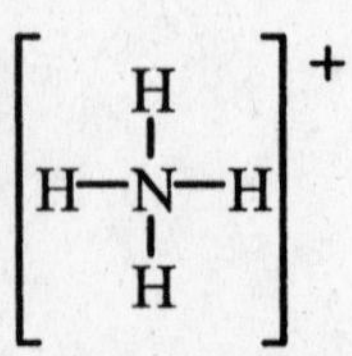

4 electron groups
4 bond pairs
tetrahedral

(d)

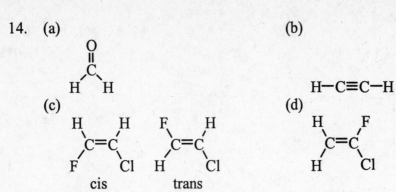

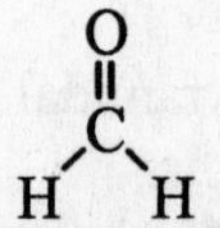

3 electron groups
3 bond pairs
trigonal planar

27. (a)

4 electron groups
2 bond pairs
2 lone pairs
bent

(b)

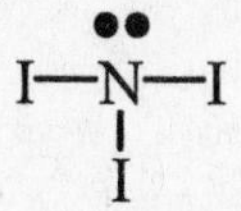

4 electron groups
3 bond pairs
1 lone pair
pyramidal

(c)

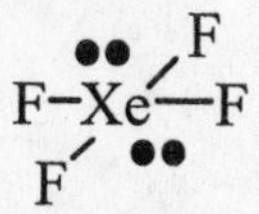

6 electron groups
4 bond pairs
2 lone pairs
square planar

(d)

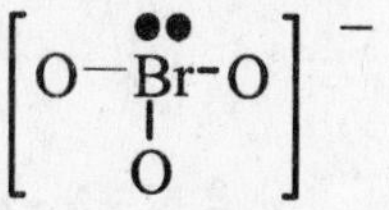

4 electron groups
3 bond pairs
1 lone pair
pyramidal

29. (a)

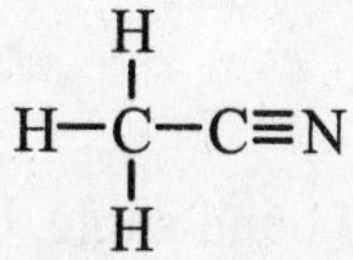

Around the left C is tetrahedral; around the right C is linear.

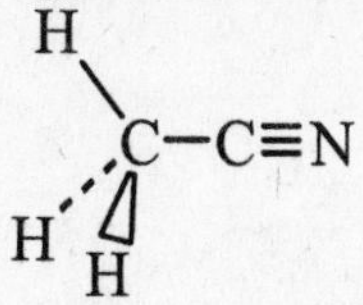

(b)

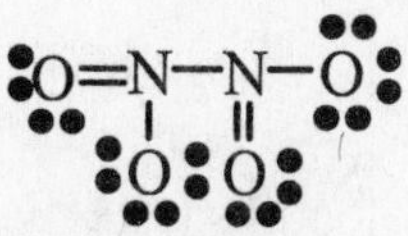

Around each N is trigonal planar.

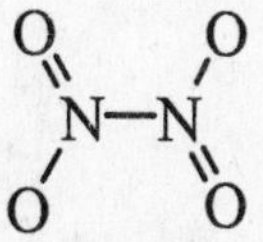

(c)

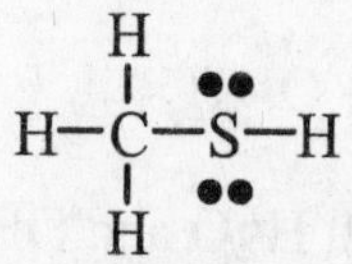

Around the C is tetrahedral; around the S is bent.

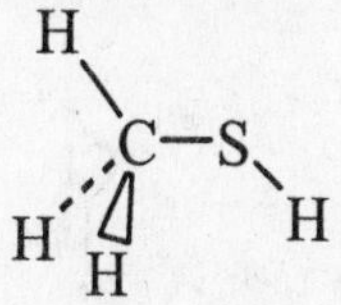

31.

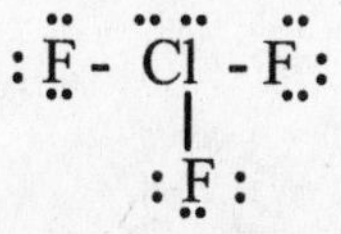

BF_3 has three electron groups, all bonding pairs, and so BF_3 is trigonal planar. ClF_3 has three bonding pairs and two lone pairs. The electron-group geometry is trigonal bipyramidal, and the molecular geometry is T-shaped.

33.

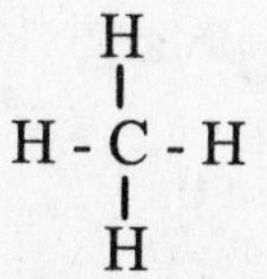

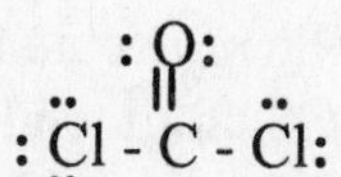

The tetrahedral bond angles are closer to the predicted angles in CH_4, as all of the bonding pairs have the same repulsion. In $COCl_2$, electron-group repulsions are strongest between the two lone pairs of electrons on the O atom, forcing the Cl-O-Cl angle to be somewhat smaller than a trigonal planar angle.

35.

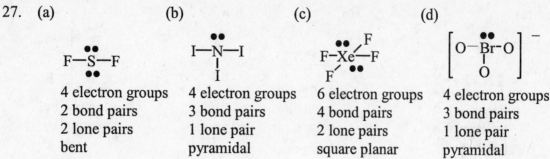

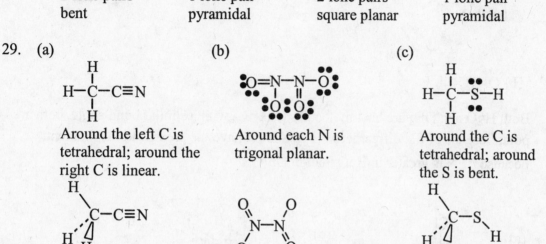

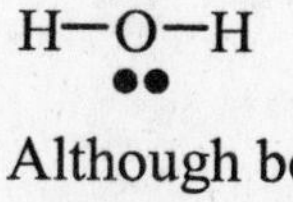

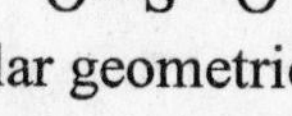

Although both molecular geometries are bent, the electron geometry of H_2O is tetrahedral but the electron geometry of SO_2 is trigonal planar.

37. (a)

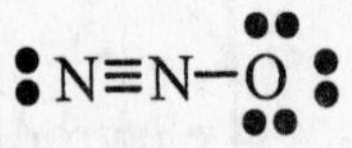

linear
different ENs
polar

(b)

T-shape
polar

(c)

bent
polar

(d)

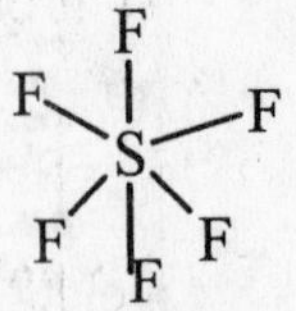

octahedral
nonpolar

39.

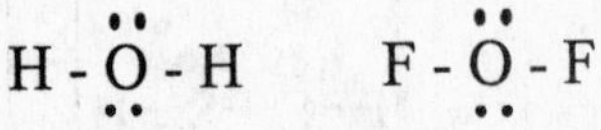

Both H_2O and OF_2 are bent molecules (AX_2E_2) with a similar bond angle, both are polar because of EN differences. H_2O should have the greater dipole moment because ΔEN is greater in H_2O than in OF_2.

41.

(a)

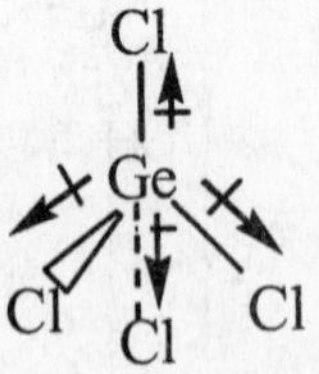

no resultant
dipole moment

(b)

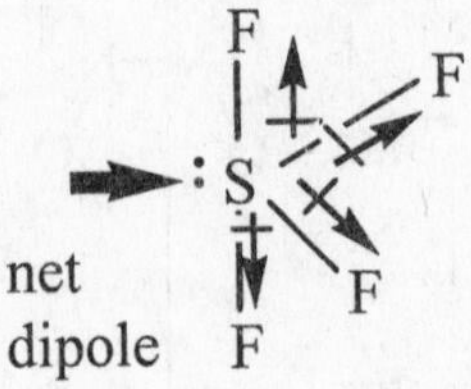

SF_4 has a seesaw shape (Table 10.1). Bond dipoles are directed toward F atoms. Two are in opposite directions along a straight line and effectively cancel. The other two are in the equatorial plane, forming a trigonal planar arrangement. The molecule is polar, but LP electrons on the S atom tend to counteract the two S-F bond dipoles in the central plane to some extent.

43. Li_2 is formed by an overlap of the 2*s* orbitals. F_2 is formed by the overlap of 2*p* orbitals along the axis between the two nuclei. The F_2 has a greater bond energy because the *p* orbitals overlap more.

45. (a)

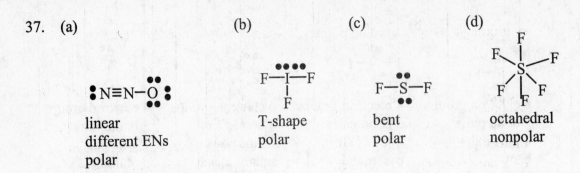

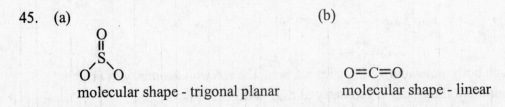

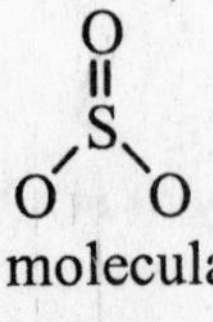

molecular shape - trigonal planar

(b)

O=C=O

molecular shape - linear

e^- group geometry - trigonal planar
sp^2 - AX_3

(c)

molecular shape - bent
e^- group geometry - tetrahedral
sp^3 - AX_2E_2

e^- group geometry - linear
sp - AX_2

(d)

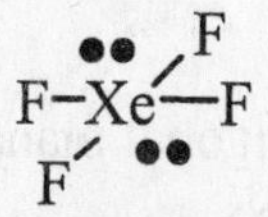

molecular shape - square planar
e^- group geometry - octahedral
d^2sp^3 - AX_4E_2

47. (a)

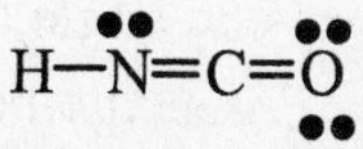

N - molecular shape - bent
e^- group geometry - trigonal planar
sp^2 - AX_2E
C-molecular shape - linear
e^- group geometry - linear
sp - AX_2

(b)

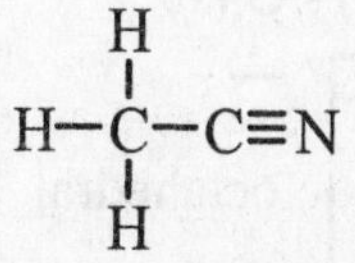

CH_3 - molecular shape - tetrahedral
e^- group geometry - tetrahedral
sp^3 - AX_4
CN - molecular shape - linear
e^- group geometry - linear
sp - AX_2

(c)

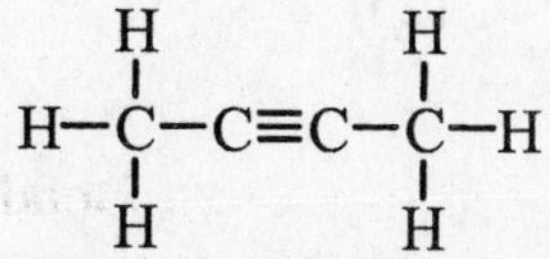

CH_3 -molecular shape - tetrahedral
e^- group geometry - tetrahedral
sp^3 - AX_4
CC - molecular shape - llinear
e^- group geometry - linear
sp - AX_2

(d)

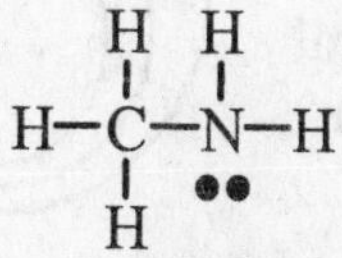

C - molecular shape - tetrahedral
e^- group geometry - tetrahedral
sp^3 - AX_4
N - molecular shape - pyramidal
e^- group geometry - tetrahedral
sp^3 - AX_3E

49.

(a) 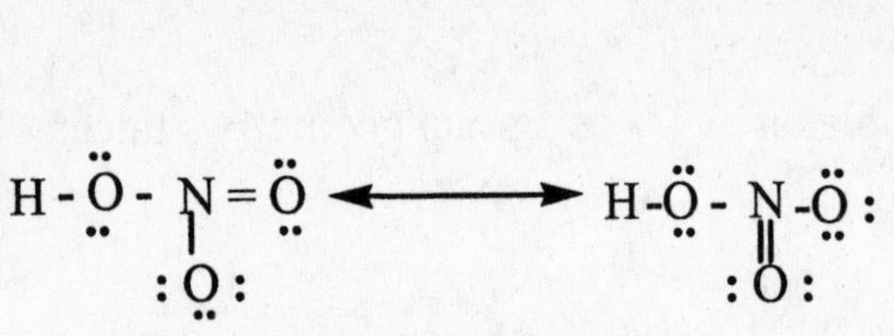

bent around O; trigonal planar around N

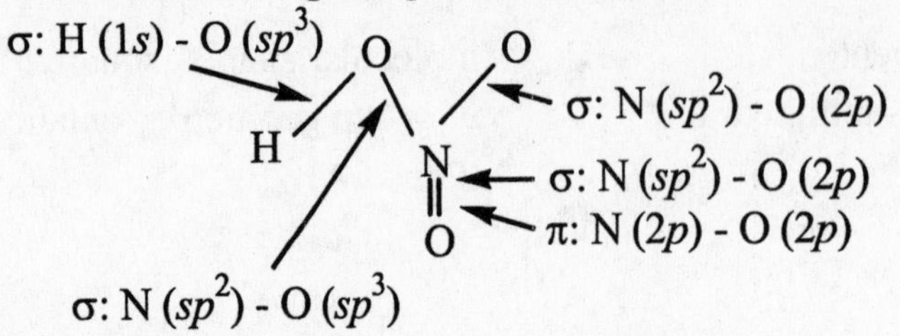

(b) 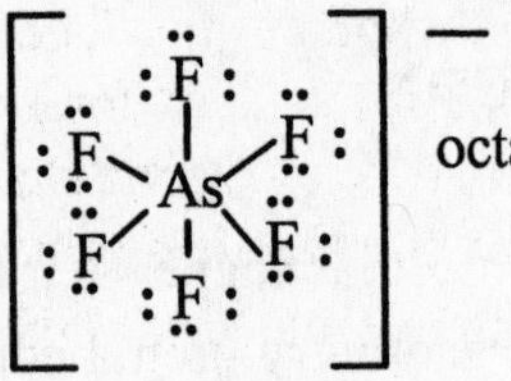

octahedral

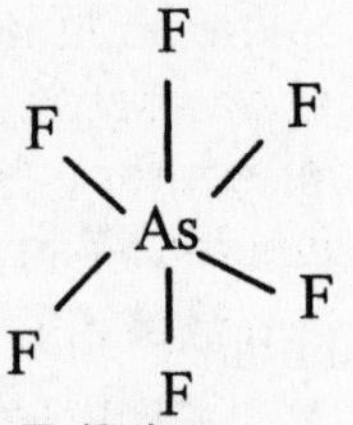

all bonds σ: As (sp^3d^2) - F (2p)

(c) 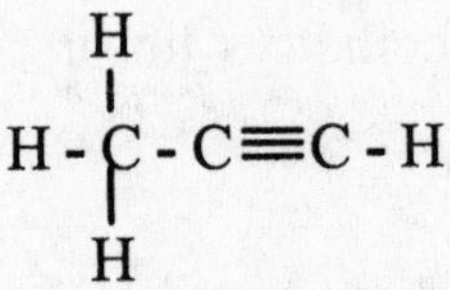

left C tetrahedral
middle C linear
right C linear

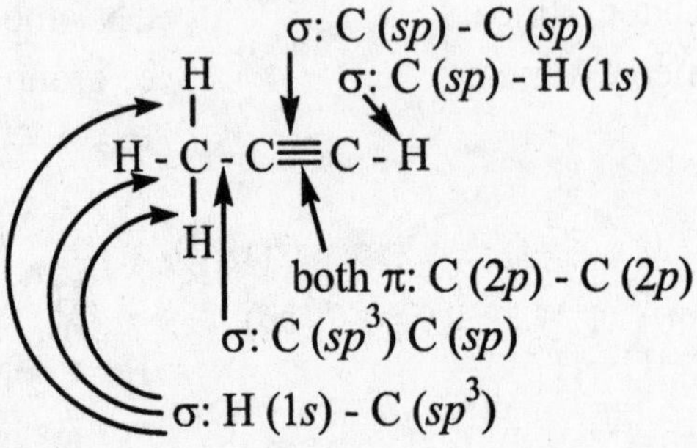

51.

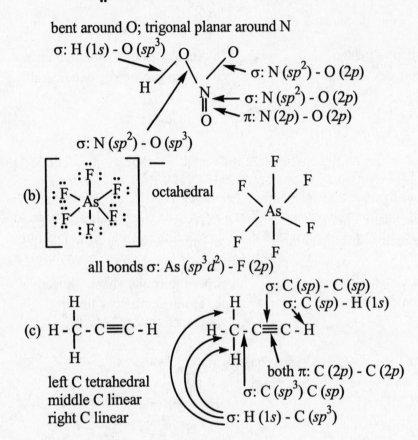

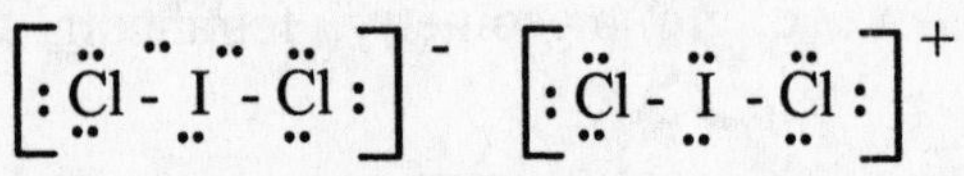

The electron-group geometry in the ICl_2^- ion is that of AX_2E_3 trigonal bipyramidal (the ion is linear). The hybridization scheme is sp^3d. The electron-group geometry in the ICl_2^+ ion is that of AX_2E_2, tetrahedral, (the ion is bent). The hybridization scheme is sp^3.

53.

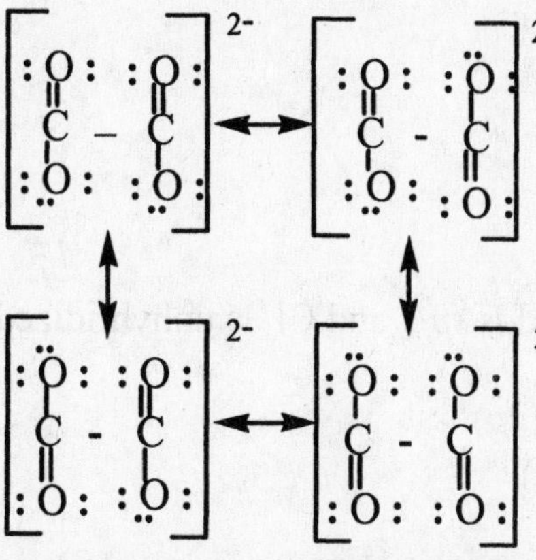

The 157 pm is the length for a single bond between the carbon atoms. The 125 pm is not short enough for a double bond between C and O, but too short for a C-O single bond. It would be correct for a delocalized pair of electrons shared by two O and C. Repulsions between lone-pair electrons on the O atoms bonded to the same C atom cause the O-C-O bonds to open up to a larger angle (126°) than the 120° angles corresponding to *sp* ^{2} hybridization. The C atoms are *sp*^{2} hybridized.

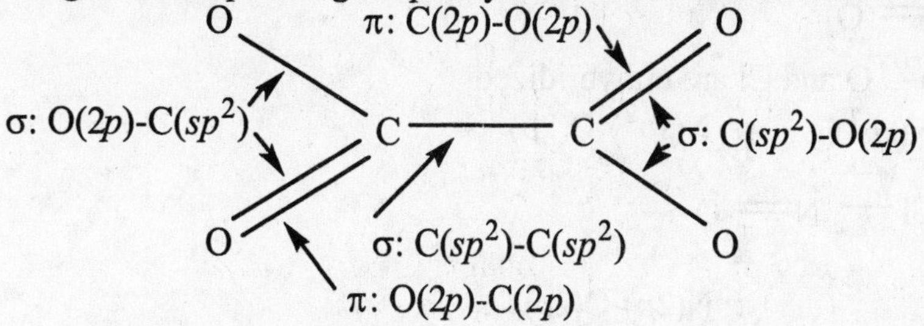

55. (a) CH_3COOCH_3

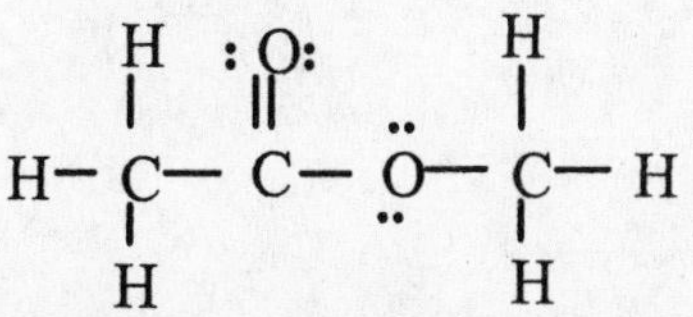

The C in each CH_3 is *sp*^{3}, the C in the CO is *sp*^{2}, the O in the CO group is unhybridized, and the O bonded to two C atoms is *sp*^{3}.

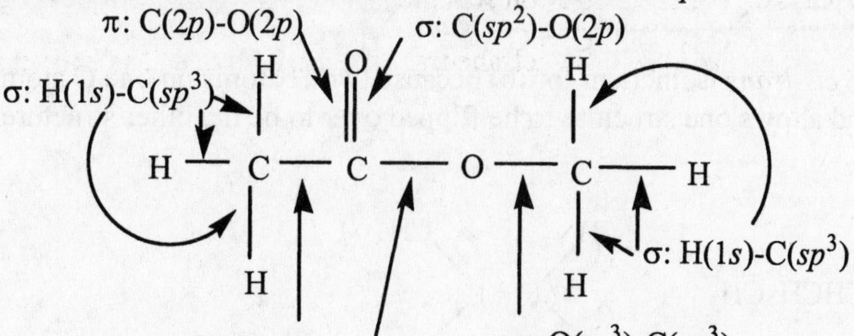

(b) *trans* $CH_3CH{=}CHCl$

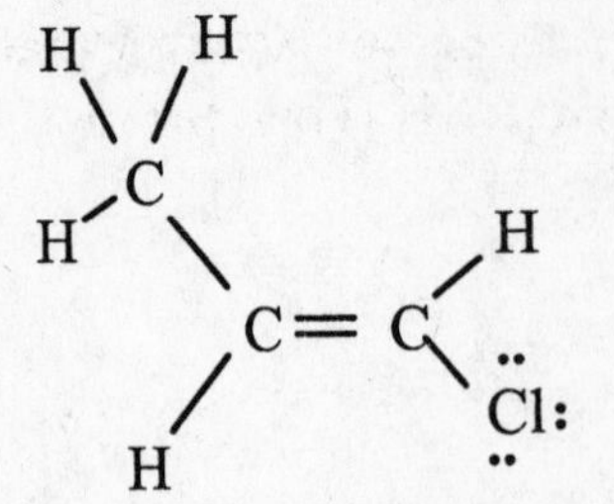

The CH$_3$ C is sp^3, each C in the double bond is sp^2, and Cl is unhybridized.

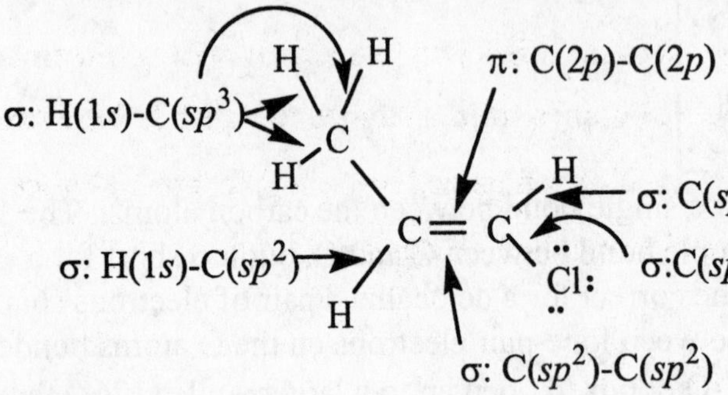

σ: H(1s)-C(sp^3)

π: C(2p)-C(2p)

σ: C(sp^2)-H(1s)

σ: H(1s)-C(sp^2)

σ:C(sp^2)- Cl(3p)

σ: C(sp^2)-C(sp^2)

(c) ClNO

: $\ddot{\underset{..}{Cl}}$ — $\ddot{N}$ = $\ddot{\underset{..}{O}}$

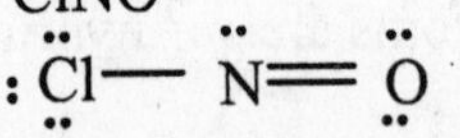

The N is sp^2. O and Cl are unhybridized.

σ: Cl(3p)-N(sp^2) σ: N(sp^2)-O(2p)

Cl — N = O

π: N(2p)-O(2p)

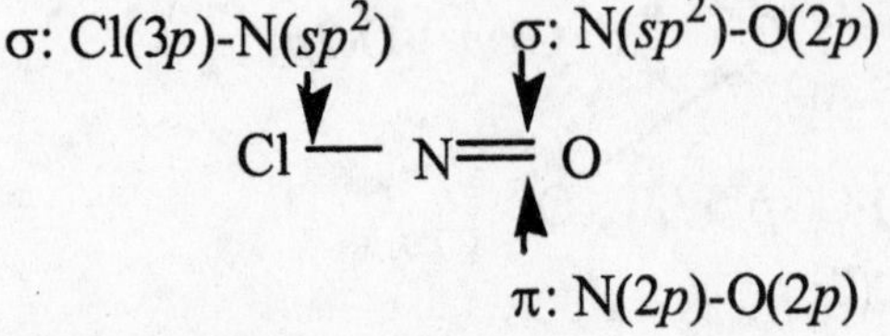

57.   (a) and (c) can be *cis-trans* isomers.

(a)

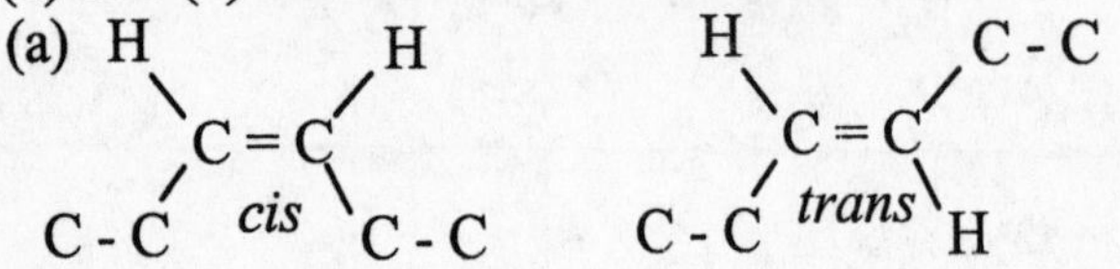

$\begin{array}{c} H \\ \diagdown \\ C=C \\ \diagup \qquad \diagdown \\ C-C \quad cis \quad C-C \end{array}$ $\begin{array}{c} H \quad C-C \\ \diagdown \quad \diagup \\ C=C \\ \diagup \quad trans \diagdown \\ C-C \qquad H \end{array}$

(c)

$\begin{array}{c} H \qquad H \\ \diagdown \quad \diagup \\ C=C \\ \diagup \quad cis \diagdown \\ C-C \qquad C \end{array}$ $\begin{array}{c} H \qquad C \\ \diagdown \quad \diagup \\ C=C \\ \diagup \quad trans \diagdown \\ C-C \qquad H \end{array}$

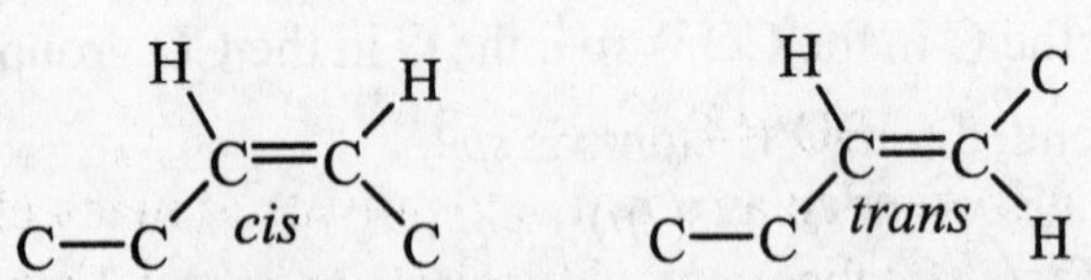

There is no *cis-trans* isomerism for (b) because two H atoms on one C atom at the double bond allows one structure to be flipped over to be the other structure.

59.

$\begin{array}{c} H \\ \diagdown \\ C=CHCH_2CH_3 \\ \diagup \\ H \end{array}$ $\begin{array}{c} H \qquad CH_3 \\ \diagdown \quad \diagup \\ C=C \\ \diagup \quad \diagdown \\ H \qquad CH_3 \end{array}$

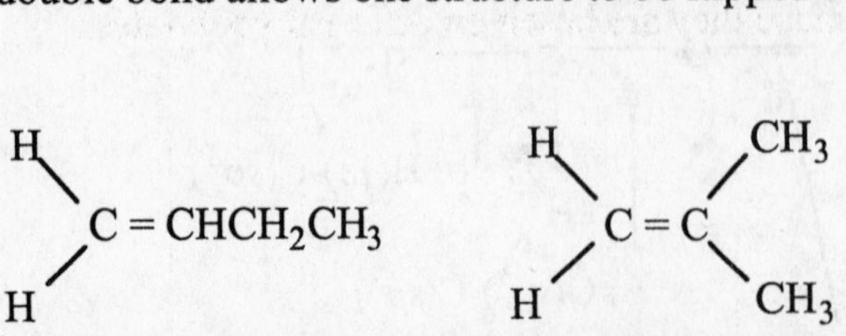

Neither would have *cis-trans* isomerism because two like groups (H) are attached to the carbons. Replacing one of the H with Cl would produce *cis-trans* isomers of 1-

butene but would not produce *cis-trans* isomers for isobutylene because the second carbon still has two identical groups (CH_3) bonded to it.

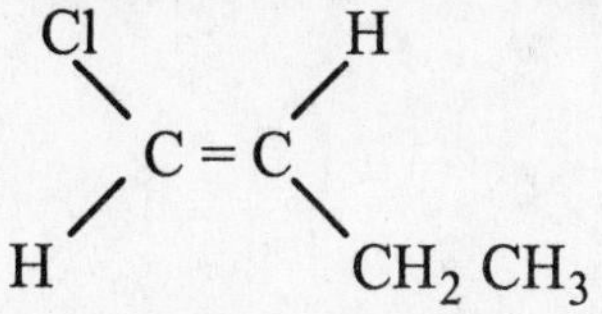

trans

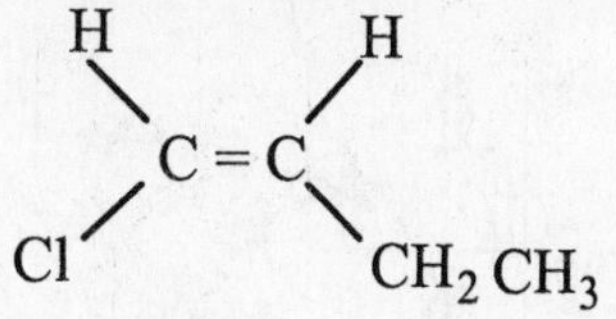

cis

61. F_2^+ has a greater bond energy because it has two fewer antibonding electrons. The electron lost by F_2 to become F_2^+ comes from an antibonding MO, therefore strengthening the F–F bond. The electron gained by F_2 to become F_2^- goes into an antibonding MO, therefore weakening the F–F bond in F_2^-.

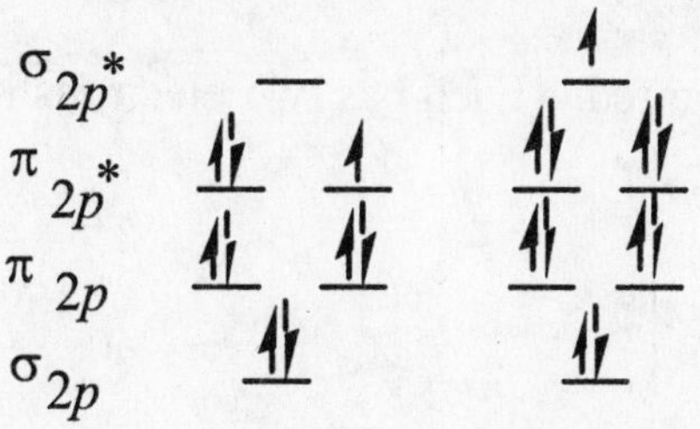

F_2^+ F_2^-

63. MO

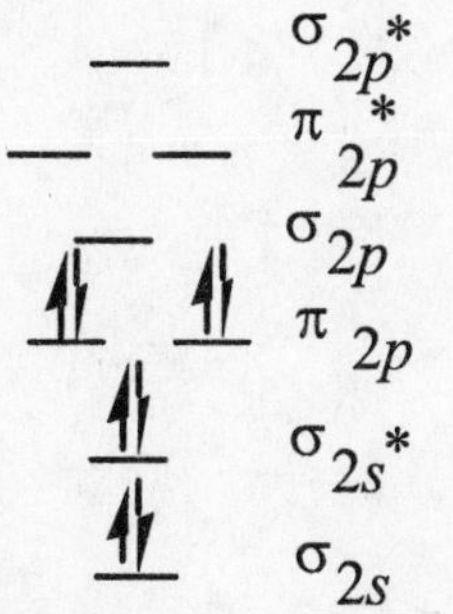

$BO = \frac{6-2}{2} = 2$

Lewis C :::: C BO = 4

The Lewis structure needs as much overlap as possible to provide as many electrons as possible to produce octets. The MO theory would indicate an σ and 2 π bonds, but it would also indicate a σ*, so the bond order would be less. The two theories would have different results because they are based on different psotulates.

65.

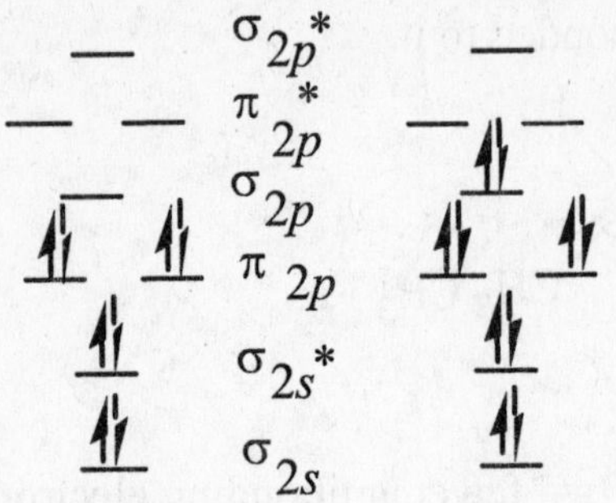

	CN^+	CN^-
BO	2	3
	diamagnetic	diamagnetic

(a) CN^- has the stronger bond because it has a higher bond order, 8 e^- in the bonding orbitals instead of six.

(b) Neither is paramagnetic because all e^- are paired. (Each has two electrons in antibonding MOs.)

67.

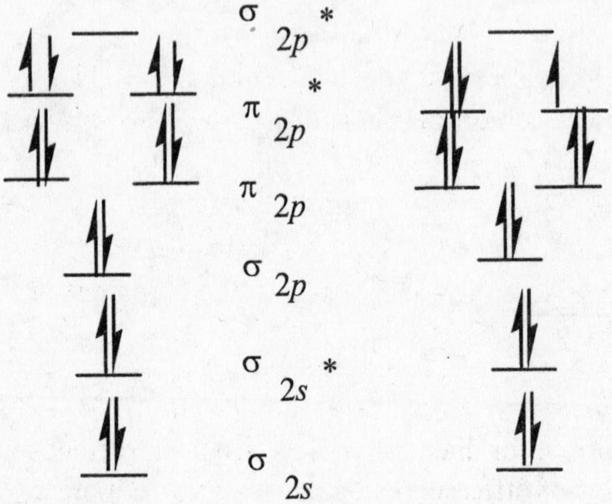

	O_2^{2-}	O_2^-
	diamagnetic	paramagnetic
BO	$\frac{8-6}{2} = 1$	$\frac{8-5}{2} = 1.5$
	peroxide ion	superoxide ion

69.

(a) 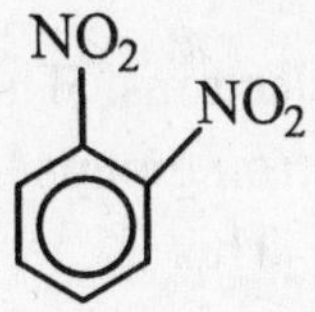 (b) 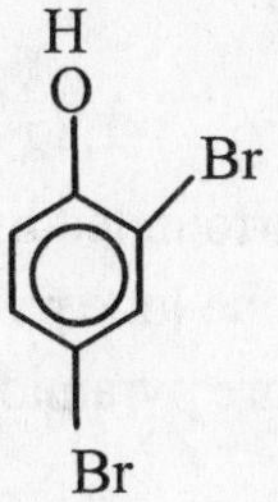 (c) 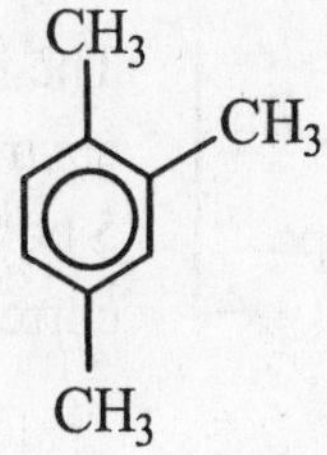

71. (a) 2,5 - diiodoaniline
(b) 1,2,4 - trifluorobenzene
(c) 3,4,5 - tribromotoluene

73.

(a) 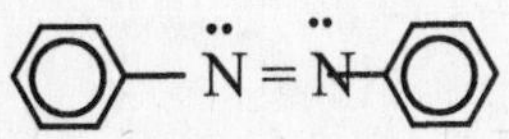

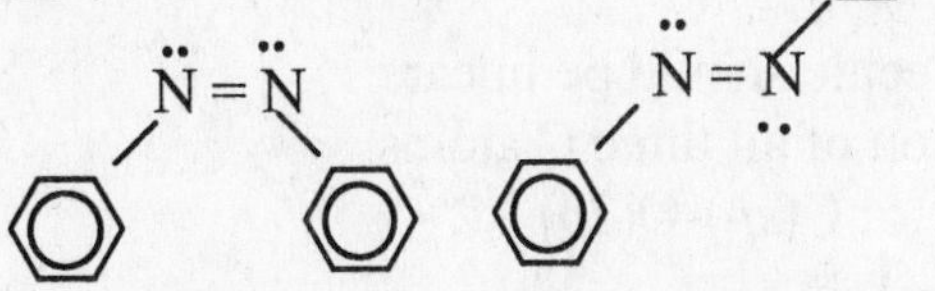

cis *trans*

(b) All C are sp^2. The delocalized e^- are in π molecular orbitals Both N are sp^2. There has to be one *p* orbital for each of the π portions of the N=N double bond.
(c) C - C - C bond angles are 120°.
C - C - N bond angles are 120°.
H-C-C- bond angles are also 120°.
C - N - N bond angles are a little less than 120°, because the lone pairs are extra repulsive.

Additional Problems

75. No. Although the statement is correct for diatomic molecules, for molecules with more than two atoms, bond dipoles of different bonds may cancel each other, resulting in a nonpolar molecule.

77. The flourine atom has a greater electronegativity than the oxygen atoms have, so the flourine end of the molecule acquires a slight negative charge, and the oxygen end a slight positive charge.

79. (a)

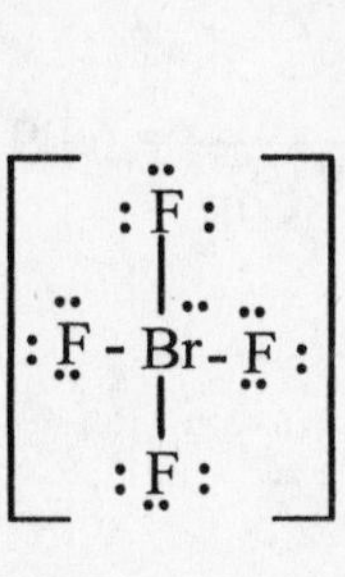

The ion has (5x7) - 1 = 34 valence electrons. The central Br atom is surrounded by five electron groups--4 bonding pairs and one lone pair: AX_4E. This corresponds to a "seesaw" shape.

(b)

$$\left[\begin{array}{c} :\ddot{F}: \\ | \\ :\ddot{F}-\overset{..}{Xe}-\ddot{F}: \\ / \quad \backslash \\ :\ddot{F}: \ :\ddot{F}: \end{array} \right]^{+}$$

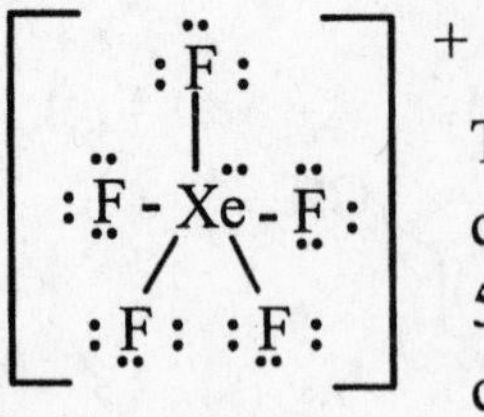

The ion has 8 + (5 x 7) - 1 = 42 valence electrons. The central Xe atom is surrounded by six electron groups-- 5 bonding pairs and one lone pair: AX_5E. This corresponds to a square pyramidal shape.

82. ethane < methylamine < methanol < fluoromethane
     C_2H_6 CH_3NH_2 CH_3OH CH_3F
     Ethane has a zero dipole moment. The other three increase as the electronegativity of N to O to F increase.

84.

$$\ddot{O} = C = C = C = \ddot{O}$$

According to VSEPR theory, the molecule should be linear, and this corresponds to sp hybridization of all three C atoms.

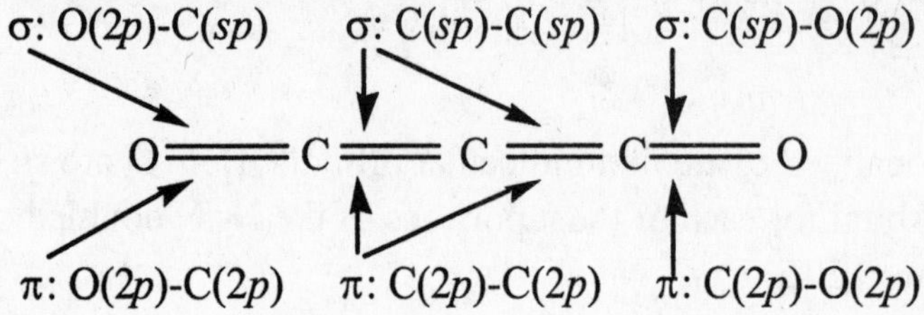

σ: O(2p)-C(sp)   σ: C(sp)-C(sp)   σ: C(sp)-O(2p)

π: O(2p)-C(2p)   π: C(2p)-C(2p)   π: C(2p)-O(2p)

87.

$$H - \overset{\overset{\displaystyle H}{|}}{\underset{1}{C}} = \underset{2}{C} = \overset{\overset{\displaystyle H}{|}}{\underset{3}{C}} - H$$

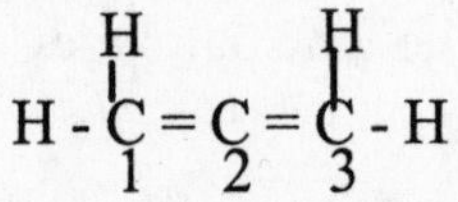

Carbon atoms 1 and 3 have a trigonal planar electron-group geometry.

The CH_2 groups are planar, with a H-C-H bond angle of about 120°

and sp^2 hybridization of the C atom. The C-C-C portion of the molecule

is linear, with sp hybridization of carbon atom 2.

90. electron charge = 1.602×10^{-19} C
     HCl bond length = Cl⁻ ionic radii = 127 pm (from Table 9.1)

$$\mu = \delta d = 1.602 \times 10^{-19} \text{ C} \times 127 \text{ pm} \times \frac{10^{-12} \text{ m}}{\text{pm}} \times \frac{\text{D}}{3.34 \times 10^{-30} \text{ C m}} = 6.09 \text{ D}$$

$$\% \text{ ionic character} = \frac{1.07}{6.09} \times 100\% = 17.6\%$$

92. (a)

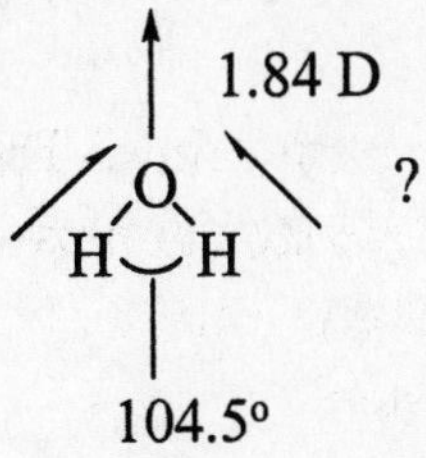

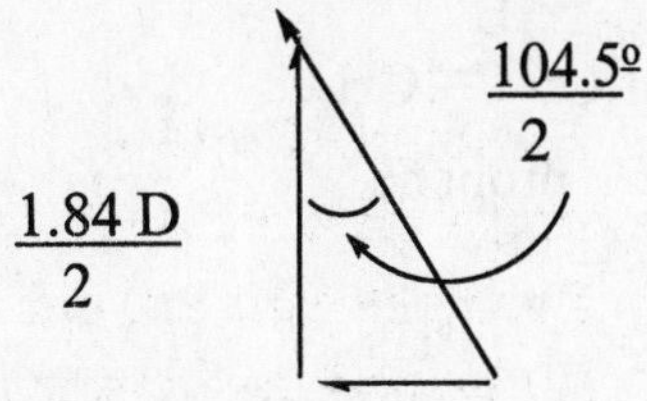

$$\cos\frac{104.5^\circ}{2} = \frac{\text{adjacent side}}{\text{hypotonuse}}$$

$$\frac{\frac{1.84\text{ D}}{2}}{\cos\frac{104.5^\circ}{2}} = 1.50\text{ D}$$

(b)

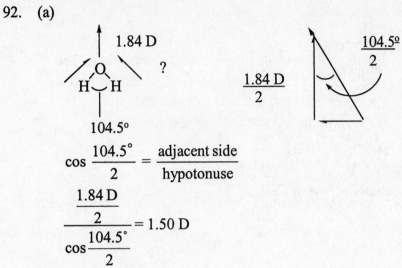

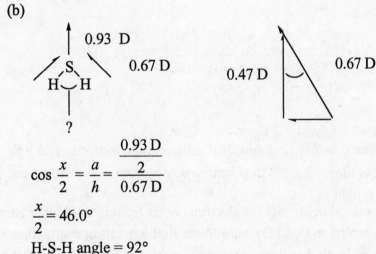

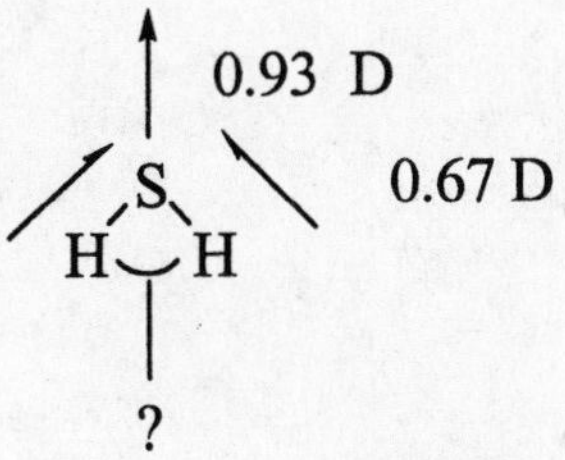

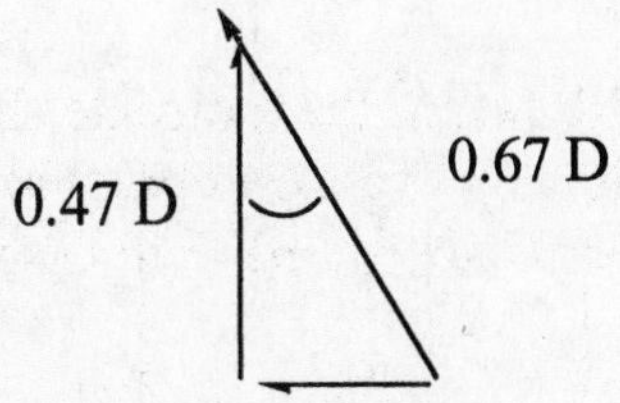

$$\cos\frac{x}{2} = \frac{a}{h} = \frac{\frac{0.93\text{ D}}{2}}{0.67\text{ D}}$$

$$\frac{x}{2} = 46.0^\circ$$

H-S-H angle = 92°

95. $85.7\text{ g C} \times \frac{\text{mol C}}{12.01\text{ g C}} = 7.14\text{ mol C} \times \frac{1}{7.14\text{ mol C}} = 1.00\,\frac{\text{mol C}}{\text{mol C}}$

$14.1\text{ g H} \times \frac{\text{mol H}}{1.008\text{ g H}} = 14.0\text{ mol H} \times \frac{1}{7.14\text{ mol C}} = 1.96\,\frac{\text{mol H}}{\text{mol C}}$

CH_2 empirical formula $14\,\frac{\text{g}}{\text{formula}}$

at 1 atm and 25° C

M = density × molar volume = 1.72 g/L × 24.45 L/mol = 42.1 g/mol

at 1 atm and 100° C

M = density × molar volume = 1.38 g/L × 30.6 L/mol = 42.2 g/mol

M = (42.1 g/mol + 42.2 g/mol)/2 =

$$M = 42.2\,\frac{\text{g}}{\text{mol}} \times \frac{\text{formula}}{14\text{ g}} = \frac{3\text{ formula}}{\text{molecule}}$$

C_3H_6

Two sets of data were given to provide an average value.

(a)

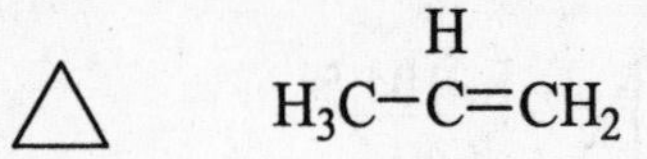

(b) cyclopropane propene

(c)

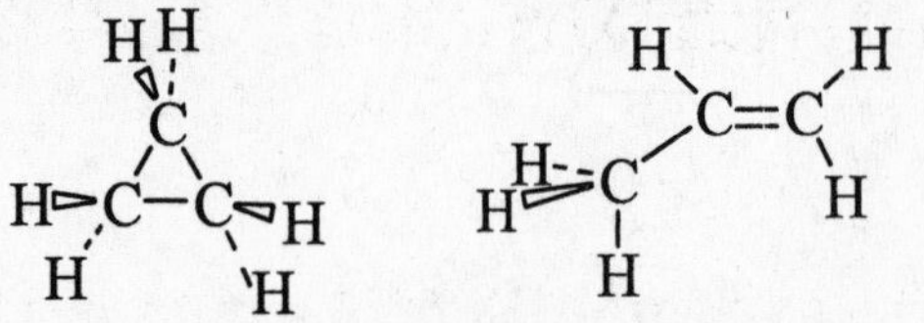

(d) There are no other combinations of C_3H_6.

96. (a) $Li_2 \rightarrow Li_2^+ + e^-$

$Be_2 \rightarrow Be_2^+ + e^-$

$B_2 \rightarrow B_2^+ + e^-$

$C_2 \rightarrow C_2^+ + e^-$

$N_2 \rightarrow N_2^+ + e^-$

$O_2 \rightarrow O_2^+ + e^-$

$F_2 \rightarrow F_2^+ + e^-$

$Ne_2 \rightarrow Ne_2^+ + e^-$

(b) Yes, Ne_2^+ has more bonding electrons than antibonding electrons. BO = ½.

(c) The antibonding orbital in F_2 that the electron comes from is at a higher energy level than the2*p* orbital in F.

(d) The antibonding orbital in O_2 that the electron comes from is at a higher energy level than the 2*p* orbital in O, so O_2 has a lower first-ionization energy than O.

(e) The bonding orbital in N_2 that the electron comes from is at a lower level than the 2*p* orbital in N, so N_2 has a higher first-ionization energy than N.

(f) The first ionization potential increases from $O_2 < O < N < N_2$. Thus molecular nitrogen has a higher ionization energy than molecular oxygen. Helium also has a higher ionization potential because of its small size.

Chapter 11

States of Matter and Intermolecular Forces

Exercises

11.1A $\Delta H_{vap} = \dfrac{652\ \text{J}}{1.50\ \text{g}\ C_6H_6} \times \dfrac{78.11\ \text{g}\ C_6H_6}{\text{mol}\ C_6H_6} \times \dfrac{\text{kJ}}{10^3\ \text{J}} = 34.0\ \text{kJ/mol}$

11.1B raise temperature

$$\Delta H = 0.750\ \text{L} \times \frac{\text{mL}}{10^{-3}\ \text{L}} \times \frac{0.789\ \text{g}}{\text{mL}} \times \frac{2.46\ \text{J}}{\text{g}\ ^\circ\text{C}} \times \frac{\text{kJ}}{10^3\ \text{J}} \times (25.0 - 0.0)\ ^\circ\text{C} = 36.4\ \text{kJ}$$

vaporize

$$\Delta H = \frac{10.0\ \text{L}}{100\ \text{L}} \times 0.750\ \text{L ethanol} \times \frac{\text{mL}}{10^{-3}\ \text{L}} \times \frac{0.789\ \text{g}}{\text{mL}} \times \frac{\text{mol}\ C_2H_5OH}{46.07\ \text{g}\ C_2H_5OH} \times \frac{43.3\ \text{kJ}}{\text{mol}} = 55.6\ \text{kJ}$$

$$\Delta H_{total} = 36.4\ \text{kJ} + 55.6\ \text{kJ} = 92.0\ \text{kJ}$$

11.2A CS_2 $1.00\ \text{kg} \times \dfrac{10^3\ \text{g}}{\text{kg}} \times \dfrac{\text{mol}}{76.13\ \text{g}} \times \dfrac{27.4\ \text{kJ}}{\text{mol}} = 360\ \text{kJ}$

CCl_4 $1.00\ \text{kg} \times \dfrac{10^3\ \text{g}}{\text{kg}} \times \dfrac{\text{mol}}{153.81\ \text{g}} \times \dfrac{37.0\ \text{kJ}}{\text{mol}} = 241\ \text{kJ}$

CH_3OH $1.00\ \text{kg} \times \dfrac{10^3\ \text{g}}{\text{kg}} \times \dfrac{\text{mol}}{32.04\ \text{g}} \times \dfrac{38.0\ \text{kJ}}{\text{mol}} = 1.19 \times 10^3\ \text{kJ}$

C_8H_{18} $1.00\ \text{kg} \times \dfrac{10^3\ \text{g}}{\text{kg}} \times \dfrac{\text{mol}}{114.24\ \text{g}} \times \dfrac{41.5\ \text{kJ}}{\text{mol}} = 363\ \text{kJ}$

C_2H_5OH $1.00\ \text{kg} \times \dfrac{10^3\ \text{g}}{\text{kg}} \times \dfrac{\text{mol}}{46.07\ \text{g}} \times \dfrac{43.3\ \text{kJ}}{\text{mol}} = 940\ \text{kJ}$

H_2O $1.00\ \text{kg} \times \dfrac{10^3\ \text{g}}{\text{kg}} \times \dfrac{\text{mol}}{18.02\ \text{g}} \times \dfrac{44.0\ \text{kJ}}{\text{mol}} = 2.44 \times 10^3\ \text{kJ}$

$C_6H_5NH_2$ $1.00\ \text{kg} \times \dfrac{10^3\ \text{g}}{\text{kg}} \times \dfrac{\text{mol}}{93.13\ \text{g}} \times \dfrac{52.3\ \text{kJ}}{\text{mol}} = 562\ \text{kJ}$

CCl_4, with a relatively low ΔH value and a large molar mass, requires the smallest quantity of heat for its vaporization.

11.2B $\Delta H = n\Delta H_{vapor}$

$$?\ \text{kJ} = 1.0\ \text{mol} \times \frac{40.7\ \text{kJ}}{\text{mol}} = 40.7\ \text{kJ}$$

$$q = mC_p\Delta T$$

$$m = \frac{40.7\,\text{kJ} \times \dfrac{10^3\ \text{J}}{\text{kJ}}}{\dfrac{4.184\ \text{J}}{\text{g}\,^\circ\text{C}} \times 100^\circ\text{C}} = 97\ \text{g}$$

11.3A $P = \dfrac{mRT}{VM} = \dfrac{1.100\ \text{g x}\ \dfrac{0.08206\ \text{L atm}}{\text{K mol}}\ \text{x}\ (273 + 39)\text{K x}\ \dfrac{760\text{mmHg}}{\text{atm}}}{335\ \text{mL x}\ \dfrac{10^{-3}\ \text{L}}{\text{mL}}\ \text{x}\ 159.80\ \text{g/mol}} = 4.00 \times 10^2\ \text{mmHg}$

11.3B $m = \dfrac{PVM}{RT} = \dfrac{19.8\ \text{mm}\ H_2O\ \text{x}\ 275\ \text{mL x}\ \dfrac{10^{-3}\ \text{L}}{\text{mL}}\ \text{x}\ 18.02\ \text{g/mol}}{\dfrac{62.36\ \text{L mmHg}}{\text{K mol}}\ \text{x}\ (273 + 22)\text{K}} = 5.33 \times 10^{-3}\ \text{g}\ H_2O$

11.4A Methane has a $T_c = -82.6$ °C at $P_c = 45.4$ atm. Room temperature is above T_c, so methane stays a gas at all pressures. The pressure gauge would work fine.

11.4B The initial volume must be enough so that there is both liquid and vapor at the critical temperature then the disappearance of the meniscus can be seen.

11.5A $P = \dfrac{nRT}{V} = \dfrac{1.05\ \text{mol x}\ \dfrac{0.08206\ \text{L atm}}{\text{K mol}}\ \text{x}\ (273.2 + 30.0)\text{K}}{2.61\ \text{L}}$

$P = 10.0$ atm

This pressure greatly exceeds the vapor pressure of H_2O at 30.0 °C, so some of the vapor condenses to liquid. The sample cannot be all liquid, however, because 1.05 mol $H_2O(l)$ occupies a volume of less than 20 mL. The final condition reached is one of liquid and vapor in equilibrium.

11.5B The solid CO_2 sublimes at the temperature of liquid water. The gaseous CO_2 is bubbling up through the water. Water vapor is condensed to droplets by the cold of the dry ice. The invisible gaseous CO_2 carries the water droplets (fog) down the side of the beaker.

11.6A IBr has the greater molecular mass and therefore the greater London forces. Both molecules are somewhat polar, but the electronegativity differences are small. IBr is a solid, and BrCl is a gas.

11.6B (a)Toluene is less polar because the carbon in a methyl group is less electronegative than the N in an amine group. It requires less energy to separate molecules of toluene, so it has a lower boiling point than aniline.

(b) The *trans*-isomer is nonpolar because the chlorine atoms are balanced on the molecules. *Trans*-1,2-dichloroethene would have the lower boiling point than the polar *cis*-1,2-dichloroethene.

(c) The *para* isomer is nonpolar because the chlorine atoms are symmetrical on the molecule. On the *ortho* isomer, the chlorine atoms are on one side, so the molecule is polar. *Para*-dichlorobenzene has the lower boiling point.

11.7A The difference in boiling point is larger as each compound has one less carbon. The differences are about 10% greater each time. That would make the next difference 32.6 °C + 3.3 °C = 35.9 °C. 36.1 °C − 35.9 °C = 0.2 °C. Butane should boil at about 0.2 °C.

11.7B $C_{10}H_{22}$ (174 °C) plus 21 °C to $C_{11}H_{24}$ (195 °C) plus 19 °C to $C_{12}H_{26}$ (214 °C) plus 17 °C to $C_{13}H_{28}$ (231 °C) plus 15 °C to $C_{14}H_{30}$ (246 °C) plus 13 °C to $C_{15}H_{32}$ (259 °C). Kerosene should be $C_{11}H_{24}$, $C_{12}H_{26}$, $C_{13}H_{28}$, $C_{14}H_{30}$, and $C_{15}H_{32}$.

11.8A NH_3, yes. H is bonded to N, a small electronegative atom.
CH_4, no. C is not electronegative enough for hydrogen bonding.
C_6H_5OH, yes. H is bonded to O, a small electronegative atom. Hydrogen bonding is somewhat overshadowed by dispersion forces associated with the large organic part of the molecule.
$CH_3\overset{\overset{O}{\|}}{C}OH$, yes. H is bonded to O, a small electronegative atom
H_2S, no. There is a little hydrogen bonding because although S is electronegative, S is too large to permit much hydrogen bonding.
H_2O_2, yes. H is bonded to O, a small electronegative atom.

11.8B (a) $(CH_3)_2CHOH$ Forces are hydrogen bonding, polar, and London.
(b) CS_2 Forces are London. 76 g/mol
(c) $HOCH_2CH_2OH$ Forces are hydrogen bonding, polar, and London.
(d) $(CH_3)_3CH$ Forces are London. 58 g/mol
CS_2 has a higher molar mass than $(CH_3)_3CH$ and $HOCH_2CH_2OH$ has more hydrogen bonding than $(CH_3)_2CHOH$.
(d) < (b) < (a) < (c)

11.9A $CsBr < KI < KCl < MgF_2$
A cesium ion is larger than a potassium ion. An iodide ion is larger than a chloride ion, which is larger than a fluoride ion. A magnesium ion is a small cation with a 2+ charge and exerts stronger interionic attractions than do either K^+ or Cs^+ ions.

11.9B CsBr, and KCl, and KI are soluble. MgF_2 is insoluble because of the stronger ionic bond caused by the 2+ charge on magnesium and the smaller interionic distance due to the smaller size of magnesium and fluoride ions.

11.10A (a) In a bcc cell the distance from one corner of the cell to the opposite corner is

4r. That distance is the hypotenuse of a right triangle whose other sides are the length (ℓ) of the cell (down one corner) and the diagonal of the bottom face of the cell. The diagonal distance from the Pythogorian theorem is $d_d = \sqrt{\ell^2 + \ell^2} = \sqrt{2\ell^2}$. The distance from one corner of the cell to the other is $d = (2\ell^2 + \ell^2)^{\frac{1}{2}}$.

Since d = 4r, then

$4r = (2\ell^2 + \ell^2)^{\frac{1}{2}} = (3\ell^2)^{\frac{1}{2}}$

$4 \times 124.1 \text{ pm} = \sqrt{3}\ \ell$

$\ell = 286.6 \text{ pm}$

11.10B $V = \ell^3 = (286.6 \text{ pm})^3 \times \left(\frac{10^{-12} \text{ m}}{\text{pm}}\right)^3 \times \left(\frac{\text{cm}}{10^{-2} \text{ m}}\right)^3 = 2.354 \times 10^{-23} \text{ cm}^3$

In a bcc cell there are two atoms per cell—one in the center and $\frac{1}{8}$ at each of the eight corners.

11.11A Avogadro's number $= \frac{2 \text{ atoms}}{\text{cell}} \times \frac{\text{cell}}{2.354 \times 10^{-23} \text{ cm}^3} \times \frac{\text{cm}^3}{7.874 \text{ g}} \times \frac{55.847 \text{ g}}{\text{mol}}$

$= 6.026 \times 10^{23}$ atoms/mol

11.11B In simple cubic, there is the equivalent of one atom in a unit cell of length $2r$.

$V_{cell} = (2r)^3 = 8r^3$

$V_{atom} = 4/3\ \pi r3 = 4.1888\ r^3$

void % $= \frac{(8.000 - 4.1888)\ r^3}{8\ r^3} \times 100\% = 47.64\%$

Review Questions

3. (a) Increases in temperature increase the vapor pressure.
   (b), (c), and (d) have no effect. The vapor pressure will be the same whatever the volume of liquid and vapor and the area of contact between them, as long as both phases exist together.

5. (d) enthalpy of sublimation is the largest of the values because it includes both enthalpy of vaporization and enthalpy of fusion.

7. (a) the solid to sublime without melting. At 1 atm and room temperature, only solid and gas phases exist.

8. All four gases have only London forces and Br_2 is heavier than the other compounds. (d) is the answer.

10. HF Forces are hydrogen bonding, polar, and London.
    CH_4 Forces are London.

CH_3OH Forces are hydrogen bonding, polar, and London.
N_2H_4 Forces are hydrogen bonding, polar, and London.
(c) all but one of there (CH_4).

13. A bcc crystal has an atom at the center with 8 closest neighbors, one at each corner.
(b) 8 is the answer.

14. (c) is true.
Ions that are at the corners or on the faces are shared with another cell.

16. N_2 has a slightly lower molar mass than O_2, so it would have weaker intermolecular forces and a lower boiling point. (b) -196°C is a reasonable choice.

Problems

17. $\Delta H = 3.530 \text{ kg } C_8H_{18} \times \frac{10^3 \text{ g}}{\text{kg}} \times \frac{\text{mol } C_8H_{18}}{114.21 \text{ g } C_8H_{18}} \times \frac{41.5 \text{ kJ}}{\text{mol}} = 1.28 \times 10^3 \text{ kJ}$

19. $\Delta H_T = \Delta H_1 + \Delta H_2$
18.0 °C → 25.0 °C $\Delta H_1 = \text{mass} \times \text{sp. ht.} \times \Delta T$

$$\Delta H_1 = 79.8 \text{ g} \times \frac{4.180 \text{ J}}{\text{g °C}} \times (25.0 - 11.3) \text{ °C} \times \frac{\text{kJ}}{1000 \text{ J}} = 4.57 \text{ kJ}$$

liquid → vapor $\Delta H_2 = n\Delta H_{vap}$

$$\Delta H_2 = 79.8 \text{ g} \times \frac{\text{mol}}{18.02 \text{ g}} \times \frac{44.0 \text{ kJ}}{\text{mol}} = 195 \text{ kJ}$$

$\Delta H_T = \Delta H_1 + \Delta H_2 = 200. \text{ kJ}$

21. $$? \text{ g } C_3H_8 = 6.75 \text{ L } H_2O \times \frac{\text{mL}}{10^{-3} \text{ L}} \times \frac{1 \text{ g}}{\text{mL}} \times \frac{\text{mol } H_2O}{18.02 \text{ g } H_2O} \times \frac{44.0 \text{ kJ}}{\text{mol}} \times \frac{\text{mol } C_3H_8}{2.22 \times 10^3 \text{ kJ}} \times \frac{44.09 \text{ g } C_3H_8}{\text{mol } C_3H_8} = 327 \text{ g } C_3H_8$$

23. In an oven the heat is a dry heat. In steam, there is the heat transfer associated with the temperature difference plus a very large quantity of heat from the condensation of steam. The transfer of heat from air in the oven to the hand occurs slowly, and so the hand warms slowly. Above the boiling water there is an almost instantaneous transfer of a large quantity of heat, from condensed steam (the heat of condensation) directly into the hand.

25. (a) about 40 mmHg
(b) about 62 °C

27. V.P. at 20° = 290 mm Hg

$$V = \frac{nRT}{P} = 1.5 \times 10^{22} \text{ molecules} \times \frac{\text{mol}}{6.022 \times 10^{23} \text{ molecules}} \times 62.4 \frac{\text{L mmHg}}{\text{K mol}} \times \frac{293 \text{ K}}{290 \text{ mmHg}} = 1.6 \text{ L}$$

29. $$P = \frac{mRT}{VM} = \frac{0.480 \text{ g x } \dfrac{0.08206 \text{ L atm}}{\text{K mol}} \text{ x } (40.0 + 273.2)\text{K}}{\dfrac{153.81 \text{ g}}{\text{mol}} \text{ x } 285 \text{ mL x } \dfrac{10^{-3} \text{ L}}{\text{mL}}} = 0.281 \text{ atm}$$

$$P = 0.281 \text{ atm} \times \frac{760 \text{ mmHg}}{\text{atm}} = 214 \text{ mmHg}$$

31. $$P = \frac{mRT}{VM} = \frac{0.195 \text{ g} \times \dfrac{0.08206 \text{ L atm}}{\text{K mol}} \times (30.0 + 273.2)\text{K}}{\dfrac{18.02 \text{ g}}{\text{mol}} \times 2.55 \text{ L}} \times \frac{760 \text{ mmHg}}{\text{atm}} = 80.2 \text{ mmHg}$$

Because the calculated vapor of $H_2O(g)$ (80.2 mmHg) greatly exceeds the vapor pressure of water at 30.0 °C (31.8 mmHg), some of the water vapor condenses. The final mixture is one of liquid and vapor. The final condition cannot be liquid exclusively because 0.195 g of $H_2O(l)$ only occupies a volume of less than 2 mL. The rest of the 2.55-L flask must be filled with $H_2O(g)$.

33. The high heat capacity of water means that the heat goes into the water instead of heating the cup above the water temperature. When the water does boil, the heat is used for the large heat of vaporization of water, and the temperature of the cup stays at the temperature of the water as long as liquid water remains. The temperature of the boiling water is below the ignition temperature of the paper cup.

35. A gas cannot be liquefied above T_c, regardless of the pressure applied. A gas can be either liquefied or solidified by a sufficient lowering of the temperature, regardless of its pressure. A gas can always be liquefied by an appropriate combination of pressure and temperature changes.

37. $$\Delta H = n\Delta H_{fus} = 4.2 \text{ cm} \times 2.1 \text{ cm} \times 2.9 \text{ cm} \times \frac{0.92 \text{ g}}{\text{cm}^3} \times \frac{\text{mol}}{18.02 \text{ g}} \times \frac{6.01 \text{ kJ}}{\text{mol}} = 7.8 \text{ kJ}$$

39. $\Delta H_{ice} = -\ \Delta H_{H2O}$

$n_{ice}\Delta H_{fus} + \text{mass}_{ice} \times \text{sp.ht.}_{ice} \times \Delta T_{ice} = -\ \text{mass}_{water} \times \text{sp.ht.}_{water} \times \Delta T_{water}$

$$\left(25.5 \text{ g} \times \frac{\text{mol}}{18.02 \text{ g}} \times \frac{6.01 \text{ kJ}}{\text{mol}}\right) + \left(25.5 \text{ g} \times \frac{4.18 \text{ J}}{\text{g °C}} \times (t - 0.0)\text{ °C}\right) \times \frac{\text{kJ}}{10^3 \text{ J}} =$$

$$-\ 125 \text{ mL} \times \frac{1.00 \text{ g}}{\text{mL}} \times \frac{4.18 \text{ J}}{\text{g °C}} \times (t - 26.5) \text{ °C} \times \frac{\text{kJ}}{10^3 \text{ J}}$$

$$8.50\text{ kJ} + 0.107\ \frac{t\text{ kJ}}{°\text{C}} = -\,0.523\ \frac{t\text{ kJ}}{°\text{C}} + 13.8\text{ kJ}$$

$$0.630\ \frac{t\text{ kJ}}{°\text{C}} = 5.3\text{ kJ}$$

$$t = 8.4\ °\text{C}$$

41.

(a) $q = 2.00\text{ kg} \times \dfrac{10^3\text{ g}}{\text{kg}} \times \dfrac{-\,2.30\text{ kJ}}{\text{mol}} \times \dfrac{\text{mol}}{201\text{ g}}$

(b) $q = 0.50\text{ mol} \times \dfrac{-44.0\text{ kJ}}{\text{mol}}$

(c) $q = 10.0\text{ g} \times \dfrac{\text{mol}}{18.0\text{ g}} \times \dfrac{(-6.0 - 44.0)\text{ kJ}}{\text{mol}}$

(d) $q = 100.0\text{ mL} \times \dfrac{1.00\text{ g}}{\text{mL}} \times \dfrac{4.18\text{ J}}{\text{g °C}} \times \dfrac{\text{kJ}}{10^3\text{ J}} \times (1.0{-}51.0)\ °\text{C}$

by estimating		actual
(a) $\dfrac{2000}{200} \times -2.3$	≈ -23	-22.9
(b) 0.5×-44	≈ -22	-22
(c) $\dfrac{10}{20} \times -50$	≈ -25	-27.8
(d) $\dfrac{100 \times 4 \times 50}{1000} = \dfrac{4 \times 50}{10}$	≈ -20	-20.9

(c) evolves the largest quantity of heat.

43.

(a)

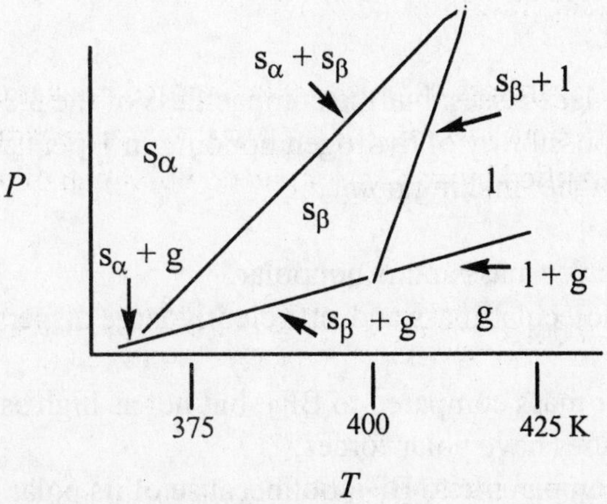

(b) The two triple points at the bottom have s_α, s_β and gas present at one. The other has s_β, g, and l present. The triple point at the top has s_α, s_β, and l all present at the same time.

(c) The sulfur gas becomes a liquid, then monoclinic sulfur and finally rhombic sulfur as the pressure increases.

45.

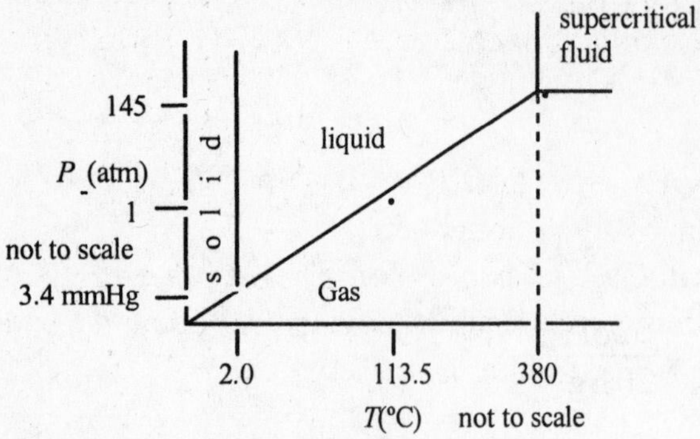

(See note in Problem 44.)

47. The solid CO_2 becomes a liquid at a temperature slightly above -56.7 °C (the triple point is 5.1 atm and -56.7 °C). The liquid then converts to gaseous CO_2 at a temperature that is probably below room temperature (the critical point is at 72.9 atm and 31 °C (304.2 K)).

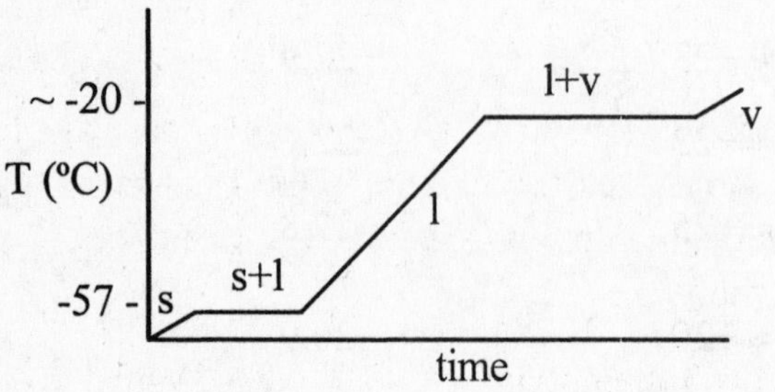

49. Both substances are nonpolar, but because of its lower molecular mass, CF_4 should have weaker dispersion forces than PBr_5. As a result, it should have the lower boiling point.

51. The substances have similar molecular masses, but the compactness of the 3,3-dimethylpentane molecule and the possibility of hydrogen bonding in 1-pentanol account for 1-pentanol having the higher melting point.

53. BF_3 is the gas. It has a small molecular mass and is nonpolar.
NI_3 is a solid because of its large molecular mass and sufficiently large dispersion and polar forces.
PCl_3 has a relatively high molecular mass compared to BF_3, but not as high as NI_3. PCl_3 is expected to be a liquid. It does have polar forces.
CH_3COOH has a molecular mass comparable to BF_3, but because of its polar character and ability to form hydrogen bonds, it is expected to be a liquid.

55. CH_4 < CH_3CH_3 < NH_3 < H_2O
−161 −89 H bond −33 H bond 100

CH_4 and CH_3CH_3 are nonpolar. NH_3 has a higher boiling point than CH_3CH_3, despite ethane's higher molecular mass, because of hydrogen bonding. In H_2O, hydrogen bonding is stronger still.

57. CH_3OH < C_6H_5OH < NaOH < LiOH
H bond –97.8 H bond 40.9 ionic 318 ionic 450
C_6H_5OH has a higher melting point than CH_3OH because of its greater molecular mass. The ionic compounds have higher melting points than the covalent compounds, and that of LiOH is greater than NaOH because of the smaller size of the Li^+ ion.

59. 1-octanol has H bonds, and octane has only dispersion forces, so 1-octanol should have the stronger intermolecular forces and the higher surface tension.

61. Water "wets" glass, because adhesive forces between water molecules and glass exceed cohesive forces in H_2O (l). The situation is the reverse for some materials, such as Teflon, and water does not "wet" them. Wetting agents lower the surface tension of water and improve its wetting ability.

63. (a) The larger, upper unit is a unit cell, but not the smaller, lower one. The larger cell can be shifted left, right, up, and down with no gaps. That is, the bottom edge of one cell coincides with the top edge of the identical cell below it. In contrast, the bottom edge of the small cell does not coincide with the top edge of the identical cell below it; there is a gap between them.

(b) $(4 \times \frac{1}{4}) + 2 = 3$ hearts $(4 \times \frac{1}{2}) + 1 = 3$ diamonds $(4 \times \frac{1}{2}) + 1 = 3$ clubs

65. (a) In a bcc cell the distance from one corner of the cell to the opposite corner is 4r. That distance is the hypotenuse of a right triangle whose other sides are the length (ℓ) of the cell (down one corner) and the diagonal of the bottom face of the cell. The diagonal distance from the Pythagorean theorem is $d_d = \sqrt{\ell^2 + \ell^2} = \sqrt{2\ell^2}$. The distance from one corner of the cell to the other is $d = (2\ell^2 + \ell^2)^{\frac{1}{2}}$.

Since $d = 4r$, then

$4r = (2\ell^2 + \ell^2)^{\frac{1}{2}} = (3\ell^2)^{\frac{1}{2}}$

$4 \times 217 \text{ pm} = \sqrt{3}\ \ell$

$\ell = 501 \text{ pm}$

(b) $V = \ell^3 = (501 \text{ pm})^3 \times \left(\frac{10^{-12}\text{ m}}{\text{pm}}\right)^3 \times \left(\frac{\text{cm}}{10^{-2}\text{ m}}\right)^3 = 1.26 \times 10^{-22} \text{ cm}^3$

(c) In a bcc unit there is one atom in the center and $8 \times \frac{1}{8}$ atom in the corner, so 2 atom/unit.

$$d = \frac{m}{V} = \frac{\frac{137.3\text{ g}}{\text{mol}} \times \frac{\text{mol}}{6.022\times10^{23}\text{ atoms}} \times \frac{2\text{ atom}}{\text{unit cell}}}{1.26\times10^{-22}\text{ cm}^3} = 3.62\text{ g/cm}^3$$

67. (a) For a Cs^+ in the center of Cl^- ions, the distance from one corner to the opposite corner is 2 × 169 pm + 2 × 181 pm = 700. pm. That line is the hypotenuse for a right triangle that has one side the length of the cell (down the corner) and one side as the diagonal of the bottom of the cell ($\sqrt{2}\ \ell$). Since $c^2 = a^2 + b^2$, the distance from one corner to the other is

$$\sqrt{\left(\sqrt{2}\ell\right)^2 + \ell^2} = 700\text{ pm}$$

$$\ell = 404\text{ pm}$$

(b) $V = \ell^3 = (404\text{ pm})^3 \times \left(\frac{10^{-10}\text{ cm}}{\text{pm}}\right)^3 = 6.59 \times 10^{-23}\text{ cm}^3$

for each cell there are one Cs and $8 \times \frac{1}{8}$ Cl^-, so $\frac{\text{molar mass}}{\text{cell}} = \frac{168.4\text{ g}}{\text{mole}}$

mass of unit cell = $6.59 \times 10^{-23}\text{cm}^3 \times 3.989\text{ g/cm}^3 = 2.63 \times 10^{-22}\text{g}$

$$N_A = \frac{168.4\text{g CsCl/mol}}{2.63 \times 10^{-22}\text{g CsCl/formula unit}} = 6.40 \times 10^{23}\ \frac{\text{formula unit}}{\text{mol}}$$

69. Ti^{4+} $8 \times \frac{1}{8} + 1 = 2$

O^{2-} $4 \times \frac{1}{2} + 2 = 4$ $Ti_2O_4 \Rightarrow TiO_2$

Ti^{4+} coordination number 6

O^{2-} coordination number 3

No. Since there are twice as many O^{2-} as Ti^{4+}, it is expected for the coordination number of Ti^{4+} to be twice that of O^{2-}.

Additional Problems

71. vapor pressure = 23.8 mmHg

$$n = \frac{PV}{RT} = \frac{23.8\text{ mmHg}\times1.00\text{ L}}{\frac{62.364\text{ mmHg L}}{\text{K mol}}\times 298\text{ K}}$$

estimate $n \approx \frac{20}{60\times300} = \frac{1}{900} = 1.1 \times 10^{-3}$

so choose 1.2×10^{-3}

actual 1.28×10^{-3} moles

73. (a) $\Delta H = n\,\Delta H_{fus} = 10.0\text{ g} \times \frac{\text{mol}}{153.81\text{ g}} \times \frac{3.28\text{ kJ}}{\text{mol}}$

$\Delta H = 0.213\text{ kJ}$

(b) $$V = \frac{nRT}{P} = \frac{1.00\text{ mol x } \frac{0.08206\text{ L atm}}{\text{K mol}} \text{ x } (77 + 273)\text{K}}{1.00\text{ atm}}$$

$V = 28.7\ \text{L}$

(c) $P = \dfrac{mRT}{VM} = \dfrac{3.5\ \text{g} \times \dfrac{0.08206\ \text{L atm}}{\text{K mol}} \times (25.0 + 273.2)\ \text{K}}{153.81\ \text{g/mol} \times 8.21\ \text{L}}$

$P = 0.06782\ \text{atm} \times \dfrac{760\ \text{mmHg}}{\text{atm}} = 51.5\ \text{mmHg}$

There will only be gas. The amount of material produces so little gas pressure that it does not reach the vapor pressure of 110 mmHg.

76. (a) CH_3CH_2OH Hydrogen bonding is most important. There are also dipole-dipole and dispersion forces.
(b) CH_3CH_2Cl Dipole-dipole is the most important force. There are also dispersion forces.
(c) HCOONa Ionic bonding is the most important force.

78. The density of liquid water rises from its value at the melting point to a maximum at 3.98 °C, because at this temperature hydrogen bonding between water molecules occurs to the greatest extent. Above this temperature its density falls with temperature in a customary fashion. Thus, for every density in a range from 0 °C to 3.98 °C and 1 atm pressure, there is another temperature in a range extending a few degrees above 3.98 °C at which the same density is observed.

80. Different printings of the text have given different formulas so there are two versions of the problem.
(a) $\log p = 7.5547 - \dfrac{1002.7}{-75.0 + 247.89} = 1.755$

$10^{\log p} = p = 10^{1.755} = 56.9\ \text{mmHg}$

(b) The normal boiling point is the boiling point at atm pressure.

$\log (760) = 7.5547 - \dfrac{1002.7}{t + 247.89}$

$2.8808 = 7.5547 - \dfrac{1002.7}{t + 247.89}$

$4.6739 = \dfrac{1002.7}{t + 247.89}$

$t + 247.89 = \dfrac{1002.7}{4.6739} = 214.53$

$t = 214.53 - 247.89 = -33.36\ °\text{C}$

(c) The critical point is the last point on the vapor-pressure curve.

$405.6\ \text{K} - 273.2 = 132.4\ °\text{C}$

$\log p = 7.5547 - \dfrac{1002.7}{132.4 + 247.89} = 4.918$

$10^{\log p} = p = P_c = 10^{4.918} = 8.279 \times 10^4\ \text{mmHg} = 108.9\ \text{atm}$

second version

(a) $\log p = 7.3605 - \dfrac{1617.9}{-75.0 + 240.17} = -2.435$

$10^{\log p} = p = 10^{-2.435} = 3.67 \times 10^{-3}$ mmHg

(b) **The normal boiling point is the boiling point at atm pressure.**

$\log(760) = 7.3605 - \dfrac{1617.9}{t + 240.17}$

$2.8808 = 7.3605 - \dfrac{1617.9}{t + 240.17}$

$4.4797 = \dfrac{1617.9}{t + 240.17}$

$t + 240.17 = \dfrac{1617.9}{4.4797} = 361.16$

$t = 361.16 - 240.17 = 120.99$ °C

(c) **The critical point is the last point on the vapor-pressure curve.**

405.6 K − 273.2 = 132.4 °C

$\log p = 7.3605 - \dfrac{1617.9}{132.4 + 240.17} = 4.342$

$10^{\log p} = p = P_c = 10^{4.342} = 2.198 \times 10^4$ mmHg = 28.9 atm

83. $2\ H_2\ (g) + O_2\ (g) \rightarrow 2\ H_2O\ (l)$

$1.00\ g\ H_2 \times \dfrac{mol}{2.016\ g} \times \dfrac{2\ mol\ H_2O}{2\ mol\ H_2} = 0.496\ mol\ H_2O$

$10.00 g\ O_2 \times \dfrac{mol\ O_2}{32.00\ g\ O_2} \times \dfrac{2\ mol\ H_2O}{mol\ O_2} = 0.625\ mol\ H_2O$ H_2 is limiting reactant

$1.00\ g\ H_2 \times \dfrac{mol}{2.016\ g} \times \dfrac{mol\ O_2}{2\ mol\ H_2} = 0.248\ mol\ O_2$ used

$10.00\ g \times \dfrac{mol}{32.00\ g} = 0.3125\ mol\ O_2$ original

0.3125 mol - 0.248 mol = 0.065 mol excess O_2

$$P_{O_2} = \frac{nRT}{V} = \frac{0.065\ \text{moles x}\ \dfrac{0.08206\ \text{L atm}}{\text{K mol}}\ \text{x}\ (25 + 273)\text{K}}{3.15\ \text{L}}$$

$P_{O_2} = 0.50$ atm $= 3.8 \times 10^2$ mmHg

The *V.P.* of H_2O at 25 °C is 23.8 mmHg.

$P_{total} = P_{O_2} + P_{H_2O}$

$P_{total} = 3.8 \times 10^2$ mmHg $+ 0.24 \times 10^2$ mmHg

$P_{total} = 4.0 \times 10^2$ mmHg

85. $P = \dfrac{mRT}{VM} = \dfrac{1.00\ \text{g} \times \dfrac{0.08206\ \text{L atm}}{\text{K mol}} \times (35.0 + 273.2)\ \text{K}}{18.02\ \text{g/mol} \times 40.0\ \text{L}} = 0.0351\ \text{atm} \times \dfrac{760\ \text{mmHg}}{\text{atm}}$

$= 26.7\ \text{mmHg}$

at 30°C, $26.7\ \text{mmHg} \times \dfrac{303\text{K}}{308\text{K}} = 26.3\ \text{mmHg}$

$<$ v.p. at 30 °C

at 28°C, $P = 26.7\ \text{mmHg} \times \dfrac{301\text{K}}{308\text{K}} = 26.1\ \text{mmHg}$

$<$ v.p. at 28 °C

at 26°C, $P = 26.7\ \text{mmHg} \times \dfrac{299\text{K}}{308\text{K}} = 25.9\ \text{mmHg}$

$>$ v.p. at 26 °C

at 27°C, $P = 26.7\text{mmHg} \times \dfrac{300\text{K}}{308\text{K}} = 26.0\ \text{mmHg}$

$<$v.p. at 27 °C

$\Delta p = 0.7\ \text{mmHg}$

The pressure = the v.p. of water between 26 and 27 °C, close to 26.5 °C. At 27 °C the pressure in the chamber will be equal to the vapor pressure of water, and liquid will begin to form.

90. diagonal of unit cell face

$4x = b$

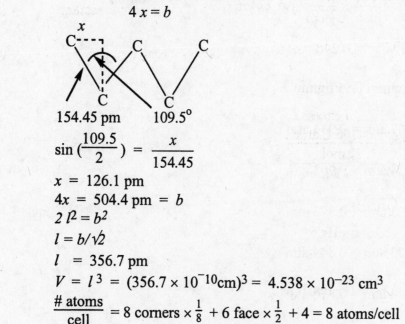

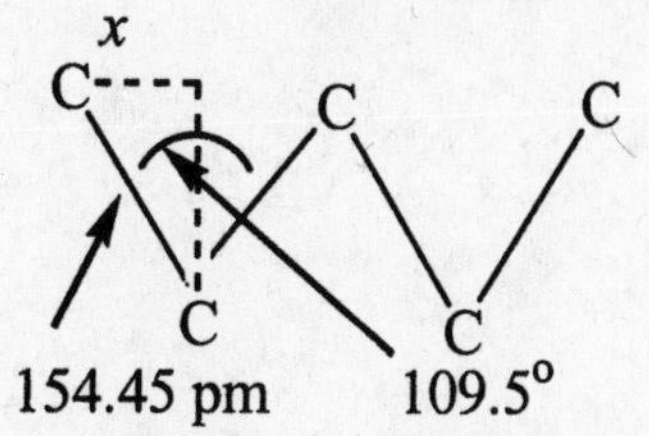

$\sin\left(\dfrac{109.5}{2}\right) = \dfrac{x}{154.45}$

$x = 126.1\ \text{pm}$

$4x = 504.4\ \text{pm} = b$

$2\,l^2 = b^2$

$l = b/\sqrt{2}$

$l = 356.7\ \text{pm}$

$V = l^3 = (356.7 \times 10^{-10}\text{cm})^3 = 4.538 \times 10^{-23}\ \text{cm}^3$

$\dfrac{\#\ \text{atoms}}{\text{cell}} = 8\ \text{corners} \times \frac{1}{8} + 6\ \text{face} \times \frac{1}{2} + 4 = 8\ \text{atoms/cell}$

$d = \dfrac{m}{V} = \dfrac{8\ \text{atoms} \times \dfrac{12.01\text{g}}{\text{mole}} \times \dfrac{\text{mole}}{6.022 \times 10^{23}\ \text{atoms}}}{4.538 \times 10^{-23}\ \text{cm}^3}$

$d = 3.516\ g/cm^3$

The tabulated value in CRC is 3.51 g/cm^3 and in Lange 14/e is 3.53 g/cm^3.

94. Hg(l) 20.0 °C → −39 °C

$\Delta H_1 = mC\Delta T$

$$\Delta H_1 = 525\ cm^3\ Hg \times \frac{13.6\ g}{cm^3} \times \frac{0.033\ cal}{g\,°C} \times (-39-20)\ °C \times \frac{kcal}{10^3\ cal} \times \frac{4.184\ kJ}{kcal} = -58\ kJ$$

Hg(l) → Hg(s) at −39 °C

$$\Delta H_2 = n\Delta H_f = 525\ cm^3 \times \frac{13.6\ g}{cm^3} \times \frac{mol}{200.0\ g} \times \frac{-2.30\ kJ}{mol} = -81.9\ kJ$$

Hg(s) −39 °C → −196 °C

$\Delta H_3 = mC\Delta T$

$$\Delta H_3 = 525\ cm^3\ Hg \times \frac{13.6\ g}{cm^3} \times \frac{0.030\ cal}{g\,°C} \times (-196-(-39))\ °C \times \frac{kcal}{10^3\ cal} \times \frac{4.184\ kJ}{kcal} = -1.4 \times 10^2\ kJ$$

$$\Delta H_{total} = \Delta H_1 + \Delta H_2 + \Delta H_3$$

$$\Delta H_{total} = -58\ kJ - 81.9\ kJ - 140\ kJ = -280\ kJ$$

$$\text{mass } N_2 = 280 kJ \times \frac{mol}{5.58\ kJ} \times \frac{28.01\ g}{mol} = 1.4 \times 10^3\ g$$

98. (a) $? \text{ m} = 900\ ft \times \frac{12\ in}{ft} \times \frac{2.54\ cm}{in} \times \frac{10^{-2}\ m}{cm} = 274\ m$

$$1917\ m \times \frac{1}{274\ m} = 7.00$$

$$1.00\ atm - \frac{1}{30} \times 1.00\ atm = 0.967\ atm$$

$$0.967\ atm - \frac{1}{30} \times 0.967\ atm = 0.934\ atm$$

$$0.934\ atm - \frac{1}{30} \times 0.934\ atm = 0.903\ atm$$

$$0.903\ atm - \frac{1}{30} \times 0.903\ atm = 0.873\ atm$$

$$0.873\ atm - \frac{1}{30} \times 0.873\ atm = 0.844\ atm$$

$$0.844\ atm - \frac{1}{30} \times 0.844\ atm = 0.816\ atm$$

$$0.816\ atm - \frac{1}{30} \times 0.816\ atm = 0.789\ atm$$

$$?\ mmHg = 0.789\ atm \times \frac{760\ mmHg}{atm} = 599\ mmHg$$

(b) from table 11.2 588.6 mmHg is 93.0 °C
610.9 mmHg is 94.0 °C

599 – 589 = 10

611 – 599 = 12

The boiling temperature is slightly closer to 93 °C than to 94 °C

100. At the triple point, the pressures should be equal.

$$6.9379 - \frac{861.34}{t + 246.33} = 9.7051 - \frac{1444.2}{t + 267.13}$$

$$-\frac{861.34}{t + 246.33} = 2.7672 - \frac{1444.2}{t + 267.13}$$

$$-861.34 = 2.7672\,t + 681.644 - \frac{(1444.2\,t + 355750)}{t + 267.13}$$

$$-1542.98 = 2.7672\,t - \frac{(1444.2\,t + 355750)}{t + 267.13}$$

$$-1542.98\,t - 412176 = 2.7672\,t^2 + 739.20\,t - 1444.2\,t - 355750$$

$$0 = 2.7672\,t^2 + 838.0\,t + 56426$$

$$0 = t^2 + 302.8\,t + 20391$$

$$t = \frac{-302.8 \pm \sqrt{(302.8)^2 - 4 \times 20391}}{2}$$

$$t = \frac{-302.8 \pm 100.6}{2}$$

$t = -101.1$ °C or -201.7 °C

(-202 °C produces a pressure of 3.22×10^{-13} mmHg, an unrealistic value.)

$T_c = -101.1$ °C

$$\log p = 6.9379 - \frac{861.34}{t + 246.33}$$

$$\log p = 6.9379 - \frac{861.34}{-101.1 + 246.33} =$$

$$10^{\log p} = p = 10^{1.007} = 10.16\ \text{mmHg} = P_c$$

Apply Your Knowledge

102. $$\gamma = \frac{hdgr}{2} = \frac{1.10\ \text{cm} \times \frac{0.789\ \text{g}}{\text{mL}} \times \frac{9.8066\ \text{m}}{\text{s}^2} \times 0.50\ \text{mm}}{2}$$

$$\gamma = \frac{1.10\ \text{cm}}{2} \times \frac{10^{-2}\ \text{m}}{\text{cm}} \times \frac{0.789\ \text{g}}{\text{mL}} \times \frac{\text{kg}}{10^3\ \text{g}} \times \frac{\text{mL}}{\text{cm}^3} \times \frac{\text{cm}^3}{(10^{-2})^3\ \text{m}} \times \frac{9.8066\ \text{m}}{\text{s}^2}$$

$$\times\ 0.50\ \text{mm} \times \frac{10^{-3}\ \text{m}}{\text{mm}} = \frac{0.021\ \text{J}}{\text{m}^2}$$

104. $P = \dfrac{nRT}{V} = \dfrac{0.40\text{ g} \times \dfrac{\text{mol}}{200.6\text{ g}} \times \dfrac{62.4\text{ L Torr}}{\text{K mol}} \times 295\text{ K}}{(10.0 \times 12.0 \times 3.0)\text{m}^3} \times \left(\dfrac{10^{-2}\text{ m}}{\text{cm}}\right)^3 \times \dfrac{\text{cm}^3}{\text{mL}} \times \dfrac{\text{mL}}{10^{-3}\text{ L}}$

$= 1.0 \times 10^{-4}$ Torr

This is less than the equilibrium vapor pressure of mercury, so all of the mercury spilled will vaporize.

$$? \frac{\text{mg Hg}}{\text{m}^3} = \frac{0.40\text{ g}}{(10.0 \times 12.0 \times 3.0)\text{ m}^3} \times \frac{\text{mg}}{10^{-3}\text{ g}} = \frac{1.1\text{ mg}}{\text{m}^3}$$

$$?\text{ times overexposure} = \frac{\dfrac{1.1\text{ mg}}{\text{m}^3}}{\dfrac{0.050\text{ mg}}{\text{m}^3}} = 22\text{ times}$$

Chapter 12

Physical Properties of Solutions

Exercises

12.1A $\frac{163 \text{ g glucose}}{(163 \text{ g} + 755 \text{ g}) \text{ solution}} \times 100\% = 17.8\%$ glucose by mass

12.1B $? \text{ g sucrose} = 225 \text{ g solution} \times \frac{6.25 \text{ g sucrose}}{100 \text{ g solution}} = 14.06 \text{ g sucrose}$

$? \text{ g sucrose} = 135 \text{ g solution} \times \frac{8.20 \text{ g sucrose}}{100 \text{ g solution}} = 11.07 \text{ g sucrose}$

$? \text{ g sucrose} = 14.06 \text{ g sucrose} + 11.07 \text{ g sucrose} = 25.13 \text{ g sucrose}$

$? \text{ g solution} = 225 \text{ g} + 135 \text{ g} = 360 \text{ g solution}$

$\text{mass } \% = \frac{25.13 \text{ g sucrose}}{360 \text{ g solution}} \times 100\% = 6.98\%$

12.2A $\text{volume } \% \text{ toluene} = \frac{40.0 \text{ mL toluene x } 100\%}{(40.0 \text{ mL} + 75.0 \text{ mL}) \text{ solution}} = 34.8 \text{ \% toluene}$

12.2B (a) $? \text{ g toluene} = \frac{0.866 \text{ g}}{\text{mL}} \times 40.0 \text{ mL toluene} = 34.6 \text{ g toluene}$

$? \text{ g benzene} = \frac{0.879 \text{ g}}{\text{mL}} \times 75.0 \text{ mL benzene} = 65.9 \text{ g benzene}$

$\text{mass } \% \text{ toluene} = \frac{34.6 \text{ toluene x } 100\%}{(34.6 \text{ g} + 65.9 \text{ g}) \text{ solution}} = 34.4 \text{ \% toluene}$

(b) $d = \frac{34.6 \text{ g} + 65.9 \text{ g}}{40.0 \text{ mL} + 75.0 \text{ mL}} = 0.874 \text{ g/mL}$

12.3A $\frac{0.1 \ \mu\text{g}}{\text{L}} = \frac{0.1 \ \mu\text{g}}{1000 \text{ g}} = \frac{0.1 \ \mu\text{g}}{1000 \text{ g x } \frac{\mu\text{g}}{10^{-6} \text{ g}}} = 0.1 \text{ ppb}$

$\frac{0.1 \ \mu\text{g}}{\text{L}} = \frac{0.1 \ \mu\text{g}}{1000 \text{ g}} = \frac{0.1 \ \mu\text{g x } 10^3}{1000 \text{ g x } \frac{\mu\text{g}}{10^{-6} \text{ g}} \text{ x } 10^3} = \frac{100 \ \mu\text{g}}{10^{12} \ \mu\text{g}} = 100 \text{ ppt}$

12.3B ppm $Na^+ = \frac{1.52 \times 10^{-3} \text{ mol } Na_2SO_4}{L} \times \frac{10^{-3} \text{ L}}{\text{mL}} \times \frac{\text{mL}}{1.00 \text{ g}} \times \frac{2 \text{ mol } Na^+}{\text{mol } Na_2SO_4}$

$\times \frac{22.99 \text{ g } Na^+}{\text{mol } Na^+} \times \frac{10^6}{10^6} = \frac{69.9 \text{ g } Na^+}{10^6 \text{ g}} = 69.9 \text{ ppm } Na^+$

12.4A ? $m = \frac{225 \text{ mg } C_6H_{12}O_6}{5.00 \text{ mL } C_2H_5OH} \times \frac{\text{mL}}{0.789 \text{ g}} \times \frac{10^3 \text{ g}}{\text{kg}} \times \frac{10^{-3} \text{ g}}{\text{mg}} \times \frac{\text{mol } C_6H_{12}O_6}{180.2 \text{ g } C_6H_{12}O_6}$

$= 0.317 \text{ m } C_6H_{12}O_6$

12.4B ? $m = \frac{25.0 \text{ mL } C_6H_5CH_3}{75.0 \text{ mL } C_6H_6} \times \frac{\text{mL}}{0.879 \text{ g}} \times \frac{10^3 \text{ g}}{\text{kg}} \times \frac{0.866 \text{ g}}{\text{mL}} \times \frac{\text{mol}}{92.13 \text{ g}} = 3.56 \text{ m}$

12.5A ? mL $C_2H_5OH = 125 \text{ mL } C_6H_6 \times \frac{0.879 \text{ g}}{\text{mL}} \times \frac{\text{kg}}{10^3 \text{ g}} \times \frac{0.0652 \text{ mol } C_2H_5OH}{\text{kg } C_6H_6}$

$\times \frac{46.07 \text{ g } C_2H_5OH}{\text{mol } C_2H_5OH} \times \frac{\text{mL}}{0.789 \text{ g}} = 0.418 \text{ mL}$

12.5B ? mL $H_2O = 25.0 \text{ g } CO(NH_2)_2 \times \frac{1 \text{ mol } CO(NH_2)_2}{60.06 \text{ g } CO(NH_2)_2} \times \frac{\text{kg } H_2O}{1.65 \text{ mol } CO(NH_2)_2}$

$\times \frac{1000 \text{ g } H_2O}{\text{kg } H_2O} \times \frac{\text{mL } H_2O}{0.998 \text{ g } H_2O} = 253 \text{ mL } H_2O$

12.6A (a) $m\ CH_3OH = \frac{7.50 \text{ g } CH_3OH}{92.5 \text{ g } C_2H_5OH} \times \frac{10^3 \text{ g } C_2H_5OH}{\text{kg } C_2H_5OH} \times \frac{\text{mol } CH_3OH}{32.04 \text{ g } CH_3OH} = 2.53\ m$

(b) mole % $CO(NH_2)_2 = \frac{1.05 \text{ mol } CO(NH_2)_2 \times 100\%}{1.05 \text{ mol } CO(NH_2)_2 + \left(1.00 \text{ kg} \times \frac{10^3 \text{ g}}{\text{kg}} \times \frac{\text{mol } H_2O}{18.02 \text{ g}}\right)}$

$= 1.86 \text{ mol \% } CO(NH_2)_2$

12.6B (a) $2.90 \text{ mol} \times 32.04 \text{ g/mol} = 92.9 \text{ g } CH_3OH$

$\frac{92.9 \text{ g}}{(1.00 \text{ kg} \times \frac{1000 \text{ g}}{\text{kg}}) + 93 \text{ g}} \times 100\% = 8.50\% \ CH_3OH$ by mass

(b) $\frac{2.90 \text{ mol}}{1093 \text{ g soln}} \times \frac{0.984 \text{ g}}{\text{mL}} \times \frac{\text{mL}}{10^{-3} \text{ L}} = 2.61 \text{ M}$

(c) $1000 \text{ g } H_2O \times \frac{\text{mol } H_2O}{18.02 \text{ g } H_2O} = 55.5 \text{ mol}$

mole % $= \frac{2.90 \text{ mol}}{55.5 \text{ mol} + 2.9 \text{ mol}} \times 100\% = 4.97\%$

12.7A (b) 5.0% C_2H_5OH by mass is the largest. The solutions are all rather dilute and have densities of approximately 1.0 g/mL. The mole percents of C_2H_5OH are not large, and are expected to be the largest in the solution having the greatest quantity of C_2H_5OH per liter or kilogram of solution. Solution (a) has 0.5 mol C_2H_5OH per liter; (b) has slightly more than 1 mol C_2H_5OH per kilogram of solution; (c) has slightly less than 0.5 mol C_2H_5OH per kilogram of solution; (d) has somewhat less than 1 mol C_2H_5OH per liter.

Another estimation method follows:

(a) is greater than (c) because (a) is 0.50 mole C_2H_5OH in 1 L solution (about 994 g solution - 23 g C_2H_5OH = 971 g H_2O), but (c) is in 1 kg of H_2O. (d) is less than (b) because the 5% mass contains a bigger percent C_2H_5OH than 5% volume, since the density of C_2H_5OH is lower than water. (b) is greater than (a) because 5 g of C_2H_5OH is about 0.1 mol. 0.1 mol is a larger percent of 5 mol than 0.5 mol is of 50 mol.

Actual calculations -

(a) $? \text{ g solution} = 1 \text{ L solution} \times \frac{\text{mL}}{10^{-3}\text{ L}} \times \frac{0.994 \text{ g solution}}{\text{mL solution}} = 994 \text{ g solution}$

$? \text{ mole } H_2O = (994 \text{ g solution} - 23 \text{ g } C_2H_5OH) \times \frac{\text{mol}}{18.01 \text{ g}} = 53.9 \text{ mol } H_2O$

$? \text{ mole \% } C_2H_5OH = \frac{0.50 \text{ mol}}{0.50 \text{ mol} + 53.9 \text{ mol}} \times 100\% = 0.92 \text{ mol\%}$

(b) $? \text{ mol } H_2O = 95 \text{ g } H_2O \times \frac{\text{mol}}{18.01 \text{ g}} = 5.27 \text{ mol } H_2O$

$? \text{ mol } C_2H_5OH = 5.0 \text{ g } C_2H_5OH \times \frac{\text{mol}}{46.0} = 0.11 \text{ mol } C_2H_5OH$

$? \text{ mol \% } C_2H_5OH = \frac{0.11 \text{ mol } C_2H_5OH}{5.38 \text{ mol solution}} \times 100\% = 2.0 \text{ mol \%}$

(c) $? \text{ mol } H_2O = 1.00 \text{ kg } H_2O \times \frac{10^3 \text{ g}}{\text{kg}} \times \frac{\text{mol}}{18.01 \text{ g}} = 55.5 \text{ mol } H_2O$

$? \text{ mol \% } C_2H_5OH = \frac{0.50 \text{ mol } C_2H_5OH}{56.0 \text{ mol solution}} \times 100\% = 0.89 \text{ mol \%}$

(d) $? \text{ g solution} = 1.00 \text{ L solution} \times \frac{\text{mL}}{10^{-3}\text{ L}} \times \frac{0.991 \text{ g}}{\text{mL}} = 991 \text{ g solution}$

$? \text{ g } C_2H_5OH = \frac{5 \text{ mL}}{100 \text{ mL}} \times 1 \text{ L} \times \frac{\text{mL}}{10^{-3}\text{ L}} \times \frac{0.789 \text{ g}}{\text{mL}} = 39 \text{ g } C_2H_5OH$

$? \text{ g } H_2O = 991 \text{ g} - 39 \text{ g} = 952 \text{ g } H_2O$

$? \text{ mol } H_2O = 952 \text{ g } H_2O \times \frac{\text{mol}}{18.02 \text{ g}} = 52.8 \text{ mol } H_2O$

$? \text{ mol } C_2H_5OH = 39 \text{ g } C_2H_5OH \times \frac{\text{mol}}{46.07 \text{ g}} = 0.85 \text{ mol}$

$? \text{ mol \% } C_2H_5OH = \frac{0.85}{52.8 + 0.85} \times 100\% = 1.6 \text{ \%}$

12.7B (d) is larger than (a) or (b) because 1.00 mol water is fewer grams of water than 1 kg or 1 L. (d) is larger than (c) because 18 grams is less than 85 grams.

(a) 0.010 M NaCl
$$? \text{ ppm} = \frac{0.010 \text{ mol}}{\text{kg solute}} \times \frac{23.0 \text{ g}}{\text{mol}} \times \frac{\text{kg}}{1000 \text{ g}} \times \frac{10^3}{10^3} = \frac{230 \text{ g Na}}{10^6 \text{ g}} = 2.3 \times 10^2 \text{ ppm Na}^+$$

(b) 0.010 M Na_2SO_4
$$? \text{ ppm} = \frac{0.010 \text{ mol}}{1.00 \text{ L solution}} \times \frac{23.0 \text{ g}}{\text{mol}} \times \frac{2 \text{ mol Na}^+}{1 \text{ mol Na}_2\text{SO}_4} \times \frac{10^{-3} \text{ L}}{\text{mL}} \times \frac{\text{mL}}{1.00 \text{ g}} \times \frac{10^6}{10^6} = \frac{230 \text{ g Na}}{10^6 \text{ g}} = 4.6 \times 10^2 \text{ ppm Na}^+$$

(c) 1.0 % $NaNO_3$
$$? \text{ ppm} = \frac{0.010 \text{ g NaNO}_3}{1.00 \text{ g solution}} \times \frac{23.0 \text{ g Na}}{89.0 \text{ g NaNO}_3} \times \frac{10^6}{10^6} = \frac{270 \text{ g Na}}{10^6 \text{ g}} = 2.7 \times 10^2 \text{ ppm Na}^+$$

(d) 0.010 mol fraction Na
$$? \text{ ppm} = \frac{0.010 \text{ mol}}{1.00 \text{ mole solution}} \times \frac{23.0 \text{ g}}{\text{mol}} \times \frac{\text{mol}}{18.0 \text{ g}} \times \frac{10^6}{10^6} = \frac{1.3 \times 10^4}{10^6 \text{ g}} = 1.3 \times 10^4 \text{ ppm}$$

12.8A Because of the large benzene part of the molecule, nitrobenzene should be more soluble in benzene.

12.8B (a) acetic acid
(b) 1-hexanol
(c) hexane
(d) butanoic acid

least soluble c < b < d < a

12.9A
$$\frac{\frac{149 \text{ mg CO}_2}{100 \text{g H}_2\text{O}}}{1 \text{ atm CO}_2} \times 0.00037 \text{ atm CO}_2 = \frac{5.5 \times 10^{-2} \text{ mg CO}_2}{100 \text{ g H}_2\text{O}}$$

12.9B
$$\chi_{N2} = \frac{P_{N_2}}{P_{\text{mixture}}} = \frac{P_{N_2}}{10.0 \text{ atm}} = 0.500$$

$$P_{N2} = 5.00 \text{ atm}$$

$$\chi_{CH4} = \frac{P_{CH_4}}{P_{mixture}} = \frac{P_{CH_4}}{10.0\text{ atm}} = 0.500$$

$$P_{CH4} = 5.00\text{ atm}$$

at 10.0 atm N_2 $\quad S = \dfrac{19\text{ mg }N2}{100\text{ g }H2O}$

CH_4 $\quad S = \dfrac{23\text{ mg }CH4}{100\text{ g }H2O}$

$$k_{N2} = \frac{S}{P_{N_2}} = \frac{\dfrac{19\text{ mg }N_2}{100\text{ g }H_2O}}{10.0\text{ atm}} = \frac{0.019\text{ mg }N2}{\text{g }H2O\text{ atm}}$$

$$k_{CH4} = \frac{S}{P_{CH_4}} = \frac{\dfrac{23\text{ mg }CH_4}{100\text{ g }H_2O}}{10.0\text{ atm}} = \frac{0.023\text{ mg }CH4}{\text{g }H2O\text{ atm}}$$

$$?\text{ mass }N2 = \frac{0.019\text{ mg }N2}{\text{g atm}} \times 5.00\text{ atm} \times 1.00\text{ L} \times \frac{\text{mL}}{10^{-3}\text{ L}} \times \frac{1.00\text{ g}}{\text{mL}} = 95\text{ mg }N2$$

$$?\text{ mass }CH4 = \frac{0.023\text{ mg }CH4}{\text{g atm}} \times 5.00\text{ atm} \times 1.00\text{ L} \times \frac{\text{mL}}{10^{-3}\text{ L}} \times \frac{1.00\text{ g}}{\text{mL}}$$

$$= 1.2 \times 10^2\text{ mg }CH4$$

total mass = 95 mg $N2$ + 1.2×10^2 mg $CH4$ = 2.1×10^2 mg mixture

12.10A $?\text{ mol }C6H5COOH = 5.05\text{ g }C6H5COOH \times \dfrac{\text{mol }C6H5COOH}{122.1\text{ g }C6H5COOH}$

$$= 0.0414\text{ mol }C6H5COOH$$

$$?\text{ mol }C6H6 = 245\text{ g }C6H6 \times \frac{\text{mol }C6H6}{78.11\text{ g }C6H6} = 3.14\text{ mol }C6H6$$

$$\chi_{C6H6} = \frac{3.14\text{ mol }C6H6}{3.18\text{ mol total}} = 0.987$$

$$P_{C6H6} = \chi_{C6H6}\, P^{\circ}_{C6H6} = 0.987 \times 95.1\text{ mmHg} = 93.9\text{ mmHg}$$

12.10B $\chi_{tol} = \dfrac{P_{tol}}{P^{\circ}_{tol}} = \dfrac{27.92\text{ mmHg}}{28.44\text{ mmHg}} = 0.9817$

$$n_{tol} = 500.0\text{ mL} \times \frac{0.862\text{ g}}{\text{mL}} \times \frac{\text{mol}}{92.13\text{ g}} = 4.68\text{ moles}$$

$$\frac{n_{tol}}{n_{tol} + n_{nap}} = 0.9817 = \frac{4.68 \text{ mol}}{4.68 \text{ mol} + n_{nap}} = 0.9817$$

$$4.68 \text{ mol} = 4.59 \text{ mol} + 0.9817\, n_{nap}$$
$$0.08 \text{ mol} = 0.9817\, n_{nap}$$
$$0.08 \text{ mol} = n_{nap}$$
$$?\ g_{nap} = 0.08 \text{ mol} \times \frac{128.2 \text{ g}}{\text{mol}} = 10 \text{ g naphthalene}$$

12.11A Assume 100.0 g

$$? \text{ mol benzene} = 50.0 \text{ g } C_6H_6 \times \frac{\text{mol } C_6H_6}{78.11 \text{ g } C_6H_6} = 0.640 \text{ mol } C_6H_6$$

$$? \text{ mol toluene} = 50.0 \text{ g } C_7H_8 \times \frac{\text{mol } C_7H_8}{92.14 \text{ g } C_7H_8} = 0.543 \text{ mol } C_7H_8$$

$$P_{benzene} = \chi_{benzene} \times P^{\circ}_{benzene} = \frac{0.640 \text{ mol}}{(0.640 + 0.543) \text{ mol}} \times 95.1 \text{ mmHg} = 51.4 \text{ mmHg}$$

$$P_{toluene} = \chi_{toluene} \times P^{\circ}_{toluene} = \frac{0.543 \text{ mol}}{(0.640 + 0.543) \text{mol}} \times 28.4 \text{ mmHg} = 13.0 \text{ mmHg}$$

$$P_{total} = P_{benzene} + P_{toluene} = 51.4 \text{ mmHg} + 13.0 \text{ mmHg} = 64.4 \text{ mmHg}$$

12.11B $P_{ben} = \chi_{ben} P^{\circ}_{ben}$

$P_{tol} = \chi_{tol} P^{\circ}_{tol}$

for mole fractions to be equal, $P_{ben} = P_{tol}$

$\chi_{ben} P^{\circ}_{ben} = \chi_{tol} P^{\circ}_{tol}$ $\qquad \chi_{ben} = 1 - \chi_{tol}$

$(1 - \chi_{tol})\, P^{\circ}_{ben} = \chi_{tol} P^{\circ}_{tol}$

$(1 - \chi_{tol})\, 95.1 \text{ mmHg} = \chi_{tol}\, 28.4 \text{ mmHg}$

$95.1 \text{ mmHg} - 95.1 \text{ mmHg}\, \chi_{tol} = 28.4 \text{ mmHg}\, \chi_{tol}$

$95.1 \text{ mmHg} = 123.5 \text{ mmHg}\, \chi_{tol}$

$$\chi_{tol} = \frac{95.1 \text{ mmHg}}{123.5 \text{ mmHg}} = 0.770$$

$\chi_{ben} = 0.230$

12.12A The greater mole fraction of benzene in the vapor occurs above the solution with the larger mole fraction of benzene. Since toluene has a larger molecular mass, equal masses of toluene and benzene means a greater mole fraction of benzene and a greater mole fraction of benzene in the vapor.

12.12B $P_{benzene} = \chi_{benzene} \times P^{\circ}_{benzene} = 0.770 \times 95.1 \text{ mmHg} = 73.2 \text{ mmHg}$

$P_{toluene} = \chi_{toluene} \times P^{\circ}_{toluene} = 0.230 \times 28.4 \text{ mmHg} = 6.53 \text{ mmHg}$

12.13A No. The process will continue until the mole fraction of water in B decreases to be equal to the mole fraction in A. That is, water vapor passes from the more dilute B to the more concentrated solution A until the concentrations are equal. The evaporation and condensation will continue, but at equal rates in both dishes. There is no net transfer, and the levels will remain constant.

12.13B If the liquid in B is more volatile than water, it will evaporate more quickly than the water. The mixture of gases will fall back into the liquid forming a solution. Eventually both containers will contain the same solution, but A will contain more as B will empty more quickly and even when the containers hold the same solution, container B will not contain as much.

12.14A $m = \dfrac{10.0 \text{ g } C_{10}H_8 \text{ x } \dfrac{\text{mol } C_{10}H_8}{128.2 \text{ g } C_{10}H_8}}{50.0 \text{ g } C_6H_6 \text{ x } \dfrac{\text{kg}}{10^3 \text{ g}}} = 1.56\ m$

$\Delta T = -K_f \times m = \dfrac{-5.12\,°\text{C}}{\text{m}} \times 1.56\ m = -7.99\ °\text{C}$

$\Delta T = T_f - T_i$

$-7.99\ °\text{C} = \text{T}_f - 5.53\ °\text{C}$

$T_f = -2.46\ °\text{C}$

12.14B $\Delta T = 100.35\ °\text{C} - 100.00\ °\text{C} = 0.35\ °\text{C}$

$\Delta T = K_b m$

$m = \dfrac{\Delta T}{K_b} = \dfrac{0.35\,°\text{C}}{0.512\,°\text{C/m}} = 0.684\ m$

$? \text{ g sucrose} = \dfrac{0.684 \text{ mol sucrose}}{\text{kg } H_2O} \times \dfrac{\text{kg}}{1000\text{g}} \times 75.0 \text{ g } H_2O \times \dfrac{342.3 \text{ g}}{\text{mol}} = 18 \text{ g sucrose}$

12.15A $\Delta T = 4.25\ °\text{C} - 5.53\ °\text{C} = -1.28\ °\text{C}$

$m = -\dfrac{\Delta T}{K_f} = \dfrac{-1.28\,°\text{C}}{-5.12\,°\text{C/m}} = 0.250\ m$

$? \text{ mol} = \dfrac{0.250 \text{ mol cpd}}{\text{kg ben}} \times \dfrac{\text{kg}}{10^3 \text{ g}} \times 30.00 \text{ g benz} = 0.00750 \text{ mol}$

$? \text{ g/mol} = \dfrac{1.065 \text{ g}}{0.00750 \text{ mol}} = 142 \text{ g/mol}$

$? \text{ mol C} = 50.69 \text{ g C} \times \dfrac{\text{mol}}{12.011} = 4.220 \text{ mol C} \times \dfrac{1}{2.818 \text{ mol}} = 1.498 \dfrac{\text{mol C}}{\text{mol O}}$

$? \text{ mol H} = 4.23 \text{ g H} + \dfrac{\text{mol}}{1.008 \text{ g}} = 4.196 \text{ mol H} \times \dfrac{1}{2.818 \text{ mol}} = 1.489 \dfrac{\text{mol H}}{\text{mol O}}$

$$? \text{ mol O} = 45.08 \text{ g O} \times \frac{\text{mol}}{15.999 \text{ g}} = 2.818 \text{ mol O} \times \frac{1}{2.818 \text{ mol}} = 1.000 \frac{\text{mol O}}{\text{mol O}}$$

empirical formula: $C_3H_3O_2$ 71 u/formula

$$\frac{142 \text{ u}}{\text{molecule}} \times \frac{\text{formula units}}{71 \text{ u}} = \frac{2 \text{ formula units}}{\text{molecule}}$$

molecular formula: $C_6H_6O_4$

12.15B $\Delta T = -\, i\, m\, K_f$

$$(-1.25°\text{C} - 0.00°\text{C}) = 1\, m\, (-1.86\ °\text{C}/m)$$

$$m = \frac{-1.25°\text{C}}{-1.86\ °\text{C/m}} = 0.672 \text{ m}$$

$$? \text{ mol} = \frac{0.672 \text{ mol}}{\text{kg}} \times \frac{\text{kg}}{10^3 \text{ g}} \times 100.0 \text{ g} = 0.0672 \text{ mol}$$

$$? \text{ mol} = 10.00 \text{ g} \times \frac{\text{mol}}{180.2 \text{ g}} = 0.05549 \text{ mol}$$

No, it is not possible. Even if the 10.0 g were pure glucose, it would only provide enough molecules to produce 0.05549 mol not the 0.0672 mol that the −1.25 °C required.

12.16A $\pi = MRT$ Replace M with $\frac{n}{V}$. Then replace n with $\frac{m}{M}$ to get $\pi = \frac{mRT}{MV}$.

Rearrange.

$$M = \frac{mRT}{\pi V} = \frac{1.08 \text{ g} \times \frac{0.08206 \text{ atm}}{\text{K mol}} \times 298 \text{ K}}{0.00770 \text{ atm} \times 50.0 \text{ mL} \times \frac{10^{-3} \text{ L}}{\text{mL}}}$$

$M = 6.86 \times 10^4$ g/mol

12.16B $\pi = MRT$ Replace M with $\frac{n}{V}$. Then replace n with $\frac{m}{M}$ to get $\pi = \frac{mRT}{MV}$.

$$\pi = \frac{mRT}{MV} = \frac{125\ \mu\text{g B-12} \times \frac{10^{-6} \text{ g}}{\mu\text{g}} \times \frac{62.36 \text{ L mm Hg}}{\text{K mol}}}{\frac{1355 \text{ g B-12}}{\text{mol B12}} \times 2.50 \text{ mL} \times \frac{10^{-3} \text{ L}}{\text{mL}}} \times (273 + 25) \text{ K}$$

$$(6.3 \times 12.01) + (88 \times 1.008) + 58.93 + (14 \times 14.01) + (14 \times 16.00) + 30.97 = 1355$$

$$\pi = 0.686 \text{ mmHg} \times \frac{13.6 \text{ g/mL}}{1.00 \text{ g/mL}} = 9.33 \text{ mm } H_2O$$

12.17A The lowest freezing point corresponds to the largest molality of ions. The solutions are dilute enough that molarity is essentially equal to molality.
0.0080 M HCl < 0.0050 m $MgCl_2$ ≈ 0.0030 M $Al_2(SO_4)_3$ < 0.010 m $C_6H_{12}O_6$

0.016 m ions lowest f.p.	0.015 m ions	0.015 m ions	0.010 m particles highest f.p.

12.17B The HCl does not dissociate in benzene. The van't Hoff factor is about 1 for HCl in benzene. $\Delta T_f = -1 \times 5.12 \times 0.01 \approx -0.05$ °C. HCl does ionize in water, producing a van't Hoff factor of about 2: $\Delta T_f = -2 \times 1.86 \times 0.01 \approx -0.04$ °C.

Review Questions

2. Volume is temperature-dependent, so molarity, mol solute/liter solution, is temperature-dependent. In molality, the amount of solute is on a mole basis, and the quantity of solvent is in kilograms. Neither quantity changes with temperature. Mole fraction depends only on the amounts of solution components, in moles. These amounts do not depend on temperature. Volume percent is temperature-dependent because the volume does change with temperature. A mass percent is not temperature-dependent because mass does not change with temperature.

5. (a) No. NaCl is ionic, and benzene has only dispersion forces as intermolecular forces.
(b) Motor oil will not dissolve in water. Motor oil has only dispersion forces as intermolecular forces, but water has hydrogen bonding. The water molecules will hold together too tightly to let the hydrocarbon molecules in.
(c) Motor oil will dissolve in benzene, as both liquids have only dispersion forces as intermolecular forces.
(d) 2-decanol has 10 carbon atoms in a chain, which only have dispersion forces. These dispersion forces are more important than the hydrogen bonding of the alcohol group. Thus, 2-decanol and water do not form a solution.
(e) Pentane and decane both only have dispersion forces and should form a solution.

6. (a) C_6H_6 forces are London
(b) C (s) forces are London
(c) NH_2OH forces are H-bond, polar and London
(d) $C_{10}H_8$ forces are London
(c) NH_2OH is more soluble in water

7. (a) phenol is soluble in both water and benzene. The C_6H_5 group is soluble in benzene and the OH is soluble in water.
(b) and (c) are soluble in benzene not water.
(c) is soluble in water not in benzene.
(a) is the answer.

9. (b) An increase in gas pressure increases solubility. Increasing temperature will decrease solubility. Increasing volume has no effect on solubility.

10. Generally, it would be true only for a solution of liquids having equal vapor pressures. Most solutions are composed of compounds with different vapor pressures and, thus, produce different vapor composition than the solution composition. Also, solutions of nonvolatile solutes have a vapor phase that is pure solvent–that is, with none of the solute component(s) at all.

13. (a) $25.0 \text{ g } CH_3OH \times \frac{\text{mol}}{32.04 \text{ g}} = 0.780 \text{ mol}$

(a) $30.0 \text{ g } CH_3CH_2OH \times \frac{\text{mol}}{46.07 \text{ g}} = 0.651 \text{ mol}$

(a) $35.0 \text{ g } C_6H_{12}O_6 \times \frac{\text{mol}}{180.2 \text{ g}} = 0.194 \text{ mol}$

The greatest depression is caused by the most moles/L.
(a) is the answer.

17. In the diffusion of gases, the net movement of molecules is from a higher to a lower pressure in a tendency to equilibrate the pressure throughout the mixture. In osmosis, the net movement of solvent molecules is from a region in which they are more abundant and have a higher vapor pressure to one in which they are less abundant and have a lower vapor pressure. This means from a lower concentration of solute (greater mole fraction of solvent) to a higher concentration of solute (smaller mole fraction of solvent). Osmotic flow is similar to gases diffusing from higher to lower pressure.

20. (a) 0.10 M $NaHCO_3$ is a higher molarity than 0.05 $NaHCO_3$.
(b) 1 M NaCl produces more particles per formula unit than 1 M glucose.
(c) 1 M $CaCl_2$ produces more ions per formula unit than 1 M NaCl.
(d) 3 M glucose has more particles per liter than 1 M NaCl.

Problems

21. $? \text{ g } NaNO_3 = \frac{4.85 \text{ g } NaNO_3}{100 \text{ g solution}} \times 2.30 \text{ kg} \times \frac{10^3 \text{ g}}{\text{kg}} = 112 \text{ g } NaNO_3$

Weigh 112 g $NaNO_3$ into a container, add 2.19 kg of water, and mix.

23. (a) $? \% = \frac{175 \text{ mg NaCl}}{\text{g solution}} \times \frac{10^{-3} \text{ g}}{\text{mg}} \times 100\% = 17.5\% \text{ by mass}$

(b) $? \text{ g} = 0.275 \text{ L} \times \frac{0.791 \text{ g}}{\text{mL}} \times \frac{\text{mL}}{10^{-3} \text{ L}} = 218 \text{ g}$

$? \% = \frac{217.5 \text{ g methanol}}{(1000 \text{ g} + 217.5 \text{ g}) \text{ solution}} \times 100\% = 17.9\% \text{ by mass}$

(c) $? \text{ g ethylene glycol} = 4.5 \text{ L} \times \frac{\text{mL}}{10^{-3} \text{ L}} \times \frac{1.114 \text{ g}}{\text{mL}} = 5.0 \times 10^3 \text{ g ethylene glycol}$

$$? \text{ g propylene glycol} = 6.5 \text{ L} \times \frac{\text{mL}}{10^{-3} \text{ L}} \times \frac{1.036 \text{ g}}{\text{mL}} = 6.7 \times 10^3 \text{ g propylene glycol}$$

$$? \% = \frac{5.0\times10^3 \text{ g ethylene glycol}}{(5.0\times10^3 \text{ g} + 6.7\times10^3 \text{ g})\text{solution}} \times 100\%$$

$$= 42 \text{ \% by mass ethylene glycol}$$

25. (a) $? \% = \frac{58.0 \text{ mL water}}{625 \text{ mL solution}} \times 100\% = 9.28\% \text{ by volume}$

(b) $? \text{ mL methanol} = 10.00 \text{ g} \times \frac{\text{mL}}{0.791 \text{ g}} = 12.64 \text{ mL methanol}$

$? \text{ mL ethanol} = 75.00 \text{ g} \times \frac{\text{mL}}{0.789 \text{ g}} = 95.06 \text{ mL ethanol}$

$? \% = \frac{1.99 \text{ g glycerol}}{(1.99 \text{ g} + 22.2 \text{ g}) \text{ solution}} \times 100\% = 11.7\% \text{ by volume}$

(c) $? \text{ mL ethanol} = 24.0 \text{ g} \times \frac{\text{mL}}{0.789 \text{ g}} = 30.4 \text{ mL ethanol}$

$? \text{ mL} = 100.0 \text{ g} \times \frac{\text{mL}}{0.963 \text{ g}} = 104 \text{ mL solution}$

$? \% = \frac{30.4 \text{ mL ethanol}}{(30.4 \text{ mL} + 104 \text{ mL})} \times 100\% = 29.3\% \text{ by volume}$

27. $? \text{ g}_{\text{chol}} = \frac{234 \text{ mg chol}}{\text{dL blood}} \times \frac{\text{dL}}{0.1 \text{ L}} \times 5.0 \text{ L} \times \frac{10^{-3} \text{ g}}{\text{mg}} = 12 \text{ g chol}$

29. (a) $? \text{ ppb trich} = \frac{5 \text{ µg trich}}{\text{L water}} \times \frac{\text{L}}{10^3 \text{ g}} \times \frac{10^{-6} \text{ g}}{\text{µg}} = \frac{5 \text{ µg}}{10^9 \text{ µg}} = 5 \text{ ppb trich.}$

(b) $? \text{ ppm KI} = \frac{0.0025 \text{ g KI}}{\text{L } H_2O} \times \frac{\text{L}}{10^3 \text{ g}} \times \frac{10^3}{10^3} = \frac{2.5 \text{ g}}{10^6 \text{ g}} = 2.5 \text{ ppm KI}$

(c) $\text{M } SO_4^{2-} = \frac{37 \text{ g } SO_4^{2-}}{10^6 \text{ g solution}} \times \frac{10^3 \text{ g solution}}{\text{L}} \times \frac{\text{mol}}{96.07 \text{ g}} = 3.9 \times 10^{-4} \text{ M } SO_4^{2-}$

31. $1\% = \frac{1 \text{ g}}{100 \text{ g}}$ $\quad \frac{1 \text{ mg}}{\text{dL}} \times \frac{10^{-3} \text{ g}}{\text{mg}} \times \frac{\text{dL}}{10^{-1} \text{ L}} \times \frac{10^{-3} \text{ L}}{\text{mL}} \times \frac{\text{mL}}{1 \text{ g}} = \frac{1 \text{ g}}{10^5 \text{ g}}$

$\text{ppt} = \frac{1 \text{ g}}{10^{12} \text{ g}}$ $\quad \text{ppm} = \frac{1 \text{ g}}{10^6 \text{ g}}$ $\quad \text{ppb} = \frac{1 \text{ g}}{10^9 \text{ g}}$

ppt < ppb < ppm < 1 mg/dL < 1%

33. $m = \frac{1.02 \text{ kg}}{554 \text{ g } H_2O} \times \frac{10^3 \text{ g}}{\text{kg}} \times \frac{10^3 \text{ g}}{\text{kg}} \times \frac{\text{mol glucose}}{342.3 \text{ g glucose}} = 5.38\ m$

35. $M = \frac{3.30 \text{ mL acetone x } 0.789 \text{ g/mL}}{75.0 \text{ mL}} \times \frac{\text{mL}}{10^{-3}\text{ L}} \times \frac{\text{mol}}{58.08 \text{ g}} = 0.598 \text{ M}$

? g acetone = 3.30 mL acetone × 0.789 g/mL = 2.60 g acetone

? g solution = 75.0 mL solution × 0.993 g/mL = 74.5 g solution

? g H_2O = 74.5 g - 2.60 g = 71.9 g H_2O

$$m = \frac{2.60 \text{ g acetone}}{71.9 \text{ g water}} \times \frac{10^3 \text{ g}}{\text{kg}} \times \frac{\text{mol}}{58.08 \text{ g}} = 0.624\ m$$

37. ? g solution $= 2.30 \text{ L} \times \frac{\text{mL}}{10^{-3}\text{ L}} \times \frac{1.035 \text{ g}}{\text{mL}} = 2.38 \times 10^3$ g solution

? g EG $= 6.27 \text{ mol} \times \frac{62.07 \text{ g}}{\text{mol}} = 389$ g EG

? % EG $= \frac{389 \text{ g EG}}{(389 \text{ g} + 1000 \text{ g}) \text{ solution}} \times 100\% = 28.0\%$ EG

? mol EG $= 2.38 \times 10^3 \text{ g solution} \times \frac{28.02 \text{ g EG}}{100 \text{ g solution}} \times \frac{\text{mol}}{62.07 \text{ g EG}} = 10.7$ mol EG

39. (a) ? mol $C_{10}H_8 = 23.5 \text{ g} \times \frac{\text{mol}}{128.16 \text{ g}} = 0.183$ mol $C_{10}H_8$

? mol $C_6H_6 = 315 \text{ g} \times \frac{\text{mol}}{78.11 \text{ g}} = 4.03$ mol C_6H_6

$$\chi = \frac{0.184 \text{ mol}}{(4.03 + 0.183) \text{ mol}} = 0.0434$$

(b) ? mol $C_6H_6 = 1 \text{ kg } C_6H_6 \times \frac{10^3 \text{ g}}{\text{kg}} \times \frac{\text{mol}}{78.11 \text{ g}} = 12.80$ mol C_6H_6

$$\chi = \frac{0.250 \text{ mol}}{(12.80 + 0.250) \text{ mol}} = 0.0192$$

41. The solution with the greatest amount of solute per unit mass of solvent has the greatest mole fraction of solute–(b). Solution (a) has 1 mol solute per 1000 g H_2O; (b) has slightly more than 1 mol solute in 950 g H_2O; (c) has about 0.3 mol solute per 900 g H_2O.

Actual -

(a) $\frac{1.00 \text{ mol}}{\text{kg } H_2O}$? mol = kg $H_2O \times \frac{10^3 \text{ g}}{\text{kg}} \times \frac{\text{mol}}{18.02 \text{ g}} = 55.49$ mol

$$\chi = \frac{1.00 \text{ mol}}{(1.00 + 55.49) \text{ mol}} = 0.0177$$

(b) ? mol $= 5.0 \text{ g } C_2H_6O \times \frac{\text{mol}}{46.07 \text{ g}} = 0.11$ mol

? mol $= 95.0 \text{ g } H_2O \times \frac{\text{mol}}{18.02 \text{ g}} = 5.27$ mol

$$\chi = \frac{0.11\,\text{mol}}{(5.27 + 0.11)\,\text{mol}} = 0.020$$

(c) $? \text{ mol} = 10.0 \text{ g } C_{12}H_{22}O_{11} \times \frac{\text{mol}}{342.3 \text{ g}} = 0.0292 \text{ mol}$

$? \text{ mol} = 90.0 \text{ g } H_2O \times \text{mol}/18.02\text{g} = 4.99 \text{ mol}$

$$\chi = \frac{0.0292\,\text{mol}}{(0.0292 + 4.99)\,\text{mol}} = 0.00582$$

43. (a) $CHCl_3$ is insoluble in water. Although $CHCl_3$ is somewhat polar, the primary intermolecular forces in water are hydrogen bonds, which are unimportant in chloroform ($CHCl_3$).
(b) Benzoic acid is slightly soluble in water. There is some hydrogen bonding to the carboxylic acid group, but the benzene ring is very unlike water.
(c) Propylene glycol is highly soluble in water; both solute and solvent have extensive hydrogen bonding.

45. Figure 12.10 shows that a saturated solution at 60 °C has about 55 g NH_4Cl per 100 g H_2O. $\frac{55 \text{ g } NH_4Cl}{100 \text{ g } H_2O} \times 55 \text{ g } H_2O = 30. \text{ g } NH_4Cl$
The given solution is unsaturated.

47. (a) $? \text{ g } H_2O = 35 \text{ g } K_2CrO_4 \times \frac{100\text{g water}}{62.\text{g } K_2CrO_4} = 56 \text{ g water at } 25\ ^\circ\text{C}$
$? \text{ g } H_2O = 56 \text{ g } H_2O - 35 \text{ g } H_2O = 21 \text{ g } H_2O$ to be added.
(b) $? \text{ g } KNO_3 = \frac{50.0\text{g } KNO_3}{75.0 \text{ g } H_2O} \times 100.0 \text{ g } H_2O = 66.7 \text{ g } KNO_3$
At temperatures above about 44 °C, the solubility of KNO_3 exceeds 66.7 g $KNO_3/100.0 \text{ g } H_2O$.

49. The solution could be left open to air so that some of the solvent evaporates. When enough solvent has evaporated, the solute will begin to crystallize.

51. (a) $? \text{ M} = \frac{4.43 \text{ mg } O_2}{100 \text{ g } H_2O} \times \frac{1.00 \text{ g}}{\text{mL}} \times \frac{10^{-3} \text{ g}}{\text{mg}} \times \frac{\text{mL}}{10^{-3} \text{ L}} \times \frac{\text{mol}}{32.00 \text{ g}} = 1.38 \times 10^{-3} \text{ M}$
NOTE: Density of water solution is approximated at 1.00 g/mL.
$? \frac{\text{mg } O_2}{H_2O} = 0.010 \text{ M} \times \frac{10^{-3} \text{ L}}{\text{mL}} \times \frac{32.00 \text{ g}}{\text{mol}} \times \frac{\text{mg}}{10^{-3} \text{ g}} \times \frac{\text{mL}}{1.00 \text{ g}} = \frac{0.32 \text{ mg } O_2}{\text{g } H_2O}$

$? \text{ mg } O_2 = 100 \text{ g } H_2O \times \frac{0.32 \text{ mg } O_2}{\text{g } H_2O} = 32 \text{ mg } O_2$

$$k = \frac{S}{P} = \frac{4.43 \text{ mg } O_2}{1.00 \text{ atm}}$$

$$P = \frac{S}{k} = 32 \text{ mg } O_2 \times \frac{1.00 \text{ atm}}{4.43 \text{ mg } O_2} = 7.2 \text{ atm}$$

53. (a) $P_P = \frac{1 \text{ mol}}{5 \text{ mol}} \times 441 \text{ mmHg} = 88.2 \text{ mmHg}$

$P_H = \frac{4 \text{ mol}}{5 \text{ mol}} \times 121 \text{ mmHg} = 96.8 \text{ mmHg}$

(b) $\chi_H = \frac{96.80 \text{ mmHg}}{(96.80 + 88.20) \text{ mmHg}} = 0.523$

$\chi_P = 1.000 - 0.523 = 0.477$

55. At 20 °C VP(H_2O) = 17.5 mmHg

$? \text{ mol} = \text{kg} \times \frac{10^3 \text{ g}}{\text{kg}} \times \frac{\text{mol}}{18.02 \text{ g}} = 55.49 \text{ mol}$

$\chi_{H_2O} = \frac{55.49 \text{ mol}}{55.49 + 0.20 \text{ mol}} = 0.9964$

$VP = 0.9964 \times 17.5 \text{ mm Hg} = 17.4 \text{ mmHg}$

57. (a) $\Delta T = -K_f \times m = \frac{-1.86\,°C}{m} \times 0.25\ I = -0.465\ °C$

$\Delta T = t_f - t_i = t_f - 0.00\ °C = -0.465\ °C$

$t_f = -0.47\ °C$

(b) $m = \frac{5.0 \text{ g } C_6H_4Cl_2}{95.0 \text{ g } H_2O} \times \frac{10^3 \text{ g}}{\text{kg}} \times \frac{\text{mol } C_6H_4Cl_2}{147.0 \text{ g } C_6H_4Cl_2} = 0.358\ m$

$\Delta T = \frac{-5.12°C}{m} \times 0.358\ m = -1.83\ °C$

$\Delta T = -1.83\ °C = T_f - 5.53\ °C$

$\Delta T = 3.70\ °C$

59. $\Delta T = -K_f \times m = \frac{-1.86°C}{m} \times 0.55\ m$

$\Delta T = -1.02\ °C = t_f - 0.00\ °C$

$t_f = -1.02\ °C$

$\Delta T = -1.02\ °C - 5.53\ °C = -6.55\ °C$

$m = \frac{\Delta T}{-K_f} = \frac{-6.55°C}{\frac{-5.12°C}{m}}$

$m = 1.28$

61. $m = \frac{2.11 \text{ g } C_{10}H_8}{35.00 \text{ g xylene}} \times \frac{10^3 \text{ g}}{\text{kg}} \times \frac{\text{mole}}{128.16 \text{ g}} = 0.470\ m$

$\Delta T = 11.25\ °C - 13.26\ °C = -2.01\ °C = -K_f \times m$

$$K_f = \frac{2.01°\text{C}}{0.4704\,m} = 4.27\ °\text{C}/m$$

63. $\Delta T = t_f - t_i = 4.70\ °\text{C} - 5.53\ °\text{C} = -0.83\ °\text{C}$

$$m = \frac{\Delta T}{-K_f} = \frac{-0.83°\text{C}}{-5.12°\text{C}/m} = 0.16\ m$$

$$?\ \text{mol} = 50.00\ \text{mL} \times \frac{0.879\ \text{g}}{\text{mL}} \times \frac{\text{kg}}{10^3\ \text{g}} \times 0.16\ \text{m} = 0.0070\ \text{mol}$$

$$M = \frac{1.505\ \text{g}}{0.0070\ \text{mol}} = 2.1 \times 10^2\ \text{g/mol}$$

$$?\ \text{mol C} = 33.81\ \text{g C} \times \frac{\text{mol}}{12.011\ \text{g}} = 2.815\ \text{mol C} \times \frac{1}{1.408\ \text{mol N}} = 1.999\ \frac{\text{mol C}}{\text{mol N}}$$

$$?\ \text{mol H} = 1.42\ \text{g H} \times \frac{\text{mol}}{1.008\ \text{g}} = 1.41\ \text{mol H} \times \frac{1}{1.408\ \text{mol N}} = 1.000\ \frac{\text{mol H}}{\text{mol N}}$$

$$?\ \text{mol O} = 45.05\ \text{g O} \times \frac{\text{mol}}{15.999\ \text{g}} = 2.81\ \text{mol O} \times \frac{1}{1.408\ \text{mol N}} = 2.000\ \frac{\text{mol O}}{\text{mol N}}$$

$$?\ \text{mol N} = 19.72\ \text{g N} \times \frac{\text{mol}}{14.006\ \text{g}} = 1.408\ \text{mol N} \times \frac{1}{1.408\ \text{mol N}} = 1.00\ \frac{\text{mol N}}{\text{mol N}}$$

empirical formula C_2HO_2N 71 g/formula

$$\frac{213\ \text{g}}{\text{mole}} \times \frac{\text{formula}}{71\ \text{g}} = 3\ \text{formula/mole}$$

molecular formula $C_6H_3O_6N_3$

65. Because the cucumbers shrivel up, water is leaving the cucumbers, so the salt solution must have a higher osmotic pressure.

67. $$\pi = \frac{mRT}{MV} = \frac{1.80\ \text{g x}\ \frac{0.08206\ \text{L atm}}{\text{K mol}}\ \text{x}\ (37 + 273)\text{K}}{46.07\ \text{g/mol x}\ 100\ \text{mL x}\ 10^{-3}\ \text{L/mL}}$$

$\pi = 9.94$ atm

$$\pi\ \text{glucose} = \frac{5.5\ \text{g x}\ \frac{0.08206\ \text{L atm}}{\text{K mole}}\ \text{x}\ (37 + 273)\text{K}}{180.16\ \text{g/mol x}\ 100\ \text{mL x}\ 10^{-3}\ \text{L/mL}}$$

$\pi = 7.77$ atm

The CH_3CH_2OH solution is hypertonic.

69. $$\text{M} = \frac{5.15\ \text{g urea}}{75.0\ \text{mL}\ H_2O} \times \frac{\text{mL}}{10^{-3}\ \text{L}} \times \frac{\text{mol urea}}{60.06\ \text{g}\ CO(NH_2)_2} = 1.14\ \text{M}$$

The flow will be to the right to the more concentrated side, that is, from A to B. Solution B has the higher osmotic pressure. More solvent molecules are able to leave the dilute side because there are less solute molecules to block the path.

71. $\Delta T = -K_f \times m = \dfrac{-1.86\,°C}{m} \times 0.10\,m = -0.19\,°C$

(a) $i = 1$ $t_f = -0.19\,°C$
(b) $i = 3$ $t_f = -0.57\,°C$
(c) i slightly greater than 1 t_f slightly less than -0.19 °C
(d) $i = 2$ $t_f = -0.38\,°C$

Answer (a) is the most precise. The solute is molecular. Unless a solution is very dilute, the ionic compounds don't completely dissociate.

73. The maximum for an infinitely dilute solution would be $i = 3$. CaO in solution forms $Ca(OH)_2$, which is partially soluble in water.

75. The lowest freezing point will have the largest ΔT. Instead of calculating ΔT, it is easier to compare $i \times m$, as all of the solutions are in the same solvent, water. The largest product ($i \times m$) produces the lowest freezing point.

(b)	$1 \times 0.15 =$	0.15
(a) slightly more than	$1 \times 0.15 =$	slightly more than 0.15
(e)	$2 \times 0.10 =$	0.20
(c) slightly more than	$2 \times 0.10 =$	slightly more than 0.20
(d)	$3 \times 0.10 =$	0.30

(b) > (a) > (e) > (c) > (d) lowest freezing point.

77. $\Delta T = 100.0\,°C - 99.4\,°C = 0.6\,°C$

$$0.6\,°C = K_b \times m \times i = \frac{0.512\,°C}{m} \times m \times 1$$

At this concentration, i is close to 1.

$m = 1.2\,m$

$$?\,g = 1.2\,m \times 3.50\,kg \times \frac{58.44\,g}{mol} = 2.5 \times 10^2\,g$$

79. The higher charge on the Al^{3+} ion would more effectively negate the negative charge on the silica and allow coagulation.

Additional Problems

81. (a) $?\,\% = \dfrac{11.3\,mL\ CH_3OH}{75.0\,mL\ solution} \times 100\% = 15.1\%$ by volume

(b) $?\,\% = \dfrac{11.3\,mL\ CH_3OH \times \dfrac{0.793\,g}{mL}}{75.0\,mL\ solution \times \dfrac{0.980\,g}{mL}} \times 100\% = 12.2\%$ by mass

(c) $? \% = \dfrac{11.3 \text{ mL CH}_3\text{OH x} \dfrac{0.793 \text{ g}}{\text{mL}}}{75.0 \text{ mL solution}} \times 100\% = 11.9 \text{ \% mass/volume}$

(d) $? \text{ g CH}_3\text{OH} = 11.3 \text{ mL} \times \dfrac{0.793 \text{ g}}{\text{mL}} = 8.961 \text{ g CH}_3\text{OH}$

$? \text{ g solution} = 75.0 \text{ mL} \times \dfrac{0.980 \text{ g}}{\text{mL}} = 73.5 \text{ g solution}$

$? \text{ g H}_2\text{O} = 73.5 \text{ g} - 9.0 \text{ g} = 64.5 \text{ g water}$

$? \text{ mol} = 8.961 \text{ g CH}_3\text{OH} \times \dfrac{\text{mol}}{32.04 \text{ g}} = 0.2797 \text{ mol}$

$? \text{ mol} = 64.5 \text{ g H}_2\text{O} \times \dfrac{\text{mol}}{18.02 \text{ g}} = 3.579 \text{ mol}$

$\text{mole \%} = \dfrac{0.2797 \text{ mol}}{(3.579 + 0.280) \text{ mol}} \times 100\% = 7.25\%$

84. In the phase: triethylamine in water:

$? \text{ g H}_2\text{O} = 40.0 \text{ g} \times \dfrac{84.5 \text{ g H}_2\text{O}}{100 \text{ g solution}} = 33.8 \text{ g H}_2\text{O}$

$? \text{ g (CH}_3\text{CH}_2)_3\text{N} = 40.0 \text{ g} - 33.8 \text{ g H}_2\text{O} = 6.2 \text{ g (CH}_3\text{CH}_2)_3\text{N}$

In the phase: water in triethylamine:

$? \text{ g H}_2\text{O} = 50.0 \text{ g} - 33.8 \text{ g} = 16.2 \text{ g H}_2\text{O}$

$? \text{ g (CH}_3\text{CH}_2)_3\text{N} = 50.0 \text{ g} - 6.2 \text{ g} = 43.8 \text{ g (CH}_3\text{CH}_2)_3\text{N}$

$? \% \text{ H}_2\text{O} = \dfrac{16.2 \text{ g H}_2\text{O}}{(16.2 + 43.8) \text{ g}} \times 100\% = 27.0 \text{ \% H}_2\text{O by mass}$

86. $? \text{ mol H}_2\text{O} = 1 \text{ L} \times \dfrac{\text{mL}}{10^{-3} \text{ L}} \times \dfrac{0.998 \text{ g}}{\text{mL}} \times \dfrac{\text{mol H}_2\text{O}}{18.01 \text{ g H}_2\text{O}} = 55.4 \text{ mol H}_2\text{O}$

$\text{mol \%} = 2.50 \% = \dfrac{X}{55.4 + X} \times 100\%$

$139 + 2.5 X = 100 X$

$97.5 X = 139$

$X = 1.42 \text{ mol}$

$? \text{ g sucrose} = 1.42 \text{ mol sucrose} \times \dfrac{342.3 \text{ g sucrose}}{\text{mol sucrose}} = 486 \text{ g sucrose}$

88. $? m = \dfrac{\Delta T_f}{-K_f} = \dfrac{-5.53 \text{ °C}}{-5.12 \dfrac{\text{°C}}{m}} = 1.08 \, m$

$? \text{ g nitrobenzene} = 1.08 \text{ mole N} \times \dfrac{123.0 \text{ g}}{\text{mole}} = 133 \text{ g nitrobenzene}$

$? \text{ \% mass nitrobenzene} = \dfrac{133 \text{ g} \times 100\%}{1000 \text{ g} + 133 \text{ g}} = 11.7\% \text{ nitrobenzene}$

$$? \; m = \frac{\Delta T_f}{-K_f} = \frac{-5.8\ ^\circ\text{C}}{-8.1\frac{^\circ\text{C}}{m}} = 0.72\ m$$

$$? \text{ g benzene} = 0.72 \text{ mole benzene} \times \frac{78.0\text{ g}}{\text{mol}} = 56 \text{ g benzene}$$

$$? \ \% \text{ mass nitobenzene} = \frac{1000\text{ g} \times 100\%}{1000\text{ g} + 56\text{ g}} = 95\ \% \text{ nitrobenzene}$$

They would not have the same boiling point, because the constants are different and the initial boiling points are very different for nitrobenzene and for benzene.

For 11.7 % nitrobenzene:

$$\Delta T_b = 1.08\ m \times 2.53\frac{^\circ\text{C}}{m} = 2.73^\circ\text{C}$$

$T_f - T_i = 2.73^\circ\text{C}$

$T_f = 2.73^\circ\text{C} + 80.10^\circ\text{C} = 82.83^\circ\text{C}$

For 95 % nitrobenzene

$$\Delta T_b = 0.72\ m \times 5.24\frac{^\circ\text{C}}{m} = 3.8^\circ\text{C}$$

$T_f - T_i = 3.8^\circ\text{C}$

$T_f = 3.8^\circ\text{C} + 210.8^\circ\text{C} = 214.6^\circ\text{C}$

91. (a) The acetone solution has the greatest vapor pressure because both the mole fraction in water is as great as in any other solution, and acetone contributes to the vapor pressure.
(b) The saturated NaCl solution has the most ions in solution, so it has the lowest freezing point.
(c) The vapor pressure of the acetone solution and the 0.10 *m* NaCl will change with time, as some of the vapor completely escapes. The vapor pressure of saturated NaCl will not change, because as water evaporates, some of the NaCl (aq) will become solid, leaving the same concentration of solute to solvent particles in the solution phase.

93. (a) $P\text{solution} = \chi_{\text{benzene}}P_{\text{benzene}} + \chi_{\text{toluene}}P_{\text{toluene}}$

$760.0 \text{ mmHg} = \chi_{\text{benzene}} \times 1351 \text{ mmHg} + (1- \chi\text{benzene}) \times 556.3 \text{ mmHg}$

$760.0 \text{ mmHg} = 1351\ \chi_{\text{ben}} + 556.3 \text{ mmHg} - 556.3\ \chi_{\text{ben}}$

$203.7 \text{ mmHg} = 795\ \chi_{\text{ben}}$

$\chi_{\text{ben}} = 0.256$

$\chi_{\text{tol}} = 1.000 - \chi_{\text{ben}} = 0.744$

(b) $P_{\text{tol}} = 0.744 \times 556.3 \text{ mmHg} = 414 \text{ mmHg}$

$P_{\text{ben}} = 0.256 \times 1351 \text{ mmHg} = 346 \text{ mmHg}$

$$\chi_{\text{ben}} = \frac{346 \text{ mmHg}}{(346 + 414) \text{ mmHg}} = 0.455$$

$$\chi_{tol} = \frac{414\text{ mmHg}}{760\text{ mmHg}} = 0.545$$

96. A: $X\text{ mol }CO(NH_2)_2 \times \frac{60.06\text{ g}}{\text{mol}} + 9X\text{ mol }H_2O \times \frac{18.02\text{ g}}{\text{mol}} = 200.0\text{ g}$

$?\text{ g} = 60.06\,X + 162.18\,X = 200.0\text{ g}$

$X = \frac{200.0\text{ g}}{222.24\text{ g/mol}} = 0.8999\text{ mol }CO(NH_2)_2$

$8.100\text{ mol }H_2O$

B: $X\text{ mol }CO(NH_2)_2 \times \frac{60.06\text{ g}}{\text{mol}} + 19\,X\text{ mol} \times \frac{18.02\text{ g}}{\text{mol}} = 100.0\text{ g}$

$?\text{ g} = 60.06\,X + 342.38\,X = 100.0\text{ g}$

$402.44\,X = 100.0$

$X = 0.2485\text{ mol }CO(NH_2)_2$

$4.712\text{ mol }H_2O$

Final solutions both have the same mole fraction.

$?\text{ mol }H_2O = 8.100 + 4.712 = 12.812\text{ mol }H_2O$

$?\text{ mol }CO(NH_2)_2 = 0.8999 + 0.2485 = 1.1484\text{ mol }CO(NH_2)_2$

$$\chi_{CO(NH_2)_2} = \frac{1.1485}{(1.1485 + 12.812)\text{ mol}} = 0.08221 \quad \chi_{H_2O} = 0.91779$$

$$\chi = 0.08221 = \frac{0.8999\text{ mol }CO(NH_2)_2}{(0.8999 + X)\text{ mol}}$$

$0.0740 + 0.0822\,X = 0.8999$

$X = 10.05\text{ mol }H_2O$

$?\text{ g }CO(NH_2)_2 = 0.900\text{ mol }CO(NH_2)_2 \times 60.06\text{ g/mol} = 54.1\text{ g }CO(NH_2)_2$ in A

$?\text{ g }H_2O = 10.05\text{ mol }H_2O \times \frac{18.02\text{ g}}{\text{mol}} = 181.1\text{ g }H_2O$ in A

$?\text{ g} = 54.1\text{ g }CO(NH_2)_2 + 181.1\text{ g }H_2O = 235.2\text{ g}$ total in A

$$\text{mass fraction} = \frac{54.1\text{ g }CO(NH_2)_2}{235.2\text{ g total}} = 0.230\ CO(NH_2)_2$$

$?\text{ g }CO(NH_2)_2 = 0.2485\text{ mol }CO(NH_2)_2 \times 60.06\text{ g/mol} = 14.9\text{ g }CO(NH_2)_2$ in B

$?\text{ g }H_2O = 300.0\text{ g} - 235.2\text{ g} - 14.9\text{ g} = 49.9\text{ g }H_2O$ in B

or

$$0.08221 = \frac{0.2485\text{ mol }CO(NH_2)_2}{(0.2485 + X)\text{ mol}}$$

$0.02043 + 0.08221X = 0.2485$

$X = 2.77\text{ mol }H_2O$

$?\text{ g }H_2O = 2.77\text{ mol }H_2O \times \frac{18.02\text{ g}}{\text{mol}} = 49.9\text{ g }H_2O$

$?\text{ g} = 14.9\text{ g }CO(NH_2)_2 + 49.9\text{ g }H_2O = 64.8\text{ g}$ total in B

$$\text{mass fraction} = \frac{14.9\text{ g CO(NH}_2)_2}{64.9\text{ g total}} = \ = 0.230\text{ CO(NH}_2)_2$$

Check: soln A + soln B = total
Initial 200.0 g + 100.0 g = 300.0 g
Final 235.2 g + 64.8 g = 300.0 g

100. $\Delta T = -m\,K_f$

$\Delta T = -i\,m\,K_f$

$(-0.795°\text{C} - 0.00°\text{C}) = -m\,(1.86\ °\text{C}/m)$

$$m = \frac{-0.795°\text{C}}{-1.86\ °\text{C/m}} = 0.427\text{m}$$

$$?\text{ mol} = \frac{0.427\text{ mol}}{\text{kg}} \times \frac{\text{kg}}{10^3\text{ g}} \times 100.0\text{ g} = 0.0427\text{ mol}$$

Let x = mass glucose, $10.0 - x$ = mass sucrose

$$x \times \frac{\text{mol glucose}}{180.2\text{ g}} + (10.0\text{ g} - x)\frac{\text{mol sucrose}}{342.3\text{ g}} = 0.0427\text{ mol}$$

$5.55 \times 10^{-3}\,x - 2.92 \times 10^{-3}\,x + 0.0292 = 0.0427$

$2.63 \times 10^{-3}\,x = 0.0135$

$x = 5.13$ g glucose

$$\%\text{ glucose} = \frac{5.13\text{ g}}{10.00\text{ g}} \times 100\% = 51.3\%$$

% sucrose = 48.7 %

102. The liquid is $CaCl_2(aq)$. $H_2O(g)$ from the air condenses on the solid to form saturated $CaCl_2(aq)$. A solid will not exhibit this phenemenon if the vapor pressure of the saturated solution exceeds the partial pressure of $H_2O(g)$ in the atmosphere. The phenomenon is deliquesence.

104. (a) $$?\text{ cm} = 1000\text{ nm} \times \frac{10^{-9}\text{ m}}{\text{nm}} \times \frac{\text{cm}}{10^{-2}\text{ m}} = 10^{-4}\text{ cm}$$

$\left(\frac{1}{2}\right)^n = 1 \times 10^{-4}$ cm

$n \log 0.5 = \log 1 \times 10^{-4} = -4$

$$n = \frac{-4}{\log 0.5} = \frac{-4}{0.3} = 13.3 \quad \text{rounds to 14 divisions}$$

(b) ? size of cube after 14 divisions $= \left(\frac{1}{2}\right)^{14} = 6.10 \times 10^{-5}$ cm

volume of each cube = $(6.10 \times 10^{-5}\text{ cm})^3 = 2.27 \times 10^{-13}\text{ cm}^3$

number of cubes = $(8)^{14} = 4.40 \times 10^{12}$ cubes

total volume of cubes = 1 cm

$$\text{Total surface area} = 4.40 \times 10^{12} \text{ cubes} \times \frac{6 \text{ sides}}{\text{cube}} \times (6.10 \times 10^{-5} \text{ cm})^2$$

$$= 9.82 \times 10^{4} \text{ cm}^2$$

surface area/volume ratio

original cube $\frac{6.0 \text{ cm}^2}{\text{cm}^3}$

after 14 subdivisions $\frac{9.82 \times 10^4 \text{ cm}^2}{\text{cm}^3}$

Colloidal particles have a much larger surface area to volume ratio than does bulk material because as the bulk material is divided, the volume is the same but the surface area increases.

Apply Your Knowledge

105. $$? \text{ Zn atoms} = 10.0 \ \mu\text{L} \times \frac{10^{-6} \text{ L}}{\mu\text{L}} \times \frac{\text{mL}}{10^{-3} \text{ L}} \times \frac{1 \text{ g}}{\text{mL}} \times \frac{0.1 \text{ g Zn}}{10^9 \text{ g sample}} \times \frac{\text{mol Zn}}{65.4 \text{ g Zn}}$$

$$\times \frac{6.022 \times 10^{23} \text{ Zn atoms}}{\text{mol Zn}} = 9 \times 10^9 \text{ Zn atoms}$$

107. (a) The weight molarity depends on two masses (moles of solute and kg of solution), which do not change with temperature. Thus, it does not depend on the temperature.

(b) $$M_{(wt)} = \frac{12.654 \text{ g NaOH}}{12.654 \text{ g NaOH} + 898.3 \text{ g } H_2O} \times \frac{\text{mol NaOH}}{39.9971 \text{ g NaOH}} \times \frac{10^3 \text{ g}}{\text{kg}}$$

$$= 0.3473 \ M_{(wt)}$$

$$? \text{ g } H_2SO_4 = 0.3473 \ M_{(wt)} \times \frac{\text{kg}}{10^3 \text{ g}} \times 69.26 \text{ g} \times \frac{\text{mol } H_2SO_4}{2 \text{ mol NaOH}}$$

$$\times \frac{98.079 \text{ g } H_2SO_4}{\text{mol } H_2SO_4} = 1.180 \text{ g } H_2SO_4$$

$$? \text{ mass \% } H_2SO_4 = \frac{1.180 \text{ g } H_2SO_4}{3.641 \text{ g acid}} \times 100 \ \% = 32.40 \ \%$$

109. $$k = \frac{S}{P_{gas}} = \frac{5.5 \times 10^{-4} \text{ M}}{0.78 \text{ atm}} = 7.1 \times 10^{-4} \ \frac{\text{M}}{\text{atm}}$$

$$S = k P_{gas} = 7.1 \times 10^{-4} \ \frac{\text{M}}{\text{atm}} \times 4.0 \text{ atm} = 2.8 \times 10^{-3} \text{ M}$$

? mol at 0.78 atm = 5.5×10^{-4} M × 5.0 L = 0.0028 mol

? mol at 4.0 atm = 2.8×10^{-3} M × 5.0 L = 0.014 mol

? mol N_2 released = (0.014 moles – 0.0028 mol) = 0.011 mol

$$? \text{ L} = 0.011 \text{ mol} \times \frac{0.08206 \text{ L atm}}{\text{K mol}} \times 310 \text{ K} \times \frac{1}{0.78 \text{ atm}} = 0.36 \text{ L}$$

$$? \text{ fl oz} = 0.36 \text{ L} \times \frac{\text{mL}}{10^{-3} \text{ L}} \times \frac{\text{fl oz}}{29.57 \text{ mL}} = 12 \text{ fl oz}$$

Gas the size of a soft drink can expanding into the body within a short period causes the bends.

Chapter 13

Chemical Kinetics: Rates and Mechanisms of Chemical Reactions

Exercises

13.1A (a) $\text{rate} = \dfrac{[D]_2 - [D]_1}{\Delta t} = \dfrac{0.3546\text{ M} - 0.2885\text{ M}}{2.55\text{ min}} = 0.0259\text{ M/min}$

(b) $\text{rate of formation of C} = 2 \times \text{rate} = 2 \times 0.0259\text{ M/min} \times \dfrac{\text{min}}{60\text{ sec}}$

$= 8.63 \times 10^{-4}\text{ M/s}$

13.1B (a) $\dfrac{-\Delta[B]}{\Delta t} = \dfrac{1}{2} \times \dfrac{-\Delta[A]}{\Delta t} = \dfrac{1}{2} \times 2.10 \times 10^{-5}\text{ M/s} = 1.05 \times 10^{-5}\text{ M/s} = \text{rate}$

(b) $\dfrac{1}{3}\dfrac{\Delta[C]}{\Delta t} = \dfrac{-1}{2}\dfrac{\Delta[A]}{\Delta t}$

$\dfrac{\Delta[C]}{\Delta t} = \dfrac{3}{2} \times 2.10 \times 10^{-5}\text{ M/s} = 3.15 \times 10^{-5}\text{ M/s}$

13.2A (a) From tangent line

$\text{rate} = \text{slope} = \dfrac{-(0 - 0.630)\text{ M}}{(570 - 0)\text{ s}} = 1.11 \times 10^{-3}\text{ M/s}$

(b) $\Delta[H_2O_2] = -10\text{ s} \times 1.11 \times 10^{-3}\text{ M/s} = -1.1 \times 10^{-2}\text{ M}$

$[H_2O_2]_{310} = [H_2O_2]_{300} + \Delta[H_2O_2]$

$= 0.298\text{ M} + (-1.1 \times 10^{-2}\text{ M}) = 0.287\text{ M}$

13.2B A tangent line has to be drawn that is parallel to the green dashed line. By eyeballing, it appears to be at about 270 s. Values from that line are used to calculate the slope and thus the rate. $\text{rate} = -\text{ slope} = \dfrac{-(0 - 0.700)\text{M}}{(520 - 0)\text{s}}$

rate = 1.3×10^{-3} M/s

Alternate: Since the green dashed line is parallel to the tangent line, its slope will be the same as the tangent line. Thus, the initial and final values in Table 13.1 can be used.

$\dfrac{-(0.094 - 0.882)\text{M}}{(600 - 0)\text{s}} = 1.3 \times 10^{-3}\text{ M s}^{-1}$ at about 270 s

Only one tangent line is possible at each point on the curved line.

13.3A rate = $k\,[NO]^2[Cl_2]$

$$\text{rate} = \frac{5.70}{M^2\ s} \times (0.200\ M)^2 \times 0.400\ M = 0.0912\ M/s$$

13.3B rate = $k[CH_3CHO]^m$

$$\frac{(\text{initial rate})_2}{(\text{initial rate})_1} = \frac{k \times [CH_3CHO]_2^m}{k \times [CH_3CHO]_1^m}$$

$$\frac{2.8 \times (\text{initial rate})_1}{(\text{initial rate})_1} = \frac{k \times 2^m \times [CH_3CHO]_1}{k \times [CH_3CHO]_1}$$

$2.8 = 2^m$

$\log 2.8 = m \log 2$

$$m = \frac{\log 2.8}{\log 2} = \frac{0.45}{0.30} = 1.5$$

Note that the order of this reaction is nonintegral. For m to be integral, the ratio of reaction rates corresponding to a doubling of the initial concentration must be an integral power of 2.

13.4A(a) $\ln \frac{[NH_2NO_2]_t}{[NH_2NO_2]_0} = -kt$

$$\ln \frac{0.0250}{0.105} = -\frac{5.62 \times 10^{-3}}{\text{min}}\, t$$

$t = 255$ min

(b) $\ln \frac{[NH_2NO_2]_t}{0.105\ M} = \frac{-5.62 \times 10^{-3}}{\text{min}} \times 6.00\ h \times \frac{60\ \text{min}}{\text{hr}}$

$$\ln \frac{[NH_2NO_2]}{0.105\ M} = -2.02$$

Raise both sides of the equation to a power of e.

$$e^{\ln \frac{[NH_2NO_2]}{0.105\ M}} = \frac{[NH_2NO_2]}{0.105\ M} = e^{-2.02} = 0.133$$

$[NH_2NO_2] = 0.0139\ M$

13.4B $\ln \frac{[NH_2NO_2]_t}{[NH_2NO_2]_0} = -kt$

$$\ln \frac{[NH_2NO_2]}{0.0750\ M} = \frac{-5.62 \times 10^{-3}}{\text{min}} \times 35.0\ \text{min}$$

$$e^{\ln \frac{[NH_2NO_2]}{0.0750\ M}} = \frac{[NH_2NO_2]}{0.0750\ M} = e^{-0.197} = 0.821$$

$[NH_2NO_2] = 0.0616\ M$

$$\text{rate} = k[NH_2NO_2] = \frac{5.62 \times 10^{-3}}{\text{min}} \times 0.0616\ M = 3.46 \times 10^{-4}\ M/\text{min}$$

13.5A (a) Because $\frac{1}{16}$ is $\left(\frac{1}{2}\right)^4$, the time required is four half-lives—that is, $4 \times t_{1/2}$.

$4 \times 120\ s = 480\ s$

or

$$\ln \frac{\frac{1}{16}}{1} = \frac{-5.78 \times 10^{-3}}{s} t$$

$$-2.77 = \frac{-5.78 \times 10^{-3}}{s} t$$

$t = 480\ s$

(b) $$\ln \frac{m}{m_o} = \frac{-5.78 \times 10^{-3}}{s} t$$

$$\ln \frac{m}{4.80\ g} = \frac{-5.78 \times 10^{-3}}{s} \times 10.0\ min \times \frac{60\ s}{min}$$

$$\ln \frac{m}{4.80\ g} = -3.47$$

Raise both sides of equation to a power of e.

$$e^{\ln \frac{m}{4.80\ g}} = \frac{m}{4.80\ g} = e^{-3.50} = 3.02 \times 10^{-2}$$

$m = 0.150\ g$

13.5B $$\ln \frac{[N_2O_5]_t}{[N_2O_5]_0} = -kt$$

$$\ln \frac{[N_2O_5]_t}{800\ mmHg} = -5.8 \times 10^{-3}\ s^{-1} \times 6.00\ min \times \frac{60\ s}{min} = -2.09$$

$$e^{\ln \frac{[N_2O_5]_t}{800\ mmHg}} = \frac{[N_2O_5]_t}{800\ mmHg} = e^{-2.09} = 0.124$$

$[N_2O_5] = 800\ mmHg \times 0.124 = 99.2\ mmHg$

	N_2O_5 →	$2\ NO_2$	$+ \frac{1}{2} O_2$
Init	800	-	-
Change	701	1402	351
Final	99	1402	351

$P\text{total} = P[N_2O_5] + P[NO] + P[O_2]$

$P\text{total} = 99\ mmHg + 1402\ mmHg + 351\ mmHg = 1852\ mmHg$

13.6A The $t_{1/2}$ is about 120 *s*. 475 *s* is 5 seconds less than four half-lives.

800 mmHg $\xrightarrow{1}$ 400 mmHg $\xrightarrow{2}$ 200 mmHg $\xrightarrow{3}$ 100 mmHg $\xrightarrow{4}$ 50 mmHg.

$P(N_2O_5)$ should be slightly more than 50 mmHg.

13.6B 320 mmHg to 80 mmHg is two half-lives.

240 min − 80 min = 160 min

$$\frac{160 \text{ min}}{2 \text{ halflives}} = \frac{80 \text{ min}}{\text{halflife}}$$

320 mmHg is at the end of one half-life so the initial pressure is 640 mmHg.

13.7A $t_{1/2} = \frac{1}{k[A]_0}$

$$55\ s = \frac{1}{k \times 0.80 \text{ M}}$$

$$k = \frac{0.023}{\text{M}\ s}$$

13.7B $\frac{1}{[A]_t} = kt + \frac{1}{[A]_0}$

$$\frac{1}{0.20 \text{ M}} = \frac{0.023}{\text{M}\ s}t + \frac{1}{0.80 \text{ M}}$$

$$\frac{5.0}{\text{M}} - \frac{1.3}{\text{M}} = \frac{0.023}{\text{M s}}t$$

$t = 1.6 \times 10^2\ s$

$$\frac{1}{0.10 \text{ M}} = \frac{0.023}{\text{M s}}t + \frac{1}{0.80 \text{ M}}$$

$$\frac{10.0}{\text{M}} - \frac{1.3}{\text{M}} = \frac{0.023}{\text{M s}}t$$

$t = 380$ sec

first $t_{1/2} = 55$ s

second $t_{1/2} = 1.6 \times 10^2 - 55 = 1.1 \times 10^2\ s$

third $t_{1/2} = 3.8 \times 10^2 - 1.6 \times 10^2 = 2.2 \times 10^2\ s$

The half-life doubles each half-life because the concentration is halved.

13.8A Method 1

$$\text{rate}_1 = -\left(\frac{0.92 \text{ M} - 1.00 \text{ M}}{100 \text{ sec}}\right) = 8.0 \times 10^{-4} \text{ M}/s$$

$$\text{rate}_2 = -\left(\frac{1.68 \text{ M} - 2.00 \text{ M}}{100 \text{ sec}}\right) = 3.2 \times 10^{-3} \text{ M}/s$$

$$\frac{\text{rate}_1}{\text{rate}_2} = \frac{k[A]_2^n}{k[A]_1^n}$$

$$\frac{\frac{3.2\times10^{-3}\text{ M}}{\text{s}}}{\frac{8.0\times10^{-4}\text{ M}}{\text{s}}} = \frac{k}{k}\left(\frac{2.00 \text{ M}}{1.00 \text{ M}}\right)^n$$

$4 = 2^n$

n = 2 second-order

Method 2

time	[A]	$\frac{1}{[A]}$	change
0	2.00	0.500	
100	1.68	0.595	.10
200	1.43	0.699	.10
300	1.26	0.794	.10
400	1.12	0.893	.10
500	1.01	0.99	.10
600	0.92	1.09	.10
700	0.84	1.19	.10
800	0.78	1.28	.09
900	0.72	1.39	.11
1000	0.67	1.49	.10

The change in $\frac{1}{[A]}$ is constant, so a plot of $\frac{1}{[A]}$ versus time would be a straight line, and the reaction is second-order.

13.8B first-order $t_{1/2} = \dfrac{0.693}{1.00 \times 10^{-3}\ s^{-1}} = 6.93 \times 10^{2}\ s$

second-order first $t_{1/2} = \dfrac{1}{1.00 \times 10^{-3}\ s^{-1} \times 1.00\ M} = 1000\ s$

second $t_{1/2} = \dfrac{1}{1.00 \times 10^{-3}\ s^{-1} \times 0.500\ M} = 2000\ s$

Because the second-order half-life increases with decreasing concentration and starts with a longer half-life, the first-order reaction concentration will decrease much more rapidly. The two concentrations willbe the same at about 1750 seconds.

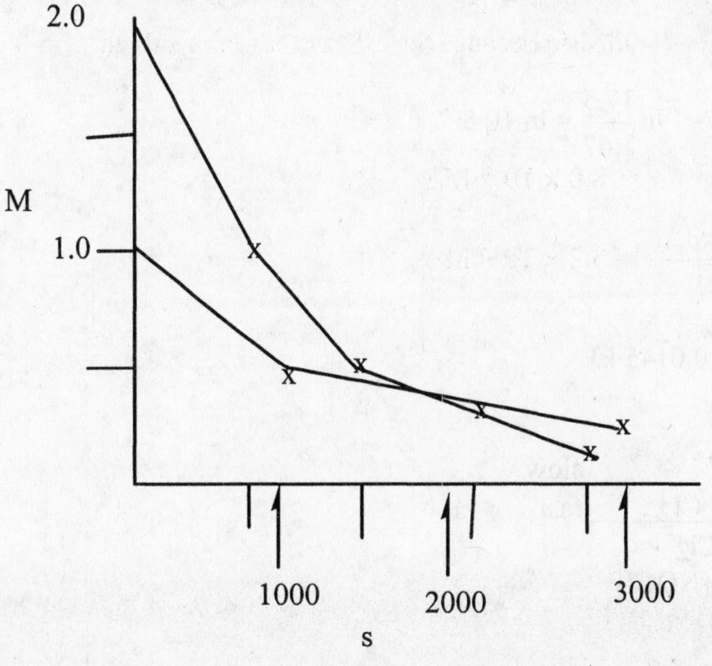

13.9A $\ln \frac{1.0 \times 10^{-5}}{2.5 \times 10^{-3}} = \frac{1.0 \times 10^{2} \text{ kJ/mol} \times 1000 \text{ J/kJ}}{\frac{8.3145 \text{ J}}{\text{mol K}}} \times \left(\frac{1}{332 \text{ K}} - \frac{1}{T_2}\right)$

$-5.52 = 1.20 \times 10^{4} \text{ K}^{-1} \times \left(\frac{1}{332 \text{ K}} - \frac{1}{T_2}\right)$

$-4.60 \times 10^{-4} \text{ K}^{-1} = \frac{1}{332 \text{ K}} - \frac{1}{T_2}$

$\frac{1}{T_2} = \frac{1}{332 \text{ K}} + 4.60 \times 10^{-4} \text{ K}^{-1} = 3.01 \times 10^{-3} \text{ K}^{-1} + 4.60 \times 10^{-4} \text{ K}^{-1}$

$= 3.47 \times 10^{-3} \text{ K}^{-1}$

$T_2 = 1/3.47 \times 10^{-3} \text{ K}^{-1} = 288 \text{ K } (15\ °\text{C})$

13.9B For first-order $k = \frac{0.693}{t_{\frac{1}{2}}}$

$k_1 = \frac{0.693}{17.5 \text{ h}} = 0.0396$ $\qquad T_1 = 125\ °\text{C} + 273 = 398 \text{ K}$

$k_2 = \frac{0.693}{1.67 \text{ h}} = 0.415$ $\qquad T_2 = 145\ °\text{C} + 273 = 418 \text{ K}$

$\ln \frac{k_2}{k_1} = \frac{E_a}{R}\left(\frac{1}{T_1} - \frac{1}{T_2}\right)$

$\ln \frac{0.415}{0.0396} = \frac{E_a \times 1000 \text{ J/kJ}}{8.3145 \text{ J mol}^{-1}\text{K}^{-1}} \left(\frac{1}{398} - \frac{1}{418}\right)$

$\ln 10.5 = 2.35 = E_a \times 0.0145 \text{ kJ}$

$E_a = 163 \text{ kJ}$

OR

$\frac{\ln \frac{0.693}{t_{\frac{1}{2}}(2)}}{\ln \frac{0.693}{t_{\frac{1}{2}}(1)}} = \ln \frac{t_{\frac{1}{2}}(1)}{t_{\frac{1}{2}}(2)} = \ln \frac{17.5}{1.67} = \ln 10.5$

$\ln 10.5 = \frac{E_a}{R}\left(\frac{1}{T_1} - \frac{1}{T_2}\right)$

$\ln 10.5 = 2.35 = E_a \times 0.0145 \text{ kJ}$

$E_a = 163 \text{ kJ}$

13.10A $NOCl \rightarrow NO + Cl$ slow

$NOCl + Cl \rightarrow NO + Cl_2$ fast

$2\ NOCl \rightarrow 2\ NO + Cl_2$

Rate of reaction $= k\,[NOCl]$

13.10B (a)

$$NO + O_2 \underset{k_{-1}}{\overset{k_1}{\rightleftharpoons}} NO_3 \quad \text{fast}$$

$$NO_3 + NO \xrightarrow{k_2} 2\,NO_2 \quad \text{slow}$$

$$2\,NO + O_2 \rightarrow 2\,NO_2$$

(b) rate = $k\,[NO]^2[O_2]$

from slow reaction rate = $k_2\,[NO_3]\,[NO]$

from fast reaction $\dfrac{k_1}{k_{-1}} = \dfrac{[NO_3]}{[O_2][NO]}$

$$[NO_3] = \frac{k_1}{k_{-1}}\,[O_2]\,[NO]$$

$$\text{rate} = k_2\frac{k_1}{k_{-1}}\,[NO]^2\,[O_2] = k\,[NO]^2\,[O_2]$$

Review Questions

3. The average rate is the rate of reaction evaluated over a period of time, not the rate at one particular instant.
The initial rate is the rate of the first few percent of the reaction. The instantaneous rate is the rate at a given time, not necessarily the beginning of the reaction.
The average and instantaneous rates of reaction are the same for a zero-order reaction, and for reactions of other orders, they are nearly the same at the very start of the reaction.
The initial rate and the instantaneous rate are the same if the beginning of the reaction is chosen as the time period.
The three rates are all equal at all times in zero-order reactions.

6. (c) zero-order produces a straight line for a concentration versus time plot.

7. (d) is the answer. rate = k[A][B] and ½rate$_A$ = rate$_B$

8. $k = \dfrac{0.693}{13.9\text{ min}} = 4.99 \times 10^{-2}\text{ min}^{-1}$

Rate = k[A] = $4.99 \times 10^{-2}\text{ min}^{-1} \times 0.40\text{ M} = 2.0 \times 10^{-2}\text{ M min}^{-1}$
(a) is the answer.

9. (b) fraction of molecules with energies in excess of the activation energy. Collision frequency increases but slowly.

12. The activation energy must be greater than the enthalpy of the reaction if the enthalpy is positive.
(c) is the answer.

15. A bimolecular reaction means that the slow step reactants are A and B, the rate = k[A][B], rate of appearance of C = rate of disappearance of A and both (b) and (c) are answers.

Problems

23. $[H_2O_2] = 0.2546\text{ M} - (9.32 \times 10^{-4}\text{ M/s} \times 35\text{ s}) = 0.222\text{ M}$

25. (a) -3.1×10^{-4} M/s $\quad -\dfrac{1}{2}\dfrac{\Delta[B]}{\Delta t} = -\dfrac{\Delta[A]}{\Delta t}$

(b) 9.3×10^{-4} M/s $\quad -\dfrac{1}{2}\dfrac{\Delta[B]}{\Delta t} = \dfrac{1}{3}\dfrac{\Delta[D]}{\Delta t}$

(c) general rate of reaction = − rate of disappearance of A = 3.1×10^{-4} M/s

27. It is not zero-order as zero-order would be a straight line.

29. (a) False. This would be true only if it is a second-order reaction and it does not have to be second-order.
(b) True. A is consumed twice as fast as B is produced.

31. (a) Assume rate = $k[S_2O_8^{2-}]^n[I^-]^m$
Determine n by using experiments 1 and 2

$$\frac{\text{rate2}}{\text{rate1}} = \frac{2.8 \text{ x } 10^{-5}}{1.4 \text{ x } 10^{-5}} = \frac{k \times (0.076\text{ M})^n \times (0.060\text{ M})^m}{k \times (0.038\text{ M})^n \times (0.060\text{ M})^m}$$

$2 = 2^n$
$n = 1$ first-order in $S_2O_8^{2-}$
Determine m by using experiments 2 and 3

$$\frac{\text{rate3}}{\text{rate2}} = \frac{5.6 \text{ x } 10^{-5}}{2.8 \text{ x } 10^{-5}} = \frac{k \times (0.076\text{ M})^n \times (0.120\text{ M})^m}{k \times (0.076\text{ M})^n \times (0.060\text{ M})^m}$$

$2 = 2^m$
$m = 1$ first-order in I^-
Second-order overall.

(b) Experiment 1
$1.4 \times 10^{-5} = k \times 0.038\text{ M} \times 0.060\text{ M}$
$k = 6.1 \times 10^{-3}\text{ M}^{-1}\text{ s}^{-1}$

(c) rate = $6.1 \times 10^{-3}\text{ M}^{-1}\text{ s}^{-1} \times 0.083\text{ M} \times 0.115\text{ M}$
rate = 5.8×10^{-5} M/sec

33. Zero-order for this reaction. For zero-order, rate is independent of condentration.

35. rate = $k\,[H_2O_2] = 3.66 \times \dfrac{10^{-3}}{\text{s}} \times 2.05\text{ M} = 7.50 \times 10^{-3}$ M/s

37. (a) $[A] = \frac{\text{rate}}{k} = \frac{0.0150\ \text{M/min}}{0.0462/\text{min}} = 0.325\ \text{M}$

(b) 88% remaining in 12 min
44% remaining in 48 min
half–life is 36 min
22% remaining in another half–life
48 min + 36 min = 84 min

39. (a) $t_{\frac{1}{2}} = 8.75\ \text{h} = \frac{0.693}{k}$

$$k = \frac{0.693}{8.75\ \text{h}} \times \frac{\text{hr}}{60\ \text{min}} \times \frac{\text{min}}{60\ \text{sec}} = 2.20 \times 10^{-5}/\text{s}$$

(b) $\ln \frac{P}{722\,\text{mmHg}} = -2.20 \times 10^{-5}/\text{s} \times 3.00\ \text{h} \times \frac{3600\ \text{s}}{\text{h}} = -0.238$

Raise both sides of equation to a power of e.

$$e^{\ln \frac{P}{722\,\text{mmHg}}} = \frac{P}{722\,\text{mmHg}} = e^{-0.238} = 0.789$$

$P = 569$ mmHg

(c) $\ln \frac{125\ \text{mmHg}}{722\ \text{mmHg}} = \frac{-2.20 \times 10^{-5}}{\text{s}}\, t$

$$-1.75 = \frac{-2.20 \times 10^{-5}}{\text{s}}\, t$$

$$t = 7.97 \times 10^{4}\ \text{s} \times \frac{\text{min}}{60\ \text{s}} = 1.33 \times 10^{3}\ \text{min} \times \frac{\text{h}}{60\ \text{min}} = 22.1\ \text{h}$$

41. $k = \frac{0.693}{32\ \text{min}} = 2.17 \times 10^{-2}/\text{min}$

$$\ln \frac{N}{2.7 \times 10^{15}} = -\ 2.17 \times 10^{-2}/\text{min} \times 2.24\ \text{h} \times \frac{60\ \text{min}}{\text{h}} = -2.92$$

$$e^{\ln \frac{N}{2.7 \times 10^{15}}} = \frac{N}{2.7 \times 10^{15}} = e^{-2.92} = 5.39 \times 10^{-2}$$

$N = 1.5 \times 10^{14}$ molecules/L

43. (a) $\frac{1}{[A]_t} = kt + \frac{1}{[A]_o}$

$$\frac{1}{[A]_t} = \frac{1.2 \times 10^{-3}}{\text{M s}} \times 2.00\ \text{h} \times \frac{60\text{min}}{\text{h}} \times \frac{60\text{s}}{\text{min}} + \frac{1}{0.56\ \text{M}}$$

$$\frac{1}{[A]_t} = \frac{8.64}{\text{M}} + \frac{1.79}{\text{M}}$$

$[A]_t = 0.096$ M

(b) $\frac{1}{0.28} - \frac{1}{0.56} = \frac{1.2 \times 10^{-3}}{\text{M s}} \times t$

$$3.57 - 1.79 = \frac{1.2 \times 10^{-3}}{\text{M s}} \times t$$

$t = 1.5 \times 10^3$ sec = 25 mins.

45. For zero order, the half-life gets longer as the initial concentration increases because the rate is constant, and the more molecules present, the longer it takes to consume half of them.
For second order, the half-life is related to the inverse of the concentration—the greater the initial concentration, the faster the initial rate and the sooner the molecules are consumed.

47.

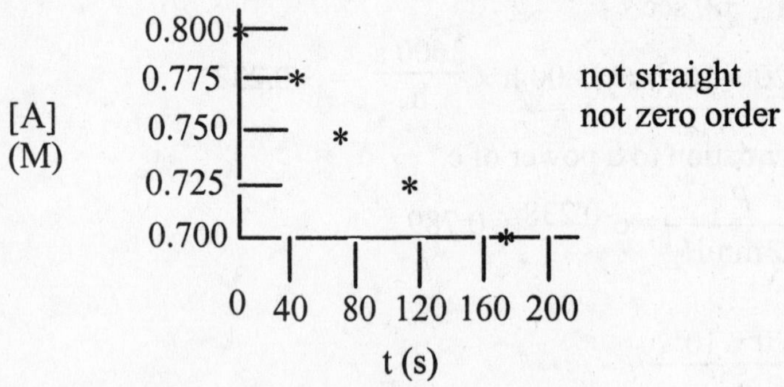

t	[A]	ln [A]	$\frac{1}{[A]}$
0	.800	-0.223	1.25
40	.775	-0.255	1.29
83	.750	-0.288	1.33
129	.725	-0.322	1.38
179	.700	-0.357	1.43

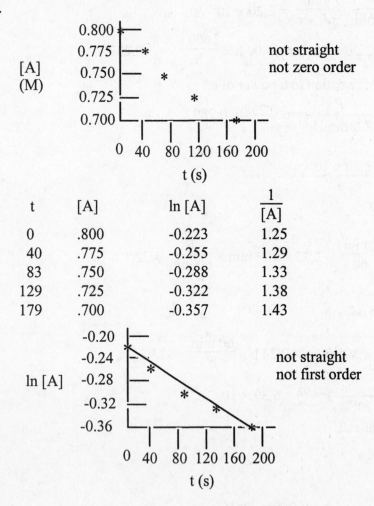

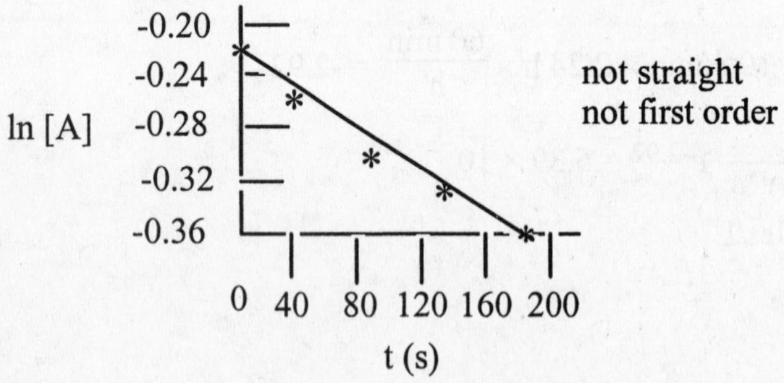

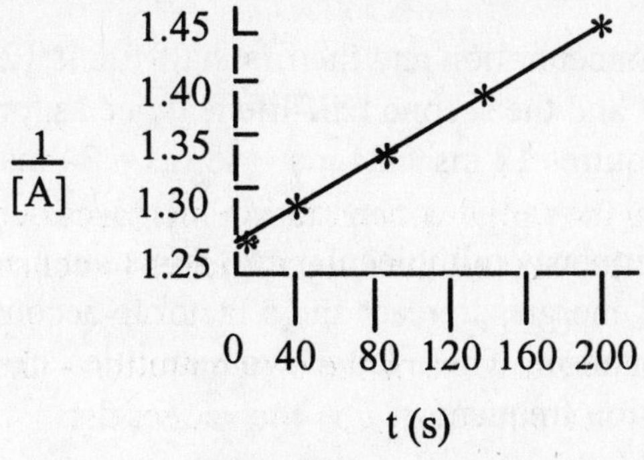

$$k = \text{slope} = \frac{1.43 - 1.25}{179 - 0} = \frac{1.0 \times 10^{-3}}{\text{M s}}$$

$$\text{rate}_1 = \frac{-(0.775 - 0.800)\text{M}}{40 \text{ s}} = \frac{6.25 \times 10^{-4} \text{ M}}{\text{s}}$$

$$\text{rate}_2 = \frac{-(0.390 - 0.400)\text{M}}{64 \text{ s}} = \frac{1.56 \times 10^{-4} \text{ M}}{2}$$

$$\frac{\text{rate}_1}{\text{rate}_2} = \frac{k\,[A]^n}{k[A]^n}$$

$$\frac{(6.25) \times 10^{-4}\text{M/s}}{1.56 \times 10^{-4}\text{M/s}} = \left(\frac{0.800}{0.400}\right)^n$$

$$4.00 = (2.00)^n$$

$n = 2$ second-order

$$\text{rate} = k\,[A]^2$$

$$k = \frac{6.25 \times 10^{-4}\text{M/s}}{(.800 \text{ M})^2} = \frac{+9.77 \times 10^{-4}}{\text{M s}}$$

49 $$t_{1/2} = \frac{1}{k[\text{Cl}]_0}$$

$$k = \frac{1}{t_{1/2}[\text{Cl}]}$$

$$k = \frac{1}{12 \text{ ms}\,[\text{Cl}]_0}$$

$$\frac{1}{[\text{Cl}]_t} = kt + \frac{1}{[\text{Cl}]_0}$$

$$\frac{1}{[\text{Cl}]_t} = \frac{1}{\frac{1}{8}[\text{Cl}]_0} = \frac{8}{[\text{Cl}]_0}$$

$$\frac{8}{[\text{Cl}]_0} - \frac{1}{[\text{Cl}]_0} = \frac{1}{[\text{Cl}]_0 12 \text{ ms}}\,t$$

$$\frac{7}{[\text{Cl}]_0} \times [\text{Cl}]_0 \; 12 \text{ ms} = t$$

$t = 84$ ms

Since the half-life is dependent on the concentration and the first half-life is 12 ms, the second half-life concentration is half and the second half-life is twice as long (24 ms). To get to 1/8, the original value requires 12 ms + 24 ms + 48 ms = 84 ms.

51. The calculation involves not only the frequency of molecular collisions but also the fraction of the molecules with sufficient energies to react and a factor to account for the collisions that have a favorable orientation. These latter two quantities are much more difficult to assess than just a collision frequency.

53. By looking at Figure 13.14, the reaction profile of an endothermic reaction, it is obvious that the activation energy must be at least as large as the enthalpy or there would be no activation. In Figure 13.13, one can easily see that the enthalpy is the difference between the initial and final states and has no relationship to how much above the initial state the activation energy is.

55. $$\ln\frac{1.63 \times 10^{-3}}{4.75 \times 10^{-4}} = \frac{E_a \times 1000\ \text{J/kJ}}{\dfrac{8.3145\ \text{J}}{\text{mol K}}} \times \left(\frac{1}{293\text{K}} - \frac{1}{303\text{K}}\right)$$

$1.233 = E_a \times 120.3\ \text{mol/kJ} \times 1.13 \times 10^{-4}$

$E_a = 91$ kJ /mol

57. (a) $$\ln\frac{0.0120}{k_1} = \frac{218\ \text{kJ mol} \times 1000\ \text{J/kJ}}{8.3145\ \text{J/mol K}} \times \left(\frac{1}{525\ \text{K}} - \frac{1}{652\ \text{K}}\right)$$

$$\ln\frac{0.0120}{k_1} = \frac{218 \times 1000}{8.3145} \times 3.71 \times 10^{-4} = 9.73$$

Raise both sides of the equation to a power of e.

$$e^{\ln\frac{0.0120}{k_1}} = \frac{0.0120}{k_1} = e^{9.73} = 1.68 \times 10^4$$

$$k_1 = \frac{0.0120}{1.68 \times 10^4} = 7.14 \times 10^{-7}/\text{min}$$

Answer should be limited to two significant figures.

(b) $$\ln\frac{0.0120}{0.0100} = \frac{218\ \text{kJ/mol} \times 1000\ \text{J/kJ}}{8.3145\ \text{J/mol K}} \times \left(\frac{1}{T_1} - \frac{1}{652\ \text{K}}\right)$$

$$6.954 \times 10^{-6} = \left(\frac{1}{T_1} - \frac{1}{652\ \text{K}}\right)$$

$T_1 = 649$ K

59. A molecule acquires enough excess energy through collisions with other molecules or the walls of its container to enable it to dissociate. Its dissociation can occur without requiring a further collision.

61. (a) $A + B \rightarrow I$

$\underline{I + B \rightarrow C + D}$

$A + 2B \rightarrow C + D$ net reaction

(b) rate = k [A][B]

63. $2\ NO_2 \rightarrow NO_3 + NO$ slow
$NO_3 + CO \rightarrow NO_2 + CO_2$ fast
$NO_2 + CO \rightarrow NO + CO_2$

65. from slow step rate = k_2 [NO][Cl]

from fast step $\frac{k_1}{k_{-1}} = \frac{[NOCl][Cl]}{[NO][Cl_2]}$ $[Cl] = \frac{k_1}{k_{-1}}\frac{[NO][Cl_2]}{[NOCl]}$

$$\text{rate} = k_2\frac{k_1}{k_{-1}}\frac{[NO]^2[Cl_2]}{[NOCl]} == k\frac{[NO]^2[Cl_2]}{[NOCl]}$$

The rate law for the mechanism does not match the exhibited rate law.

67. rate = $k_2[Hg][Tl^{3+}]$ from slow step

$\frac{k_1}{k_{-1}} = \frac{[Hg][Hg^{2+}]}{[Hg_2^{2+}]}$ from fast equilibrium step

$$[Hg] = \frac{k_1[Hg_2^{2+}]}{k_{-1}[Hg^{2+}]}$$

$$\text{rate} = k_2\frac{k_1}{k_{-1}}\frac{[Hg_2^{2+}][Tl^{3+}]}{[Hg^{2+}]} = k\frac{[Hg_2^{2+}][Tl^{3+}]}{[Hg^{2+}]}$$

69. Because $[I^-]$ remains constant (it is a catalyst), the rate law, rate = $k[I^-][H_2O_2]$, simplifies to rate = $k'[H_2O_2]$. The value of k' depends on $[I^-]$ chosen, but once $[I^-]$ is fixed, the value of k' is fixed.

71. An inhibitor may block the active site of the enzyme, or it may react with the enzyme to change the shape of the active site.

73. Both an enzyme and the surface on which a surface-catalyzed reaction occurs require that reaction occurs at active sites. The kinetics of each type of reaction, then, is governed by the availability of these active sites.

Additional Problems

75. From the data pairs:

t = 0 s	[A] = 0.88 M	t = 100 s	[A] = 0.44 M
t = 25 s	[A] 0.74 M	t = 125 s	[A] = 0.37 M
t = 50 s	[A] = 0.62 M	t = 150 s	[A] = 0.31 M

The half-life is 100 s.

$$k = \frac{0.693}{100\ s} = 6.93 \times 10^{-3}\ s^{-1}$$

$$\text{rate} = k\,[A]_{125} = 6.93 \times 10^{-3}\ s^{-1} \times 0.37\ M = 2.6 \times 10^{-3}\ M/s$$

76. $$0.00500\ L \times 0.882\ M\ H_2O_2 \times \frac{2\ \text{mol}\ MnO_4^-}{5\ \text{mol}\ H_2O_2} \times \frac{1\ L}{0.0500\ \text{mol}\ MnO_4} \times \frac{1000\ mL}{1\ L}$$

$$= 35.3\ mL$$

0 s, 35.3 mL; 60 s, 27.9 mL; 120 s, 22.6 mL; 180 s, 18.3 mL,
240 s, 14.9 mL; 300 s, 11.9 mL; 360 s, 9.44 mL; 420 s, 7.52 mL;
480 s, 6.08 mL; 540 s, 4.80 mL; 600 s, 3.76 mL

79. (a) $$P = \frac{mRT}{VM} = \frac{4.50\ g \times \frac{0.08206\ L\ atm}{K\ mol} \times (147 + 273)K}{1.00\ L \times 146.2\ g/mol} = 1.06\ atm$$

(b) after 1 half-life $P_{DTBP} = 0.53$ atm

$P_{acetone} = 1.06$ atm $P_{ethane} = 0.53$ atm
$P_{total} = 2.12$ atm

(c) $$\ln\frac{P_t}{P_o} = -kt = -\frac{0.693}{t_{1/2}}\,t$$

$$\ln\frac{P_t}{1.06} = -\frac{0.693}{80.0\ min} \times 125\ min = -1.08$$

Raise both sides of equation to power of e.

$$e^{\ln\frac{P_t}{1.06}} = \frac{P_t}{1.06\ atm} = e^{-1.08} = 0.340$$

$P_t = 0.340 \times 1.06\ atm = 0.36\ atm = P_{DTBP}$
$P_{ethane} = 1.06 - 0.36 = 0.70$ atm
$P_{acetone} = P_{ethane} \times 2 = 1.40$ atm
$P_{total} = 0.36 + 0.70 + 1.40 = 2.46$ atm

82. $$\ln\frac{k_2}{k_1} = \frac{+Ea}{R}\left(\frac{1}{T_1} - \frac{1}{T_2}\right)$$

$$k_1 = \frac{0.693}{14.1\ h \times \frac{60\ min}{h}} = 8.19 \times 10^{-4}\ min^{-1}$$

$$k_2 = \frac{0.693}{48.8\ min} = 1.42 \times 10^{-2}\ min^{-1}$$

$$\ln\frac{1.42 \times 10^{-2}}{8.19 \times 10^{-4}} = \frac{E_a \times \frac{10^3\ J}{kJ}}{\frac{8.3145\ J}{mol\ K}} \times \left(\frac{1}{25 + 273} - \frac{1}{50 + 273}\right) K^{-1}$$

$2.85 = Ea \times 3.12 \times 10^{-2}$ mol/kJ

$$Ea = \frac{91.2\text{ kJ}}{\text{mol}}$$

Answer should be limited to two significant figures.

$$\ln\frac{1.42\times10^{-2}\text{ min}^{-1}}{k_1} = \frac{\frac{91.2\text{ kJ}}{\text{mol}}\times\frac{10^3\text{ J}}{\text{kJ}}}{\frac{8.3145\text{ J}}{\text{mol K}}}\times\left(\frac{1}{65+273}-\frac{1}{50+273}\right)\text{K}^{-1} = -1.5$$

Raise both sides of the equation to a power of e.

$$e^{\ln\frac{1.42\times10^{-2}\text{ min}^{-1}}{k_1}} = \frac{1.42\times10^{-2}\text{ min}^{-1}}{k_1} = e^{-1.5} = 0.22$$

$$k = \frac{6.41 \times 10^{-2}}{\text{min}}$$

$$\ln\frac{10}{100} = \frac{-6.41 \times 10^{-2}\text{ t}}{\text{min}}$$

$$-2.30 = \frac{-6.41 \times 10^{-2}\text{ t}}{\text{min}}$$

t = 35.9 min

Answer should be limited to two significant figures.

85. (a) For [OCl^-] use rate1/rate 3

$$\frac{\text{rate1}}{\text{rate3}} = \frac{k}{k}\left(\frac{0.0040}{0.0020}\right)^n \times \left(\frac{0.0020}{0.0020}\right)^m \times \frac{[1.00]^p}{[1.00]^p}$$

$$\frac{4.8 \times 10^{-4}}{2.4 \times 10^{-4}} = \left(\frac{0.0040}{0.0020}\right)^n$$

$2 = 2^n$

$n = 1$ first-order in OCl^-

For [I^-] use rate2/rate 3

$$\frac{\text{rate2}}{\text{rate3}} = \frac{k\,[0.0020]}{k\,[0.0020]} \times \left(\frac{0.0040}{0.0020}\right)^m \times \frac{[1.00]^p}{[1.00]^p}$$

$$\frac{5.0 \times 10^{-4}}{2.4 \times 10^{-4}} = \left(\frac{0.0040}{0.0020}\right)^m$$

$2.08 = 2^m$

$m = 1$ first-order in I^-

For [OH^-] use rate4/rate 3

$$\frac{\text{rate4}}{\text{rate3}} = \frac{k\,[0.0020]}{k\,[0.0020]} \times \frac{[0.0020]}{[0.0020]} \times \left(\frac{0.50}{1.00}\right)^p$$

$$\frac{4.6 \times 10^{-4}}{2.4 \times 10^{-4}} = \left(\frac{0.50}{1.00}\right)^p$$

$$1.92 = \left(\frac{1}{2}\right)^p$$

$\log 1.92 = p \log 0.5$

$p = \frac{\log 1.92}{\log 0.5} = -0.94$

$p = -1$ negative first-order

Overall order of 1

(b) $\text{rate} = k\frac{[OCl^-][I^-]}{[OH^-]}$

$$\frac{4.8 \times 10^{-4}\ M}{s} = k\frac{(0.0040\ M)(0.0020\ M)}{1.00\ M}$$

$$k = \frac{60}{s}$$

(c) $OCl^- + H_2O \rightleftharpoons HOCl + OH^-$

$I^- + HOCl \rightarrow HOI + Cl^-$

$HOI + OH^- \rightarrow H_2O + OI^-$

$OCl^- + I^- \rightarrow Cl^- + OI^-$

The first step is a fast reversible reaction. The rapid equilibrium assumption leads to:

$$\frac{k_1}{k_{-1}} = \frac{[HOCl][OH^-]}{[OCl^-][H_2O]}$$

$$[HOCl] = \frac{k_1}{k_{-1}}\frac{[OCl^-][H_2O]}{[OH^-]}$$

The second step is the rate-determining step:

$\text{rate} = k_2[I^-][HOCl]$

$$\text{rate} = k_2\frac{k_1}{k_{-1}}\frac{[I^-][OCl^-][H_2O]}{[OH^-]}$$

Note that $[H_2O]$ is a constant.

$$\text{rate} = k_2\frac{k_1[H_2O]}{k_{-1}}\frac{[I^-][OCl^-]}{[OH^-]} = k_{total}\frac{[I^-][OCl^-]}{[OH^-]}$$

The third step is a rapid neutralization of the acid HOI by the base OH^-.

(d) Since OH^- is made in the first step and consumed in the third step, it is an intermediate.

86. $O_3 \rightleftharpoons O_2 + O$ fast

$O + O_3 \rightarrow 2\ O_2$ slow

$2O_3 \rightarrow 3\ O_2$

from slow step $\text{rate} = k_2[O][O_3]$

from fast step $\frac{k_1}{k_{-1}} = \frac{[O_2][O]}{[O_3]}$

$$[O] = \frac{k_1}{k_{-1}}\frac{[O_3]}{[O_2]}$$

rate = $k_2 \frac{k_1}{k_{-1}} \frac{[O_3]^2}{[O_2]}$

89.

Time, hr	mass, g	log mass	1/mass
0	1.71	0.233	0.585
3.58	1.15	0.0607	0.870
7.67	0.87	-0.060	1.15
24.33	0.46	-0.337	2.17
44.58	0.28	-0.553	3.57

Log mass can be used as well as ln mass.

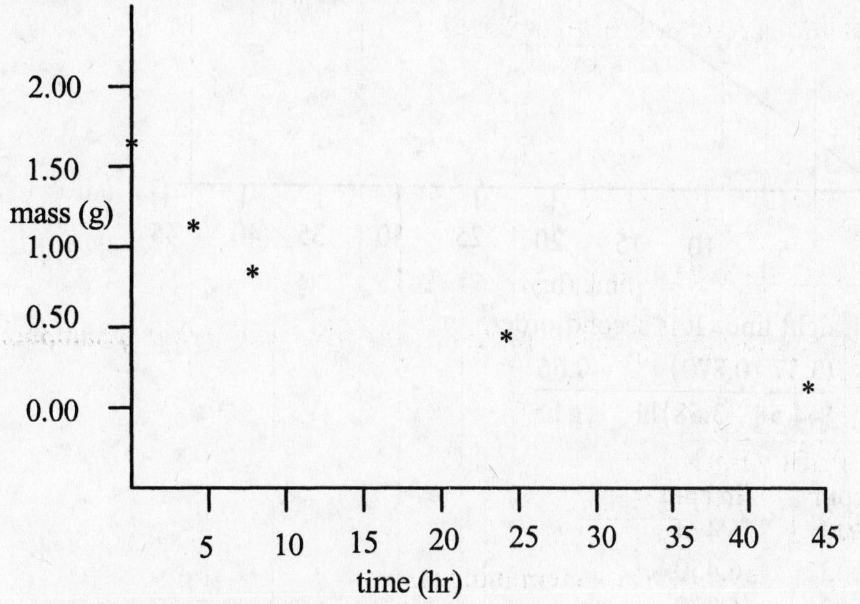

This is not a straight line. It is not zero-order.

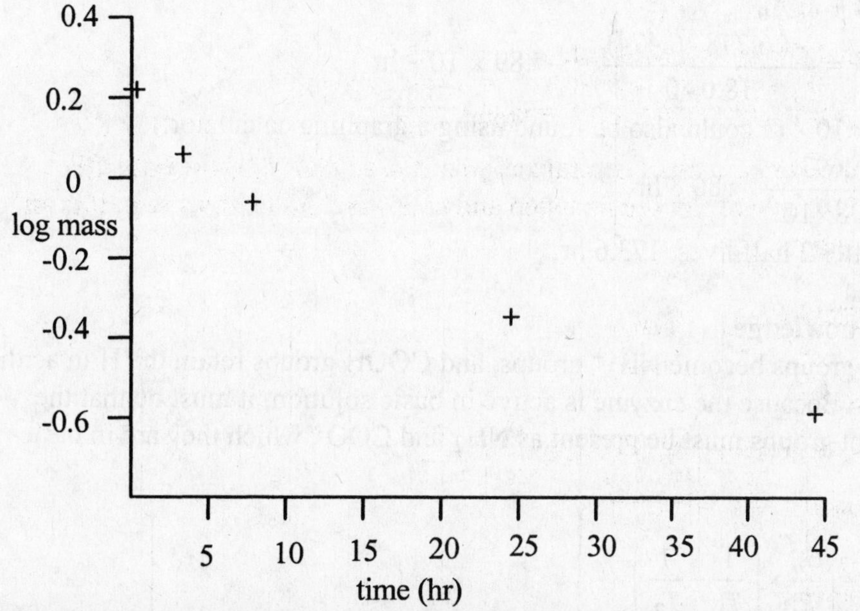

This is not a straight line It is not first-order.

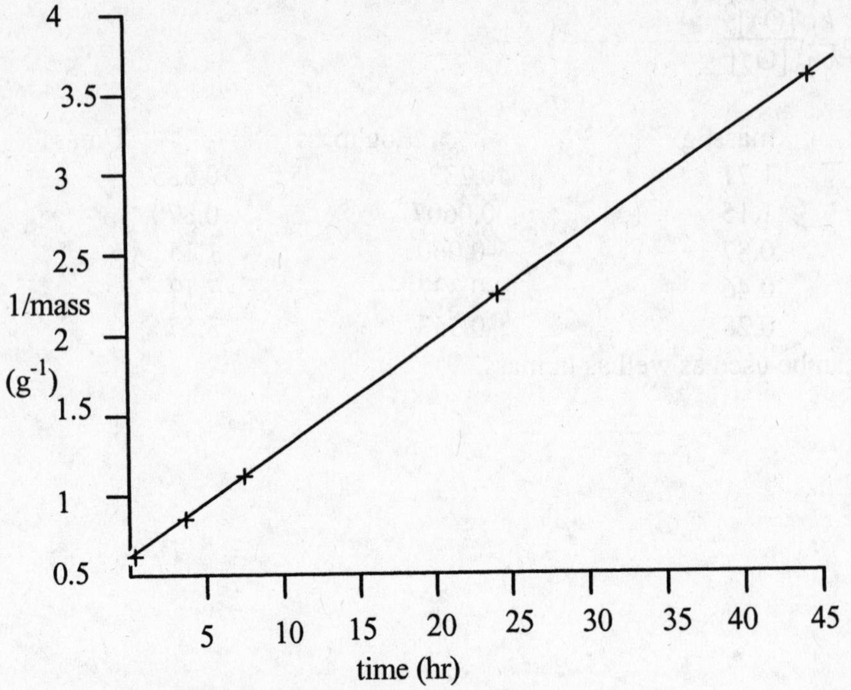

This is a straight line. It is second-order.

$$k = \text{slope} = \frac{(3.57 - 0.870)\,g^{-1}}{(44.58 - 3.58)\,hr} = \frac{0.66}{g\,hr}$$

90.

	cpm	ln cpm
0	648	6.474
3.00	633	6.450
12.0	589	6.378
18.0	562	6.332

$\ln A = -kt + \ln A_0$

$$\text{slope} = -k = \frac{-(6.378 - 6.474)}{18.0 - 0} = -7.89 \times 10^{-3}\ hr^{-1}$$

$k = 7.89 \times 10^{-3}$ (k could also be found using a graphing calculator.)

$$t_{1/2} = \frac{.693}{7.98 \times 10^{-3}} = 86.8\ hr$$

25 % is after 2 half lives, 173.6 hr.

Apply Your Knowledge

92. The NH_2 groups become NH_3^+ groups, and COOH groups retain the H in acidic solutions. Because the enzyme is active in basic solution, it must be that the substituent groups must be present as NH_2 and COO^-, which they are in basic solutions.

94. (a) $\ln\frac{r2}{r1} = \frac{E_a}{R} \times \left(\frac{1}{T_1} - \frac{1}{T_2}\right)$

$$\ln\frac{142}{179} = \frac{E_a}{R} \times \left(\frac{1}{25.0 + 273.2} - \frac{1}{21.7 + 273.2} \right)$$

$-0.232 = Ea \times -4.51 \times 10^{-3}$

$$Ea = \frac{51.3 \text{ kJ}}{\text{mol}}$$

(b) $$\ln\frac{r2}{r1} = \ln\frac{r2}{179} = \frac{\frac{51.3 \text{ kJ}}{\text{mol K}} \times \frac{10^3 \text{ J}}{\text{kJ}}}{\frac{8.3145 \text{ J}}{\text{mol K}}} \times \left(\frac{1}{25.0 + 273.2} - \frac{1}{20.0 + 273.2} \right) = -0.353$$

Raise both sides of the equation to a power of e.

$$e^{\ln\frac{r2}{179}} = \frac{r2}{179} = e^{-0.353} = 0.702$$

$$r_2 = \frac{126 \text{ chirps}}{\text{min}}$$

(c) $$\frac{126 \text{ chirps}}{\text{min}} = \frac{31.5 \text{ chirps}}{15 \text{ s}}$$

$40 + 31.4 = 71.4$ °F

$$°\text{F} = (20.0\ °\text{C} \times \frac{9}{5}) + 32° = 68\ °\text{F}$$

The rule of thumb is close but not exact.

95. (a) $$\text{Rate} = \frac{\Delta[I_3^-]}{\Delta t} = \frac{[I_3^-]_{\text{final}} - [I_3^-]_{\text{Init}}}{t_{\text{final}} - t_{\text{init}}}$$

$t_{\text{inti}} = 0$

$\Delta[I_3^-]$ = the constant value of $S_2O_3^{2-}$ initially added.

Thus, t_{final} is the only variable, and the rate is inversely proportional to the time. That is, the more quickly the blue color appears, the faster the I^- is consumed, and the faster the reaction: rate $\propto$ 1/time.

(b) For $S_2O_8^{2-}$

$$\frac{\text{Rate}_1}{\text{Rate}_2} = \frac{t_2}{t_1} = \left(\frac{[S_2O_8]_1}{[S_2O_8]_2} \right)^x = \frac{42\text{s}}{21\text{s}} = \left(\frac{0.20}{0.10} \right)^x$$

$2 = 2^x$ $\quad x = 1$

For I^-

$$\frac{\text{Rate}_1}{\text{Rate}_4} = \frac{t_4}{t_1} = \left(\frac{[I^-]_1}{[I^-]_4} \right)^y = \frac{42\text{s}}{21\text{s}} = \left(\frac{0.20}{0.10} \right)^y$$

$2 = 2^y$ $\quad y = 1$

first-order in $S_2O_8^{2-}$

first-order in I^-

second-order overall

(c) $[S_2O_8^{2-}] = \dfrac{25.0\text{ mL} \times 0.20\text{ M}}{(25.0+25.0+10.0+5.0)\text{mL}} = 0.077\text{ M}$

$[I^-] = \dfrac{25.0\text{ mL} \times 0.20\text{ M}}{65.0\text{ mL}} = 0.077\text{ M}$

$[S_2O_3^{2-}] = \dfrac{10.0\text{ mL} \times 0.010\text{ M}}{65.0\text{ mL}} = 0.00154\text{ M}$

For I_3^- to react with starch, all the $S_2O_3^{2-}$ must be used up. The $S_2O_8^{2-}$ that is necessary to make the I_3^- use up the $S_2O_3^{2-}$ is:

$$0.00154\text{ M} \times \frac{\text{mol } I_3^-}{2\text{ mol } S_2O_3^{2-}} \times \frac{\text{mol } S_2O_8^{2-}}{\text{mol } I_3^-} = 7.7 \times 10^{-4}\text{ M}$$

$$\text{rate} = \frac{-\Delta[S_2O_8^{2-}]}{t} = \frac{-(-7.7 \times 10^{-4})}{21s} = 3.7 \times 10^{-5}\text{ M/s}$$

(d) Rate = $k[S_2O_8^{2-}][I^-]$

$$k = \frac{3.7 \times 10^{-5}\text{ M/s}}{(0.077\text{ M})^2} = 0.0062\text{ M}^{-1}\text{s}^{-1}$$

(e) $S_2O_8^{2-} + I^- \rightarrow IS_2O_8^{3-}$ slow

$IS_2O_8^{3-} \rightarrow 2\ SO_4^{2-} + I^+$ fast

$I^+ + I^- \rightarrow I_2$ fast

$I_2 + I^- \rightarrow I_3^-$ fast

$S_2O_8^{2-} + 3I^- \rightarrow 2\ SO_4^{2-} + I_3^-$

The first step is slow because it requires collisions between like-charged particles—that is, between negative ions. The third step is very fast because it is aided by the attraction between oppositely charged ions. The reaction has a slow first step followed by fast steps, so the rate is determined by the first step alone. Thus, the rate law is rate = $k[S_2O_8^{2-}][I^-]$.

Chapter 14

Chemical Equilibrium

Exercises

14.1A $K_c = \frac{[COCl_2]}{[CO][Cl_2]} = \frac{[COCl_2]}{[CO]^2}$

There would not be just one value for $[COCl_2]$. There would be a different value of $[COCl_2]$ for each value of CO.

14.1B $[SO_2]$ and $[SO_3]$ do not have unique values, but the ratios $[SO_2]/[SO_3]$, $[SO_2]^2/[SO_3]^2$, and their inverses do have unique values.

$$1.00 \times 10^2 = \frac{[SO_3]^2}{[SO_2]^2[O_2]}$$

When $[O_2]$ is set, then the ratio $\frac{[SO_3]^2}{[SO_2]^2}$ must have a set value. For $[O_2] = 1.00$ M,

$$\frac{[SO_3]^2}{[SO_2]^2} = 1.00 \times 10^2 \text{ and } \frac{[SO_3]}{[SO_2]} = 10.0$$

14.2A $K_c' = \left(\frac{1}{K_c}\right)^2 = \left(\frac{1}{20.0}\right)^2 = 2.50 \times 10^{-3}$

14.2B The reaction is the reverse and four times the first reaction.

$$K_c = \frac{1}{(1.97 \times 10^{-20})^4} = 6.64 \times 10^{78}$$

14.3A $K_c = \left(\frac{1}{1.8x10^{-6}}\right)^{\frac{1}{2}}$ for $NO_2(g) \rightleftharpoons NO(g) + \frac{1}{2}\ O_2(g)$

$K_c = 7.5 \times 10^2$

$K_p = K_c\,(RT)^{\Delta n} = 7.5 \times 10^2 \times (0.08206 \text{ x } 457)^{1/2}$

$K_p = 4.6 \times 10^3$

14.3B $K_c = \frac{K_p}{(RT)^{\frac{9}{4}-\frac{5}{2}}}$ for $NO(g) + \frac{3}{2}\ H_2O(g) \rightleftharpoons NH_3(g) + \frac{5}{4}\ O_2(g)$

$$K_c = \frac{2.6 \times 10^{-16}}{(0.0821 \times 900)^{-\frac{1}{4}}} = (2.6 \times 10^{-16}) \times (0.0821 \times 900)^{\frac{1}{4}} = 7.6 \times 10^{-16}$$

For $4\ NH_3(g) + 5\ O_2(g) \rightleftharpoons 4\ NO(g) + 6\ H_2O(g)$, the chemical equation is four times the reverse of the starting equation.

$$K_c = \frac{1}{(7.6 \times 10^{-16})^4} = 3.0 \times 10^{60}$$

14.4A $K_p = \dfrac{P_{CO}\,P_{H_2}}{P_{H_2O}}$

14.4B $K_c = \dfrac{[H_2\,(g)]^4}{[H_2O\,(g)]^4}$ $\quad K_p = \dfrac{P_{H_2}^{\ 4}}{P_{H_2O}^{\ 4}}$

14.5A A value of K greater than 1 (1.2×10^3) means that the forward reaction is favored. Because K is not extremely large, the reaction will not go to the point where the reactant concentrations are essentially zero.

14.5B A significant amount of both products and reactants means that K_P must be close to 1. 10.0 is the answer.

14.6A $Q_c = \dfrac{[H_2][I_2]}{[HI]^2} = \dfrac{0.100 \times 0.100}{1.00^2} = 1.00 \times 10^{-2}$

$Q_c < K_c$ $\quad 1.00 \times 10^{-2} < 1.84 \times 10^{-2}$ $\quad$ The reaction will proceed to the right.

14.6B $Q_p = \dfrac{P_{HI}^{2}}{P_{H_2S}} = \dfrac{(0.0010)^2}{(0.010)} = 1.0 \times 10^{-4}$

$Q_p > K_p$ The reaction will proceed to the left.

At equilibrium P_{H_2S} will be greater, P_{HI} will be less; I_2 (s) amount will increase and S (s) amount will decrease.

14.7A (a) Reaction would go to the right, using up some N_2 and producing more NH_3. The amount of H_2, however, would be greater than in the original equilibrium. The amount of N_2 would be less than in the original equilibrium.

(b) Reaction would go to the left, using up some NH_3 and producing more H_2.

(c) Reaction would go to the right by consuming some N_2 and some H_2. The amount of NH_3 would be less than in the original equilibrium.

14.7B The direction depends on the concentration since adding an aqueous solution of acetic acid is adding both a reactant and a product. For almost pure acetic acid, the reaction goes to the right. A dilute solution has more water than acetic acid, it will cause the reaction direction to shift back to the left. Intermediate solutions will depend on the exact concentration to determine reaction direction.

14.8A There is no change in the equilibrium amount of HI(g) by changing the pressure or volume, because there are the same number of moles of gas on both sides of the equation.

14.8B Additional $NO_2(g)$ will cause the reaction to go to the left to use up the excess $NO_2(g)$. The volume increase causes a pressure decrease; the system reacts to the left to generate more moles of gas and, thus, more pressure. Both the volume change and NO_2 addition push the reaction to the left.

14.9A At low temperature. Because $\Delta H°$ is negative, the forward reaction is exothermic. Thus at low temperature the reaction shifts to the right to make more heat and also to increase conversion to the product.

14.9B $NO(g) + NO_2(g) \rightleftharpoons N_2O_3(g)$

$$\Delta H^\circ_{rxn} = \Sigma\Delta H^\circ_f(\text{prod}) - \Sigma\Delta H^\circ_f(\text{react})$$

$$\Delta H^\circ_{rxn} = 1\text{ mol} \times \frac{83.72\text{ kJ}}{\text{mol}} - 1\text{ mol} \times \frac{33.18\text{ kJ}}{\text{mol}} - 1\text{ mol} \times \frac{90.25\text{ kJ}}{\text{mol}} = -39.71\text{ kJ}$$

Since the reaction is exothermic, lowering the temperature will cause the reaction to shift toward the right, making more N_2O_3. The partial pressure will be greater at the freezing point of water.

14.10A (a) The reverse reaction occurs, using up CO_2 and making CO and H_2O. There will be more CO and H_2O and less CO_2. There will be more H_2 because the extra is not all used up.

(b) Because both a reactant (H_2O) and a product (H_2) are added to the equilibrium mixture, we cannot make a qualitative prediction of whether the equilibrium will shift to the left or to the right.

(c) Adding more H_2O favors the forward reaction, as does lowering the temperature (the forward reaction is exothermic). The new equilibrium will have more CO_2, and H_2, and less CO than the original equilibrium. The amount of H_2O is in doubt, because it is not known if the combined effects of the two changes will consume more or less than the 1.0 mol H_2O added.

14.10B (a) $CH_4(g) + 2\ H_2O(g) \rightleftharpoons CO_2(g) + 4\ H_2(g)$

$$\Delta H^\circ_{rxn} = \Sigma\Delta H^\circ_f(\text{prod}) - \Sigma\Delta H^\circ_f(\text{react})$$

$$\Delta H^\circ_{rxn} = 1\text{ mol} \times \frac{-393.5\text{ kJ}}{\text{mol}} + 4\text{ mol} \times \frac{0\text{ kJ}}{\text{mol}} - 2\text{ mol} \times \frac{-241.8\text{ kJ}}{\text{mol}} - 1\text{ mol} \times \frac{-74.81\text{ kJ}}{\text{mol}} = 164.9\text{ kJ}$$

Since the reaction is an endothermic reaction, products will be favored at a higher temperature.

(b) Since there are five moles of gas on the right and three moles of gas on the left, products will be favored at low pressure.

14.11A $K_c = \dfrac{[PCl_5]}{[PCl_3][Cl_2]}$

	$PCl_3(g)$	+	$Cl_2(g)$	$\rightleftharpoons$	$PCl_5(g)$
Initial, M:	1.00		1.00		0
Changes, M:	-0.82		-0.82		+0.82
Equilibrium, M:	0.18		0.18		0.82

$K_c = \dfrac{0.82}{(0.18)^2} = 25$

14.11B $[H_2] = \dfrac{10.0\text{ g}}{25.0\text{ L}} \times \dfrac{\text{mol}}{2.016\text{ g}} = 0.198\text{ M}$

$[H_2S] = \dfrac{72.6\text{ g}}{25.0\text{ L}} \times \dfrac{\text{mol}}{34.09\text{ g}} = 0.085\text{ M}$

	$Sb_2S_3(s) + 3\,H_2(g)$	$\rightleftharpoons$	$2\,Sb(s) + 3\,H_2S(g)$
Initial, M:	0.198		--
Changes, M:	–0.085		0.085
Equilibrium, M:	0.113		0.085

$[H_2]_{eq} = 0.1984\text{ M} - [H_2S]_{eq} = 0.198\text{ M} - 0.085\text{ M} = 0.113\text{ M}$

$K_c = \dfrac{[H_2S]^3}{[H_2]^3} = \dfrac{(0.0852)^3}{(0.113)^3}$

$K_c = 0.429$

$K_p = K_c(RT)^{\Delta n} = K_c(RT)^0 = 0.429$

14.12A (a)

	$CO(g)$	+	$H_2O(g)$	$\rightleftharpoons$	$CO_2(g)$ +	$H_2(g)$
Initial, M:	$\dfrac{0.100\text{ mol}}{5.00\text{ L}}$		$\dfrac{0.100\text{ mol}}{5.00\text{ L}}$			
Change, M:	$-X$		$-X$		X	X
Equilibrium, M:	$0.0200 - X$		$0.0200 - X$		X	X

$\dfrac{[CO_2][H_2]}{[CO][H_2O]} = 23.2 = \dfrac{X^2}{(0.0200 - X)^2}$

$\sqrt{23.2} = 4.82 = \dfrac{X}{0.0200 - X}$

$0.0963 - 4.82\,X = X$
$0.0963 = 5.82\,X$
$X = 0.0165$ M
? mol H_2 = 0.0165 M × 5.00 L = 0.0827 mol H_2

(b) $$P = \frac{nRT}{V} = \frac{0.0827\,\text{mol} \times \dfrac{0.08206\,\text{L atm}}{\text{K mol}} \times 600\,\text{K}}{5.00\,\text{L}}$$

$P = 0.814$ atm

14.12B $$K_c = \frac{[HI]^2}{[H_2][I_2]} = \frac{\left(\dfrac{\text{mol HI}}{V}\right)^2}{\left(\dfrac{\text{mol } H_2}{V}\right) \text{x} \left(\dfrac{\text{mol } I_2}{V}\right)} = \frac{(\text{mol HI})^2}{\text{mol } H_2 \text{ x mol } I_2}$$

This would be true only for those reactions that have the same number of moles in the numerator and the denominator, so the volumes cancel.

14.13A

	$CO(g)$	+	$Cl_2(g)$	⇌	$COCl_2(g)$
Initial, M:	$\frac{0.100\text{ mol}}{25.0\text{ L}}$		$\frac{0.200\text{ mol}}{25.0\text{ L}}$		--
Change, M:	$-X$		$-X$		X
Equilibrium, M:	$0.00400 - X$		$0.00800 - X$		X

$$K_c = 1.2 \times 10^3 = \frac{X}{(0.00400 - X) \times (0.00800 - X)} = \frac{X}{3.20 \times 10^{-5} - 0.01200X + X^2}$$

$1.2 \times 10^3\, X^2 - 14.4\,X + 3.84 \times 10^{-2} = X$
$1.2 \times 10^3\, X^2 - 15.4\,X + 3.84 \times 10^{-2} = 0$
$X^2 - 1.28 \times 10^{-2}\, X + 3.2 \times 10^{-5} = 0$

$$X = \frac{-b \pm \sqrt{b^2 - 4ac}}{2a}$$

$$= \frac{1.28 \times 10^{-2} \pm \sqrt{(1.28 \times 10^{-2})^2 - 4 \times (1) \times 3.2 \times 10^{-5}}}{2}$$

$$= \frac{1.28 \times 10^{-2} \pm 5.99 \times 10^{-3}}{2}$$

$X = 3.4 \times 10^{-3}$ or 9.4×10^{-3}
too large, not valid

? mol $COCl_2$ = 3.4×10^{-3} M × 25.0 L = 8.5×10^{-2} mol

14.13B

	$N_2(g)$ +	$O_2(g)$ ⇌	$2\,NO(g)$
Initial, mol:	0.78	0.21	--
Change, mol:	$-X$	$-X$	$2X$
Equilibrium, mol:	$0.78 - X$	$0.21 - X$	$2X$

$$\frac{(2X)^2}{(0.78 - X)(0.21 - X)} = 2.1 \times 10^{-3}$$

$$\frac{4X^2}{0.164 - 0.99X + X^2} = 2.1 \times 10^{-3}$$

$4X^2 = 2.1 \times 10^{-3} (0.164 - 0.99X + X^2)$

$4X^2 = 3.44 \times 10^{-4} - 2.08 \times 10^{-3}X + 2.1 \times 10^{-3}X^2$

$4X^2 - 2.1 \times 10^{-3}X^2 \approx 4X^2$

$4X^2 + 2.08 \times 10^{-3}X - 3.44 \times 10^{-4} = 0$

$$X = \frac{-2.08 \text{ x } 10^{-3} \pm \sqrt{(2.08 \text{ x } 10^{-3})^2 + 4 \text{ x } 4 \text{ x } 3.44 \text{ x } 10^{-4}}}{2 \text{ x } 4}$$

$$X = \frac{-2.08 \text{ x } 10^{-3} \pm 7.42 \text{ x } 10^{-2}}{8} = 9.0 \times 10^{-3}$$

at equilibrium moles NO = $2 \times 9.0 \times 10^{-3}$ mol = 0.018 mol

moles N_2 = 0.78 mol – 0.009 mol = 0.77 mol

moles O_2 = 0.21 mol – 0.009 mol = 0.20 mol

$$\text{mol fraction of NO} = \frac{0.018 \text{ mol NO}}{(0.77 + 0.20 + 0.018) \text{ mol}} = 0.018$$

The volume does not matter, because it will cancel out in the equilibrium constant expression. To make concentrations and numbers of moles of reactant numerically equal, a volume of 1.00 L can be assumed.

14.14A

	$H_2(g)$ +	$I_2(g)$ ⇌	2 HI(g)
Initial, mol:	0.0100	--	0.100
Change, mol:	X	X	$-2X$
Equilibrium, mol:	$0.0100+X$	X	$0.100-2X$

The reaction must go to the left because there is no I_2 present initially.

$$\frac{(0.100-2X)^2}{(0.0100+X)X} = 54.3$$

$0.0100 - 0.400X + 4X^2 = 54.3X^2 + 0.543X$

$50.3X^2 + 0.943X - 0.0100 = 0$

$$X = \frac{-0.943 \pm \sqrt{(0.943)^2 - 4 \text{ x } 50.3 \text{ x } (-0.0100)}}{2 \text{ x } 50.3}$$

$X = 7.6 \times 10^{-3}$ or -0.026 not valid

$H_2 = 1.76 \times 10^{-2}$ moles

$I_2 = 7.6 \times 10^{-3}$ moles

HI $= 0.100 - 2 \times 7.6 \times 10^{-3} = 0.085$ moles

In this reaction the same answer will be obtained regardless of the volume of the system. To make concentrations and numbers of moles of reactant numerically equal, a volume of 1.00 L can be assumed.

14.14B $Q = \dfrac{[HI]^2_{init}}{[H_2]_{init} \text{ x } [I_2]_{init}}$

$$Q = \frac{(0.100/5.25)^2}{(0.0100/5.25)(0.0100/5.25)} = 100$$

Because $Q > K_c$ (that is, 100 > 54.3), a net reaction must occur in the reverse direction. At equilibrium, the amounts of H_2 and I_2 will be greater than initially, and the amount of HI will be less. This guides us in labeling the changes in amounts as positive or negative.

The reaction:	$H_2(g)$	+	$I_2(g)$	$\rightleftharpoons$	2 HI(g)
Initial amounts:	0.0100 mol		0.0100 mol		0.100 mol
Changes:	$+X$ mol		$+X$ mol		$-2X$ mol
Equil. amounts:	$(0.0100 + X)$ mol		$(0.0100 + X)$ mol		$(0.100 - 2X)$ mol
Equil. concns., M:	$\frac{(0.0100 + X)}{5.25}$		$\frac{(0.0100 + X)}{5.25}$		$\frac{(0.0100 + X)}{5.25}$

$$K_c = \frac{[HI]^2}{[H_2][I_2]} = \frac{\left(\frac{0.100 - 2X}{5.25}\right)^2}{\left(\frac{0.0100 + X}{5.25}\right)\left(\frac{0.0100 + X}{5.25}\right)} = 54.3$$

$$K_c = \frac{(0.100 - 2X)^2}{(0.0100 + X)^2} = 54.3$$

$$\left(\frac{(0.100 - 2X)^2}{(0.0100 + X)^2}\right)^{1/2} = \frac{0.100 - 2X}{0.0100 + X} = (54.3)^{1/2}$$

$0.100 - 2X = (54.3)^{1/2} \times (0.0100 + X)$

$0.100 - 2X = 0.0737 + 7.37X$

$9.37X = 0.0263$

$X = 0.00281$

The equilibrium amounts are:

H_2: $0.0100 + X = 0.0100 + 0.00281 = 0.0128$ mol H_2

I_2: $0.0100 + X = 0.0100 + 0.00281 = 0.0128$ mol I_2

HI: $0.100 - 2X = 0.100 - (2 \times 0.00281) = 0.094$ mol HI

14.15A

	CO(g)	+	$Cl_2(g)$	$\rightleftharpoons$	$COCl_2(g)$	$K_p = 22.5$
Initial, atm:	1.00		1.00		--	
Change, atm:	$-X$		$-X$		X	
Equilibrium, atm:	$1.00 - X$		$1.00 - X$		X	

$$K_p = \frac{P_{COCl_2}}{P_{CO}P_{Cl_2}} = \frac{X}{(1.00 - X)^2} = 22.5$$

$$22.5 = \frac{X}{1.00 - 2.00X + X^2}$$

$22.5 - 45.0X + 22.5X^2 = X$

$22.5X^2 - 46.0X + 22.5 = 0$

$X^2 - 2.04X + 1 = 0$

$$X = \frac{-b \pm \sqrt{b^2 - 4ac}}{2a} = \frac{2.04 \pm \sqrt{(2.04)^2 - 4 \times 1 \times 1}}{2}$$

$$X = \frac{2.04 \pm 0.40}{2}$$

$X = 0.82$ or 1.22 too large, not valid

$P_{COCl_2} = 0.82$ atm

$P_{CO} = 1.00 - 0.82 = 0.18$ atm

$P_{Cl_2} = 1.00 - 0.82 = 0.18$ atm

$P_{total} = P_{COCl_2} + P_{CO} + P_{Cl_2} = 1.18$ atm

14.15B $K_p = 0.108 = P_{NH_3} \times P_{H_2S}$

$P_{NH_3} = P_{H_2S}$

$P_{NH_3} = \sqrt{0.108} = 0.329$ atm

$P_T = P_{NH_3} + P_{H_2S} = 0.329 \text{ atm} + 0.329 \text{ atm} = 0.658$ atm

Review Questions

2. $K_c = \dfrac{[NO]^3}{[NO_2][N_2O]} = \dfrac{(0.0015)^3}{(1.25)(1.80)} = 1.5 \times 10^{-9}$
   (d) is the answer.

3. $K_p = K_c(RT)^{\Delta n}$ $\quad\quad \Delta n = 2 - 3 = -1$
   (b) is the answer.

4. $K_c = 10.0 = \dfrac{[C][D]}{[A][B]}$

   $[A][B] = \dfrac{[C][D]}{10.0} = 0.10 \times [C][D]$
   (c) is the answer.

5. $Q = \dfrac{[I_2][Cl_2]}{[ICl]^2} = \dfrac{(0.00125)(0.0075)}{(0.15)^2} = 4.2 \times 10^{-4}$
   $Q < K$ The reaction is not at equilibrium. Reaction is toward the right and the Cl_2 concentration will increase.
   (d) is the answer.

7. (a) False. The ratio of the products of the concentrations of products of a reaction, raised to appropriate powers, to the product of the concentrations of the reactants, also raised to appropriate powers, is equal to the equilibrium constant.
   (b) False. See (a).
   (c) False. See (a).
   (d) True.

(e) False. There are many different sets of values that equal K.
(f) True.

9. $K_c = 100 = \dfrac{[SO_3]^2}{[SO_2]^2[O_2]}$

If $[SO_3] = [SO_2]$, then $100 = \dfrac{1}{[O_2]}$

$[O_2] = 0.01$ M
(c) is the answer.

10. (a) $K_p = \dfrac{P_{CO_2}P_{H_2}}{P_{CO}P_{H_2O}}$ (b) $K_p = \dfrac{P^2_{NH_3}}{P_{N_2}P^3_{H_2}}$ (c) $K_p = P_{NH_3}P_{H_2S}$

11. (a) $1/2\ N_2(g) + 1/2\ O_2(g) \rightleftharpoons NO(g)$, $K_p = \dfrac{P_{NO}}{(P_{N_2})^{1/2}(P_{O_2})^{1/2}}$

(b) $1/2\ N_2(g) + 3/2\ H_2(g) \rightleftharpoons NH_3(g)$, $K_p = \dfrac{P_{NH_3}}{(P_{N_2})^{1/2}(P_{H_2})^{3/2}}$

(c) $1/2\ N_2(g) + 1/2\ O_2(g) + 1/2\ Cl_2(g) \rightleftharpoons NOCl\ (g)$,

$$K_p = \frac{P_{NOCl}}{(P_{N_2})^{1/2}(P_{O_2})^{1/2}(P_{Cl_2})^{1/2}}$$

12. (a) $2\ CO(g) + 2\ NO(g) \rightleftharpoons N_2(g) + 2\ CO_2(g)$ $K_c = \dfrac{[N_2][CO_2]^2}{[CO]^2[NO]^2}$

(b) $5\ O_2(g) + 4\ NH_3(g) \rightleftharpoons 4\ NO(g) + 6\ H_2O(g)$ $K_c = \dfrac{[NO]^4[H_2O]^6}{[NH_3]^4[O_2]^5}$

(c) $2\ NaHCO_3(s) \rightleftharpoons Na_2CO_3(s) + H_2O(g) + CO_2(g)$ $K_c = [H_2O][CO_2]$

14. (a) Yes, there is a smaller number of moles of gas on the right.
(b) No, there are the same number of moles of gases on both sides of the equation, so a pressure change will have no effect.
(c) Yes, there is a smaller number of moles of gas on the right.

17. $K_c = 10.0 = \dfrac{[CO_2][H_2]}{[CO][H_2O]}$

	$CO(g)$ +	$H_2O(g)$	$\rightleftharpoons$	$CO_2(g)$ +	$H_2(g)$
Init	1.00	1.00		1.00	-

Eq	<1.00	<1.00	>1.00	<1.00

(b) is the answer.

18.

	$Cl_2(g)$ +	$I_2(g)$ ⇌	2 ICl(g)
Init	3.00	2.10	-
Change	−1.16	−1.16	+2.32
Eq	1.84	0.94	2.32

$$K_c = \frac{[ICl]^2}{[Cl_2][I_2]} = \frac{(2.32)^2}{(1.84)(0.94)} = 3.11$$

(d) is the answer.

Problems

19. (a) $K_c = K_p/(RT)^{\Delta n}$

$$K_c = \frac{2.9 \text{ x } 10^{-2}}{(0.08206 \text{ x } 303)^1} = 1.2 \times 10^{-3}$$

(b) $K_c = \frac{0.275}{(RT)^1} = \frac{0.275}{(0.08206 \text{ x } 700)^1} = 4.79 \times 10^{-3}$

(c) $K_c = \frac{22.5}{[0.08206 \times (395 + 273)]^{-1}} = 1.23 \times 10^3$

21. $K_c = \left(\frac{1}{16.7}\right)^2 = 3.59 \times 10^{-3}$

23. $K_c = \frac{K_p}{(RT)^{\Delta n}}$ for $2\,Cl_2(g) + 2\,H_2O(g) \rightleftharpoons 4\,HCl(g) + O_2(g)$

$$K_c = \frac{0.0366}{\left(\frac{0.08206 \text{ L atm}}{\text{K mol}} \times (273 + 450)\text{K}\right)^{5-4}} = 6.17 \times 10^{-4}$$

New equation is one-fourth and the reverse of original.

$$K_c = \frac{1}{(6.17 \times 10^{-4})^{1/4}} = 6.35$$

25. $CH_4(g) + H_2O(g) \rightleftharpoons CO(g) + 3\,H_2(g)$ $\qquad \frac{1}{K_{p1}} = \frac{1}{8.00 \times 10^{24}}$

$\underline{CO(g) + H_2O(g) \rightleftharpoons CO_2(g) + H_2(g)}$ $\qquad \frac{1}{K_{p2}} = \frac{1}{9.80 \text{ x } 10^{-6}}$

$CH_4(g) + 2\,H_2O(g) \rightleftharpoons CO_2(g) + 4\,H_2(g)$ $\qquad K_{p3} = \frac{1}{K_{p1} \times K_{p2}}$

$$K_{p3} = \frac{1}{9.80 \times 10^{-6} \times 8.00 \times 10^{24}}$$

$K_{p3} = 1.28 \times 10^{-20}$

Coefficients of the desired equation are ½ of those in the above equation.

$K_p = \sqrt{K_{p3}} = (1.28 \times 10^{-20})^{1/2} = 1.13 \times 10^{-10}$

27. $2\ CH_4(g) + 2\ H_2O(g) \rightleftharpoons 2\ CO(g) + 6\ H_2(g)$ $\qquad K_{p1}^2 = (1.2 \times 10^{-25})^2$

$2\ CO(g) + H_2O(g) \rightleftharpoons \frac{3}{2}\ O_2(g) + C_2H_2(g)$ $\qquad \dfrac{1}{K_{p2}^{1/2}} = \dfrac{1}{(1.1 \times 10^{2})^{1/2}}$

$3\ H_2(g) + \frac{3}{2}\ O_2(g) \rightleftharpoons 3\ H_2O(g)$ $\qquad K_{p3}^3 = (1.1 \times 10^{40})^3$

$2\ CH_4(g) \rightleftharpoons C_2H_2(g) + 3\ H_2(g)$ $\qquad K_p = \dfrac{K_{p1}^2 \times K_{p3}^3}{K_{p2}^{1/2}}$

$$K_p = \frac{(1.2 \times 10^{-25})^2 \times (1.1 \times 10^{40})^3}{(1.1 \times 10^{2})^{1/2}}$$

$K_p = 1.8 \times 10^{69}$

$$K_c = \frac{K_p}{(RT)^{\Delta n}} = \frac{1.8 \times 10^{69}}{(RT)^{4-2}} = \frac{1.8 \times 10^{69}}{(0.08206 \times 298\ \text{K})^2} = 3.0 \times 10^{66}$$

29. $K_c = (K_c')^2$ K_c will be greater only as long as K_c' is greater than 1. When K_c' is less than one, K_c will be smaller than K_c'.

31. $K_c = \dfrac{[C]}{[A]^2[B]} = \dfrac{6.2 \times 10^{-3}}{(2.4 \times 10^{-2})^2(4.6 \times 10^{-3})} = 2.3 \times 10^3$

33. $K_c = 6.28 \times 10^3 = \dfrac{[H_2S]^2}{[H_2]^2[S_2]}$

$$6.28 \times 10^3 = \frac{([S_2]^{\frac{1}{2}})^2}{[H_2]^2[S_2]} = \frac{1}{[H_2]^2}$$

$[H_2] = 0.0126\ M$

35. $K_p = 1.33 \times 10^{-5} = \dfrac{P_{HI}^2}{P_{H_2S}} = \dfrac{(0.010\ P_{H_2S})^2}{P_{H_2S}}$

$P_{H_2S} = 0.133$ atm

$P_{HI} = 1.33 \times 10^{-3}$ atm

$P_{total} = 0.134$ atm

37. $K_p = \dfrac{P_{NO_2}^2}{P_{N_2O_4}} = \dfrac{(3(P_{N_2O_4})^{1/2})^2}{P_{N_2O_4}} = 9$

39. (a) Not correct. Changing the volume, which changes the pressure, has no effect on K_c because the number of moles of gas are the same on both sides of the equation.
(b) Correct. $Q_c = 1 > K_c$, so the reaction will go to the left.
(c) Not correct. $Q_c = 1 > K_c$, so the reaction will go to the left and produce graphite.
(d) Not correct. They will be the same value, because Δn gas is equal to zero. ($K_p = K_c(RT)^{\Delta n} = K_c \times (0.0821 \times 2025\text{ K})^{2-2}$)
(e) Correct. The change is twice as great for HCN as for H_2 and N_2 individually, because the mole HCN to mole H_2 to mole N_2 is 2 to 1 to 1.

41. $2NO(g) + Br_2(g) \rightleftharpoons 2NOBr(g)$

4 2 20 $Q = \dfrac{(20)^2}{(2)(4)^2} = 12.5$

Reaction goes to left.
+2x +x -2x
If we try integer number, x = 1

$Q = \dfrac{(18)^2}{(3)(6)^2} = 3$

6 3 18 Drawing c

43. (a) More $H_2(g)$ will be produced as the reaction goes to the right to use up some of the $H_2O(g)$.
(b) There is no effect on the equilibrium amount of $H_2(g)$.
(c) The equilibrium amount of $H_2(g)$ will increase as the reaction goes to the right to produce more moles of gas and thus more pressure.
(d) The $H_2(g)$ concentration will decrease as the reaction goes in the direction that uses up some of the CO(g).
(e) The partial pressure of each gas does not change, so there is no change in the equilibrium amount of $H_2(g)$.
(f) The $H_2(g)$ concentration increases as the reaction goes in the direction that uses up the heat.
(g) Removing part of the C(s) will not change the concentration of $H_2(g)$, as the equilibrium condition is independent of the amount of a solid reactant.

45. $Cl_2(g) \rightleftharpoons 2\,Cl(g)$
$S_2(g) \rightleftharpoons 2\,S(g)$
$H_2(g) \rightleftharpoons 2\,H(g)$

These dissociation reactions are endothermic, because they require the breaking of bonds with no new bonds formed. The forward reaction is favored with increasing temperature–dissociation is more extensive when equilibrium is reached.

47. (a) $N_2(g) + O_2(g) \rightleftharpoons 2\ NO(g)$
There is no effect, because the same number of moles of gas appear on each side of the equation.
(b) $N_2(g) + 3\ H_2(g) \rightleftharpoons 2\ NH_3(g)$
Because the number of molecules of gas produced is less than the number of molecules of reactant gases, this reaction occurs to a greater extent with increased pressure.
(c) $H_2(g) + I_2(g) \rightleftharpoons 2\ HI(g)$
There is no effect because the same number of moles of gas appear on each side of the equation.
(d) $2\ H_2(g) + S_2(g) \rightleftharpoons 2\ H_2S(g)$
Because fewer molecules of gas are produced, this reaction occurs to a greater extent at high pressure.

49. $$Q_c = \frac{[CO_2][H_2]}{[H_2O][CO]} = \frac{\left(25.0\,g \times \frac{mol}{44.01\,g}\right) \times \left(25.0\,g \times \frac{mol}{2.016\,g}\right)}{\left(20.0\,g \times \frac{mol}{28.01\,g}\right) \times \left(20.0\,g \times \frac{mol}{18.01\,g}\right)} = 8.88$$

$Q_c < K_c$ The reaction will shift toward the right to increase the amount of $CO_2(g)$ and $H_2(g)$.

51. $$Q_c = \frac{\left(\frac{2.10\,mol\,CH_4}{7.25\,L}\right) \times \left(\frac{3.15\,mol\,H_2O}{7.25\,L}\right)}{\left(\frac{0.103\,mol\,CO}{7.25\,L}\right) \times \left(\frac{0.205\,mol\,H_2}{7.25\,L}\right)} = 3.92 \times 10^5$$

$Q_p = Q_c(RT)^{\Delta n} = 3.92 \times 10^5 \times (0.0821 \times 773\ K)^{2-4} = 97.4$
The reaction will shift slightly to the right to increase the amount of $CH_4(g)$ and $H_2O(g)$ because Q_p (97.4) is slightly less than K_p (102).

53. $$[NO_2] = \frac{1.353\ g}{10.5\ L} \times \frac{mol}{46.01\ g} = 2.80 \times 10^{-3}\ M$$

$$[NO] = \frac{0.0960\ g}{10.5\ L} \times \frac{mol}{30.01\ g} = 3.05 \times 10^{-4}\ M$$

$$[O_2] = \frac{0.0512\ g}{10.5\ L} \times \frac{mol}{32.00\ g} = 1.52 \times 10^{-4}\ M$$

$$K_c = \frac{[NO]^2[O_2]}{[NO_2]^2} = \frac{(3.05 \times 10^{-4})^2 \times 1.52 \times 10^{-4}}{(2.80 \times 10^{-3})^2} = 1.80 \times 10^{-6}$$

$K_p = K_c\,(RT)^1 = 1.80 \times 10^{-6} \times (0.08206 \times (184 + 273))$
$K_p = 6.76 \times 10^{-5}$

55. $NH_2COONH_4(s) \rightleftharpoons 2\,NH_3(g) + CO_2(g)$

$P_{total} = P_{NH_3} + P_{CO_2}$

$P_{NH_3} = 2\,P_{CO_2}$

$P_{total} = 2\,P_{CO_2} + P_{CO_2} = 0.164$ atm

$P_{CO_2} = 0.164$ atm/3 = 0.0547 atm

$P_{NH_3} = 0.110$ atm

$K_p = P_{NH_3}^2\,P_{CO_2} = (0.110)^2(0.0547) = 6.7 \times 10^{-4}$

57.

	$2\,ICl(g)$	$\rightleftharpoons$	$I_2(g)$	+	$Cl_2(g)$
Initial, M:	6.72×10^{-3}		--		--
Changes, M:	$-2\times2.41\times10^{-4}$		$+2.41 \times 10^{-4}$		$+2.41 \times 10^{-4}$
Equilibrium, M:	6.24×10^{-3}		2.41×10^{-4}		2.41×10^{-4}

$$[ICl]_{init} = \frac{0.682\text{ g ICl}}{625\text{ mL}} \times \frac{\text{mol}}{162.4\text{ g}} \times \frac{\text{mL}}{10^{-3}\text{ L}} = 6.72 \times 10^{-3}\text{ M}$$

$$[I_2]_{eq} = \frac{0.0383\text{ g } I_2}{625\text{ mL}} \times \frac{\text{mol}}{253.8\text{ g}} \times \frac{\text{mL}}{10^{-3}\text{ L}} = 2.41 \times 10^{-4}\text{ M}$$

$$K_c = \frac{[I_2][Cl_2]}{[ICl]^2} = \frac{(2.41\times10^{-4})^2}{(6.24\times10^{-3})^2} = 1.49 \times 10^{-3}$$

59.

	C_6H_{12}	$\rightleftharpoons$	$C_5H_9CH_3$
Initial, M:	$\frac{100\text{ g}}{84.16\text{ g/mol}} = 1.19$ moles		--
Change, M:	$-X$		X
Equilibrium, M:	$1.19 - X$		X

$$\frac{X}{1.19 - X} = 0.143$$

$X = 0.170 - 0.143\,X$
$1.143\,X = 0.170$
$X = 0.149$ moles

$$0.149\text{ moles} \times \frac{84.16\text{ g}}{\text{mole}} = 12.5\text{ g methylcyclopentane}$$

(As long as the same number of concentration terms appear in the numerator and denominator, the volume does not matter.)

61.

	C(s) +	$H_2O(g) \rightleftharpoons$	CO(g) +	$H_2(g)$
Initial, M:		0.100	--	0.100
Changes, M:		$-X$	X	X
Equilibrium, M:		$0.100 - X$	X	$0.100 + X$

A volume of 1.00 L has been assumed.

$$K_c = 0.111 = \frac{X(0.100 + X)}{(0.100 - X)}$$

$$0.111 = \frac{0.100\,X + X^2}{0.100 - X}$$

$$0.0111 - 0.111\,X = 0.100\,X + X^2$$

$$X^2 + 0.211\,X - 0.0111 = 0$$

$$X = \frac{-0.211 \pm \sqrt{(0.211)^2 - 4 \times (-0.0111)}}{2}$$

$X = 0.0436$ or -0.254 not valid

$X = 0.0436\text{ M} \times 1.00\text{ L} = 0.0436\text{ mol CO}$

63. $$[CO]_{init} = \frac{20.0\text{ g CO}}{8.05\text{ L}} \times \frac{\text{mol}}{28.01\text{ g}} = 0.0887\text{ M}$$

$$[Cl_2]_{init} = \frac{35.5\text{ g }Cl_2}{8.05\text{ L}} \times \frac{\text{mol}}{70.90\text{ g}} = 6.22 \times 10^{-2}\text{ M}$$

	CO(g) +	$Cl_2(g) \rightleftharpoons$	$COCl_2(g)$
Initial, M:	8.87×10^{-2}	6.22×10^{-2}	--
Changes, M:	$-X$	$-X$	X
Equilibrium, M:	$8.87 \times 10^{-2} - X$	$6.22 \times 10^{-2} - X$	X

$$\frac{X}{(8.87 \times 10^{-2} - X) \times (6.22 \times 10^{-2} - X)} = 1.2 \times 10^3$$

$$X = 1.2 \times 10^3\,X^2 - 1.81 \times 10^2\,X + 6.62$$

$$0 = 1.2 \times 10^3\,X^2 - 1.82 \times 10^2\,X + 6.62$$

$$X = \frac{-b \pm \sqrt{b^2 - 4ac}}{2a}$$

$$X = \frac{1.82 \times 10^2 \pm \sqrt{(1.82 \times 10^2)^2 - 4 \times 1.2 \times 10^3 \times 6.62}}{2 \times 1.2 \times 10^3}$$

$X = 0.0911$ M or 0.0605 M

too large, not valid

$? \text{ mol } COCl_2 = 0.0605\text{ M} \times 8.05\text{ L} = 0.487\text{ mol } COCl_2$

$$? \text{ g } COCl_2 = 0.487\text{ mol } COCl_2 \times \frac{98.91\text{ g}}{\text{mol}} = 48.2\text{ g } COCl_2$$

65. $$[PCl_3] = [Cl_2] = \frac{0.100\text{ mol}}{6.40\text{ L}} = 0.01563\text{ M}$$

$$[PCl_5] = \frac{0.0100\text{ mol}}{6.40\text{ L}} = 0.001563\text{ M}$$

$$Q_c = \frac{(0.001563)}{(0.01563)(0.01563)} = 6.398$$ Since $Q_c < K_c$, the net reaction must shift to the right to establish equilibrium.

	$PCl_3(g)$	+	$Cl_2(g)$	$\rightleftharpoons$	$PCl_5(g)$
Initial, M:	0.01563		0.01563		0.001563
Changes, M:	$-X$		$-X$		X
Equilibrium, M:	$0.01563 - X$		$0.01563 - X$		$0.001563 + X$

$$\frac{0.001563 + X}{(0.01563 - X)^2} = 26$$

$0.001563 + X = 26X^2 - 0.8128X + 0.006352$

$0 = 26X^2 - 1.8128X + 0.004789$

$$X = \frac{-b \pm \sqrt{b^2 - 4ac}}{2a}$$

$$X = \frac{1.8128 \pm \sqrt{(1.8128)^2 - 4 \times 26 \times 0.004789}}{52}$$

$X = 0.0670$ M or 0.00275 M

not valid

PCl_3 and Cl_2 $(0.01563 - 0.00275) \times 6.40\text{ L} = 0.082$ mol

PCl_5 $(0.001563 + 0.00275) \times 6.40\text{ L} = 0.028$ mol

67.

	$2\,NH_3(g)$	$\rightleftharpoons$	$N_2(g)$	+	$3\,H_2(g)$
Initial, M:	$\frac{1.00\text{ mol}}{10.0\text{ L}}$		--		--
Changes, M:	$-X$		$+\frac{1}{2}X$		$\frac{3}{2}X$
Equilibrium, M:	$0.100 - X$		$\frac{1}{2}X$		$\frac{3}{2}X$

$$K_c = \frac{(\frac{1}{2}X)(\frac{3}{2}X)^3}{(0.100 - X)^2} = \frac{[N_2][H_2]^3}{[NH_3]^2} = 6.46 \times 10^{-3}$$

$1.69X^4 = 6.46 \times 10^{-5} - 1.29 \times 10^{-3}X + 6.46 \times 10^{-3}X^2$

$X^4 - 3.82 \times 10^{-3}X^2 + 7.63 \times 10^{-4}X = 3.82 \times 10^{-5}$

Assume $X \approx 0.0500$

$(0.0500)^4 - 3.82 \times 10^{-3} \times (0.0500)^2 + 7.63 \times 10^{-4} \times 0.0500$

$= 3.49 \times 10^{-5} < 3.82 \times 10^{-5}$

Make X larger $X = 0.0520$

$(0.0520)^4 - 3.82 \times 10^{-3} \times (0.0520)^2 + 7.63 \times 10^{-4} \times 0.0520$

$= 3.67 \times 10^{-5} < 3.82 \times 10^{-5}$

Make X larger $X = 0.0529$

$(0.0529)^4 - 3.82 \times 10^{-3} \times (0.0529)^2 + 7.63 \times 10^{-4} \times 0.0529$

$= 3.75 \times 10^{-5} < 3.82 \times 10^{-5}$

Make X larger $X = 0.0538$

$(0.0538)^4 - 3.82 \times 10^{-3} \times (0.0538)^2 + 7.63 \times 10^{-4} \times 0.0538$

$= 3.84 \times 10^{-5} > 3.82 \times 10^{-5}$

Make X smaller $X = 0.0537$

$(0.0537)^4 - 3.82 \times 10^{-3} \times (0.0537)^2 + 7.63 \times 10^{-4} \times 0.0537$
$= 3.83 \times 10^{-5} > 3.82 \times 10^{-5}$

Make X smaller $X = 0.0536$
$(0.0536)^4 - 3.82 \times 10^{-3} \times (0.0536)^2 + 7.63 \times 10^{-4} \times 0.0536$
$= 3.82 \times 10^{-5} = 3.82 \times 10^{-5}$

$[NH_3] = 0.100\text{ M} - 0.0536\text{ M} = 0.046\text{ M}$

$\% = \frac{0.046}{0.100} \times 100\% = 46\%$ undissociated

As a check:

$K_c = \frac{(0.0268)(0.0804)^3}{(0.046)^2} = 6.6 \times 10^{-3}$

69. (a) $K_P = 5.60 = \frac{P_{CS_2}}{P_{S_2}} = \frac{0.152}{P_{S_2}}$

$P_{S_2} = 0.0271$ atm

(b) $P_{total} = P_{S_2} + P_{CS_2} = 0.179$ atm

71. $P_{CH_4} = \frac{nRT}{V} = \frac{0.100\text{ mol} \times \frac{0.08206\text{ L atm}}{\text{K mol}} \times 1273\text{ K}}{4.16\text{ L}}$

$P_{CH_4} = 2.511$ atm

	C (s) +	$2\,H_2(g)$ $\rightleftharpoons$	$CH_4(g)$
Initial, atm:		--	2.511 atm
Change, atm:		$2X$	$-X$
Equilibrium, atm:		$2X$	$2.511 - X$

$K_P = 0.263 = \frac{P_{CH_4}}{P^2_{H_2}} = \frac{2.511 - X}{(2X)^2}$

$1.052\,X^2 = 2.511 - X$

$1.052\,X^2 + X - 2.511 = 0$

$X = \frac{-b \pm \sqrt{b^2 - 4ac}}{2a}$

$X = \frac{-1 \pm \sqrt{1^2 - 4 \times 1.052 \times (-2.511)}}{2.104}$

$X = 1.141$ atm or -2.09 not valid

$P_{total} = P_{H_2} + P_{CH_4} = 2 \times 1.141 + (2.511 - 1.141) = 3.65$ atm

73. $12.0\text{ atm} = 2P + P = 3P \qquad P = 4$ atm

	$CO(g)$ +	$2\,H_2(g)$ $\rightleftharpoons$	$CH_3OH(g)$
Initial, atm:	4.0	8.0	--
Changes, atm:	$-P$	$-2P$	P
Equilibrium, atm:	$4.0 - P$	$8.0 - 2P$	P

$K_p = \frac{[CH_3OH]}{[CO][H_2]^2} = \frac{P}{(4.0-P)(8.0-2P)^2} = 9.23 \times 10^{-3}$

$$\frac{P}{(4.0-P)(64 - 32P + 4P^2)} = 9.23 \times 10^{-3}$$

$$\frac{P}{(256 - 192P + 48P^2 - 4P^3)} = 9.23 \times 10^{-3}$$

$P = 2.36 - 1.77P + 0.44P^2 - 0.037P^3$

$0.037P^3 - 0.44P^2 + 2.8P = 2.36$

$X^3 - 12X^2 + 76P = 64$

P must be less than 4, so the important term is $76P$, thus $76.0P = 64$.

$P \approx 0.84$

$(0.84)^3 - 12 \times (0.84)^2 + 76 \times 0.84 = 56 < 64$

Make P larger $P = 0.90$

$(0.90)^3 - 12 \times (0.90)^2 + 76 \times 0.90 = 59 < 64$

Make P larger $P = 1.00$

$(1.00)^3 - 12 \times (1.00)^2 + 76 \times 1.00 = 65 > 64$

Make P smaller $P = 0.99$

$(0.99)^3 - 12 \times (0.99)^2 + 76 \times 0.99 = 64$

$P = P_{CH_3OH} = 0.99$ atm

To check:

$$\frac{(0.99)}{(4.0 - 0.99)(8.0 - 2 \times 0.99)^2} = \frac{(0.99)}{(3.0)(6.)^2} = 9.2 \times 10^{-3}$$

An alternate method is to plug an estimate of P into the K_p equation.

$P = 3.00$ atm

$$\frac{3.00}{(1.00)(2.00)^2} = 0.75 >> 9.23 \times 10^{-3}$$

$P = 2.00$ atm

$$\frac{2.00}{(2.00)(4.00)^2} = 0.0625 > 9.23 \times 10^{-3}$$

$P = 1.00$ atm

$$\frac{1.00}{(3.00)(6.00)^2} = 9.26 \times 10^{-3} > 9.23 \times 10^{-3}$$

$P = 0.99$ atm

$$\frac{0.99}{(3.01)(6.02)^2} = 9.08 \times 10^{-3} < 9.23 \times 10^{-3}$$

$P = P_{CH_3OH} = 1.00$ atm is closer than 0.99 atm, so the answer is 1.00 atm.

Additional Problems

75. (a) to the left
    (b) 7 moles gas on the left, and 8 moles gas on the right, so to the left.
    (c) Reacts in direction to heat system, so to the right.
    (d) to the right

(e) Cooling will cause the reaction to generate heat, so it will go to the right. At the temperature at which water is a liquid, an increase in pressure will cause the reaction to go to the right.
(f) Cooling causes a shift to the right, and at temperatures where water and ammonia are solids, there are 2 moles of gas on the right and 3 moles on the left, so the reaction goes to the right.

77. Need to establish the value of K_c from equilibrium concentrations first.

$$K_c = \frac{0.179 \text{ x } 0.079}{0.021 \text{ x } 0.121} = 5.6$$

(a)	$CO(g)$ +	$H_2O(g)$	$\rightleftharpoons$	$CO_2(g)$ +	$H_2(g)$
init	0.021	0.121		0.179	0.179

$$Q_c = \frac{0.179 \text{ x } 0.179}{0.021 \text{ x } 0.121} = 12.6$$

$Q_c > K_c$, so a net reaction occurs to the left.
The same conclusion is reached by considering Le Chatelier's principle. Adding H_2 to a mixture at equilibrium will cause the reaction to go in the direction that uses up H_2–that is, to the left.

(b) $0.021 + X \quad 0.121 + X \quad 0.179 - X \quad 0.179 - X$

$$K_c = 5.6 = \frac{(0.179 - X)^2}{(0.021 + X)(0.121 + X)}$$

$$5.6\, X^2 + 0.795\, X + 0.01423 = X^2 - 0.358\, X + 0.03204$$

$$4.6\, X^2 + 1.153\, X - 0.01781 = 0$$

$$X = \frac{-1.153 \pm \sqrt{(1.153)^2 - 4 \text{ x } 4.6 \text{ x } (-0.01781)}}{9.2}$$

$X = 0.01460$ mol or -0.26525
not valid

$n_{CO_2} = n_{H_2} = 0.164$ mol

$n_{CO} = 0.036$ mol

$n_{H_2O} = 0.136$ mol

79. Base the calculation of 100.0 g of the gaseous phase at equilibrium

$$13.71 \text{ g C} \times \frac{\text{mol}}{12.011 \text{ g}} = 1.1415 \text{ mol C}$$

All the C in the gas phase is present as $CS_2(g)$.

The number of moles S in $CS_2 = 1.142 \text{ mol } CS_2 \times \frac{2 \text{ mol S}}{1 \text{ mol } CS_2} = 2.284$ mol S.

$$86.29 \text{ g S} \times \frac{\text{mol}}{32.066 \text{ g}} = 2.691 \text{ mol S}$$

The number of moles S_2 = [2.691 mol S (total) - 2.284 mol S in CS_2]

$$\times \frac{1 \text{ mol } S_2}{2 \text{ mol S}} = 0.204 \text{ mol } S_2$$

$$K_c = \frac{n_{CS_2}}{n_{S_2}} = \frac{1.142 \text{ mol } CS_2}{0.204 \text{ mol } S_2} = 5.60$$

80.

	$N_2O_4(g)$ $\rightleftharpoons$	$2\ NO_2(g)$
Initial, mol:	0.100	--
Changes, mol:	$-\frac{1}{2}X$	$+X$
Equilibrium, mol:	$0.100 - \frac{1}{2}X$	X

$$0.746 = \text{mol fraction } NO_2 = \frac{X}{0.100 - \frac{1}{2}X + X} = \frac{X}{0.100 + \frac{1}{2}X}$$

$X = 0.0746 + 0.373\,X$

$0.0746 = 0.627\,X$

$X = 0.119$ mol = mol NO_2

mol $N_2O_4 = 0.100 - \frac{1}{2} \times (0.119) = 0.041$ mol

$$P_{NO_2} = \frac{nRT}{V} = \frac{0.119 \text{ mol x } \frac{0.08206 \text{ L atm}}{\text{K mol}} \text{ x } 348 \text{ K}}{2.50 \text{ L}} = 1.36 \text{ atm}$$

$$P_{N_2O_4} = \frac{0.041 \text{ mol x } \frac{0.08206 \text{ L atm}}{\text{K mol}} \text{ x } 348 \text{ K}}{2.50 \text{ L}} = 0.47 \text{ atm}$$

$$Kp = \frac{P^2_{NO_2}}{P_{N_2O_4}} = \frac{(1.36)^2}{0.47} = 3.9$$

$\Delta H^\circ_{rxn} = 2 \text{ mol } NO_2 \times 33.18 \text{ kJ/mol} - 1 \text{ mol } N_2O_4 \times 9.16 \text{ kJ/mol} = 57.20 \text{ kJ}$

High pressure and low temperature favor purer N_2O_4.

82. $P_{O_2} = \chi P^\circ = 0.2095 \times 1.0000 = 0.2095$ atm

	$2\ CaSO_4\ (s) \rightleftharpoons 2\ CaO\ (s) +$	$2\ SO_2\ (g)$	$+\ O_2\ (g)$
Initial, atm:		--	0.2095
Changes, atm:		X	$+\frac{1}{2}X$
Equilibrium, atm:		X	$0.2095 + \frac{1}{2}X$

$1.45 \times 10^{-5} = P^2_{SO_2} P_{O_2} = X^2 \times (0.2095 + \frac{1}{2}X)$

assume $\frac{1}{2}X << 0.2095$

$1.45 \times 10^{-5} = X^2 \times 0.2095$

$P_{SO_2} = X = 8.32 \times 10^{-3}$ atm

$(8.32 \times 10^{-3}\text{ atm})^2 \times (0.2095 + 8.32 \times 10^{-3}) = 1.51 \times 10^{-5}$ The approximation produced an error of about 4% so the approximation is justified.

85.

	$COCl_2(g)$	$\rightleftharpoons$	$CO(g)$	+	$Cl_2(g)$
Initial, atm:	Y		--		--
Changes,atm:	$-X$		X		X
Equilibrium, atm:	$Y-X$		X		X

$$K_p = \frac{P_{CO}P_{Cl_2}}{P_{COCl_2}} = \frac{X^2}{Y\text{-}X} = 4.44 \times 10^{-2}$$

$P_T = P_{COCl_2} + P_{CO} + P_{Cl_2} = 3.00$ atm

$P_T = Y - X + X + X = 3.00$ atm

$Y - X = 3.00 - 2X$

$$K_p = 4.44 \times 10^{-2} = \frac{X^2}{3.00 - 2X}$$

$0.133 - 8.88 \times 10^{-2} X = X^2$

$X^2 + 8.88 \times 10^{-2} X - 0.133 = 0$

$$X = \frac{-8.88\times10^{-2} \pm \sqrt{(8.88\times10^{-2})^2 - 4\times1\times(-0.133)}}{2}$$

$$X = \frac{-8.88 \times 10^{-2} \pm 0.735}{2}$$

$X = 0.323$ atm or -0.412 atm Not valid, cannot be a negative pressure.

$P_{CO} = 0.323$ atm

$P_{Cl_2} = 0.323$ atm

$P_T = 3.00\text{ atm} = P_{CO} + P_{Cl_2} + P_{COCl_2}$

$3.00\text{ atm} = 0.323\text{ atm} + 0.323\text{ atm} + P_{COCl_2}$

$P_{COCl_2} = 2.35$ atm

$$\text{mol }\% = \frac{\text{moles }COCl_2}{\text{moles }T} \times 100\% = \frac{P_{COCl_2}}{P_T} \times 100\% = \frac{2.35\text{ atm}}{3.00\text{ atm}} \times 100\% = 78.3\%$$

88. $CH_4(g) + H_2O(g) \rightleftharpoons 3\,H_2(g) + CO(g)$ $\Delta H° = 230$ kJ $k = 1/190$

$CO(g) + H_2O(g) \rightleftharpoons CO_2(g) + H_2(g)$ $\Delta H° = -40$ kJ $k = 1.4$

$CH_4(g) + 2\,H_2O(g) \rightleftharpoons CO_2(g) + 4\,H_2(g)$ $k = 1.4/190 = 7.4 \times 10^{-3}$

	$CH_4(g)$	+	$2\,H_2O(g)$	$\rightleftharpoons$	$CO_2(g)$	+	$4\,H_2(g)$	
Initial:mol	0.100		0.100		--		--	$\Delta H° = 190$ kJ

Changes: mol	−x	−2x	x	4x
Equil: mol	−0.100 − x	0.100 − 2x	x	4x

$$7.4 \times 10^{-3} = \frac{[CO_2][H_2]^4}{[CH_4][H_2O]^2} = \frac{(x/10.0)(4x/10.0)^4}{[(1.00-x)/10.0][(1.00-2x)/10.0]^2}$$

$$7.4 \times 10^{-3} = \frac{x(4x)^4}{100(1.00-x)(1.00-2x)^2}$$

$256x^5 = 0.74\,[(1.00 - x)(1.00 - 2x)^2]$

$256x^5 - 0.74\,[(1.00 - x)(1.00 - 4x + 4x^2)]$

$256x^5 - 0.74\,(1.00 - 5x + 8x^2 - 4x^3) = 0$

$256x^5 + 2.96x^3 - 5.92x^2 + 3.70x = 0.74$

$$x \approx \sqrt[5]{\frac{0.74}{256}} = 0.3$$

$256 \times (0.3)^5 + 2.96 \times (0.3)^3 - 5.92 \times (0.3)^2 + 3.70 \times (0.3) = 1.28 > 0.74$

x is too large; assume x = 0.2

$256 \times (0.2)^5 + 2.96 \times (0.2)^3 - 5.92 \times (0.2)^2 + 3.70 \times (0.2) = 0.61 < 0.74$

x is too small; assume x = 0.22

$256 \times (0.22)^5 + 2.96 \times (0.22)^3 - 5.92 \times (0.22)^2 + 3.70 \times (0.22) = 0.69 < 0.74$

x is too large; assume x = 0.23

$256 \times (0.23)^5 + 2.96 \times (0.23)^3 - 5.92 \times (0.23)^2 + 3.70 \times (0.23) = 0.74 = 0.74$

x = 0.23 moles

moles $H_2 = 4 \times 0.23 = 0.92$ moles

Because raising the temperature of a reaction shifts the equilibrium toward the endothermic direction, the amount of H_2 will increase.

Apply Your Knowledge

90. $? \text{ mol } CH_3CO_2H = 22.44 \text{ mL} \times \frac{10^{-3}\text{ L}}{\text{mL}} \times 0.1025 \text{ M } Ba(OH)_2 \times \frac{2 \text{ mol } CH_3CO_2H}{\text{mol } Ba(OH)_2}$

$= 4.6002 \times 10^{-3} \text{ mol } CH_3CO_2H$

$4.6002 \times 10^{-3} \text{ mol} \times 100 = 0.46002 \text{ mol } CH_3CO_2H$

	$CH_3CO_2H(l)$ +	$CH_3CH_2OH(l)$ $\rightleftharpoons$	$CH_3CO_2CH_2CH_3(l)$ +	$H_2O(l)$
Initial: mol:	1.51 mol	1.66 mol	--	--
Changes: mol	−1.05	−1.05	1.05	1.05
Equilibrium: mol	0.46 mol	0.61	1.05	1.05

$$K_c = \frac{[CH_3CO_2CH_2CH_3][H_2O]}{[CH_3CO_2H][CH_3CH_2OH]} = \frac{1.05 \text{ x } 1.05}{0.46 \text{ x } 0.61}$$

$K_c = 3.9$

92. bulb 5

$$28.68 \text{ mL} \times \frac{10^{-3}\text{ L}}{\text{mL}} \times 0.0150 \text{ M } Na_2S_2O_3 \times \frac{\text{mol } I_2}{2 \text{ mol } Na_2S_2O_3} = 2.15 \times 10^{-4} \text{ mol } I_2$$

$$? \text{ moles HI initially} = 0.280 \text{ g HI} \times \frac{\text{mol HI}}{127.9 \text{ g HI}} = 0.00219 \text{ mol HI}$$

	$2\ HI(g)$	$\rightleftharpoons$	$H_2(g)$	+	$I_2(g)$
Initial: mol	0.00219 mol		--		--
Changes: mol	$-2 \times 2.15 \times 10^{-4}$		$+2.15 \times 10^{-4}$		$+2.15 \times 10^{-4}$
Equil: mol	1.76×10^{-3}		2.15×10^{-4}		2.15×10^{-4}

$$P_{HI} = \frac{nRT}{V} = \frac{1.76\times10^{-3} \text{ mole} \times \frac{0.08206 \text{ L atm}}{\text{K mol}} \times 623 \text{ K}}{400 \text{ cm}^3 \times \frac{\text{mL}}{\text{cm}^3} \times \frac{10^{-3} \text{ L}}{\text{mL}}} = 0.225 \text{ atm}$$

$$P_{I_2} = P_{H_2} = \frac{2.15\times10^{-4} \text{ mole} \times \frac{0.08206 \text{ L atm}}{\text{K mol}} \times 623 \text{ K}}{400 \text{ cm}^3 \times \frac{\text{mL}}{\text{cm}^3} \times \frac{10^{-3} \text{ L}}{\text{mL}}} = 0.0275 \text{ atm}$$

$$K_P = \frac{P_{H_2} P_{I_2}}{P_{HI}^2} = \frac{(0.0275 \text{ atm})^2}{(0.225 \text{ atm})^2} = 0.0149$$

OR

Since there are two moles of gas on both sides of the chemical equation, $K_P = K_c$.

$$K_c = \frac{\left(2.15\times10^{-4}\right)^2}{\left(1.76\times10^{-3}\right)^2} = 0.0149$$

Bulb 1 $K_c = \frac{\left(1.57\times10^{-4}\right)^2}{\left(2.03\times10^{-3}\right)^2} = 0.00599$

Bulb 2 $K_c = \frac{\left(2.09\times10^{-4}\right)^2}{\left(2.08\times10^{-3}\right)^2} = 0.0101$

Bulb 3 $K_c = \frac{\left(2.42\times10^{-4}\right)^2}{\left(1.98\times10^{-3}\right)^2} = 0.0150$

Bulb 4 $K_c = \frac{\left(3.11\times10^{-4}\right)^2}{\left(2.55\times10^{-3}\right)^2} = 0.0149$

Bulb 5 $K_c = \frac{\left(2.15\times10^{-4}\right)^2}{\left(1.76\times10^{-3}\right)^2} = 0.0149$

Since the last three values are the same, the system is at equilibrium.

Chapter 15

Acids, Bases, and Acid-Base Equilibria

Exercises

15.1A (a) NH_3 + HCO_3^- $\rightleftharpoons$ NH_4^+ + CO_3^{2-}

base (1) acid (2) acid (1) base (2)

where 1 is for one acid-base pair and 2 for the other

(b) H_3PO_4 + H_2O $\rightleftharpoons$ H_3O^+ + $H_2PO_4^-$

acid (1) base (2) acid (2) base (1)

15.1B HCO_3^- acted as an acid in Exercise 15.1A.

HCO_3^- + HCl → H_2CO_3 + Cl^-

base acid

H_2O acted as a base in Exercise 15.1A.

H_2O + NH_3 $\rightleftharpoons$ NH_4^+ + OH^-

acid base

$H_2PO_4^-$ acted as a base from the right side of the equation in Exercise 15.1A.

$H_2PO_4^-$ + NH_3 $\rightleftharpoons$ NH_4^+ + HPO_4^{2-}

15.2A (a) H_2Te is the stronger acid. Because the Te atom is larger than the S atom, it is expected that the H-Te bond energy will be less than the H-S bond energy, and the H-Te bond will be more easily broken than the H-S bond.

(b) $CH_3CH_2CH_2CHBrCOOH$ is the stronger acid, because the Br on the second carbon is more electron-withdrawing than the Cl on the fifth carbon.

15.2B (a) < (d) < (b) < (c)

(a) has no extra electron-withdrawing group. (d) has the Br on the far side of the benzene, so it is stronger than (a). (b) has the Cl next to the COOH group and is a stronger withdrawing group than Br. (c) is the strongest because it has two close withdrawing groups.

15.3A (d) < (a) < (c) < (b)

(d) is the weakest because it is aromatic and has two electron-withdrawing groups. (a) is aromatic and has one electron-withdrawing group. (c) has a electron-withdrawing group, so it is weaker than (b), the strongest of these bases.

15.3B (a) The most strongly basic is $-CH_2CH_2CH_3$ as it has no electron-withdrawing groups.

(b) The most weakly basic is $-C_6H_3Br_2$ as it has two electron-withdrawing groups and is aromatic.

weakly basic $-C_6H_3Br_2$, $-C_6H_4Cl$, $-C_6H_5$, $-CH_2CHClCH_3$, $-CH_2CH_2CH_3$
strongly basic

15.4A $[H^+] = \dfrac{0.0105 \text{ mol } HNO_3}{225 \text{ L}} = 4.67 \times 10^{-5}$ M
pH = -log[H^+] = 4.331

15.4B $[OH^-] = \dfrac{2.65 \text{ g}}{735 \text{ mL}} \times \dfrac{1 \text{ mol } Ba(OH)_2}{171.3 \text{ g } Ba(OH)_2} \times \dfrac{1000 \text{ mL}}{\text{L}} \times \dfrac{2 \text{ mol } OH^-}{\text{mol } Ba(OH)_2} = 0.0421$ M
pOH = - log [OH^-] = - log (0.0421) = 1.376
pH = 14.000 – pOH = 12.624

15.5A The solution is basic. The concentration of OH^- comes from the dissociation of NaOH *and* the self-ionization of water.

15.5B The NaOH and HCl react to neutralize each other. The solution is neutral.

15.6A $CH_3CH_2COOH + H_2O \rightleftharpoons CH_3CH_2COO^- + H_3O^+$

$K_a = 1.3 \times 10^{-5} = \dfrac{[H_3O^+][CH_3CH_2COO^-]}{[CH_3CH_2COOH]}$

$[CH_3CH_2COOH] = 0.250 - [H_3O^+]$
Assumption: Self-ionization of water is negligible, so that $[H_3O^+]$
= $[CH_3CH_2COO^-]$.

$K_a = 1.3 \times 10^{-5} = \dfrac{[H_3O^+]^2}{0.250 - [H_3O^+]}$

Assume 0.250 >> $[H_3O^+]$.
$[H_3O^+] = 1.8 \times 10^{-3}$ M
Assumption is good.
pH = 2.74

15.6B $[C_6H_4NO_2OH] = \dfrac{2.1 \text{ g}}{\text{L}} \times \dfrac{\text{mole}}{139 \text{ g}} = 0.015$ M

$C_6H_4NO_2OH + H_2O \rightleftharpoons H_3O^+ + C_6H_4NO_2O^-$

	$C_6H_4NO_2OH$	H_3O^+	$C_6H_4NO_2O^-$
Initial, M:	0.015	≈ 0	–
Change, M:	-y	+y	+y
Equilibrium, M:	0.015 -y	y	y

$Ka = \dfrac{[H_3O^+][C_6H_4NO_2O^-]}{[C_6H_4NO_2OH]} = \dfrac{y^2}{0.015 - y} = 6.0 \times 10^{-8}$

Assume 0.015 >> y.
$y^2 = 9.0 \times 10^{-10}$
$y = 3.0 \times 10^{-5} = [H_3O^+]$ Assumption is good.
pH = -log[H_3O^+] = -log(3.0×10^{-5}) = 4.52

15.7A $HClO_2 + H_2O \rightleftharpoons H_3O^+ + ClO_2^-$

	$HClO_2$	H_3O^+	ClO_2^-
Initial, M:	0.0100	≈0	0
Changes, M:	-y	+y	+y
Equilibrium, M:	0.0100 - y	y	y

$$\frac{[H_3O^+][ClO_2^-]}{[HClO_2]} = \frac{y\,y}{[0.0100 - y]} = 1.1 \times 10^{-2}$$

$$\frac{M_{acid}}{K_a} = \frac{0.0100}{1.1 \times 10^{-2}} = 0.91\text{, much less than 100.}$$

The assumption is expected to fail, and the quadratic equation must be used.

$y2 = 1.1 \times 10^{-4} - 1.1 \times 10^{-2}\ y$

$y2 + 1.1 \times 10^{-2}\ y - 1.1 \times 10^{-4} = 0$

$$y = \frac{-1.1 \times 10^{-2} \pm \sqrt{(1.1 \times 10^{-2})^2 - 4 \times 1 \times (-1.1 \times 10^{-4})}}{2}$$

$$y = \frac{-1.1 \times 10^{-2} \pm 2.4 \times 10^{-2}}{2}$$

$y = 6.3 \times 10^{-3}$ $\qquad -1.7 \times 10^{-2}$

Negative values are not valid.

$[H_3O^+] = 6.3 \times 10^{-3}$

$pH = -\log[H_3O^+] = -\log[6.3 \times 10^{-3}] = 2.20$

15.7B The two relationships that must be simultaneously satisfied are $\frac{x^2}{M} = 1.4 \times 10^{-3}$

and $\frac{x}{M} = 0.050$.

$$\frac{(0.050M)^2}{M} = 1.4 \times 10^{-3}$$

$2.5 \times 10^{-3}\ M = 1.4 \times 10^{-3}$

$M = 0.56$ M is minimum molarity.

15.8A $CH_3NH_2 + H_2O \rightleftharpoons CH_3NH_3^+ + OH^-$

	CH_3NH_2	$CH_3NH_3^+$	OH^-
Initial, M:	0.200	0	≈0
Changes, M:	− y	+y	+y
Equilibrium, M:	0.200 − y	y	y

$$\frac{[CH_3NH_3^+][OH^-]}{[CH_3NH_2]} = \frac{y^2}{0.200 - y} = 4.2 \times 10^{-4}$$

assume 0.200 >> y

$y^2 = 8.4 \times 10^{-5}$

$y = [OH^-] = 9.2 \times 10^{-3}$

y is 4.6% of 0.200, and the assumption is just within the 5% rule.

$pOH = -\log[OH^-] = -\log[9.2 \times 10^{-3}] = 2.04$

$pH = 14.00 - pOH = 14.00 - 2.04 = 11.96$

15.8B $[NH_3] = \dfrac{5.00 \text{ mL x } 0.0100\text{M}}{1.000\text{L}} \times \dfrac{10^{-3}\text{L}}{\text{mL}} = 5.00 \times 10^{-5}$ M

	$NH_3 + H_2O \rightleftharpoons$	$NH_4^+ +$	OH^-
Initial, M:	5.00×10^{-5}	0	≈0
Changes, M:	-y	y	y
Equilibrium, M:	5.00×10^{-5} - y	y	y

$$\frac{M_{\text{base}}}{K_b} = \frac{5.00 \times 10^{-5}}{1.8 \times 10^{-5}} = 2.7\text{, much less than 100}$$

The assumption will fail, so the quadratic must be used.

$$\frac{[NH_4^+][OH^-]}{[NH_3]} = \frac{y^2}{5.00 \times 10^{-5} - y} = 1.8 \times 10^{-5}$$

$y^2 = 9.00 \times 10^{-10} - 1.8 \times 10^{-5}\ y$

$y^2 + 1.8 \times 10^{-5}\ y - 9.00 \times 10^{-10} = 0$

$$y = \frac{-1.8 \times 10^{-5} \pm \sqrt{(1.8 \times 10^{-5})^2 - 4 \times 1 \times (-9.00 \times 10^{-10})}}{2}$$

$$y = \frac{-1.8 \times 10^{-5} \pm 6.3 \times 10^{-5}}{2}$$

$y = 2.2 \times 10^{-5}$ $\quad\quad$ -4.0×10^{-5}

Negative values are not valid.

pH = -log[OH⁻] = -log(2.2×10^{-5}) = 4.66

pH = 14.00 - pOH = 14.00 - 4.66 = 9.34

15.9A $K_a = \dfrac{X^2}{0.0100 - X}$

$X = 10^{-3.12} = 7.59 \times 10^{-4}$ M

$$K_a = \frac{(7.59 \times 10^{-4})^2}{0.0100 - 7.59 \times 10^{-4}} = 6.2 \times 10^{-5}$$

$pK_a = 4.21$

15.9B The same pH means the same [OH⁻].

$[OH^-] = \sqrt{x} = [(CH_3)_2NH] \times K_{b[(CH_3)_2NH]}$

$[OH^-] = \sqrt{x} = [NH_3] \times K_{b(NH_3)}$

$[(CH_3)_2NH] \times K_{b[(CH_3)_2NH]} = [NH_3] \times K_{b\ (NH_3)}$

$$[NH_3] = \frac{[(CH_3)_2NH] \text{ x } K_b((CH_3)_2NH)}{K_b(NH_3)} = \frac{[(CH_3)_2NH] \times K_{b((CH_3)_2NH)}}{K_{b(NH_3)}}$$

$$[NH_3] = \frac{0.200 \text{ M x } 5.9 \text{ x } 10^{-4}}{1.8 \text{ x } 10^{-5}} = 6.56 \text{ M}$$

15.10A Since both K_b and molarity are larger, it should be obvious that methylamine is more basic and would have the higher pH.

$$K_b = \frac{[OH^-][NH_4^+]}{[NH_3]} = \frac{[OH^-]^2}{[NH_3]}$$

$$[OH^-] = \sqrt{K_b [NH_3]} \qquad\qquad [OH^-] = \sqrt{K_b [CH_3NH_2]}$$

$$[OH^-] = \sqrt{1.8 \times 10^{-5} \times 0.025} \qquad [OH^-] = \sqrt{4.2 \times 10^{-4} \times 0.030}$$

$$[OH^-] = \sqrt{4.5 \times 10^{-7}} \qquad\qquad [OH^-] = \sqrt{1.26 \times 10^{-5}}$$

pOH = -log[OH$^-$] ≈ 3 actual(3.17) $\qquad$ pOH ≈ 2 actual (2.45)

pH = 10.83 $\qquad\qquad\qquad\qquad\qquad$ pH = 11.55

15.10B HCl is a strong acid so the [H$_3$O$^+$] = 0.0010 M, pH = 3. For 0.10 M acetic acid,

$$[H_3O^+] = \sqrt{0.10 \times K_a} = \sqrt{1.8 \times 10^{-6}} > 1.0 \times 10^{-3} \text{ so pH} < 3.$$

15.11A $\qquad\qquad\qquad$ C$_4$H$_4$O$_4$ + H$_2$O $\rightleftharpoons$ H$_3$O$^+$ + C$_4$H$_3$O$_4^-$

Initial, M:	0.125	≈0	0
Changes, M:	-y	+y	+y
Equilibrium, M:	0.125 - y	y	y

$$\frac{[H_3O^+][C_4H_3O_4^-]}{[C_4H_4O_4]} = \frac{y^2}{0.125 - y} = 1.2 \times 10^{-2}$$

$$\frac{M_{acid}}{K_a} = \frac{0.125}{1.2 \times 10^{-2}} = 10.4 \qquad \text{Assumption fails, use quadratic.}$$

$$y^2 = 1.5 \times 10^{-3} - 1.2 \times 10^{-2} y$$

$$y^2 + 1.2 \times 10^{-2} y - 1.5 \times 10^{-3} = 0$$

$$y = \frac{-1.2 \times 10^{-2} \pm \sqrt{(1.2 \times 10^{-2})^2 - 4 \times 1 \times (-1.5 \times 10^{-3})}}{2}$$

$$y = \frac{-1.2 \times 10^{-2} \pm 7.8 \times 10^{-2}}{2}$$

$$y = 3.3 \times 10^{-2} \qquad\qquad\qquad -4.5 \times 10^{-3}$$

$\qquad\qquad\qquad\qquad\qquad\qquad$ Negative values are not valid.

pH = $-$log [H$_3$O$^+$] = $-$log (3.3 × 10^{-2}) = 1.48

Because $K_{a1} \gg K_{a2}$, essentially all of the H$_3$O$^+$ is produced in the K_{a1} reaction.

15.11B $[H_3PO_4] = \dfrac{0.057 \text{ g acid solution}}{100 \text{ g cola drink}} \times \dfrac{75 \text{ g H}_3\text{PO}_4}{100 \text{ g acid solution}} \times \dfrac{\text{mol H}_3\text{PO}_4}{98.0 \text{ g H}_3\text{PO}_4}$

$\qquad\qquad\qquad \times \dfrac{1 \text{ g cola drink}}{\text{mL cola drink}} \times \dfrac{\text{mL cola drink}}{10^{-3} \text{ L cola drink}} = 4.4 \times 10^{-3} \text{ M}$

$[H_3PO_4] = \dfrac{0.084 \text{ g acid solution}}{100 \text{ g cola drink}} \times \dfrac{75 \text{ g H}_3\text{PO}_4}{100 \text{ g acid solution}} \times \dfrac{\text{mol H}_3\text{PO}_4}{98.0 \text{ g H}_3\text{PO}_4}$

$\qquad\qquad\qquad \times \dfrac{1 \text{ g cola drink}}{\text{mL cola drink}} \times \dfrac{\text{mL cola drink}}{10^{-3} \text{ L cola drink}} = 6.4 \times 10^{-3} \text{ M}$

$\frac{4.4 \times 10^{-3}}{7.1 \times 10^{-3}} << 100$ Approximations will not work, so the quadratic equation is needed.

Assume all H_3O^+ is produced in the first acid-dissociation reaction.

$$H_3PO_4 + H_2O \rightleftharpoons H_3O^+ + H_2PO_4^-$$

	4.4×10^{-3} M	≈ 0	0
	$-y$	$+y$	$+y$
	$4.4 \times 10^{-3} - y$	y	y

$$K_a = \frac{[H_3O^+][H_2PO_4^-]}{[H_3PO_4]} = \frac{y^2}{4.4 \times 10^{-3} - y} = 7.1 \times 10^{-3}$$

$y^2 = 3.1 \times 10^{-5} \quad -7.1 \times 10^{-3}\, y$

$y^2 + 7.1 \times 10^{-3}\, y - 3.1 \times 10^{-5} = 0$

$$y = \frac{-7.1 \times 10^{-3} \pm \sqrt{(7.1 \times 10^{-3})^2 - 4 \times 1 \times (-3.1 \times 10^{-5})}}{2}$$

$$y = \frac{-7.1 \times 10^{-3} \pm 1.3 \times 10^{-2}}{2}$$

$y = 3.1 \times 10^{-3} = [H_3O^+]$ or -1.0×10^{-2}

Negative values are not valid.

$pH = -\log[H_3O^+] = -\log(3.0 \times 10^{-3}) = 2.52$

for $[H_3PO_4] = 6.4 \times 10^{-3}$, pH = 2.39.

15.12A For very dilute solutions, reaction goes essentially to completion.

$[H_3O^+] = 2 \times 8.5 \times 10^{-4} = 1.7 \times 10^{-3}$ M

pH = 2.77

15.12B Ionization of 0.020 M H_2SO_4 is complete in the first step and partial in the second: $0.020\text{ M} < [H_3O^+] < 0.040$ M. Only response (b) fits this requirement adequately: $[H_3O^+] = 0.025$ M. Response (c) would require almost complete ionization in both steps. A more exact calculation is below.

$$1.1 \times 10^{-2} = \frac{[H_3O^+][SO_4^{2-}]}{[HSO_4^-]} = K_{a2}$$

$$1.1 \times 10^{-2} = \frac{(0.020 + X)\, X}{0.020 - X}$$

$X^2 + 0.031\, X - 2.2 \times 10^{-4} = 0$

$$X = \frac{-0.031 \pm \sqrt{(0.031)^2 - 4 \times 1 \times (-2.2 \times 10^{-4})}}{2}$$

$X = 0.0060$ M or -0.037 M

not valid

$[H_3O^+] = 0.020 + 0.006 = 0.026$ M

15.13A (a) $NaNO_3$ is neutral. Na^+ is from the strong base, NaOH; NO_3^- is from the strong acid, HNO_3.

(b) $CH_3CH_2CH_2COOK$ is basic. K^+ is from the strong base, KOH. $CH_3CH_2CH_2COO^-$ is from the weak acid, $CH_3CH_2CH_2COOH$.

$$CH_3CH_2CH_2COO^- + H_2O \rightleftharpoons CH_3CH_2CH_2COOH + OH^-$$

15.13B HCl (aq), NH_4Br (aq), NaCl (aq), KNO_2(aq), NaOH(aq)

HCl is a strong acid. NH_4Br is the salt of a weak base, so it is slightly acidic. NaCl is a salt of a strong acid–strong base, so it is neutral. KNO_2 is the salt of a weak acid, so it is slightly basic. NaOH is a strong base.

15.14A $NH_4^+ + H_2O \rightleftharpoons H_3O^+ + NH_3$

$$K_a = \frac{[H_3O^+][NH_3]}{[NH_4^+]} = \frac{X \cdot X}{0.052 - X} = \frac{K_w}{K_b} = \frac{10^{-14}}{1.8 \times 10^{-5}} = 5.6 \times 10^{-10}$$

Assume $X << 0.052$.

$X^2 = 0.052 \times 5.6 \times 10^{-10} = 2.9 \times 10^{-11}$

$X = 5.4 \times 10^{-6}$ M (assumption valid)

pH = 5.27

15.14B

	CH_3COOH +	OH^- $\rightleftharpoons$	$CH_3COO^- + H_2O$
	50.00 mL × 0.120 M	18.75 mL × 0.320 M	–
	=6.00 mmol	=6.00 mmol	
	-6.00mmol	-6.00 mmol	+6.00
	—	—	6.00 mmol

$$[CH_3COO^-] = \frac{6.00 \text{ mmol}}{(50.00 + 18.75)\text{mL}} = 0.0873 \text{ M}$$

$$CH_3COO^- + H_2O \rightleftharpoons CH_3COOH + OH^-$$

	CH_3COO^-	CH_3COOH	OH^-
Initial, M:	0.0873	0	≈0
Changes, M:	-y	+y	+y
Equilibrium, M:	0.0873 - y	y	y

$$K_b = \frac{K_w}{K_a} = \frac{[CH_3COOH][OH^-]}{[CH_3COO^-]} = \frac{y^2}{(0.0873 - y)} = \frac{10^{-14}}{1.8 \times 10^{-5}} = 5.6 \times 10^{-10}$$

assume 0.0873 >> y

$y^2 = 4.85 \times 10^{-11}$

$y = 6.96 \times 10^{-6}$ (assumption valid)

pOH = $-\log[OH^-] = -\log(6.96 \times 10^{-6}) = 5.16$

pH = 14.00 - pOH = 8.84

15.15A $CH_3COO^- + H_2O \rightleftharpoons CH_3COOH + OH^-$

$$K_b = \frac{K_w}{K_a} = \frac{10^{-14}}{1.8 \times 10^{-5}} = \frac{[OH^-]^2}{[CH_3COO^-]}$$

pH = 9.10 so pOH = 4.90, and $[OH^-] = 10^{-4.90} = 1.26 \times 10^{-5}$

$$\frac{10^{-14}}{1.8 \times 10^{-5}} = \frac{(1.26 \times 10^{-5})^2}{[CH_3COO^-]}$$

$[CH_3COO^-] = 0.29$ M

15.15B The K_a values are $K_a(HNO_2) = 7.2 \times 10^{-4}$ and $K_a(HCN) = 6.2 \times 10^{-10}$. In comparing K_b values for the conjugate base ions, that of CN^- is greater than that of NO_2^-, making NH_4CN (aq) more basic than NH_4NO_2 (aq).

15.16A $NH_3 + H_2O \rightleftharpoons NH_4^+ + OH^-$

$0.15 - X \qquad 0.35 + X \qquad X$

$$K_b = 1.8 \times 10^{-5} = \frac{(0.35 + X)X}{0.15 - X}$$

Assume $X << 0.15$.

$$1.8 \times 10^{-5} = \frac{0.35\,X}{0.15}$$

$X = 7.71 \times 10^{-6}$ M = $[OH^-]$ (assumption valid)

pOH = 5.11

pH = 8.89

15.16B The common ion is the H^+.

$$CH_3COOH + H_2O \rightleftharpoons CH_3COO^- + H_3O^+$$

	CH_3COOH	CH_3COO^-	H_3O^+
Initial, M:	0.10	-	0.10
Changes, M:	-y	+y	+y
Equilibrium, M:	0.10 - y	y	–0.10 + y

$$K_a = \frac{[CH_3COO^-][H_3O^+]}{[CH_3COOH]} = \frac{y\,(0.10 + y)}{0.10 - y} = 1.8 \times 10^{-5}$$

Assume 0.10 >> y.

$$\frac{y\,(0.10)}{0.10} = 1.8 \times 10\text{-}5 \text{ M} = [CH_3COO^-]$$ (assumption valid)

$[H_3O^+] = 0.10$ M + 1.8×10^{-5} M = 0.10 M

15.17A $NH_3 + H^+ \rightleftharpoons NH_4^+$

NH_3	H^+	NH_4^+
0.24 M	$\frac{0.03}{0.50} = 0.06$ M	0.20M
0.24 - 0.06	--	0.20 + 0.06

$$K_b = 1.8 \times 10^{-5} = \frac{[NH_4^+][OH^-]}{[NH_3]}$$

$$K_b = 1.8 \times 10^{-5} = \frac{0.26}{0.18} \times [OH^-]$$

$[OH^-] = 1.25 \times 10^{-5}$ M

pOH = 4.90

pH = 9.10

or

pK_a for $NH_4^+ = 14.00 - pK_b = 14.00 - (-\log 1.8 \times 10^{-5}) = 9.26$

$$pH = pK_a + \log\frac{[\text{base}]}{[\text{conj. acid}]}$$

$$pH = 9.26 + \log\frac{0.18}{0.26} = 9.10$$

15.17B $9.50 = 9.26 + \log\frac{[NH_3]}{[NH_4^+]}$

$$0.24 = \log\frac{[NH_3]}{[NH_4^+]}$$

$$10^{0.24} = 10^{\log\frac{[NH_3]}{[NH_4^+]}} = \frac{[NH_3]}{[NH_4^+]}$$

$$1.74 = \frac{[NH_3]}{[NH_4^+]} = \frac{0.24 + y}{0.20 - y}$$

$0.348 - 1.74\ y = 0.24 + y$

$0.108 = 2.74\ y$

$0.039\ M = y = [OH^-]$ added

?mol OH^- added = 0.039 M × 0.500 L = 0.020 mol NaOH added

15.18A $pH = pKa + \log\frac{[CH_3COO^-]}{[CH_3COOH]}$

$$4.50 = 4.74 + \log\frac{0.250}{[CH_3COOH]}$$

$$-0.24 = \log\frac{0.250}{[CH_3COOH]}$$

Raise both sides of equation to a power of 10.

$$10^{-0.24} = 10^{\log\frac{0.250}{[CH_3COOH]}} = \frac{0.250}{[CH_3COOH]}$$

$$[CH_3COOH] = \frac{0.250\ M}{10^{-0.24}} = \frac{0.250\ M}{0.58}$$

$[CH_3COOH] = 0.43$ M

15.18B $[H_3O^+] = \frac{K_w}{K_b} \times \frac{[NH_4^+]}{[NH_3]}$

$[H_3O^+] = 10^{-9.05} = 8.91 \times 10^{-10}$ M

$$[NH_4^+] = \frac{8.91 \times 10^{-10} \times 1.8 \times 10^{-5} \times 0.150}{1 \times 10^{-14}}$$

$[NH_4^+] = 0.241$ M

mass = 0.241 M × 0.250 L × 53.49 g/mol = 3.22 g NH_4Cl

An alternate way to work this problem is shown below:

$$pH = pK_a + \log\frac{[\text{base}]}{[\text{conj. acid}]}$$

$$9.05 = 9.26 + \log\frac{0.150}{M}$$

$$\log\frac{0.150}{M} = -0.21$$

$$\frac{0.150}{M} = 10^{-0.21} = 0.62$$

$M = 0.24$

#g NH_4Cl = 0.24 M × 0.250 L × 53.49 g/mol = 3.2 g NH_4Cl

15.19A (a) NH_4Cl pH = 4.63
(b) NH_4Cl - NH_3 pH = 9.26
(c) HCl - HNO_3 pH = - 0.30
(d) CH_3COOH - CH_3COO^- pH = 4.74

The indicator color shows that the pH is in the range of about 4 to 5.5. Solutions (b) and (c) have pH values outside this range. The 1.00 M NH_4Cl would have a pH in this range due to hydrolysis of NH_4^+. The 1.00 M CH_3COOH-CH_3COONa is a buffer with pH = 4.74 (pK_a of acetic acid). To distinguish between (a) and (d), add a small amount of either an acid or a base. The pH of the buffer solution (d) would not change, and that of the 1.00 M NH_4Cl (solution b) would.

15.19B Thymol blue changes to blue at about pH 9.5. NaOH is required to neutralize buffer and then to raise pH to 9.5.

$$? \text{ mol NaOH} = 100.0 \text{ mL} \times 0.200\text{M} \times \frac{10^{-3}\text{L}}{\text{mL}} = 2.00 \times 10^{-2} \text{ mol NaOH to neutralize}$$

buffer.

$[H^+] = 10^{-9.5} = 3.16 \times 10^{-10}$ M

$$? \text{ mol NaOH in excess} = 3.16 \times 10^{-10} \text{ M} \times 100 \text{ mL} \times \frac{10^{-3}\text{L}}{\text{mL}}$$

$$= 3.16 \times 10^{-11} \text{ mol NaOH}$$

?mol $NaOH_{total}$ = 3.16×10^{-11} mol + 2.00×10^{-2} mol = 2.00×10^{-2} mol NaOH

15.20A $H_3O^+ + OH^- \rightleftharpoons H_2O$

(a) $$[H_3O^+] = \frac{20.00 \text{ mL x } 0.500 \text{ M} - 19.90 \text{ mL x } 0.500 \text{ M}}{(20.00 + 19.90) \text{ mL}}$$

$[H_3O^+] = 1.253 \times 10^{-3}$ pH = 2.90

(b) $$[H_3O^+] = \frac{20.00 \text{ mL x } 0.500 \text{ M} - 19.99 \text{ mL x } 0.500 \text{ M}}{(20.00 + 19.99) \text{ mL}}$$

$[H_3O^+] = 1.250 \times 10^{-4}$ pH = 3.90

(c) $$[OH^-] = \frac{20.01 \text{ mL x } 0.500 \text{ M} - 20.00 \text{ mL x } 0.500 \text{ M}}{(20.00 + 20.01) \text{ mL}}$$

$[OH^-] = 1.250 \times 10^{-4}$ pOH = 3.90 pH = 10.10

(d) $[OH^-] = \dfrac{20.10\ mL \times 0.500\ M - 20.00\ mL \times 0.500\ M}{(20.00 + 20.10)\ mL}$

$[OH^-] = 1.247 \times 10^{-3}$ $\quad$ $pOH = 2.90$ $\quad$ $pH = 11.10$

15.20B $NaOH + HCl \rightleftharpoons H_2O + Na^+ + Cl^-$

NaOH	HCl
25.00 mL × 0.220	8.10 mL × 0.252
5.50mmol	2.04 mmol
-2.04	-2.04
3.46 mmol	—

$[OH^-] = \dfrac{3.46\ mmol}{33.10\ mL} = 0.105\ M$

$pOH = -\log[OH] = 0.981$

$pH = 14.00 - pOH = 13.02$

15.21A (a) $CH_3COOH + OH^- \rightleftharpoons H_2O + CH_3COO^-$

CH_3COOH	OH^-	CH_3COO^-
10.00 mmol	12.50 ml × 0.500 M	
-6.25 mmol	-6.25 mmol	+ 6.25 mmol
3.75 mmol	--	6.25 mmol

This is a buffer solution.

$[H_3O^+] = Ka\dfrac{[acid]}{[base]} = 1.8 \times 10^{-5} \times \dfrac{3.75\ mmol/32.50\ mL}{6.25\ mmol/32.50\ mL}$

Notice that since both acid and base are in the same solution, the volume is the same, does cancel out, and can be left out of the calculation.

$[H_3O^+] = 1.08 \times 10^{-5}$

$pH = 4.97$

(b) $CH_3COOH + OH^- \rightleftharpoons H_2O + CH_3COO^-$

CH_3COOH	OH^-	CH_3COO^-
10.00 mmol	20.10 mL × 0.500 M	
-10.00 mmol	-10.00 mmol	+10.00 mmol
0	0.05 mmol	+10.00 mmol

$[OH^-] = \dfrac{0.05\ mmol}{(20.00 + 20.10)\ mL} = 1.25 \times 10^{-3}\ M$

$pOH = 2.90$

$pH = 11.10$

15.21B initial $[OH^-] = 0.500\ M$

$pOH = 0.30$, pH - 13.70

(a)

	OH^-	+	CH_3COOH	→	CH_3COO^-	+ H_2O
Init	20.00 mL × 0.500 M		10.00 mL × 0.500 M		-	-
Eq	5.00 mmol		-		5.00 mmol	

$[OH^-] = \dfrac{5.00\ mmol}{30.00\ mL} = 0.167\ M$

$pOH = 0.78$, pH - 13.22

(b)

	OH^-	+	CH_3COOH	→	CH_3COO^-	+ H_2O
Init	10.0 mmol		10.0 mmol		-	-

Eq	-	-	10.0 mmol

$[CH_3COO^-] = \dfrac{10.0\text{ mmol}}{40.00\text{ mL}} = 0.250\text{ M}$

$CH_3COO^- + H_2O \rightleftharpoons CH_3COOH + OH^-$

$\dfrac{K_w}{K_a} = \dfrac{10^{-14}}{1.8\times10^{-5}} = \dfrac{[CH_3COOH][OH^-]}{[CH_3COO^-]}$

$\dfrac{10^{-14}}{1.8\times10^{-5}} = \dfrac{[OH^-]^2}{0.25\text{ M}}$

$[OH^-]^2 = 1.39 \times 10^{-10}$

$[OH^-] = 1.18 \times 10^{-5}$

pOH = 4.93, pH - 9.07

(c)

	OH^-	+	CH_3COOH	$\rightarrow$	$CH_3COO^- + H_2O$	
Init	10.0 mmol		15.0 mmol		-	-
Eq	-		5.0 mmol		10.0 mmol	

$[H^+] = K_a \times \dfrac{[\text{acid}]}{[\text{base}]} = 1.8 \times 10^{-5} \times \dfrac{(5.0\text{ mmol})}{(10.0\text{ mmol})}$

$[H^+] = 9.0 \times 10^{-6}$

pH = 5.04

(d)

	OH^-	+	CH_3COOH	$\rightarrow$	$CH_3COO^- + H_2O$	
Init	10.0 mmol		20.0 mmol		-	-
Eq	-		10.0 mmol		10.0 mmol	

$[H^+] = K_a \times \dfrac{[\text{acid}]}{[\text{base}]} = 1.8 \times 10^{-5} \times \dfrac{(10.0\text{ mmol})}{(10.0\text{ mmol})}$

$[H^+] = 1.8 \times 10^{-5}$

pH = 4.74

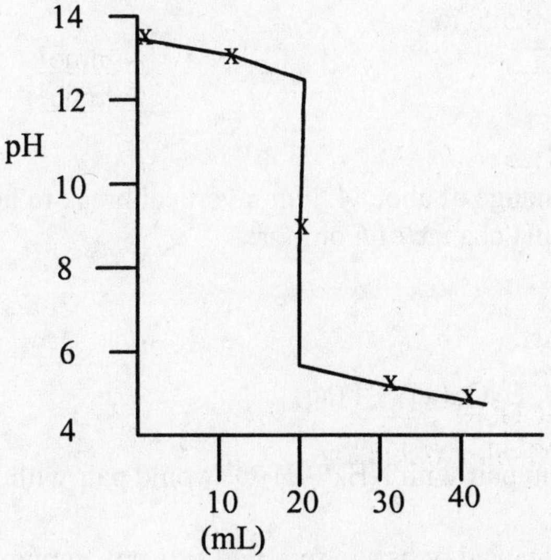

15.22A (a) At half-neutralization, $[H_3O^+] = \frac{K_w}{K_b}$. In this case, this occurs near pH = 9 or

$[H^+] = 10^{-9}$

$10^{-9} = \frac{10^{-14}}{K_b}$

$K_b = 1.0 \times 10^{-5}$

or

$pH = pK_a \approx 9$

$pK_b = 14.00 - pK_a \approx 5$

$pK_b \approx 1 \times 10^{-5}$

(b) From the graph, pH ≈ 5 at equivalence.
At equivalence, hydrolysis of the cation of the weak base occurs.

$[\text{cation}] = \frac{1.0\text{ M base} \times V}{2\text{ V}} = 0.50\text{ M}$

$K_a = \frac{K_w}{K_b} = \frac{10^{-14}}{10^{-5}} = \frac{[H_3O^+]^2}{[\text{cation}]} = \frac{[H_3O^+]^2}{0.50\text{ M}}$

$[H_3O^+] = 2.24 \times 10^{-5}$

$pH = 4.7$

15.22B

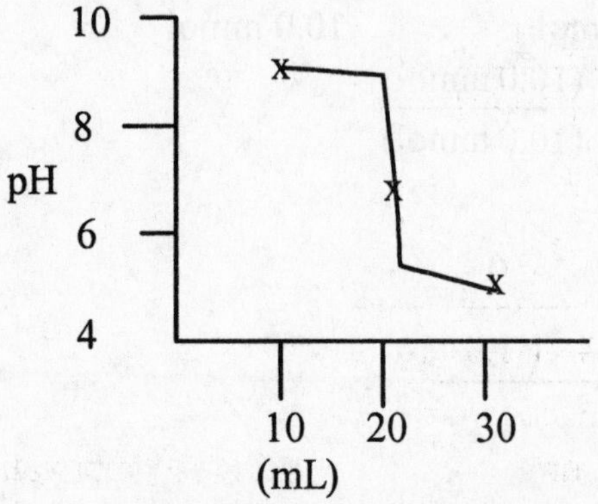

The vertical break is only a pH change of about 4. For a vertical break to be easily discernible, it should be a pH change of 6 or more.

Review Questions

1. Arrhenius; $HI(aq) \rightarrow H^+(aq) + I^-(aq)$
   Brønsted-Lowry; $HI(aq) + H_2O(l) \rightarrow H_3O^+(aq) + I^-(aq)$

3. (c) is not correctly paired. NH_3 would pair with NH_4^+. H_3O^+ would pair with H_2O.

4. An amphiprotic species can react as an acid or as a base. H_3PO_4 is not amphiprotic; it can only act as an acid. $H_2PO_4^-$ and H_2O are amphiprotic; they can react as an acid or as a base. NH_4^+ can only lose H^+; it cannot gain H^+.

5. (a) $HClO_2 + H_2O \rightleftharpoons H_3O^+ + ClO_2^-$ $\qquad K_a = \dfrac{[H_3O^+][ClO_2^-]}{[HClO_2]}$

(b) $CH_3CH_2COOH + H_2O \rightleftharpoons CH_3CH_2COO^- + H_3O^+$

$$K_a = \frac{[H_3O^+][CH_3CH_2COO^-]}{[CH_3CH_2COOH]}$$

(c) $HCN + H_2O \rightleftharpoons H_3O^+ + CN^-$ $\qquad K_a = \dfrac{[H_3O^+][CN^-]}{[HCN]}$

(d) $C_6H_5OH + H_2O \rightleftharpoons H_3O^+ + C_6H_5O^-$ $\qquad K_a = \dfrac{[H_3O^+][C_6H_5O^-]}{[C_6H_5OH]}$

6. For a weak base, the pH must be >7. (d) is the answer.

9. $[H^+] = 10^{-pH} = 10^{-5.00}$

$[H^+][OH^-] = 10^{-14}$

$$[OH^-] = \frac{10^{-14}}{10^{-5}} = 10^{-9}$$

(a) is the answer.

12. $H_2SO_3 + H_2O \rightleftharpoons H_3O^+ + HSO_3^-$

$$1.3 \times 10^{-2} = \frac{[H_3O^+][HSO_3^-]}{[H_2SO_3]} = \frac{[H_3O^+]^2}{0.10 - [H_3O^+]}$$

$$1.3 \times 10^{-3} - 1.3 \times 10^{-2}[H_3O^+] = [H_3O^+]^2$$

$$[H_3O^+]^2 + 1.3 \times 10^{-2}[H_3O^+] - 1.3 \times 10^{-3} = 0$$

$$[H_3O^+] = \frac{-1.3\times10^{-2} \pm \sqrt{(1.3\times10^{-2})^2 - 4\times(-1.3\times10^{-3})}}{2}$$

$[H_3O^+] = 3.0 \times 10^{-2}$ or -4.3×10^{-2} not valid

	HSO_3^- +	$H_2O \rightleftharpoons$	H_3O^+ +	SO_3^{2-}
I	3.0×10^{-2}		3.0×10^{-2}	-
C	$-x$		$+x$	x
E	$3.0 \times 10^{-2} - x$		$3.0 \times 10^{-2} + x$	x

$$6.3 \times 10^{-8} = \frac{[H_3O^+][SO_3^{2-}]}{[HSO_3^-]} = \frac{(3.0\times10^{-2} + x)\,x}{3.0\times10^{-2} - x}$$

Assume $x \ll 3.0 \times 10^{-2}$

$x = [SO_3^{2-}] = 6.3 \times 10^{-8}$

(d) is the answer.

13. K_a (acid) × K_b (conjugate base) = K_w

16. The equivalence point in an acid-base titration is the point where the acid and base are in the exact stoichiometric proportions. The end point in an acid-base titration is the point where the indicator color changes. The indicator is selected so that the endpoint occurs near the equivalence point.

17. To have the indicator change in the 8-10 range, it must have a $K_a \approx 10^{-9}$.
(b) is the answer.

18. The pH is the highest before the acid is added. The last acid added produces the lowest pH.

19. (a) above 7 weak acid with strong base
(b) below 7 strong acid with weak base
(c) 7 strong acid with strong base

Problems

21. (a) $HIO_4 + NH_3 \rightarrow NH_4^+ + IO_4^-$
(b) $H_2O + NH_2OH \rightarrow NH_2OH_2^+ + OH^-$
(c) $H_3BO_3 + NH_2^- \rightarrow NH_3 + H_2BO_3^-$

23. (a) $HOClO_2 + H_2O \rightleftharpoons H_3O^+ + OClO_2^-$
acid 1 base 2 acid 2 base 1
(b) $HSeO_4^- + NH_3 \rightleftharpoons NH_4^+ + SeO_4^{2-}$
acid 1 base 2 acid 2 base 1
(c) $HCO_3^- + OH^- \rightleftharpoons CO_3^{2-} + H_2O$
acid 1 base 2 base 1 acid 2
(d) $C_5H_5NH^+ + H_2O \rightleftharpoons C_5H_5N + H_3O^+$
acid 1 base 2 base 1 acid 2

25. $H_2PO_4^- + H_2O \rightarrow H_3O^+ + HPO_4^{2-}$
$H_2PO_4^- + H_2O \rightarrow OH^- + H_3PO_4$

27. HCl is the strongest acid; it will lose its H^+ more easily than the others. All of the others are weak acids.

29. CH_3COOH is a stronger acid than is H_2O. Because aniline is more able to accept a proton from CH_3COOH than from H_2O, it is a stronger base in $CH_3COOH(l)$ than it is in $H_2O(l)$.

31. (a) H_2Se is stronger because it has a larger anion radius and lower bond dissociation energy than H_2S, so the H^+ is easier to remove.
(b) $HClO_3$ is stronger because Cl is more electronegative than I. The electron density is pulled alway from the —OH bond more in $HClO_3$ than in HIO_3.

(c) H_3AsO_4 is stronger than $H_2PO_4^-$ because the H^+ does not have to be removed from a -2 ion , only a -1 ion.
(d) HBr is stronger because Br is to the right of Se in the periodic table, so the ΔEN is greater between H and Br than between H and Se.
(e) HN_3 is stronger because the ΔEN is greater between N and H than between C and H.
(f) HNO_3 is stronger. The H^+ is separated from NO_3^- against a pull of -1, not -2, as in SO_4^{2-}.

33. (a) $1.4 \times 10^{-3} < K_a < 8.7 \times 10^{-3}$
Less than 2,2-dichloropropanoic acid but more than 2-chloropropanoic acid (p. 623).
(b) The K_a of 4-chlorobutanoic acid should be less than 3-chloropropanoic acid (1.0×10^{-5}) because the electron-withdrawing group is farther away from the carboxylic acid group. The K_a value should be slightly greater than that of unsubstituted acids such as butyric acid (also known as butanoic acid) ($K_a = 1.5 \times 10^{-5}$). K_a (4-chlorobutanoic acid) $\approx 5 \times 10^{-5}$.

35. (a) (b) (c) (d)

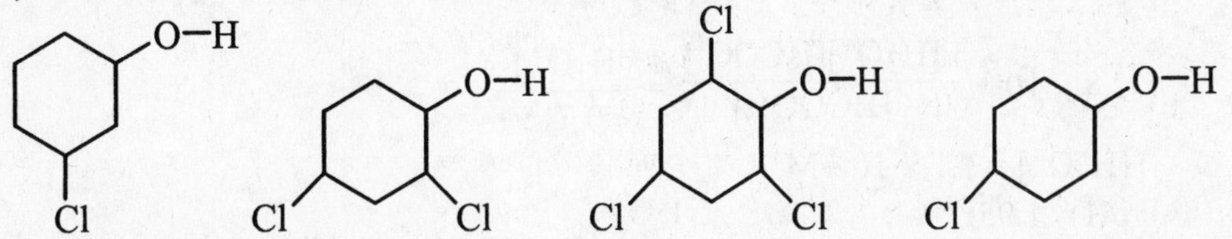

(d) < (a) < (b) < (c)
Phenol should be weaker than (d). This ranking is determined by the number of Cl atoms in the molecules and their proximity to the -OH group.

37. (a) $[H_3O^+] = 10^{-pH} = 10^{-2.91} = 1.2 \times 10^{-3}$ M
(b) $[H_3O^+] = 10^{-9.26} = 5.5 \times 10^{-10}$ M
(c) $[H_3O^+] = 10^{-4.35} = 4.5 \times 10^{-5}$ M
(d) $[H_3O^+] = 10^{-3.94} = 1.1 \times 10^{-4}$ M

39. (a) $[H_3O^+] =$ 0.039 M pH = 1.41
(b) $[OH^-] =$ 0.070 M pOH = 1.15 pH = 12.85
(c) $[H_3O^+] =$ 0.65 M pH = 0.19
(d) $[OH^-] = 2 \times 2.5 \times 10^{-4}$ M $= 5.0 \times 10^{-4}$ M pOH = 3.30 pH = 10.70

41. (a) $[OH^-] =$ 0.073 M pOH = 1.14
(b) $[OH^-] =$ 1.75 M pOH = −0.24
(c) $[OH^-] = 2 \times 0.045$ M = 0.090 M pOH = 1.05

(d) $[H_3O^+] = 9.1 \times 10^{-2}$ pH = 1.04 pOH = 12.96

43. $[OH^-] = 10^{-pOH} = 10^{-(14 - pH)} = 10^{-3.30} = 5.0 \times 10^{-4}$

$$V_{NaOH} = \frac{5.00\ L \times 5.0 \times 10^{-4}\ M}{0.250\ M} \times \frac{mL}{10^{-3}\ L} = 10\ mL$$

Dilute 10 mL of 0.250 M NaOH to 5.00 L.

45. $[OH^-] = 2 \times 0.0062 = 0.0124$ M pOH = 1.91 pH = 12.09
The $Ba(OH)_2$ is more basic and, thus, has the higher pH..

47. (a) $K_a = 4.9 \times 10^{-5} = \frac{[H^+][C_8H_7O_2^-]}{[HC_8H_7O_2]}$

$$4.9 \times 10^{-5} = \frac{[H^+]^2}{0.22}$$

$[H^+]^2 = 1.08 \times 10^{-5}$
$[H^+] = 3.28 \times 10^{-3}$ M
pH = 2.48

(b) $M = \frac{32.9\ g\ HCOOH}{L} \times \frac{mol\ HCOOH}{46.03\ g\ HCOOH} = 0.715\ M$

$$1.8 \times 10^{-4} = \frac{[H_3O^+][HCOO^-]}{[HCOOH]} = \frac{[H_3O^+]^2}{0.715}$$

$[H_3O^+] = 1.1 \times 10^{-2}$ M
pH = 1.95

49. $[H_3O^+] = 10^{-4.90} = 1.26 \times 10^{-5}$ M

$$K_a = \frac{[H_3O^+]^2}{[C_3H_5OH]} = 1.0 \times 10^{-10}$$

$$[C_3H_5OH] = \frac{(1.26 \times 10^{-5})^2}{1.0 \times 10^{-10}} = 1.6\ M$$

51. $[aspirin] = \frac{1.00\ g}{0.300\ L} \times \frac{mol}{180.15\ g} = 1.850 \times 10^{-2}\ M$

$[H_3O^+] = 10^{-2.62} = 2.40 \times 10^{-3}$ M

$$K_a = \frac{[H_3O^+]^2}{M_{aspirin} - [H_3O^+]} = \frac{(2.40 \times 10^{-3})^2}{(1.85 \times 10^{-2} - 2.40 \times 10^{-3})} = 3.6 \times 10^{-4}$$

53. $CCl_3COOH + H_2O \rightleftharpoons H_3O^+ + CCl_3COO^-$

$K_a = 10^{-pKa} = 10^{-0.52} = 3.02 \times 10^{-1}$

K_a is too large for the assumptions to work.

$$0.302 = \frac{[H_3O^+]^2}{0.105\text{-}[H_3O^+]}$$

$[H_3O^+]^2 + 0.302\ \{H_3O^+] - 0.0317 = 0$

$$[H_3O^+] = \frac{-0.302 \pm \sqrt{(0.302)^2 - 4 \text{ x } 1\text{x } (-0.0317)}}{2}$$

$[H_3O^+] = 0.0825$ M or -0384 M Negative values are not valid.

$[H_3O^+] = 0.0825$ M

pH= 1.08

55. For 0.150 M CH_3 COOH

$$K_a = 1.8 \times 10^{-5} = \frac{[H_3O^+]^2}{0.150 \text{ M}}$$

$[H_3O^+] = 1.6 \times 10^{-3}$ M

pH = 2.78

For HCOOH,

$$K_a = 1.8 \times 10^{-4} = \frac{[H_3O^+]^3}{[HCOOH] - [H_3O^+]}$$

$$1.8 \times 10^{-4} = \frac{(1.6 \text{ x } 10^{-3})^2}{[HCOOH] - 1.6 \text{ x } 10^{-3}}$$

$[HCOOH] - 1.6 \times 10^{-3} = 0.014$ M

$[HCOOH] = 0.016$ M

57. 0.0045 M H_2SO_4 has the lower pH. H_2SO_4 is a strong acid in its first ionization step; moreover, even K_{a2} of H_2SO_4 is larger than K_{a1} of H_3PO_4.

59. (a) $? \text{ M} = \frac{83 \text{ g}}{\text{L}} \times \frac{\text{mol}}{90.04 \text{ g}} = 0.922 \text{ M}$

$$5.4 \times 10^{-2} = \frac{[H_3O^+]^2}{0.922 \text{ M} - [H_3O^+]}$$

$[H_3O^+]^2 + 5.4 \times 10^{-2}\ [H_3O^+] - 4.98 \times 10^{-2} = 0$

$$[H_3O^+] = \frac{-5.4 \text{ x } 10^{-2} \pm \sqrt{(5.4 \text{ x } 10^{-2})^2 - 4 \text{ x } 1 \text{ x } (-4.98 \text{ x } 10^{-2})}}{2}$$

$[H_3O^+] = 0.198$ M or - 0.251 M

Because $K_{a1} >> K_{a2}$, all H_3O^+ is formed in the first ionization step.

$[H_3O^+] = 0.198$ M pH = 0.70

(b) $HOOCCOOH + H_2O \rightleftharpoons H_3O^+ + HOOCCOO^-$

$0.922 - X \qquad X \qquad X$

$HOOCCOO^- + H_2O \rightleftharpoons H_3O^+ + {}^-OOCCOO^-$

$X - Y \qquad X + Y \qquad Y$

$$K_{a2} = 5.3 \times 10^{-5} = \frac{(X+Y)(Y)}{X-Y}$$

Assume $X >> Y$.

$K_{a2} = 5.3 \times 10^{-5} = \dfrac{XY}{X}$

$Y = [^{-}OOCCOO^{-}] = 5.3 \times 10^{-5}$ M

61. (a) $H_3PO_4 + H_2O \rightarrow H_3O^+ + H_2PO_4^-$

$Ka = 7.1 \times 10^{-3} = \dfrac{[H_3O^+][H_2PO_4^-]}{[H_3PO_4]} = \dfrac{[H_3O^+]^2}{0.15 - [H_3O^+]}$

$[H_3O^+]^2 + 7.1 \times 10^{-3}[H_3O^+] - 1.07 \times 10^{-3} = 0$

$[H_3O^+] = \dfrac{-7.1 \times 10^{-3} \pm \sqrt{(7.1 \times 10^{-3})^2 - 4(1)(-1.07 \times 10^{-3})}}{2}$

$[H_3O^+] = 0.029$ M or -0.036 Negative values are not valid.

$[H_3O^+] = 0.029\text{ M} = [H_2PO_4^-]$

pH = 1.53 M

(b) $[H_3PO_4] = 0.15 - 0.029 = 0.12$ M

(c) $H_3PO_4 + H_2O \rightarrow H_3O^+ + H_2PO_4^-$

See (a) above. $[H_2PO_4^-] = 0.029$ M

(d) $H_2PO_4^- + H_2O \rightarrow H_3O^+ + HPO_4^{2}$

$K_{a2} = 6.3 \times 10^{-8} = \dfrac{[H_3O^+][HPO_4^{2-}]}{[H_2PO_4^-]} = 6.3 \times 10^{-8}$

Assume that $[H_3O^+] \approx [H_2PO_4^-]$ (p. 638), each cancels in the equation and $[HPO_4^{2-}] = K_{a2} = 6.3 \times 10^{-8}$ M.

(e) $HPO_4^{2-} + H_2O \rightleftharpoons H_3O^+ + PO_4^{3-}$ (see p. 639)

$4.3 \times 10^{-13} = K_{a3} = \dfrac{[H_3O^+][PO_4^{3-}]}{[HPO_4^{2-}]}$

$[HPO_4^{2-}] = K_{a2} = 6.3 \times 10^{-8}$ M and $[H_3O^+] = 0.029$ M

$4.3 \times 10^{-13} = \dfrac{0.029 \times [PO_{4^{3-}}]}{6.3 \times 10^{-8}}$

$9.3 \times 10^{-19} = [PO_4^{3-}]$

OR

$K_{a3} = \dfrac{[H_3O^+][PO_4^{3-}]}{[HPO_4^{2-}]}$

$K_{a3} = \dfrac{0.029[PO_4^{3-}]}{K_{a_2}}$

$[PO_4^{3-}] = \dfrac{K_{a_2}K_{a_3}}{0.029} = \dfrac{6.3 \times 10^{-8} \times 4.3 \times 10^{-13}}{0.029} = 9.3 \times 10^{-19}$

63. (a) $RbClO_4$(aq) should be neutral because Rb^+ would be from a strong base

and ClO_4^- would be from a strong acid.

(b) $CH_3CH_2NH_3^+ + H_2O \rightleftharpoons CH_3CH_2NH_2 + H_3O^+$

Solution is acidic.

(c) $HCOO^- + H_2O \rightleftharpoons HCOOH + OH^-$

$NH_4^+ + H_2O \rightleftharpoons NH_3 + H_3O^+$

$$K_a(NH_4^+) = \frac{K_w}{K_b} = \frac{1 \times 10^{-14}}{1.8 \times 10^{-5}} = 5 \times 10^{-10}$$

$$K_b(HCOO^-) = \frac{K_w}{K_a} = \frac{1 \times 10^{-4}}{1.8 \times 10^{-4}} = 5 \times 10^{-11}$$

Because K_b ($HCOO^-$) is smaller than K_a (NH_4^+), hydrolysis of NH_4^+ occurs more extensively than that of $HCOO^-$, and the solution will be slightly acidic.

65. (c) NH_4I. NH_4^+ is the conjugate acid of the base NH_3 and produces an acidic solution by hydrolysis. Of the other solutions, (a) is neutral and (b) and (d) are basic due to hydrolysis of the anions.

67. (a) $OCl^- + H_2O \rightleftharpoons HOCl + OH^-$

(b) $$K_b = \frac{K_w}{K_a} = \frac{10^{-14}}{2.9 \times 10^{-8}} = 3.4 \times 10^{-7}$$

(c) $$K_b = \frac{[HOCl][OH^-]}{[OCl^-]} = \frac{[OH^-]^2}{0.080 - [OH^-]} \approx \frac{[OH^-]^2}{0.080} = \frac{K_w}{K_a} = \frac{10^{-14}}{2.9 \times 10^{-8}} = 3.4 \times 10^{-7}$$

$$\frac{[OH^-]^2}{0.080} = 3.4 \times 10^{-7}$$

$[OH^-] = 1.66 \times 10^{-4}$ M

pOH = 3.78

pH = 10.22

69. $[OH^-] = 10^{-(14.00 - 9.05)} = 1.12 \times 10^{-5}$

$CH_3CO_2^- + H_2O \rightleftharpoons CH_3CO_2H + OH^-$

$$K_b = \frac{K_w}{K_a} = \frac{10^{-14}}{1.8 \times 10^{-5}} = \frac{[OH^-]^2}{[CH_3CO_2^-]} = \frac{(1.12 \times 10^{-5})^2}{[CH_3CO_2^-]}$$

$[CH_3CO_2^-] = 0.23$ M

71. $HCOOH + H_2O \rightleftharpoons HCOO^- + H_3O^+$

(c) or (d). The H_3O^+ ion from $HClO_4$ or the $HCOO^-$ ion from $(HCOO)_2Ca$ will suppress the reaction.

73.

	$NH_3 + H_2O \rightleftharpoons$	NH_4^+ +	OH^-
Initial:	0.15 M	0.015 M	
Equil:	$0.15 - X$	$0.015 + X$	X

$$K_b = \frac{[NH_4^+][OH^-]}{[NH_3]} = 1.8 \times 10^{-5}$$

$$1.8 \times 10^{-5} = \frac{X(0.015 + X)}{0.15 - X}$$

Assuming that 0.015 >> X, then

$$1.8 \times 10^{-5} = \frac{X0.015}{0.15}.$$

$X = 1.8 \times 10^{-4}$ M = $[NH_4^+]$

75. $CH_3CH_2COOH + H_2O \rightleftharpoons CH_3CH_2COO^- + H_3O^+$

	CH_3CH_2COOH	$CH_3CH_2COO^-$	H_3O^+
Initial:	0.350	0.0786	
Equil:	$0.350 - X$	$0.0786 + X$	X

$$K_a = \frac{[H_3O^+][CH_3CH_2COO^-]}{[CH_3CH_2COOH]}$$

Assume $[H_3O^+] << CH_3CH_2COO^-$.

$$1.3 \times 10^{-5} = \frac{X\,0.0786}{0.350}$$

$X = 5.79 \times 10^{-5} = [H_3O^+]$

pH = 4.24

77. (a) C_6H_5COOH is an acid. pH < 7
(b) NH_3 is a weak base. pH > 7
(c) CH_3NH_2 is a weak base; $CH_3NH_3^+$ is its conjugate acid. pH < 7
(d) $K_{b1} = \frac{K_w}{K_{a1}} = \frac{10^{-14}}{7.1 \times 10^{-3}} = 1.4 \times 10^{-12}$. $K_{a2} = 6.3 \times 10^{-8}$ Because K_{a2} is larger than K_{b1}, KH_2PO_4 is more likely to lose a H^+ than to gain one. The solution is acidic. pH < 7
(e) $Ba(OH)_2$ (aq) will be basic.
(f) NO_2^- is the conjugate base of a weak acid. The solution is basic.

79. $HPO_4^{2-} + H_2O \rightleftharpoons PO_4^{3-} + H_3O^+$ K_{a3}

$HPO_4^{2-} + H_2O \rightleftharpoons H_2PO_4^- + OH^-$ K_b

$K_{a3} = 4.3 \times 10^{-13}$ $K_b = \frac{K_w}{K_{a2}} = \frac{10^{-14}}{6.3 \times 10^{-8}} = 1.6 \times 10^{-7}$

Basic, because hydrolysis of HPO_4^{2-} occurs more extensively than its ionization as an acid.

81. $[H_3O^+] = K_a \times \frac{[acid]}{[base]} = 1.8 \times 10^{-4} \times \frac{0.405}{0.326}$

$[H_3O^+] = 2.24 \times 10^{-4}$ M

pH = 3.65

83. $[H_3O^+] = 10^{-10.05} = 8.91 \times 10^{-11}$ M

$$8.91 \times 10^{-11} = \frac{10^{-14}}{1.8 \times 10^{-5}} \frac{[\text{acid}]}{[0.350]}$$

[acid] $= 5.6 \times 10^{-2}$ M

$$? \text{ g}(NH_4)_2SO_4 = 5.6 \times 10^{-2} \text{ M} \times 0.100\text{L} \times \frac{\text{mol }(NH_4)_2SO_4}{2 \text{ mol } NH_4^+} \times \frac{132 \text{ g }(NH_4)_2SO_4}{\text{mol }(NH_4)_2SO_4}$$

$= 0.37$ g$(NH_4)_2SO_4$

The solution is so basic that a negligible amount of SO_4^{2-} will become HSO_4^-, and that reaction can be ignored.

85. $HCOO^- + H_3O^+ \rightleftharpoons HCOOH + H_2O$

	$HCOO^-$	H_3O^+	$HCOOH$
	0.326 M × 50.0 mL	1.00 mL × 0.250 M	0.405 M × 50.0 mL
initial:	16.3 mmol	0.250 mmol	20.3 mmol
equil:	16.0	--	20.6

$$[H_3O^+] = 1.8 \times 10^{-4} \times \frac{20.6}{16.0} = 2.32 \times 10^{-4}$$

pH = 3.63

87. No. The components need to be a base and its *conjugate* acid or an acid and its *conjugate* base. The HCl and NaOH would neutralize one another, leaving NaCl (aq) with either excess HCl or excess NaOH, not a buffer solution.

89. The pH range from just slightly before to just slightly beyond the equivalence point is longer for a strong acid–strong base titration than for a weak acid–strong base titration. More indicators are available that change color in this longer pH range.

91. (a) NH_4Cl: $NH_4^+ + H_2O \rightleftharpoons NH_3 + H_3O^+$

$$K_a = \frac{K_w}{K_b} = \frac{1 \times 10^{-14}}{1.8 \times 10^{-5}} = \frac{[H_3O^+]^2}{[NH_4^+]} = \frac{[H_3O^+]^2}{(0.10 \text{ M})}$$

$[H_3O^+] = 7.45 \times 10^{-6}$ M

pH = 5.13 yellow

(b) CH_3COOK: $CH_3COO^- + H_2O \rightleftharpoons CH_3COOH + OH^-$

$$K_b = \frac{K_w}{K_a} = \frac{10^{-14}}{1.8 \times 10^{-5}} = \frac{[OH^-]^2}{[CH_3COO^-]} = \frac{[OH^-]^2}{(0.10 \text{ M})}$$

$[OH^-] = 7.45 \times 10^{-6}$ M

pOH = 5.13

pH = 8.87 blue-violet

(c) Na_2CO_3: $CO_3^{2-} + H_2O \rightleftharpoons HCO_3^- + OH^-$

$$K_b = \frac{K_w}{K_a} = \frac{10^{-14}}{4.7 \times 10^{-11}} = 2.1 \times 10^{-4} = \frac{[OH^-]^2}{[CO_3^{2-}]} = \frac{[OH^-]^2}{(0.10 \text{ M})}$$

$[OH^-] = 4.58 \times 10^{-3}$ M

pOH = 2.34

pH = 11.66 red

(d) CH_3COOH: $CH_3COOH + H_2O \rightleftharpoons CH_3COO^- + H_3O^+$

$$K_a = 1.8 \times 10^{-5} = \frac{[H_3O^+]^2}{[CH_3COOH]} = \frac{[H_3O^+]^2}{(0.10\ M)}$$

$[H_3O^+] = 1.34 \times 10^{-3}$ M

pH = 2.87 violet

93. The new solution will be pH = 7. The new color will be yellow because the phenolphthalein will be colorless and the thymol blue will be yellow.

95. In a strong base–strong acid titration; (1) the initial pH is high because the base is completely ionized; (2) at the half-neutralization point, the pH depends on the concentration of the base remaining (half has been neutralized); (3) at the equivalence point, the pH is 7.00 because neither cation nor anion ionize; (4) the steep portion of the curve is over a wide range; (5) the choice of indicators is extensive. Any indicator with a color change in the pH range of 4 to 10 will work. The same indicator can be used for either titration because the region of rapid pH change (steep portion of titration curve) is the same in either case; it is just appoached from a different direction. (Something not discussed in the text is that since it is easier to see the color change from a light to a dark color, usually the same indicator is not used for both types of titrations.)

97.

CH_3COOH	+ OH^-	$\rightleftharpoons$ $CH_3COO^- + H_2O$
20.00 mL × 0.500 M	7.45 mL × 0.500 M	--
10.00 mmol	3.73 mmol	
-3.73 mmol	-3.73 mmol	+3.73 mmol
6.27 mmol	--	3.73 mmol

$$[H_3O^+] = K_a \times \frac{[acid]}{[base]} = 1.8 \times 10^{-5} \times \frac{6.27}{3.73} = 3.03 \times 10^{-5}\ M$$

pH = 4.52

99. (a) $NH_3 + HCl \rightarrow NH_4^+ + Cl^-$

$$?\ mL\ HCl = 40.00\ mL \times 0.200\ M\ NH_3 \times \frac{1\ mol\ HCl}{1\ mol\ NH_3} \times \frac{L}{0.500\ mol\ HCl}$$

= 16 mL HCl

(b) $NH_3 + H_2O \rightleftharpoons NH_4^+ + OH^-$

$$K_b = 1.8 \times 10^{-5} = \frac{[OH^-]^2}{0.200}$$

$[OH^-] = 1.90 \times 10^{-3}$ M pOH = 2.72 pH = 11.28

(c)

	NH_3	+ H^+	$\rightarrow$ NH_4^+
	40.00 mL 0.200 M	5.00 mL × 0.500 M	
I	8.00 mmol	2.50 mmol	
C	- 2.50	- 2.50	+ 2.50 mmol
E	5.50 mmol	0	2.50 mmol

$$[H_3O^+] = \frac{10^{-14}}{1.8 \times 10^{-5}} \times \frac{(2.50)}{(5.50)} = 2.53 \times 10^{-10} \text{ M}$$

pH = 9.60

(d) $$[H_3O^+] = \frac{10^{-14}}{1.8 \times 10^{-5}} = 5.56 \times 10^{-10} \text{ M}$$

pH = 9.26

or

$pH = pK_a(NH_4^+) = 14.00 - pK_b(NH_3)$

pH = 14.00- 4.74 = 9.26

(e)

	NH_3	+	H^+	→	NH_4^+
I	8.00 mmol		10.00 mL × 0.500 M		
C	- 5.00		- 5.00		+ 5.00
E	3.00				5.00

$$[H_3O^+] = \frac{1 \times 10^{-14}}{1.8 \times 10^{-5}} \times \frac{(5.00)}{(3.00)} = 9.26 \times 10^{-10} \text{ M}$$

pH = 9.03

(f) $$K_a = \frac{K_w}{K_b} = \frac{10^{-14}}{1.8 \times 10^{-5}} = \frac{[H_3O^+]^2}{[NH_4^+]}$$

$$[NH_4^+] = \frac{8.00 \text{ mmol}}{(40.00 + 16.00) \text{ mL}} = 0.143 \text{ M}$$

$$K_a = \frac{10^{-14}}{1.8 \times 10^{-5}} = \frac{[H_3O^+]^2}{0.143}$$

$[H_3O^+] = 8.91 \times 10^{-6}$

pH = 5.05

(g)

NH_3	+	H^+	→	NH_4^+
8.00 mmol		20.00 mL × 0.500 M		
		10.00 mmol		
- 8.00		- 8.00		+ 8.00
--		2.00 mmol		8.00

$$[H_3O^+] = \frac{2.00 \text{ mmol}}{60.00 \text{ mL}} = 3.33 \times 10^{-2} \text{ M}$$

pH = 1.48

101.

	Acid	Base
	e^- pair acceptor	e^- pair donor
(a)	$Al(OH)_3$	OH^-
(b)	Cu^{2+}	NH_3
(c)	CO_2	OH^-

Additional Problems

103. (a) $[H_3O^+] = K_a \dfrac{[\text{acid}]}{[\text{base}]}$

$$[H_3O^+] = 10^{-4.0} = 1.4 \times 10^{-4} \frac{[\text{acid}]}{[\text{base}]}$$

$$\frac{1.0\times10^{-4}}{1.4\times10^{-4}} = 0.71 = \frac{[\text{acid}]}{[\text{base}]}$$

$$? \text{ moles NaOH} = 2.0 \text{ g NaOH } \frac{\text{mol}}{40.0 \text{ g}} = 0.0500 \text{ mol NaOH}$$

$CH_3CHOHCOOH + OH^- \rightarrow CH_3CHOHCOO^- + H_2O$

0.050 mol NaOH will make 0.0500 mol of conjugate base.

? mol acid = 0.71 × 0.0500 mol NaOH = 0.036 mol acid

? mol acid total = mol reacted with NaOH + mol as acid

? mol acid = 0.050 mol + 0.036 mol = 0.086 mol

$$? \text{ g lactic acid} = 0.086 \text{ mol} \times \frac{90.0 \text{ g}}{\text{mol}} = 7.7 \text{ g lactic acid}$$

(b) $$\text{pH} = \text{p}K\text{a} + \log \frac{[\text{base}]}{[\text{acid}]}$$

$$3.80 = 4.74 + \log \frac{[\text{OAc}^-]}{[\text{HOAC}]}$$

$$-0.94 = \log \frac{[\text{OAc}^-]}{[\text{HOAC}]}$$

$$0.115 = \frac{[\text{OAc}^-]}{[\text{HOAC}]}$$

Also $[\text{HOAc}] + [\text{OAc}^-] = 0.0500$ M

$$0.115 = \frac{[\text{OAc}^-]}{[0.0500 - [\text{OAc}]}$$

$0.00575 - 0.115\,[\text{OAc}^-] = [\text{OAc}^-]$

$1.115\,[\text{OAc}^-] = 0.00575$

$[\text{OAc}^-] = 0.0052$ M

$[\text{HOAc}] = 0.0448$ M

$$0.0052 \text{ mol OAc}^- \times \frac{\text{mol NaOH}}{\text{mol OAc}^-} \times \frac{40.00 \text{ g}}{\text{mol}} = 0.21 \text{ g NaOH}$$

$$0.0500 \text{ M} \times 1.00 \text{ L} \times \frac{60.05 \text{ g}}{\text{mol}} = 3.00 \text{ g acetic acid}$$

106. $$[\text{HZ}] = \frac{10.0 \text{ mL} \times 0.0900 \text{ M}}{10.0 \text{ M} + 20.0 \text{ mL}} = 0.0300 \text{ M}$$

$$[\text{Z}^-] = \frac{20.0 \text{ mL} \times 0.150 \text{ M}}{30.0 \text{ mL}} = 0.100 \text{ M}$$

$$[H_3O^+] = 10^{-5.90} = 1.3 \times 10^{-6} = K_a \frac{[0.0300]}{[0.100]}$$

$K_a = 4.3 \times 10^{-6}$
$pK_a = 5.36$

108. Yes, as long as $[H_3O^+][OH^-] = K_w$.
$2[OH^-][OH^-] = 10^{-14}$
$[OH^-]^2 = 5 \times 10^{-15}$
$[OH^-] = 7.07 \times 10^{-8}$ M
$[H_3O^+] = 1.41 \times 10^{-7}$ M
Yes, as long as pH + pOH = 14.
2pOH + pOH = 14
3pOH = 14
pOH = 4.67
$[OH^-] = 2.14 \times 10^{-5}$ M
pH = 9.33
$[H_3O^+] = 4.68 \times 10^{-10}$ M
The solutions having $[H_3O^+] = 2 \times [OH^-]$ and pH = 2 × pOH are not the same. The first is nearly pH neutral, whereas the second is more basic.

110. $H_2SO_4 + H_2O \longrightarrow H_3O^+ + HSO_4^-$

$HSO_4^- + H_2O \longrightarrow H_3O^+ + SO_4^-$

	HSO_4^-	H_3O^+	SO_4^-
I:	0.020	0.020	—
C:	$-y$	$+y$	$+y$
E:	$0.020 - y$	$0.020 + y$	y

$$K_{a2} = 1.1 \times 10^{-2} = \frac{(0.020 + y)(y)}{0.020 - y}$$

$2.2 \times 10^{-4} - 1.1 \times 10^{-2}\, y = 0.020\, y + y^2$
$y^2 + 0.031\ y - 2.2 \times 10^{-4} = 0$

$$y = \frac{-0.031 \pm \sqrt{(0.031)^2 - 4 \times 1 \times (-2.2 \times 10^{-4})}}{2}$$

$$y = \frac{-0.031 \pm 0.043}{2}$$

$y = 6.0 \times 10^{-3}$ M $\qquad$ -3.7×10^{-3} M Negative values are not valid.
$[H_3O^+]_{total} = [H_3O^+]$ from K_{a1} + $[H_3O^+]$ from K_{a2}
$[H_3O^+]_{total} = 0.020\ \text{M} + 6.0 \times 10^{-3}\ \text{M} = 0.026$ M
This compares closely to the estimated answer of 0.025 M.

113 $NaHSO_4 + NaOH \rightarrow Na_2SO_4 + HOH$

$$?\ \text{g } NaHSO_4 = 36.56\ \text{mL} \times \frac{10^{-3}\ \text{L}}{\text{mL}} \times 0.225\ \text{M} \times \frac{\text{mol } NaHSO_4}{\text{mol NaOH}} \times \frac{120.1\ \text{g}}{\text{mol}} = 0.988\ \text{g}$$

$$? \% \, NaHSO_4 = \frac{0.988 \text{ g}}{1.016 \text{ g}} \times 100 \% = 97.2 \% \, NaHSO_4$$

? % NaCl = 100.0 % – 97.2 % = 2.8 % NaCl

With a K_a of 1.1×10^{-2} a titration of HSO_4^- will reach the equivalence point close to 7 as shown below.

$$?\text{mol } HSO_4^- = \text{mol } SO_4^{2-} = \frac{0.988 \text{ g}}{120.1 \text{ g/mol}} = 8.22 \times 10^{-3} \text{ mol}$$

$$V_{eq.pt.} = 2 \times V_{base} = 2 \times 36.56 \text{ mL} = 73.12 \text{ mL}$$

$$[SO_4^{2-}] = \frac{8.22 \times 10^{-3} \text{ mol}}{73.12 \text{ mL}} \times \frac{\text{mL}}{10^{-3} \text{ L}} = 0.112 \text{ M}$$

$$K_b = \frac{K_w}{K_a} = \frac{1.0 \times 10^{-14}}{1.1 \times 10^{-2}} = 9.1 \times 10^{-13} = \frac{X^2}{0.112}$$

$X = 3.2 \times 10^{-7} = [OH^-]$

pOH = 6.49

pH = 7.51

Bromthymol blue, phenol red, and other indicators that change close to pH 7 are suitable.

116. (a) $$n = \frac{PV}{RT} = \frac{316 \text{ Torr} \times 275 \text{ mL}}{\frac{62.36 \text{ L Torr}}{\text{K mol}} \times 298 \text{ K}} \frac{10^{-3} \text{ L}}{\text{mL}} = 4.68 \times 10^{-3} \text{ moles}$$

$$[\text{1-propylamine}] = \frac{4.68 \times 10^{-3} \text{ M}}{0.500 \text{ L}} = 9.35 \times 10^{-3} \text{ M}$$

$$C_3H_7NH_2 + H_2O \rightarrow C_3H_7NH_3^+ + OH^-$$

$$K_b = 10^{-3.43} = \frac{[OH^-][C_3H_7NH_3^+]}{[C_3H_7NH_2]}$$

$$3.72 \times 10^{-4} = \frac{[OH^-]^2}{9.35 \times 10^{-3} - [OH^-]}$$

$$[OH^-]^2 + 3.72 \times 10^{-4} [OH^-] - 3.48 \times 10^{-6}$$

$$[OH^-] = \frac{-3.72 \times 10^{-4} \pm \sqrt{(3.72 \times 10^{-4})^2 - 4 \times (-3.48 \times 10^{-6})}}{2}$$

$$[OH^-] = \frac{-3.72 \times 10^{-4} \pm 3.75 \times 10^{-3}}{2}$$

$[OH^-] = 1.69 \times 10^{-3}$ M or -2.06×10^{-3} M Not Valid

pOH = 2.77

pH = 11.23

(b) $$? \text{ mg NaOH} = 1.7 \times 10^{-3} \text{ M} \times 0.500 \text{ L} \times \frac{40.00 \text{ g}}{\text{mole}} \times \frac{\text{mg}}{10^{-3} \text{ g}} = 34 \text{ mg NaOH}$$

117. (a) The pH starts high and decreases during the titration. The solution being titrated is a base, and the titrant is an acid. The base must be a strong base; the initial pH is near 14. The equivalence point is pH $\approx$ 8.8, so the acid is a weak acid.
Also, the final pH remains fairly high (about 4), indicating that the titrant is a weak acid.
(b) At twice the equivalence point the concentration of weak acid and conjugate base will be equal and pH = p$K_a \approx$ 3.8. $K_a \approx 1.6 \times 10^{-4}$.
(c) From the graph the 2< pH <9 at the equivalence point. By calculation the pH = 8.8 (assuming pK_a = 3.8 for the titrant).

$$A^- + H_2O \rightleftharpoons HA + OH^-$$

$$\frac{K_w}{K_a} = \frac{[OH^-]^2}{[A^-]}$$

$$\frac{10^{-14}}{1.6 \times 10^{-4}} = \frac{[OH^-]^2}{0.50}$$

$[A^-]$ = 0.50 M because equal volumes of 1.00 M acid and 1.00 M base are required to reach the equivalence point.

$[OH^-] = 5.6 \times 10^{-6}$

pOH = 5.25

pH = 14 - 5.25 = 8.75

120. $$[\text{o-phthalic}] = \frac{0.6\text{ g}}{100\text{ mL}} \times \frac{\text{mL}}{10^{-3}\text{L}} \times \frac{\text{mol}}{166\text{ g}} = 3.6 \times 10^{-2}\text{ M}$$

$$\text{o-phthalic} + H_2O \rightleftharpoons H_3O^+ + \text{hydrogen o-phthalate}^-$$

$$K_{w1} = \frac{[H_3O^+][\text{Ho-phthalic}^-]}{[\text{o-phthalic}]} = \frac{[H_3O^+]^2}{[\text{o-phthalic}]}$$

$$K_{a1} = \frac{(10^{-2.33})^2}{3.6\times10^{-2} - 10^{-2.33}} = \frac{(10^{-2.33})^2}{3.6\times10^{-2} - 4.7\times10^{-3}} = 7.1 \times 10^{-4}$$

$$pK_{a1} = 3.15$$

$$pH = \frac{1}{2}(pK_{a1} + pK_{a2})$$

$$4.19 = \frac{1}{2}(3.34 + pK_{a2})$$

$$pK_{a2} = 5.23$$

122. (a) 2 moles acetic acid and 1 mole of sodium acetate in 2.00 L of solution.
(b) 3 moles acetic acid and 1 mole of NaOH in 2.00 L of solution.
(c) 3 moles sodium acetate and 2 moles hydrochloric acid in 2.00 L of solution.
In (a) and (b) there are acetic acid molecules, acetate ions, sodium ions, and water. In (c) there is also 1 M NaCl, and the high ionic strength may have a small effect on the pH.

125. $CH_3COOH + H_2O \rightleftharpoons CH_3COO^- + H_3O^+$

$$K_a = 1.8 \times 10^{-5} = \frac{[CH_3COO^-][H_3O^+]}{0.250\text{ M}}$$

$HCOOH + H_2O \rightleftharpoons HCOO^- + H_3O^+$

$$K_a = 1.8 \times 10^{-4} = \frac{[HCOO^-][H_3O^+]}{0.150\text{ M}}$$

$$\frac{4.5 \times 10^{-6}}{[H_3O^+]} = [CH_3COO^-]$$

$$\frac{2.7 \times 10^{-5}}{[H_3O^+]} = [HCOO^-]$$

$$[CH_3COO^-] + [HCOO^-] = [H_3O^+] = \frac{4.5 \times 10^{-6}}{[H_3O^+]} + \frac{2.7 \times 10^{-5}}{[H_3O^+]}$$

$[H_3O^+]^2 = 3.15 \times 10^{-5}$

$[H_3O^+] = 5.61 \times 10^{-3}$

pH = 2.25

Apply Your Knowledge

127. $$[HCl] = \frac{0.10\text{ mL} \times 0.05\text{ M}}{100.0\text{ mL}} = 5 \times 10^{-5}$$

pH = 4.3 instead of 7.

Any residual drops of solution on the electrode would change the pH of the water. A buffer having pH = 7 would react with the residual drops and keep the pH at 7.

One appropriate buffer would be $H_2PO_4^-$ and HPO_4^{2-} in the ratio of $\frac{[HPO_4^{2-}]}{[H_2PO_4^-]} =$ 0.63.

$$pH = pK_{a2} + \log\frac{[HPO_4^{2-}]}{[H_2PO_4^-]} = 7.00$$

$$pH = 7.20 + \log\frac{[HPO_4^{2-}]}{[H_2PO_4^-]} = 7.00$$

$$\frac{[HPO_4^{2-}]}{[H_2PO_4^-]} = 10^{-0.20} = 0.63$$

129. $CaO\ (s) + H_2O\ (l) \rightarrow Ca(OH)_2\ (aq) \rightarrow Ca^{2+}\ (aq) + 2\ OH^-\ (aq)$

$$1.0\text{ ton} \times \frac{2000\text{ lb}}{\text{ton}} \times \frac{454\text{ g}}{\text{lb}} \times \frac{72\text{ g}}{100\text{ g}} \times \frac{\text{mL}}{\text{g}} \times \frac{\text{L}}{1000\text{ mL}} = 654\text{ L solution}$$

$[H_3O^+] = 1 \times 10^{-7}$ in pure water. The amount of H_3O^+ that must be neutralized to raise the pH from 6.0 to 7.0 is:

$$654\text{ L} \times \left(\frac{1 \times 10^{-6}\text{ mol } H_3O^+}{\text{L}} - \frac{1 \times 10^{-7}\text{ mol } H_3O^+}{\text{L}}\right) = 5.9 \times 10^{-4}\text{ mol } H_3O^+$$

(a negligible amount)

To raise the pH of water from 7.0 to 12.0:

$[OH^-] = 10^{-(14.00 - 12.00)} = 10^{-2}$ M

$$10^{-2}\text{ M} \times 654\text{ L} = 6.54\text{ moles } OH^- \times \frac{\text{mole CaO}}{2\text{ mole } OH^-} \times \frac{56.08\text{ g}}{\text{mole}} = 1.8 \times 10^2\text{ g CaO}$$

130. At the equivalence point. The solution consists of the salt (NaP) of the weak monoprotic acid (HP). The value of $[P^-]$ at the equivalence point could be calculated from the titration and dilution data, but this is not necessary. Addition of a second 25.00-mL portion of the weak acid at the equivalence point produces a solution in which $[HP] = [P^-]$, no matter what their particular value. In a solution with $[HP] = [P^-]$, $pH = pK_a + \log\frac{[P^-]}{[HP]} = pK_a + \log 1 = pK_a = 3.84$. Thus, $[H_3O^+] = K_a = 10^{-3.84} = 1.4 \times 10^{-4}$.

133. $$\left(\frac{0.759\text{ mL } CO_2}{\text{mL } H_2O} = \frac{0.759\text{ L } CO_2}{\text{L } H_2O}\right)$$

$$?\text{ M} = \frac{0.759\text{ L } CO_2}{\text{L } H_2O} \times \frac{\text{mol}}{22.4\text{L}} = 0.034\text{ M}$$

$$k = \frac{S}{P_{gas}} = \frac{\text{concentration}}{P} = \frac{0.034\text{ M}}{1\text{ atm}} = \frac{[H_2CO_3]}{0.00037\text{ atm}}$$

$[H_2CO_3] = 1.26 \times 10^{-5}$ M

$$K_{a1} = 4.4 \times 10^{-7} = \frac{[H_3O^+]^2}{1.26 \times 10^{-5} - [H_3O^+]}$$

$[H_3O^+]^2 + 4.4 \times 10^{-7}[H_3O^+] - 5.5 \times 10^{-12} = 0$

$$[H_3O^+] = \frac{-4.4 \times 10^{-7} \pm \sqrt{(4.4 \times 10^{-7})^2 - 4 \times 1 \times (-5.5 \times 10^{-12})}}{2}$$

$[H_3O^+] = 2.1 \times 10^{-6}$ M or -2.58×10^{-6} not valid

pH = 5.68

Chapter 16

More Equilibria in Aqueous Solutions: Slightly Soluble Salts and Complex Ions

Exercises

16.1A (a) $MgF_2(s) \rightleftharpoons Mg^{2+}(aq) + 2\,F^-(aq)$ $\quad K_{sp} = [Mg^{2+}][F^-]^2$

(b) $Li_2CO_3(s) \rightleftharpoons 2\,Li^+(aq) + CO_3^{2-}(aq)$ $\quad K_{sp} = [Li^+]^2[CO_3^{2-}]$

(c) $Cu_3(AsO_4)_2(s) \rightleftharpoons 3\,Cu^{2+}(aq) + 2\,AsO_4^{3-}(aq)$ $\quad K_{sp} = [Cu^{2+}]^3[AsO_4^{3-}]^2$

16.1B (a) $Mg(OH)_2(s) \rightleftharpoons Mg^{+2}(aq) + 2\,OH^-(aq)$ $\quad K_{sp} = [Mg^{2+}][OH^-]^2$

(b) $ScF_3(s) \rightleftharpoons Sc^{3+}(aq) + 3\,F^-(aq)$ $\quad K_{sp} = [Sc^{3+}][F^-]^3$

(c) $Zn_3(PO_4)_2(s) \rightleftharpoons 3\,Zn^{2+}(aq) + 2\,PO_4^{3-}(aq)$ $\quad K_{sp} = [Zn^{2+}]^3[PO_4^{3-}]^2$

16.2A $s = 1.7 \times 10^{-4}$ M $Mg(OH)_2$

$Mg(OH)_2(s) \rightleftharpoons Mg^{+2}(aq) + 2\,OH^-(aq)$

$\qquad\qquad s \qquad\qquad 2s$

$K_{sp} = [Mg^{+2}][OH^-]^2 = s\,(2s)^2 = 4s^3$

$K_{sp} = 4 \times (1.7 \times 10^{-4})^3 = 2.0 \times 10^{-11}$

16.2B $[Ag^+] = \dfrac{14\text{ g }Ag^+}{10^6\text{g solution}} \times \dfrac{1\text{ g}}{\text{mL}} \times \dfrac{\text{mL}}{10^{-3}\text{L}} \times \dfrac{\text{mol}}{108\text{ g}} = 1.3 \times 10^{-4}$ M

$Ag_2CrO_4(s) \rightleftharpoons 2\,Ag^+(aq) + CrO_4^{2-}(aq)$

$[CrO_4^{2-}] = \frac{1}{2}[Ag^+] = 6.5 \times 10^{-5}$

$K_{sp} = [Ag^+]^2[CrO_4^{2-}] = (1.3 \times 10^{-4})^2(6.5 \times 10^{-5})$

$K_{sp} = 1.1 \times 10^{-12}$

16.3A $Ag_3AsO_4(s) \rightleftharpoons 3Ag^+(aq) + AsO_4^{3-}(aq)$

$\qquad\qquad 3s \qquad\qquad s$

$K_{sp} = [Ag^+]^3[AsO_4^{3-}] = 1.0 \times 10^{-22}$

$27\,s^4 = 1.0 \times 10^{-22}$

$s = 1.4 \times 10^{-6}$ M

16.3B $PbI_2(s) \rightleftharpoons Pb^{2+}(aq) + 2\,I^-(aq)$

$\frac{1}{2}X \qquad X$

$K_{sp} = 7.1 \times 10^{-9} = \frac{1}{2}X(X)^2 = \frac{1}{2}X^3$

$X = 2.4 \times 10^{-3}$ M I^-

$$? \text{ ppm } I^- = \frac{2.4 \times 10^{-3} \text{ mol}}{\text{L}} \times \frac{10^{-3} \text{ L}}{\text{mL}} \times \frac{\text{mL}}{1.00 \text{ g}} \times \frac{127 \text{ g } I^-}{\text{mol}} \times \frac{10^6 \text{ g ppm}}{\text{g}}$$

$$= 3.1 \times 10^2 \text{ ppm}$$

OR

$$? \text{ ppm } I^- = \frac{2.4 \times 10^{-3} \text{ mol}}{\text{L}} \times \frac{127 \text{ g } I^-}{\text{mol}} \times \frac{10^{-3} \text{ L}}{\text{mL}} \times \frac{\text{mL}}{1.00 \text{ g}} \times \frac{1.00 \times 10^6}{1.00 \times 10^6}$$

$$= \frac{3.1 \times 10^2 \text{ g } I^-}{1000000 \text{ g}} = 3.1 \times 10^2 \text{ ppm}$$

16.4A The solutes are all of the same type, MX_2. Their molar solubilities parallel their K_{sp} values:

CaF_2	<	PbI_2	<	MgF_2	<	$PbCl_2$
5.3×10^{-9}		7.1×10^{-9}		3.7×10^{-8}		1.6×10^{-5}

16.4B The square root of $K_{sp\ BaSO_4} \approx 10^{-5}$. The third root of both CaF_2 and PbI_2 should be $\approx 10^{-3}$. The molar solubility of PbI_2 is larger than that of CaF_2 because the K_{sp} is greater.

$PbI_2 > CaF_2 > BaSO_4$

$BaSO_4 \rightleftharpoons Ba^{2+}(aq) + SO_4^{2-}(aq)$

$s \qquad s \qquad s$

$s^2 = 1.1 \times 10^{-10}$

$s = 1.0 \times 10^{-5}$ M

$CaF_2 \rightleftharpoons Ca^{2+}(aq) + 2F^-(aq)$

$s \qquad s \qquad 2s$

$4s^3 = 5.3 \times 10^{-9}$

$s = 1.1 \times 10^{-3}$ M

$PbI_2 \rightleftharpoons Pb^{2+}(aq) + 2I^-(aq)$

$s \qquad s \qquad s$

$4s^3 = 7.1 \times 10^{-9}$

$s = 1.2 \times 10^{-3}$ M

16.5A $Ag_2SO_4(s) \rightleftharpoons 2\,Ag^+(aq) + SO_4^{2-}(aq)$

$1.00 + 2s \qquad s$

$K_{sp} = 1.4 \times 10^{-5} = (1.00 + 2s)^2\,s$

assume 1.00 >> 2s

$1.4 \times 10^{-5} = 1.00\, s$
$s = 1.4 \times 10^{-5}$ M (Assumption is valid.)

16.5B $Ag_2SO_4(s) \rightleftharpoons 2\, Ag^+(aq) + SO_4^{2-}(aq)$
$2\,s + y \qquad +s$

$s = 1.0 \times 10^{-3}$ M
$K_{sp} = 1.4 \times 10^{-5} = (2.0 \times 10^{-3} + y)^2 (1.0 \times 10^{-3})$
$1.4 \times 10^{-5} = (4.0 \times 10^{-6} + 4.0 \times 10^{-3}\, y + y^2)(1.0 \times 10^{-3})$
$1.4 \times 10^{-5} = 4.0 \times 10^{-9} + 4.0 \times 10^{-6}\, y + 1.0 \times 10^{-3}\, y^2$
$1.0 \times 10^{-3}\, y^2 + 4.0 \times 10^{-6}\, y - 1.4 \times 10^{-5} = 0$
$y^2 + 4.0 \times 10^{-3}\, y - 1.4 \times 10^{-2} = 0$

$$y = \frac{-4.0\times10^{-3} \pm \sqrt{(4.0\times10^{-3})^2 - 4\times1\times(-1.4\times10^{-2})}}{2}$$

$$y = \frac{-4.0\times10^{-3} \pm 0.237}{2}$$

y = 0.117 M or -0.12 M (Negative values are not valid.)

$$?\text{ g } AgNO_3 = 0.117\text{ M} \times 250.0\text{ mL} \times \frac{10^{-3}\text{ L}}{\text{mL}} \times \frac{170\text{ g}}{\text{mol}} = 5.0\text{ g } AgNO_3$$

16.6A $$[Pb^{2+}] = \frac{1.00\text{ g}}{1.50\text{ L}} \times \frac{\text{mol}}{331.2\text{ g}} \times \frac{1\text{ mol } Pb^{2+}}{\text{mol } Pb(NO_3)_2} = 2.01 \times 10^{-3}\text{ M}$$

$$[I^-] = \frac{1.00\text{ g}}{1.50\text{ L}} \times \frac{\text{mol}}{278.1\text{ g}} \times \frac{2\text{ mol } I^-}{\text{mol } MgI_2} = 4.79 \times 10^{-3}\text{ M}$$

$Q_{ip} = (2.01 \times 10^{-3})(4.79 \times 10^{-3})^2 = 4.61 \times 10^{-8}$
$Q_{ip} > K_{sp}$ A precipitate should form.

16.6B $[OH^-] = 10^{-(14.00 - 10.35)} = 2.2 \times 10^{-4}$ M

$$[Mg^{2+}] = \frac{2.5\text{ g } MgCl_2}{10^6\text{ g solution}} \times \frac{1.00\text{ g}}{\text{mL}} \times \frac{\text{mL}}{10^{-3}\text{ L}} \times \frac{\text{mol } MgCl_2}{95.2\text{ g } MgCl_2} \times \frac{\text{mol } Mg^{2+}}{\text{mol } MgCl_2}$$
$$= 2.6 \times 10^{-5}\text{ M}$$

$Q_{ip} = [Mg^{2+}][OH^-]^2 = (2.6 \times 10^{-5})(2.2 \times 10^{-4})^2$
$Q_{ip} = 1.3 \times 10^{-12}$
$Q_{ip} < K_{sp}$ No precipitate.

16.7A Add KI(aq) dropwise from a buret to a known volume of solution of known $[Pb^{2+}]$. Stir after each drop is added, observing first the appearance and then disappearance of $PbI_2(s)$. Continue until a single drop produces a lasting precipitate. At this point, $Q_{iP} = K_{sp}$. Calculate K_{sp} from the $[Pb^{2+}]$ and $[I^-]$ accounting for dilution effects.

16.7B (a) for 4 drops:

$$[I^-] = \frac{4\text{ drops} \times \frac{\text{mL}}{20\text{ drops}} \times 0.10\text{ M}}{100.0\text{ mL} + 4\text{ drops} \times \frac{\text{mL}}{20\text{ drops}}} = 2.0 \times 10^{-4}\text{ M}$$

$$[Pb^{2+}] = \frac{100.0\text{ mL} \times 1 \times 10^{-4}\text{ M}}{100.2\text{ mL}} = 1 \times 10^{-4}\text{ M}$$

$Q_{ip} = [Pb^{2+}][I^-]^2 = (1 \times 10^{-4})(2.0 \times 10^{-4})^2 = 4 \times 10^{-12}$

$Q_{ip} < K_{sp}$

Yes, the observations will resemble Figure 64. There will be a yellow precipitate at the point of impact which will later dissolve.

(b) $$[Pb^{2+}] = \frac{4\text{ drops} \times \frac{\text{mL}}{20\text{ drops}} \times 0.10\text{ M}}{100.0\text{ mL} + 4\text{ drops} \times \frac{\text{mL}}{20\text{ drops}}} = 2.0 \times 10^{-4}\text{ M}$$

$$[I^-] = \frac{100.0\text{ mL} \times 1 \times 10^{-4}\text{ M}}{100.2\text{ mL}} = 1 \times 10^{-4}\text{ M}$$

$Q_{ip} = [Pb^{2+}][I^-]^2 = (2 \times 10^{-4})(1 \times 10^{-4})^2 = 2 \times 10^{-12}\text{ M}$

$Q_{ip} < K_{sp}$ The same observations are still made.

16.8A Adding equal volumes will mean that the ion concentrations are halved.

$$[Mg^{2+}] = \frac{1}{2} \times \frac{0.0010\text{ mol } MgCl_2}{\text{L}} \times \frac{1\text{ mol } Mg^{2+}}{1\text{ mol } MgCl_2} = 5.0 \times 10^{-4}\text{ M}$$

$$[F^-] = \frac{1}{2} \times \frac{0.020\text{ mol NaF}}{\text{L}} \times \frac{1\text{ mol } F^-}{1\text{ mol NaF}} = 1.0 \times 10^{-2}\text{ M}$$

$Q_{ip} = (5 \times 10^{-4})(1 \times 10^{-2})^2 = 5 \times 10^{-8} > 3.7 \times 10^{-8}\ K_{sp}$

Yes, precipitation should just occur.

16.8B $K_{sp} = [Pb^{2+}][I^-]^2 = 7.1 \times 10^{-9} = 2.00 \times 10^{-3} \times [I^-]^2$

$[I^-] = 1.9 \times 10^{-3}\text{ M}$

$$?\text{ g KI} = 10.5\text{ L} \times 1.9 \times 10^{-3}\text{ M} \times \frac{\text{mol KI}}{\text{mol } I^-} \times \frac{166\text{ g}}{\text{mol}} = 3.3\text{ g KI}$$

16.9A

	$Ca^{2+}(aq)$ +	$C_2O_4^{2-}(aq)$ ⇌	$CaC_2O_4(s)$
Init.	0.0050 M	0.0100 M	
Change	– 0.0050 M	– 0.0050 M	
Eq.	--	0.0050 M	

	$CaC_2O_4(s)$ ⇌	$Ca^{2+}(aq)$ +	$C_2O_4^{2-}(aq)$
Init.		0	0.0050 M
Change		s	s
Eq.		s	$0.0050 + s$

$K_{sp} = 2.7 \times 10^{-9} = s \times (0.0050 + s)$
assume $s << 0.0050$
$2.7 \times 10^{-9} = s \times 0.0050$
$s = 5.4 \times 10^{-7}$ M (Assumption is valid.)
$\frac{5.4 \times 10^{-7}}{0.0050} \times 100\% = 0.011\%$
Yes, precipitation is complete.

16.9B An estimate of the completeness of precipitation can be made by comparing Q_{ip} and K_{sp}.
(a) $Q_{ip} = 0.110 \text{ M} \times 0.090 \text{ M} = 9.9 \times 10^{-3}$ $K_{sp} = 9.1 \times 10^{-6}$
(b) $Q_{ip} = 0.12 \text{ M} \times (0.25 \text{ M})^2 = 7.5 \times 10^{-3}$ $K_{sp} = 1.6 \times 10^{-5}$
(c) $MgSO_4$ is soluble and will be the farthest from completeness.
(d) $Q_{ip} = 0.050 \text{ M} \times 0.055 \text{ M} = 2.8 \times 10^{-3}$ $K_{sp} = 1.8 \times 10^{-10}$
For (d), $Q_{ip} >> K_{sp}$ and it is the closest to completeness. Comparing (a) and (b), the ratio of $\frac{Q_{ip}}{K_{sp}}$ is greater for (a) is greater than (b).
(c) Mg^{2+} > (b) Pb^{2+} > (a) Ca^{2+} > (d) Ag^{+}
Exact calculations are below.

(a) $Ca^{2+}(aq) + SO_4^{2-}(aq) \rightarrow CaSO_4(s)$

	Ca^{2+}	SO_4^{2-}
Init:	0.110 M	0.090 M
Change:	0.090 M	0.090 M
After:	0.020 M	–

$CaSO_4(s) \rightleftharpoons Ca^{2+}(aq) + SO_4^{2-}(aq)$

	Ca^{2+}	SO_4^{2-}
	0.020 M	–
	s	s
	$0.020 + s$	s

$9.1 \times 10^{-6} = (0.020 + s)s$
$s = 4.6 \times 10^{-4}$ $\frac{4.6 \times 10^{-4}}{0.110 \text{ M}} \times 100\% = 0.41\%$

(b) $Pb^{2+}(aq) + 2Cl^{-}(aq) \rightarrow PbI_2$

	Pb^{2+}	Cl^{-}
Init:	0.12 M	0.25 M
Change:	−0.12 M	−0.24 M
After:	–	0.01 M

$PbI_2 \rightleftharpoons Pb^{2+}(aq) + 2Cl^{-}(aq)$

	Pb^{2+}	Cl^{-}
	–	0.01 M
	s	$2s$
	s	$2s + 0.01$

$1.6 \times 10^{-5} = (s)(2s + 0.01)^2 \approx s(0.01)^2$
$s = 0.16$ $\frac{0.16}{0.12 \text{ M}} \times 100\% = 1.3 \times 10^{2}\%$
(c) $Mg^{2+}(aq) + SO_4^{2-}(aq) \rightarrow MgSO_4(aq)$

$MgSO_4$ is soluble

(d)	$Ag^+(aq)$ +	$Cl^-(aq) \rightarrow AgCl(s)$	
Init:	0.050 M	0.055 M	
Change:	0.050 M	0.050 M	
After:	–	0.005 M	
$AgCl(s)$	$\rightleftharpoons$	$Ag^+(aq)$+	$Cl^-(aq)$
		–	0.005 M
		s	s
		s	$0.005 + s$

$1.8 \times 10^{-10} = (s)(0.005 + s)$

$s = 3.6 \times 10^{-8}$ $\quad \dfrac{3.6\times10^{-8}}{0.050\text{ M}} \times 100\% = 7.2 \times 10^{-5}\,\%$

(c) Mg^{2+} > (b) Pb^{2+} > (a) Ca^{2+} > (d) Ag^+

16.10A $AgCl(s) \rightleftharpoons Ag^+(aq) + Cl^-(aq)$ $\quad K_{sp} = 1.8 \times 10^{-10}$

$AgBr(s) \rightleftharpoons Ag^+(aq) + Br^-(aq)$ $\quad K_{sp} = 5.0 \times 10^{-13}$

(a) Br- precipitates first.

$$[Ag^+] = \frac{K_{sp}}{[Cl^-]} = \frac{1.8\times10^{-10}}{0.0100} = 1.8 \times 10^{-8}\text{ M}$$

$$[Ag^+] = \frac{K_{sp}}{[Br^-]} = \frac{5.0\times10^{-13}}{0.0100} = 5.0 \times 10^{-11}\text{ M}$$

(b) Cl^- begins to precipitate when

$[Ag^+] = 1.8 \times 10^{-8}$ M

$$[Br^-] = \frac{K_{sp}}{[Ag^+]} = \frac{5.0\times10^{-13}}{1.8\times10^{-8}} = 2.8 \times 10^{-5}\text{ M}$$

(c) Percent of Br left in solution

$$?\ \%\ Br^- = \frac{2.8\times10^{-5}}{0.0100} \times 100\% = 0.28\%$$

Because 0.28% is greater than the 0.1% criterium, the separation is not complete, but it is close.

16.10B $AgBr(s) \rightleftharpoons Ag^+(aq) + Br^-(aq)$ $\quad K_{sp} = 5.0 \times 10^{-13}$

$AgI(s) \rightleftharpoons Ag^+(aq) + I^-(aq)$ $\quad K_{sp} = 8.5 \times 10^{-17}$

The selective precipitation of Br^- and Cl^- barely failed. There is a greater difference between the K_{sp} values of AgBr and AgI than between AgCl and AgBr; thus, the selective precipitation of iodide ion as AgI(s) will probably work.

16.11A $Fe(OH)_2(s) \rightleftharpoons Fe^{2+}(aq) + 2\ OH^-(aq)$

s $\qquad 10^{-(14.00\ -\ 6.50)}$

$K_{sp} = 8.0 \times 10^{-16} = (s)(3.16 \times 10^{-8})^2$

$s = [Fe^{2+}] = 0.80\ M$ = molar solubility of $Fe(OH)_2(s)$

16.11B $[H^+] = K_a \dfrac{\text{acid}}{\text{base}} = 1.8 \times 10^{-5} \times \dfrac{0.520}{0.180} = 5.2 \times 10^{-5}\ M$

$$[OH^-] = \frac{K_w}{[H^+]} = \frac{10^{-14}}{5.2 \times 10^{-5}} = 1.9 \times 10^{-10}\ M$$

$Q_{ip} = [Fe^{3+}][OH^-]^3 = (1.0 \times 10^{-3}) \times (1.9 \times 10^{-10})^3$

$Q_{ip} = 6.9 \times 10^{-33} > 4 \times 10^{-38} = K_{sp}$

No, Fe^{3+} cannot exist in the buffer solution without precipitation occurring.

16.12A $Mg(OH)_2(s) \rightarrow Mg^{2+}(aq) + 2\ OH^-(aq)$

$2\ NH_4^+(aq) + 2\ H_2O(l) + 2\ Cl^-(aq) \rightarrow 2\ NH_3(aq) + 2\ H_3O^+(aq) + 2\ Cl^-(aq)$

$\underline{2\ H_3O^+(aq) + 2\ OH^-(aq) \rightarrow 4\ H_2O(l)}$

$Mg(OH)_2(s) + 2NH_4Cl(aq) \rightarrow MgCl_2(aq) + 2H_2O(l) + 2NH_3(aq)$

16.12B in 1.00 M Na_2SO_3:

$SrSO_3\ (s) \rightleftharpoons Sr^{2+} + SO_3^{2-}$ $\quad K_{sp} = 4 \times 10^{-8}$

$\quad s \quad (s + 1.00)$

$4 \times 10^{-8} = (s)(s + 1.00)$

$s = 4 \times 10^{-8}\ M$

in water:

$s^2 = 4 \times 10^{-8}$

$s = 2 \times 10^{-4}\ M$

in 1.00 M NH_3:

$NH_3 + H_2O \rightleftharpoons NH_4^+ + OH^-$

$$1.8 \times 10^{-5} = \frac{[OH^-]^2}{1.00}$$

$[OH^-]^2 = 1.8 \times 10^{-5}$

$[OH^-] = 4.2 \times 10^{-3}\ M$

$$[H^+] = \frac{10^{-14}}{4.2 \times 10^{-3}} = 2.4 \times 10^{-12}\ M$$

$SrSO_3(s) \rightleftharpoons Sr^{2+} + SO_3^{2-}$ $\quad 4 \times 10^{-8}$

$\underline{SO_3^{2-} + H^+ \rightleftharpoons HSO_3^-}$ $\quad \underline{\dfrac{1}{6.2 \times 10^{-8}}}$

$SrSO_3\ (s) + H^+ \rightleftharpoons HSO_3^- + Sr^{2+}$ $\quad K = 6.5 \times 10^{-1}$

$$\frac{s^2}{[H^+]} = \frac{s^2}{2.4 \times 10^{-12}} = 6.5 \times 10^{-1}$$

$s = 1.2 \times 10^{-6}\ M$

in 1.00 M NH_4NO_3:

$$\frac{10^{-14}}{1.8\times10^{-5}} = \frac{[H^+]^2}{1.00} \qquad [H^+] = 2.4 \times 10^{-5}\ M$$

$$\frac{s^2}{[H^+]} = \frac{s^2}{2.4\times10^{-5}} = 6.5 \times 10^{-1}$$

$s = 3.9 \times 10^{-3}$ M

Strontium sulfite is more soluble in NH_4NO_3, less in H_2O, even less in NH_3, and least in Na_2SO_3. NH_4NO_3 is acidic and the reaction to form HSO_3^- causes $SrSO_3$ to be more soluble in NH_4NO_3 than in water. Solubility is decreased in Na_2SO_3 by the common ion effect. Solubility is decreased in NH_3 because NH_3 is basic.

16.13A $Ag^+(aq) + 2\ S_2O_3^{2-}(aq) \rightleftharpoons [Ag(S_2O_3)_2]^{3-}(aq) \qquad K_f = 1.7 \times 10^{13}$

$X \qquad 1.0 - 2(0.10 - X) \qquad 0.10 - X$

Setting $[Ag^+] = X$ is the equivalent of the formation of $[Ag(S_2O_3)_2]^{3-}$ going to completion and the amount X dissociating.

$$K_f = 1.7 \times 10^{13} = \frac{0.10-X}{X(0.80 + 2X)^2}$$

Assume $X << 0.10$

$$1.7 \times 10^{13} = \frac{0.10}{X(0.80)^2}$$

$X = [Ag^+] = 9.2 \times 10^{-15}$ M

16.13B $K_f = \dfrac{[Ag(NH_3)_2]^+}{[Ag^+][NH_3]^2} = 1.6 \times 10^7$

$$\frac{(0.050 - 1.0\times10^{-8})}{(1.0\times10^{-8})[NH_3]^2} = 1.6 \times 10^7$$

$[NH_3] = 0.56$ M

$[NH_3]_{total} = [NH_3]_{free} + [NH_3]_{complex}$

$$[NH_3]_{total} = 0.56\ M + 0.050\ M \times \frac{2\ mol\ NH_3}{mol\ complex} = 0.66\ M$$

16.14A $[I^-] = \dfrac{1.00\ g}{1.00\ L} \times \dfrac{mol}{166.0\ g} = 6.024 \times 10^{-3}$ M

$Q_{ip} = [Ag^+][I^-] = 9.2 \times 10^{-15} \times 6.024 \times 10^{-3} = 5.5 \times 10^{-17}$

$K_{sp} = 8.5 \times 10^{-17} > Q_{ip}$ No, precipitation should not occur.

16.14B $K_f = \dfrac{[Ag(NH_3)_2]^+}{[Ag^+][NH_3]^2} = 1.6 \times 10^7 = \dfrac{0.050}{[Ag^+](1.00)^2}$

$[Ag^+] = 3.1 \times 10^{-9}$ M

$$[Br^-] = \frac{K_{sp}}{[Ag^+]} = \frac{5.0 \times 10^{-13}}{3.1 \times 10^{-9}} = 1.6 \times 10^{-4}\ M$$

$$?\ g\ KBr = 250.0\ mL \times \frac{10^{-3}\ L}{mL} \times 1.6 \times 10^{-4}\ M \times \frac{mol\ KBr}{mol\ Br^-} \times \frac{119\ g\ KBr}{mol\ KBr}$$

$$= 4.8 \times 10^{-3}\ g\ KBr$$

16.15A $AgBr(s) + 2\ S_2O_3^{2-}(aq) \rightleftharpoons [Ag(S_2O_3)_2]^{3-}(aq) + Br^-(aq)$

	$S_2O_3^{2-}$	$[Ag(S_2O_3)_2]^{3-}$	Br^-
	0.500 M	0	0
	-2s	+s	+s
	0.500-2s	s	s

$$K_c = 8.5 = \frac{s^2}{(0.500 - 2s)^2}$$

$$2.9 = \frac{s}{0.500 - 2s}$$

1.5 - 5.8 s = s

1.5 = 6.8 s

s = 0.22 M molar solubility

16.15B Because $[Ag(CN)_2]^-$ has a larger K_f (5.6×10^{18}) than the K_f of $[Ag(S_2O_3)_2]^{3-}$ (1.7×10^{13}) or $[Ag(NH_3)_2]^+$ (1.6×10^7), AgI is most soluble in 0.100 M NaCN.

16.16A No. NH_4NO_3 does not produce the acid to destroy the complex ion, $[Ag(NH_3)_2]^+$, and the concentration of free Ag^+ remains too low for AgCl(s) to precipitate.

16.16B (a) $Zn^{2+}(aq) + NH_3(aq) + 2\ H_2O(l) \rightarrow Zn(OH)_2(s) + 2\ NH_4^+(aq)$

(b) $Zn(OH)_2(s) + 4\ NH_3(aq) \rightarrow [Zn(NH_3)_4]^{2+}(aq) + 2\ OH^-(aq)$

(c) $[Zn(NH_3)_4]^{2+}(aq) + 2\ OH^-(aq) + 6\ HC_2H_3O_2(aq) \rightarrow$

$Zn^{2+}(aq) + 4\ NH_4^+(aq) + 6\ C_2H_3O_2^-(aq)$

Review Questions

1. $K_{sp} = [Fe^{3+}][OH^-]^3$

   $K_{sp} = [Au^{3+}]^2[C_2O_4^{2-}]^3$

4. (a) $PbSO_4$ is more soluble because its K_{sp} value is larger and the two solutes are of the same type: MX.

   (b) PbI_2 is more soluble because its K_{sp} is larger and the two compounds are of the same type: MX_2.

5. Adding a common ion of SO_4^{2-} causes reduction of the solubility of $BaSO_4$.

   (a) $[Ba^{2+}]$ is reduced.

6. Raising the pH (increasing $[OH^-]$) reduces H^+ to force more $[S^{2-}]$ to be formed.
(c) raise the pH.

9. The solubility of $Fe(OH)_3(s)$ is increased by HCl(aq) and $CH_3COOH(aq)$, both acids, because the H_3O^+ would react with the OH^- from the $Fe(OH)_3(s)$. Its solubility is lowered by NaOH(aq) and $NH_3(aq)$, because of the common ion OH^-.

11. (a) $K_f = \dfrac{[[Ag(NH_3)_2]^+]}{[Ag^+][NH_3]^2}$

(b) $K_f = \dfrac{[[Zn(NH_3)_4]^{2+}]}{[Zn^{2+}][NH_3]^4}$

(c) $K_f = \dfrac{[[Ag(S_2O_3)_2]^{3-}]}{[Ag^+][S_2O_3{}^{2-}]^2}$

12. $Pb(NO_3)_2(aq)$, through the common ion Pb^{2+}, reduces the solubility of $PbCl_2(s)$; but because Pb^{2+} forms the complex ion $[PbCl_3]^-$, HCl(aq) increases the solubility of $PbCl_2(s)$.

13. HCl, HNO_3, and NH_3 will dissolve $Cu(OH)_2$. H^+ from HCl or HNO_3 will react with OH^- to form H_2O, and the NH_3 will complex the Cu^{2+} to form $[Cu(NH_3)_4]^{2+}$. $Cu(OH)_2$ is slightly soluble in H_2O.
(a) is the answer.

14. (b) Adding HCl will allow separation of Cu^{2+} and Ag^+. AgCl will precipitate and $CuCl_2$ will not. H_2S will precipitate both. HNO_3 will precipitate neither. NH_4NO_3 may form complexes but not precipitates.

15. $[Al(H_2O)_6]^{3+} + H_2O \rightarrow H_3O^+ + [Al(OH)(H_2O)_5]^{2+}$

Problems

19. (a) $Hg_2(CN)_2(s) \rightleftharpoons Hg_2{}^{2+}(aq) + 2\ CN^-(aq)$
(b) $Ag_3AsO_4(s) \rightleftharpoons 3\ Ag^+(aq) + AsO_4{}^{3-}(aq)$

21. (a) $K_{sp} = 5 \times 10^{-40} = [Hg_2{}^{2+}][CN^-]^2$
(b) $K_{sp} = 1.0\times 10^{-22} = [Ag^+]^3[AsO_4{}^{3-}]$

23. No, the molar solubility and K_{sp} cannot have the same value. The molar solubility must be raised to a power and often times multiplied by a factor to obtain K_{sp}. The molar solubility is larger than K_{sp}, because the solubility is generally much smaller

than 1M, and raising such a number to a power (2, 3, 4 ...) produces a result that is smaller still.

25. (a) $CaF_2(s) \rightleftharpoons Ca^{2+}(aq) + 2F^-(aq)$

$s \quad 2s$

$K_{sp} = (s)(2s)^2 = 4s^3$

$K_{sp} = 4 \times (3.32 \times 10^{-4})^3$

$K_{sp} = 1.46 \times 10^{-10}$

(b) (b) $Al(OH)_3\ (s) \leftrightarrow Al^{3+}(aq) + 3OH^-(aq)$

$s \quad 3s$

$K_{sp} = (s)(3s)^3 = 27s^4 = 27(2.6 \times 10^{-9})^4$

$K_{sp} = 1.2 \times 10^{-33}$

(c) $Cd(IO_3)_2(s) \rightleftharpoons Cd^{2+}(aq) + 2\,IO_3^-(aq)$

$s \quad 2s$

$K_{sp} = s(2s)^2 = 4s^3$

$$s = \frac{0.097\text{ g}}{100\text{ mL}} \times \frac{\text{mL}}{10^{-3}\text{ L}} \times \frac{\text{mol}}{462.21\text{ g}} = 2.10 \times 10^{-3}\text{ M}$$

$K_{sp} = 4 \times (2.10 \times 10^{-3})^3 = 3.7 \times 10^{-8}$

27. (a) $AgCl(s) \rightleftharpoons Ag^+(aq) + Cl^-(aq)$

$s \quad s$

$s^2 = K_{sp} = 1.8 \times 10^{-10}$

$s = 1.3 \times 10^{-5}$ M

$Ag_2CrO_4(s) \rightleftharpoons 2\,Ag^+(aq) + CrO_4^{2-}(aq)$

$2s \quad s$

$4s^3 = K_{sp} = 1.1 \times 10^{-12}$

$s = 6.5 \times 10^{-5}$ M $\quad$ Ag_2CrO_4 has a higher molar solubility than AgCl.

(b) $Mg(OH)_2(s) \rightleftharpoons Mg^{2+}(aq) + 2\,OH^-(aq)$

$s \quad 2s$

$4s^3 = C = 1.8 \times 10^{-11}$

$s = 1.7 \times 10^{-4}$ M

$MgCO_3(s) \rightleftharpoons Mg^{2+}(aq) + CO_3^{2-}(aq)$

$s \quad s \quad s$

$s^2 = K_{sp} = 3.5 \times 10^{-8}$

$s = 1.9 \times 10^{-4}$ M $\quad$ The molar solubility of $MgCO_3$ is greater than that of $Mg(OH)_2$.

29. $$[S_2O_3^{2-}] = 6.4\text{ mg S} \times \frac{\text{mol S}}{32.1\text{ g S}} \times \frac{1\text{ mol } S_2O_3^{2-}}{1\text{ mol S}} \times \frac{1}{50.0\text{ mL}} = 4.0 \times 10^{-3}\text{ M}$$

$K_{sp} = [Ba^{2+}][S_2O_3^{2-}] = [S_2O_3^{2-}]^2 = (4.0 \times 10^{-3})^2 = 1.6 \times 10^{-5}$

31.

	$CaCO_3$	$CaSO_4$	CaF_2
K_{sp}	2.8×10^{-9}	9.1×10^{-6}	5.3×10^{-9}

$CaSO_4$ produces more $[Ca^{2+}]$ than $CaCO_3$ because of K_{sp} values. To determine whether CaF_2 or $CaSO_4$ has a higher $[Ca^{2+}]$ requires at least an estimation. That is, $s^2 = 9.1 \times 10^{-6}$ and $s \approx 3 \times 10^{-3}$ for $CaSO_4$, and $s^3 \approx 1 \times 10^{-9}$ and $s \approx 1 \times 10^{-3}$ for CaF_2.

$CaSO_4$ also produces more $[Ca^{2+}]$ than CaF_2.

actual

$CaSO_4(s) \rightleftharpoons Ca^{2+}(aq) + SO_4^{2-}(aq)$

$s \quad s$

$s^2 = 9.1 \times 10^{-6}$

$s = 3.0 \times 10^{-3}$ M

$CaCO_3(s) \rightleftharpoons Ca^{2+}(aq) + CO_3^{2-}(aq)$

$s \quad s$

$s^2 = 2.8 \times 10^{-9}$

$s = 5.3 \times 10^{-5}$ M

$CaF_2(s) \rightleftharpoons Ca^{2+}(aq) + 2\,F^-(aq)$

$s \quad 2s$

$4s^3 = 5.3 \times 10^{-9}$

$s = 1.1 \times 10^{-3}$ M

33. $MgF_2(s) \rightleftharpoons Mg^{2+}(aq) + 2\,F^-(aq)$

$\frac{1}{2}X \quad X$

$K_{sp} = \frac{1}{2}X(X)^2 = \frac{1}{2}X^3 = 3.7 \times 10^{-8}$

$[F^-] = X = 4.2 \times 10^{-3}$ M

$$4.2 \times 10^{-3}\text{ M} \times \frac{10^{-3}\text{ L}}{\text{mL}} \times \frac{19.00\text{ g}}{\text{mol}} \times \frac{\text{mL}}{\text{g}} \times \frac{10^3}{10^3} = \frac{80\text{ g F}^-}{10^6\text{ g solution}} = 80\text{ ppm F}^-$$

Yes, it is soluble enough.

35. $PbCl_2(s) \rightleftharpoons Pb^{2+}(aq) + 2\,Cl^-(aq)$

$s \quad 2s$

K_{sp} (80 °C) $= 3.3 \times 10^{-3} = 4s^3$

$s = 0.094$ M

K_{sp} (25 °C) $= 1.6 \times 10^{-5} = 4s^3$

$s = 0.016$ M

$0.094 - 0.016 = 0.078$ M

$$\frac{0.078\text{ mmol}}{\text{mL}} \times 1.00\text{ mL} \times \frac{278.1\text{ mg}}{\text{mmol}} = 22\text{ mg}$$

Yes, it is visible.

37. AgBr will be most soluble in water, as both of the other solutions contain common ions that reduce solubility.

39. (a) $Mg(OH)_2(s) \rightleftharpoons Mg^{2+}(aq) + 2\ OH^-(aq)$

$$0.10 + s \qquad 2s$$

$K_{sp} = 1.8 \times 10^{-11} = [Mg^{2+}][OH^-]^2 = (0.10 + s)(2s)^2$

Assume $0.10 >> s$

$$4s^2 = \frac{1.8 \times 10^{-11}}{0.10} = 1.8 \times 10^{-10}$$

$s = 6.7 \times 10^{-6}$ M (Assumption is valid.)

(b) $MgNH_4PO_4(s) \rightleftharpoons Mg^{2+}(aq) + NH_4^+(aq) + PO_4^{3-}(aq)$

$$0.10 + s \qquad s \qquad s$$

$K_{sp} = [Mg^{2+}][NH_4^+][PO_4^{3-}]$

$2.5 \times 10^{-13} = (0.10 + s)(s)(s)$

Assume $0.10 >> s$

$2.5 \times 10^{-13} = 0.10s^2$

$s = 1.6 \times 10^{-6}$ M (Assumption is valid.)

41. $$[I^-] = \frac{15.0\ g}{0.250\ L} \times \frac{mol}{149.89\ g} = 0.400\ M$$

$PbI_2(s) \rightleftharpoons Pb^{2+}(aq) + 2\ I^-(aq)$

$$s \qquad 0.400 + 2s$$

$K_{sp} = 7.1 \times 10^{-9} = s\,(0.400 + 2s)^2$

Assume $0.400 >> 2s$

$s \times (0.400)^2 = 7.1 \times 10^{-9}$

$s = 4.4 \times 10^{-8}\ M = [Pb^{2+}]$ (Assumption is valid.)

$[I^-] = 0.400$ M

43. $Ag_2SO_4(s) \rightleftharpoons 2\ Ag^+(aq) + SO_4^{2-}(aq)$

$$X \qquad \frac{1}{2}X$$

$$K_{sp} = 1.4 \times 10^{-5} = X^2\left(\frac{1}{2}X\right) = \frac{1}{2}X^3$$

(a) $[Ag^+] = X = 3.0 \times 10^{-2}$ M

$[SO_4^{2-}] = \frac{1}{2}X = 1.5 \times 10^{-2}$ M

(b) $$[SO_4^{2-}] = \frac{K_{sp}}{[Ag^+]^2} = \frac{1.4 \times 10^{-5}}{(4.0 \times 10^{-3}\ M)^2} = 0.88\ M$$

0.88 M needed in sol. - 1.5×10^{-2} M already in sol.= 0.86 M to be added

$$0.86\ M\ SO_4^{2-} \times 0.500\ L \times \frac{mol\ Na_2SO_4}{mol\ SO_4^{2-}} \times \frac{142.05\ g}{mol\ Na_2SO_4} = 61\ g\ Na_2SO_4\ \text{added}$$

An alternate method to reach this number is shown below.

$Ag_2SO_4(s) \rightleftharpoons 2\ Ag^+(aq) + SO_4^{2-}(aq)$

Initial	$2s$	s	
Change		M	
Equil	$2\,s$	$s + M$	$2s = 4.0 \times 10^{-3}$
	4.0×10^{-3}	$2.0 \times 10^{-3} + M$	

$1.4 \times 10^{-5} = (4.0 \times 10^{-3})^2 \times (2.0 \times 10^{-3} + M)$

$0.88 = 2.0 \times 10^{-3} + M$

$M = 0.88$ M SO_4^{2-} needed in solution.

0.88M	$-\ 1.5 \times 10^{-2}$ M	$= 0.86$ M
needed in sol.	already in sol.	to be added

$$0.86\text{ M } SO_4^{2-} \times 0.500\text{ L} \times \frac{\text{mol } Na_2SO_4}{\text{mol } SO_4^{2-}} \times \frac{142.05\text{ g}}{\text{mol } Na_2SO_4} = 61\text{ g } Na_2SO_4 \text{ added}$$

45. $BaS_2O_3(s) \rightleftharpoons Ba^{2+}(aq) + S_2O_3^{2-}(aq)$

$\qquad\qquad s \qquad\qquad s + 0.0033$

$K_{sp} = 1.6 \times 10^{-5} = (s)(s + 0.0033)$

$1.6 \times 10^{-5} = s^2 + 0.0033\,s$

$s^2 + 0.0033\,s - 1.6 \times 10^{-5} = 0$

$$s = \frac{-0.0033 \pm \sqrt{(0.0033)^2 - 4 \times 1 \times (-1.6 \times 10^{-5})}}{2}$$

$$s = \frac{-0.0033 \pm 8.7 \times 10^{-3}}{2}$$

$s = 2.7 \times 10^{-3}$ M or -6.0×10^{-3} Negative values are not valid.

47. $$[CrO_4^{2-}] = \frac{K_{sp}}{[Ag^+]^2} = \frac{1.1 \times 10^{-12}}{(1.05 \times 10^{-3})^2} = 1.0 \times 10^{-6}\text{ M}$$

49. (a) NaCl is soluble, but MgF_2 is slightly soluble. $(K_{sp} = 3.7 \times 10^{-8})$

$$[Mg^{2+}] = \frac{235\text{ mL x } 0.0022\text{ M}}{235\text{ mL} + 485\text{ mL}} = 7.2 \times 10^{-4}\text{ M}$$

$$[F^-] = \frac{485\text{ mL x } 0.0055\text{ M}}{235\text{ mL} + 485\text{ mL}} = 3.7 \times 10^{-3}\text{ M}$$

$Q_{ip} = [Mg^{2+}][F^-]^2 = 7.2 \times 10^{-4} \times (3.7 \times 10^{-3})^2$

$Q_{ip} = 9.9 \times 10^{-9} < 3.7 \times 10^{-8} = K_{sp}$ No precipitate.

(b) $NaNO_3$ is soluble, but $Pb_3(AsO_4)_2$ is slightly soluble. $K_{sp} = 4.0 \times 10^{-36}$

$$[Pb^{2+}] = \frac{136\text{ mL x } 0.00015\text{ M}}{(136 + 234)\text{ mL}} = 5.5 \times 10^{-5}\text{ M}$$

$$[AsO_4^{3-}] = \frac{234\text{ mL x } 0.00028\text{ M}}{(136 + 234)\text{ mL}} = 1.8 \times 10^{-4}\text{ M}$$

$Q_{ip} = [Pb^{2+}]^3[AsO_4]^2 = (5.5 \times 10^{-5})^3(1.8 \times 10^{-4})^2$

$= 5.4 \times 10^{-21} > 4.0 \times 10^{-36} = K_{sp}$

A precipitate should form.

51. $[F^-] = \dfrac{1.0\text{ g}}{1.0 \times 10^3\text{ L}} \times \dfrac{\text{mol}}{19.00\text{ g}} = 5.26 \times 10^{-5}\text{ M}$

$Q_{ip} = [Ca^{2+}][F^-]^2 = 2.0 \times 10^{-3} \times (5.26 \times 10^{-5})^2$

$Q_{ip} = 5.5 \times 10^{-12}$

$K_{sp} = 5.3 \times 10^{-9}$ No precipitation occurs.

53. $CaCO_3$ $K_{sp} = 2.8 \times 10^{-9}$ $CaSO_4$ $K_{sp} = 9.1 \times 10^{-6}$

$CaCO_3$ will precipitate first.

$[Ca^{2+}] = \dfrac{K_{sp}}{[SO_4^{2-}]} = \dfrac{9.1 \times 10^{-6}}{0.010} = 9.1 \times 10^{-4}\text{ M}$

$[CO_3^{2-}] = \dfrac{K_{sp}}{[Ca^{2+}]} = \dfrac{2.8 \times 10^{-9}}{9.1 \times 10^{-4}} = 3.1 \times 10^{-6}\text{ M}$

55. (a) $[Mg^{2+}] = \dfrac{K_{sp}}{[OH^-]^2} = \dfrac{1.8 \text{ x } 10^{-11}}{(2.0 \text{ x } 10^{-3})^2} = 4.5 \times 10^{-6}\text{ M}$

(b) $\dfrac{4.5 \text{ x } 10^{-6}}{0.059} \times 100\% = 7.6 \times 10^{-3}\% < 0.01\%$

Yes, precipitation is complete.

57. $[Mg^{2+}] = \dfrac{0.010\text{ mol}}{100\text{ mol}} \times 0.360\text{ M} = 3.6 \times 10^{-5}\text{ M}$

$[OH^-] = \sqrt{\dfrac{K_{sp}}{[Mg^{2+}]}} = \sqrt{\dfrac{1.8 \text{ x } 10^{-11}}{3.6 \text{ x } 10^{-5}}} = 7.1 \times 10^{-4}\text{ M}$

pOH = 3.15

pH = 10.85

59. CaF_2 MgF_2

K_{sp} 5.3×10^{-9} K_{sp} 3.7×10^{-8}

(a) CaF_2 will precipitate first, as it has a smaller K_{sp}.

(b) $[F^-] = \sqrt{\dfrac{K_{sp}}{[Mg^{2+}]}} = \sqrt{\dfrac{3.7 \text{ x } 10^{-8}}{0.010\text{ M}}} = \sqrt{3.7 \text{ x } 10^{-6}\text{ M}} = 1.9 \times 10^{-3}\text{ M}$

(c) No, the K_{sp} values are too close together.

$[Ca^{2+}] = \dfrac{K_{sp}}{[F^-]^2} = \dfrac{5.3 \text{ x } 10^{-9}}{(1.9 \text{ x } 10^{-3})^2} = 1.47 \times 10^{-3}\text{ M}$

$$\% \text{ Ca remaining} = \frac{1.47\times10^{-3}\text{ M}}{0.010\text{ M}} \times 100\% = 15\%$$

To be a successful separation, the % remaining should be < 0.1%.

61. (a) $[Ag^+] = \dfrac{10.00 \text{ mL x } 0.50 \text{ M}}{110.0 \text{ mL}} = 4.5 \times 10^{-2}$ M

$[Cl^-] = \dfrac{100.0 \text{ mL x } 0.010 \text{ M}}{110.0 \text{ mL}} = 9.1 \times 10^{-3}$ M

$Q_{ip} = [Ag^+][Cl^-] = 4.5 \times 10^{-2} \times 9.1 \times 10^{-3} = 4.1 \times 10^{-4} > K_{sp} = 1.8 \times 10^{-10}$

Precipitation will occur.

$Ag^+(aq) \quad + \quad Cl^-(aq) \rightleftharpoons AgCl(s)$

4.5×10^{-2} M 9.1×10^{-3} M

-9.1×10^{-3} M -9.1×10^{-3} M

3.6×10^{-2} M $\qquad \approx 0$

? mol AgCl $= 9.1 \times 10^{-3} \text{ M} \times 110 \text{ mL} \times \dfrac{10^{-3}\text{ L}}{\text{mL}} = 1.0 \times 10^{-3}$ moles

1.0×10^{-3} M AgCl precipitates; the remaining solution has $[Ag^+] = 0.036$ M.

(b) $[SO_4^{2-}] = \dfrac{100.0 \text{ mL x } 0.010 \text{ M}}{110.0 \text{ mL}} = 9.1 \times 10^{-3}$ M

$Q_{ip} = [Ag^+]^2[SO_4^{2-}] = (3.6 \times 10^{-2})^2(9.1 \times 10^{-3})$

$Q_{ip} = 1.2 \times 10^{-5} < 1.4 \times 10^{-5} = K_{sp}$

Precipitation of Ag_2SO_4 will not occur after the AgCl precipitates.

63. $NaHSO_4$ is acidic. The H_3O^+ will react with CO_3^{2-} to form HCO_3^- and increase the molar solubility $CaCO_3$.

65. $[OH^-] = 10^{-(14.00 - 4.60)} = 4.0 \times 10^{-10}$ M

$Co(OH)_3(s) \rightleftharpoons Co^{3+}(aq) + 3\ OH^-(aq)$

$s \qquad 4.0 \times 10^{-10}$ M

In a buffer solution, the $[OH^-]$ is constant.

$K_{sp} = 1.6 \times 10^{-44} = s\ (4.0 \times 10^{-10})^3$

$s = 2.5 \times 10^{-16}$ M

67. $CaCO_3(s) + 2\ H_3O^+(aq) \rightarrow Ca^{2+}(aq) + H_2CO_3\ (aq) + 2\ H_2O(l) \rightarrow$
$Ca^{2+}(aq) + 3\ H_2O(l) + CO_2(g)$

$CaCO_3(s) + 2\ CH_3COOH(\text{vinegar}) \rightarrow$
$Ca^{2+}(aq) + 2\ CH_3COO^-(aq) + H_2O(l) + CO_2(g)$

69. $[CrCl_2(NH_3)_4]^+$

71. Both H_3O^+ from HCl (d) and HSO_4^- from $NaHSO_4$ (b) can donate protons to NH_3 in the complex ion, causing $[Zn(NH_3)_4]^{2+}$ to dissociate and the concentration of free Zn^{2+} to increase.

73. $SO_4^{2-}(aq) + 2\ Ag^+(aq) \rightleftharpoons Ag_2SO_4(s)$
$Ag_2SO_4(s) + 4\ NH_3(aq) \rightleftharpoons 2[Ag(NH_3)_2]^+(aq) + SO_4^{2-}(aq)$
$4\ HNO_3(aq) + 2\ [Ag(NH_3)_2]^+(aq) + SO_4^{2-}(aq) \rightarrow Ag_2SO_4(s) + 4\ NH_4^+(aq) + 4\ NO_3^-(aq)$

75. (a) $Fe(OH)_3(s) + 3\ H_3O^+(aq) \rightarrow Fe^{3+}(aq) + 6\ H_2O(l)$
(b) $Na^+(aq) + Cl^-(aq) + Na^+(aq) + OH^-(aq) \rightarrow$ N.R.
(c) $Cr^{3+}(aq) + 4\ OH^-(aq) \rightarrow [Cr(OH)_4]^-(aq)$

77. $K_f = 4.1 \times 10^8 = \dfrac{0.25}{[Zn^{2+}](1.50)^4}$
$[Zn^{2+}] = 1.2 \times 10^{-10}$ M

79. $K_f = 2.4 \times 10^1 = \dfrac{[[PbCl_3]^-]}{[Cl^-]^3[Pb^{2+}]} = \dfrac{0.010}{(1.50)^3[Pb^{2+}]}$
$[Pb^{2+}] = 1.2 \times 10^{-4}$ M
$[I^-] = \dfrac{2.00 \text{ mL x } 2.00 \text{ M}}{0.302 \text{ L}} \times \dfrac{10^{-3} \text{ L}}{\text{mL}} = 1.3 \times 10^{-2}$ M
$[Pb^{2+}][I^-]^2 = (1.2 \times 10^{-4}) \times (1.3 \times 10^2)^2 = 2.1 \times 10^{-8} > 7.1 \times 10^{-9}$
Yes, a precipitate should form.

81. $K_f = 1.6 \times 10^7 = \dfrac{[[Ag(NH_3)_2]^+]}{[NH_3]^2[Ag^+]} = \dfrac{0.220}{(0.805)^2[Ag^+]}$
$[Ag^+] = 2.1 \times 10^{-8}$ M
$[Br^-] = \dfrac{K_{sp}}{[Ag^+]} = \dfrac{5.0 \text{ x } 10^{-13}}{2.1 \text{ x } 10^{-8}} = 2.4 \times 10^{-5}$ M
$? \text{ mg KBr} = 2.4 \times 10^{-5} \text{ M} \times 1.42 \text{ L} \times \dfrac{119 \text{ g KBr}}{\text{mol KBr}} \times \dfrac{\text{mg}}{10^{-3} \text{ g}} = 4.0$ mg KBr

83. $AgBr(s) \rightleftharpoons Ag^+(aq) + Br^-(aq)$ $\quad K_{sp} = 5.0 \times 10^{-13}$

$\underline{Ag^+(aq) + 2\, S_2O_3^{2-}(aq) \rightleftharpoons [Ag(S_2O_3)_2]^{3-}(aq)}$ $\quad K_f = 1.7 \times 10^{13}$

$AgBr(s) + 2\, S_2O_3^{2-}(aq) \rightleftharpoons [Ag(S_2O_3)_2]^{3-}(aq) + Br^-(aq)$ $\quad K_c = K_{sp}K_f = 8.5$

	$S_2O_3^{2-}$	$[Ag(S_2O_3)_2]^{3-}$	Br^-
	0.100 M		
	- 2s	+s	+s
	0.100 - 2s	s	s

$$8.5 = \frac{s^2}{(0.100 - 2s)^2}$$

$$2.9 = \frac{s}{0.100 - 2s}$$

$0.29 - 5.8s = s$

$0.29 = 6.8s$

$s = 4.3 \times 10^{-2}$ M = molar solubility

85. $PbCl_2$ is soluble enough that $[Pb^{2+}]$ remaining in solution is sufficiently high that K_{sp} of PbS is exceeded in Group 2 after the Group 1 precipitation. AgCl is so insoluble that $[Ag^+]$ remaining in solution after the Group 1 precipitation is not enough to yield a detectable precipitate of Ag_2S in Group 2.

87. Only Hg_2^{2+} is proven to be present, based on the gray color produced when the Group 1 precipitate is treated with $NH_3(aq)$. The presence of Pb^{2+} and Ag^+ remains uncertain. Treatment of the Group 1 precipitate with hot water and a subsequent test for Pb^{2+} was not performed, and the $NH_3(aq)$ was not tested for the presence of Ag^+.

Additional Problems

91. $$[Pb^{2+}] = \frac{50.0\text{ mg}}{0.500\text{ L}} \times \frac{10^{-3}\text{ g}}{\text{mg}} \times \frac{\text{mole}}{331.2\text{ g}} = 3.02 \times 10^{-4}\text{ M}$$

$H_2C_2O_4(aq) + H_2O(l) \rightleftharpoons HC_2O_4^-(aq) + H_3O^+(aq)$

Because for a weak diprotic acid, H_2A, $[A^{2-}] = K_{a2}$, it can be written that

$[C_2O_4^{2-}] = 5.3 \times 10^{-5}$ M

$Q_{ip} = [Pb^{2+}][C_2O_4^{2-}] = 3.02 \times 10^{-4} \times 5.3 \times 10^{-5} = 1.6 \times 10^{-8} > 4.8 \times 10^{-10} = K_{sp}$

Yes, a precipitate will form.

92. $5\, C_2O_4^{2-}(aq) + 2\, MnO_4^-(aq) + 16\, H^+(aq) \rightarrow 10\, CO_2(g) + 2\, Mn^{2+}(aq) + 8\, H_2O(l)$

$$\frac{0.00500\text{ M } MnO_4^- \times 3.22\text{ mL}}{100\text{ mL}} \times \frac{5\text{ mmol } C_2O_4^{2-}}{2\text{ mmol } MnO_4^-} = 4.03 \times 10^{-4}\text{ M } C_2O_4^{2-}$$

or

$$\frac{\dfrac{5.00\times10^{-3}\ \text{mmol MnO}_4^-}{\text{mL}}\times 3.22\ \text{mL}\times\dfrac{5\ \text{mmol C}_2\text{O}_4^{2-}}{2\ \text{mmol MnO}_4^-}}{100\ \text{mL}} = 4.03\times10^{-4}\ \text{M C}_2\text{O}_4^{2-}$$

$K_{sp} = [Sr^{2+}][C_2O_4^{2-}] = [C_2O_4^{2-}]^2 = (4.03\times10^{-4})^2$

$K_{sp} = 1.62\times10^{-7}$

94. $Pb(N_3)_2(s) \rightleftharpoons Pb^{2+}(aq) + 2\ N_3^-(aq)$ $\qquad K_{sp} = 2.5\times10^{-9}$

$\underline{2\ N_3^-(aq) + 2\ H_3O^+(aq) \rightleftharpoons 2\ HN_3(aq) + 2\ H_2O(l)}$ $\qquad \dfrac{1}{K_a^{\,2}} = \dfrac{1}{(1.9\times10^{-5})^2}$

$Pb(N_3)_2(s) + 2\ H_3O^+(aq) \rightleftharpoons Pb^{2+}(aq) + 2\ HN_3(aq) + H_2O(l)$ $\quad Kc = \dfrac{Ksp}{(Ka)^2} = 6.9$

$\qquad\qquad\qquad\qquad s \qquad\qquad 2s$

$[H_3O^+] = 10^{-2.85} = 1.4\times10^{-3}\ M$

$$Kc = \frac{[Pb^{2+}][HN_3]^2}{[H_3O^+]^2} = \frac{s\,(2s)^2}{(1.4\times10^{-3})^2} = \frac{4\,s^3}{(1.4\times10^{-3})^2}$$

$s = 0.015$ M molar solubility

98. $K_f = 1.6\times10^7 = \dfrac{[[Ag(NH_3)_2]^+]}{[NH_3]^2[Ag^+]} = \dfrac{0.62}{(0.50)^2[Ag^+]}$

$[Ag^+] = 1.6\times10^{-7}\ M$

$$[Br^-] = \frac{Ksp}{[Ag^+]} = \frac{5.0\times10^{-13}}{1.6\times10^{-7}} = 3.1\times10^{-6}\ M$$

$$?\ \text{mL NaBr} = \frac{3.1\times10^{-6}\ \text{mol Br}^-}{\text{L}}\times 165\ \text{mL}\times\frac{\text{L NaBr(aq)}}{0.0050\ \text{mol Br}^-} = 0.10\ \text{mL NaBr(aq)}$$

101. To simplify this solution, the possible formation of the complex ion $[Pb(S_2O_3)_3]^{4-}$ is ignored based on $K_{overall} = K_{sp1}\times K_{sp2}\times K_f$

$= 1.6\times10^{-8}\times4.0\times10^{-7}\times2.2\times10^{6} = 1.4\times10^{-8}$

The small value of $K_{overall}$ means that the reaction of $PbSO_4(s)$ and $PbS_2O_3(s)$ to produce $[Pb(S_2O_3)_3]^{4-}$ is unlikely.

The expressions required are the two K_{sp} expressions:

$PbSO_4(s) \rightleftharpoons Pb^{2+}(aq) + SO_4^{2-}(aq)$ $\qquad K_{sp1} = [Pb^{2+}][SO_4^{2-}] = 1.6\times10^{-8}$

$PbS_2O_3(s) \rightleftharpoons Pb^{2+}(aq) + S_2O_3^{2-}(aq)$ $\qquad K_{sp2} = [Pb^{2+}][S_2O_3^{2-}] = 4.0\times10^{-7}$

together with the charge balance condition, which requires the total concentration of positive and negative charge to be equal.

Positive charge = negative charge

$2\times[Pb^{2+}] = 2\times[SO_4^{2-}] + 2\times[S_2O_3^{2-}]$

$[Pb^{2+}] = [SO_4^{2-}] + [S_2O_3^{2-}]$

(Note: The concentrations of H_3O^+ and OH^- are assumed to be neglible compared to the other species, that is, the pH is close to 7.)

$$[Pb^{2+}] = \frac{1.6\times10^{-8}}{[Pb^{2+}]} + \frac{4.0\times10^{-7}}{[Pb^{2+}]}$$

$[Pb^{2+}]^2 = 1.6 \times 10^{-8} + 4.0 \times 10^{-7} = 4.2 \times 10^{-7}$

$[Pb^{2+}] = 6.5 \times 10^{-4}$ M

103. (a) $Ca(palm)_2(s) \rightleftharpoons Ca^{2+}(aq) + 2\ palm^-(aq)$

$s \qquad 2s$

$K_{sp} = [Ca^{2+}][palm^-]^2 = s\,(2s)^2 = 4\,s^3$

$$\text{molar solubility} = \frac{0.003\text{ g}}{100\text{ mL}} \times \frac{\text{mL}}{10^{-3}\text{ L}} \times \frac{\text{mol}}{550.9\text{ g}} = 5 \times 10^{-5}\text{ M}$$

$K_{sp} = (5 \times 10^{-5})(2 \times 5 \times 10^{-5})^2 = 5 \times 10^{-13}$

$$[Ca^{2+}] = \frac{25\text{ g}}{10^6\text{ g}} \times \frac{\text{mol}}{40.0\text{ g}} \times \frac{\text{g}}{\text{mL}} \times \frac{\text{mL}}{10^{-3}\text{ L}} = 6.3 \times 10^{-4}\text{ M}$$

$Q_{ip} = [Ca^{2+}][palm]^2 = (6.3 \times 10^{-4})(0.10)^2 = 6 \times 10^{-6} > K_{sp} = 5 \times 10^{-13}$

Yes, soap scum will form.

(b) ? g scum = 6.3×10^{-4} M × 6.5 L × 550.9 g/mol = 2.3 g calcium palmitate

Apply Your Knowledge

104. (a) $? \text{ g } Ca^{2+} = 0.6248\text{ g } CaCO_3 \times \frac{\text{mole } CaCO_3}{100.090\text{ g}} \times \frac{\text{mole } Ca^{2+}}{\text{mole } CaCO_3} \times \frac{40.078\text{ g}}{\text{mole}}$

$= 0.2502\text{ g } Ca^{2+}$

$$? \%\ Ca^{2+} = \frac{0.2502\text{ g } Ca^{2+}}{0.7291\text{ g sample}} \times 100\ \% = 34.32\ \%\ Ca$$

(b) $K_{sp} = [Ca^{2+}][C_2O_4^{2-}] = 2.7 \times 10^{-9}$

$[Ca^{2+}] = \sqrt{K_{sp}} = 5.2 \times 10^{-5}$ M

$$? \text{ moles } Ca^{2+} = 5.2 \times 10^{-5}\text{ M} \times 325\text{ mL} \times \frac{10^{-3}\text{ L}}{\text{mL}}$$

$= 1.7 \times 10^{-5}$ moles Ca^{2+} in solution

? moles Ca^{2+} precipitated = total moles – moles in solution

$= 6.242 \times 10^{-3} - 1.7 \times 10^{-5} = 6.225 \times 10^{-3}$ moles.

$$? \text{ g } CaCO_3 = 6.225 \times 10^{-3}\text{ moles } Ca^{2+} \times \frac{\text{mole } CaCO_3}{\text{mole } Ca^{2+}} \times \frac{100.09\text{ g}}{\text{mole } CaCO_3}$$

$= 0.6231\text{ g } CaCO_3$

$$? \text{ g } Ca^{2+} = 0.6231 \text{ g } CaCO_3 \times \frac{40.078 \text{ g } Ca^{2+}}{100.09 \text{ g } CaCO_3} = 0.2495 \text{ g } Ca^{2+}$$

$$? \text{ \% } Ca^{2+} = \frac{0.2495 \text{ g } Ca^{2+}}{0.7291 \text{ g sample}} \times 100 \text{ \%} = 34.22 \text{ \%}$$

Failure to use an excess of oxalate ion produces an absolute error of 0.10% and a relative error of $\frac{34.32 - 34.22}{34.32} \times 100\% = 0.29\%$.

105. (a) $BaSO_4$ is insoluble enough that little dissolves and therefore will not be dangerous.

(b) $BaSO_4(s) \rightleftharpoons Ba^{2+}(aq) + SO_4^{2-}(aq)$

$s \qquad s \qquad s$

$K_{sp} = 1.1 \times 10^{-10} = s^2$

$$s = 1.05 \times 10^{-5} \text{ M} \times \frac{137.33 \text{ g}}{\text{mol}} \times \frac{\text{mg}}{10^{-3} \text{ g}} = 1.4 \text{ mg/L } Ba^{2+}$$

(c) SO_4^{2-} from $MgSO_4$ reduces the solubility of $BaSO_4$ through the common-ion effect.

108. (a) $K_{sp} = 7.9 \times 10^{-9} = [Pb^{2+}][I^-]^2$

$$= \left(\frac{0.020 \text{ M} \times 10.0 \text{ mL}}{10.0 \text{ mL} + X \text{ drops}\left(\frac{1.12 \text{ mL}}{100 \text{ drops}}\right)}\right)\left(\frac{0.020 \text{ M} \times X \text{ drops}\frac{1.12 \text{ mL}}{100 \text{ drops}}}{10.0 \text{ mL} + X \text{ drops}\frac{1.12 \text{ mL}}{100 \text{ drops}}}\right)^2$$

$$7.9 \times 10^{-9} = \left(\frac{0.20 \text{ mmol}}{10.0 \text{ mL} + 0.0112X \text{ mL}}\right)\left(\frac{2.24 \times 10^{-4} X \text{ mmol}}{10.0 \text{ mL} + 0.0112X \text{ mL}}\right)^2$$

$$7.9 \times 10^{-9} = \left(\frac{0.20 \text{ mmol}}{(10.0 + 0.0112X)\text{mL}}\right)\left(\frac{5.02 \times 10^{-8} X^2}{100 + 0.224X + 1.25 \times 10^{-4} X^2}\right)$$

$$7.9 \times 10^{-9} = \frac{1.00 \times 10^{-8} X^2}{1000 + 3.36X + 3.76 \times 10^{-3} X^2 + 1.4 \times 10^{-6} X^3}$$

$1.11 \times 10^{-14} X^3 - 9.97 \times 10^{-9} X^2 + 2.65 \times 10^{-8} X + 7.9 \times 10^{-6} = 0$

Using successive iterations.

$$\text{Estimate } X = \sqrt{\frac{7.9 \times 10^{-6}}{9.97 \times 10^{-8}}} = 28$$

$2.43 \times 10^{-10} - 7.84 \times 10^{-6} + 2.97 \times 10^{-7} + 7.9 \times 10^{-6} = 3.57 \times 10^{-7}$

Try 29

$2.71 \times 10^{-10} - 8.41 \times 10^{-6} + 3.07 \times 10^{-7} + 7.9 \times 10^{-6} = 2.02 \times 10^{-7}$

Try 30

$3.00 \times 10^{-10} - 9.00 \times 10^{-6} + 3.18 \times 10^{-7} + 7.9 \times 10^{-6} = -6.86 \times 10^{-11}$
Between 29 and 30 drops are required.

(b) In (a) the $[Pb^{2+}]$ is essentially set at about 0.020 M and the $[I^-]$ is increased until the K_{sp} value is met. In (b) the $[I^-]$ is set at about 0.020 M and the $[Pb^{2+}]$ is increased until the K_{sp} value is met. Because the $[I^-]$ is squared and the $[Pb^{2+}]$ is to the first power, the number of drops to meet the K_{sp} value is different depending on which ion is being added.

$? [Pb^{2+}] = \dfrac{K_{sp}}{[I^-]^2} = \dfrac{7.9\times10^{-9}}{(0.020)^2} = 2.0 \times 10^{-5}$ M required to precipitate PbI_2

$? \text{ mol } Pb^{2+} = 2.0 \times 10^{-5} \text{ M} \times 10.0 \text{ mL} \times \dfrac{10^{-3}\text{ L}}{\text{mL}} = 2.0 \times 10^{-7}$ mol Pb^{2+} required

$? \text{ drops } Pb^{2+} \text{ solution} = 2.0 \times 10^{-7} \text{ mol} \times \dfrac{\text{L}}{0.020 \text{ mol}} \times \dfrac{10^{-3}\text{ L}}{\text{mL}} \times \dfrac{100 \text{ drops}}{1.12 \text{ mL}}$
$= 8.9 \times 10^{-7}$ drops Pb^{2+} solution

One drop of Pb^{2+} solution is required.

110. $CaCO_3(s) \rightleftharpoons Ca^{2+}(aq) + CO_3^-(aq)$ $\quad K_{sp} = 2.8 \times 10^{-9}$

$CO_3^-(aq) + H_3O^+(aq) \rightleftharpoons HCO_3^-(aq) + H_2O(aq)$ $\quad K = \dfrac{1}{K_{a_1}} = \dfrac{1}{4.7\times10^{-11}}$

$CaCO_3(s) + H_3O^+(aq) \rightleftharpoons Ca^{2+}(aq) + HCO_3^-(aq) + H_2O(aq)$ $\quad K_c = \dfrac{K_{sp}}{K_{a_1}}$

$$K_c = \frac{K_{sp}}{K_{a_1}} = \frac{2.8\times10^{-9}}{4.7\times10^{-11}} = 60$$

$CaCO_3(s) + H_3O^+(aq) \rightleftharpoons Ca^{2+}(aq) + HCO_3^-(aq) + H_2O(aq)$
$\qquad\qquad\qquad\qquad s \qquad\qquad s$

(a) $6.0 \times 10^1 = K_c = \dfrac{[Ca^{2+}][HCO_3^-]}{[H_3O^+]} = \dfrac{s^2}{[H_3O^+]} = \dfrac{s^2}{10^{-5.6}} = \dfrac{s^2}{2.51 \text{ x } 10^{-6}}$

$s = 1.2 \times 10^{-2}$ M

(b) $6.0 \times 10^1 = \dfrac{s^2}{[H_3O^+]} = \dfrac{s^2}{10^{-4.22}} = \dfrac{s^2}{6.03\times10^{-5}}$

$s = 6.0 \times 10^{-2}$ M

Chapter 17

Thermodynamics: Spontaneity, Entropy, and Free Energy

Exercises

17.1A (a) Spontaneous. The molecules in the wood (principally cellulose, a carbohydrate) would eventually oxidize to CO_2 and H_2O. The decay is greatly enhanced by microorganisms.

(b) Nonspontaneous. Stirring NaCl(aq) cannot supply the energy input required to dissociate NaCl into its elements.

(c) Spontaneous. In water, hydrogen chloride becomes a strong acid and completely dissociates.

17.1B (a) Spontaneous. The motion of the molecules of ethanol and water will cause mixing of the two compounds.

(b) Indeterminate. $CaCO_3$(s) should decompose on heating, but whether the decomposition is sufficient to produce CO_2(g) at 1 atm at 600 °C cannot be determined without more information.

(c) Indeterminate. Whether the reaction is spontaneous depends on the temperature. The temperature must be lower than the temperature at which the partial pressure is equal to 0.50 atm.

(d) Nonspontaneous. Copper's positive electrode potential (or its position in the activity series) indicates that it does not react with acid to produce H_2(g). Most other metals do react with acid to produce H_2(g).

17.2A (a) This is a decrease in entropy because two moles of gas go to one mole of a solid.

(b) This is an increase in entropy because two moles of solid go to five moles (two moles of solid and three moles of gas).

(c) There is no prediction here. All compounds are gaseous. The same number of moles are on both sides of the equation, and the molecules have the same total number of atoms.

17.2B The dish of water is enclosed. Once enough molecules of water have vaporized, there will be an equilibrium between the vapor molecules and the liquid molecules. That is, vapor molecules will become liquid at the same rate that liquid molecules become vapor. The molecules in the vapor phase produce the vapor pressure. If the dish were not enclosed, the liquid would disappear because the vapor molecules would escape from the vicinity of the dish.

17.3A $\Delta S^\circ = \Sigma \nu_p \times S^\circ$ products - $\Sigma \nu_r \times S^\circ$ reactants

$\Delta S^\circ =$ 1 mol $CO_2 \times 213.6$ J /mol K + 1 mol $H_2 \times 130.6$ J/mol K - 1 mol CO × 197.6 J /mol K - 1 mol $H_2O \times 188.7$ J /mol K

$\Delta S^\circ = -42.1$ J /K

17.3B 4 $NH_3(g) + 3\ O_2(g) \rightarrow 2\ N_2(g) + 6\ H_2O(g)$

$\Delta S^\circ = \Sigma \nu_p \times S^\circ$ products - $\Sigma \nu_r \times S^\circ$ reactants

$\Delta S^\circ = 2$ mol $N_2 \times 191.5$ J /mol K + 6 mol $H_2O \times 188.7$ J/mol K – 4 mole $NH_3 \times 192.3$ J/mol K – 3 mol $O_2 \times 205.0$ J/mol K = 131.0 J/K

17.4A (a) $\Delta S < 0, \Delta H < 0$, case 2
(b) $\Delta S > 0, \Delta H > 0$, case 3

17.4B $\Delta H^\circ_{rxn} = \Sigma \nu_p \times \Delta H^\circ_f$ product - $\Sigma \nu_r \times \Delta H^\circ_f$ reactant

$$\Delta H^\circ_{rxn} = 3 \text{ mol BrF} \times \frac{-93.85 \text{ kJ}}{\text{mol}} - 1 \text{ mol Br}_2 \times \frac{30.91 \text{ kJ}}{\text{mol}} - 1 \text{ mol BrF}_3 \times \frac{-255.6 \text{ kJ}}{\text{mol}} = -56.9 \text{ kJ}$$

Because 2 moles of gas go to 3 moles of gas, $\Delta S > 0$ is expected. For a negative ΔH and positive ΔS, the reaction should be spontaneous at all temperatures. Case 1.

17.5A The phase diagram (Figure 11.12) shows that at one atm the equilibrium of $CO_2(s) \rightleftharpoons CO_2(g)$ occurs at -78.5 °C. This is the temperature at which the line, $P = 1$ atm, intersects the sublimation curve of CO_2 (s). At higher T, vaporization goes to completion. At lower T, the vaporization does not occur at 1 atm pressure.

17.5B 2 $NO_2(g) \rightarrow N_2(g) + 2\ O_2(g)$

$\Delta H^\circ_{rxn} = \Sigma \nu_p \times \Delta H^\circ_f$ product - $\Sigma \nu_r \times \Delta H^\circ_f$ reactant

$$\Delta H^\circ_{rxn} = 1 \text{ mol N}_2 \times \frac{0 \text{ kJ}}{\text{mol}} + 2 \text{ mol O}_2 \times \frac{0 \text{ kJ}}{\text{mol}} - 2 \text{ mol NO}_2 \times \frac{33.18 \text{ kJ}}{\text{mol}} = -66.36 \text{ kJ}$$

ΔS° should be positive because the reaction goes from 2 moles to 3 moles. The actual calculation is shown below.

$\Delta S^\circ = \Sigma \nu_p \times S^\circ$ products - $\Sigma \nu_r \times S^\circ$ reactants

$\Delta S^\circ = 1$ mol $N_2 \times 191.5$ J /mol K + 2 mol $O_2 \times 205.0$ J/mol K – 2 mole $NO_2 \times 240.0$ J/mol K = 121.5 J/K

A negative ΔH° and positive ΔS° is spontaneous at all temperatures.

2 $NH_3(g) \rightarrow N_2(g) + 3\ H_2(g)$

$\Delta H^\circ_{rxn} = \Sigma \nu_p \times \Delta H^\circ_f$ product - $\Sigma \nu_r \times \Delta H^\circ_f$ reactant

$$\Delta H^\circ_{rxn} = 1 \text{ mol N}_2 \times \frac{0 \text{ kJ}}{\text{mol}} + 3 \text{ mol H}_2 \times \frac{0 \text{ kJ}}{\text{mol}} - 2 \text{ mol NH}_3 \times \frac{-46.11 \text{ kJ}}{\text{mol}} = 92.22 \text{ kJ}$$

ΔS° should be positive because the reaction goes from 2 moles to 4 moles. The actual calculation is shown below.

$\Delta S^\circ = \Sigma \nu_p \times S^\circ$ products $- \Sigma \nu_r \times S^\circ$ reactants

$$\Delta S° = 1 \text{ mol } N_2 \times 191.5 \text{ J/mol K} + 3 \text{ mol } H_2 \times 130.6 \text{ J/mol K} - 2 \text{ mole } NH_3 \times 192.3 \text{ J/mol K} = 198.7 \text{ J/K}$$

A positive $\Delta H°$ and positive $\Delta S°$ is spontaneous only at high temperatures.

17.6A (a) $\Delta G° = \Delta H° - T\Delta S°$

$$\Delta G° = -114.1 \text{ kJ} - \left(298 \text{ K} \times (-146.2 \text{ J/K}) \times \frac{\text{kJ}}{10^3 \text{ J}}\right)$$

$$\Delta G° = -70.5 \text{ kJ}$$

(b) $\Delta G° = \Delta H° - T\Delta S°$

$$\Delta G° = 409.0 \text{ kJ} - \left(298 \text{ K} \times \left(\frac{-129.1 \text{ J}}{\text{mol K}}\right) \times \frac{\text{kJ}}{10^3 \text{ J}}\right)$$

$$\Delta G° = 447.5 \text{ kJ}$$

17.6B (a) $\Delta G° = \Sigma\nu_p \times \Delta G°_f \text{ products} - \Sigma\nu_r \times \Delta G°_f \text{ reactants}$

$$\Delta G° = 1 \text{ mol } CCl_4 \text{ (l)} \times (-65.27 \text{ kJ/mol}) + 6 \text{ mol S} \times 0 - 1 \text{ mol } CS_2 \times (+65.27 \text{ kJ/mol}) - 2 \text{ mol } S_2Cl_2 \times (-31.8 \text{ kJ/mol})$$

$$\Delta G° = -66.9 \text{ kJ}$$

(b) $4\ NH_3(g) + 3\ O_2(g) \rightarrow 2\ N_2(g) + 6\ H_2O(g)$

$$\Delta G° = \Sigma\nu_p \times \Delta G°_f \text{ products} - \Sigma\nu_r \times \Delta G°_f \text{ reactants}$$

$$\Delta G° = 2 \text{ mol } N_2 \times 0 \text{ kJ/mol} + 6 \text{ mol } H_2O \times (-228.6 \text{ kJ/mol}) - 3 \text{ mol } O_2 \times 0 \text{ kJ/mol} - 4 \text{ mol } NH_3 \times (-16.48 \text{ kJ/mol})$$

$$\Delta G° = -1305.7 \text{ kJ}$$

17.7A $\Delta S°_{vapor} = \dfrac{\Delta H°_{vap}}{(145.1 + 273.2)\text{K}} \approx \dfrac{87 \text{ J}}{\text{mol K}}$

$$\Delta H°_{vapor} = \frac{3.6 \times 10^4 \text{ J}}{\text{mol}} \times \frac{\text{kJ}}{10^3 \text{ J}} = \frac{36 \text{ kJ}}{\text{mol}}$$

17.7B Methanol will hydrogen-bond extensively. Vaporization will involve breaking the hydrogen bonds and creating a disordered vapor from an ordered liquid. The $\Delta S°$ vapor should be much larger than 87 J/mol K.

$CH_3OH(l) \rightarrow CH_3OH(g)$

$$\Delta S°_{vapor} = \frac{239.7 \text{ J}}{\text{mol K}} - \frac{126.8 \text{ J}}{\text{mol K}} = \frac{112.9 \text{ J}}{\text{mol K}}$$

or

$$\Delta H°_{vapor} = \frac{-200.7 \text{ kJ}}{\text{mol}} - \frac{(-238.7 \text{ kJ})}{\text{mol}} = \frac{38.0 \text{ kJ}}{\text{mol}}$$

$$\Delta S°_{vapor} = \frac{\Delta H°_{vap}}{T_{bp}} = \frac{\dfrac{38.0 \text{ kJ}}{\text{mol}} \times \dfrac{10^3 \text{ J}}{\text{kJ}}}{(64.7° + 273.2°)\text{K}} = \frac{112.5 \text{ J}}{\text{mol}}$$

17.8A $2\ Al(s) + 6\ H^+(aq) \rightarrow\ 2\ Al^{3+}(aq) + 3\ H_2(g)$

$$K_{eq} = \frac{P^3_{H_2}\ [Al^{3+}]^2}{[H^+]^6}$$

17.8B $Mg(OH)_2(s) + 2\ H_3O^+(aq) \rightarrow\ Mg^{2+}(aq)\ + 4\ H_2O(l)$

$$K_{eq} = \frac{[Mg^{2+}]}{[H_3O^+]^2}$$

17.9A $\Delta G° = 2\ \text{mol Hg} \times 0 + 1\ \text{mol}\ O_2 \times 0 - 2\ \text{mol HgO} \times -\ 58.56\ \text{kJ /mol}$

$\Delta G° = 117.12\ \text{kJ/mol}$

$$\ln K_{eq} = \frac{\Delta G°}{-RT} = \frac{117.12\ \text{kJ/mol} \times 10^3\ \text{J/kJ}}{-8.3145\ \text{J/mol K} \times (273.15 + 25)\ \text{K}} = -\ 47.25$$

$e^{\ln Keq} = K_{eq} = e^{-47.25} = 3.02 \times 10^{-21}$

17.9B $\Delta G° = \Sigma\nu_p \times \Delta G°_f\ \text{products} - \Sigma\nu_r \times \Delta G°_f\ \text{reactants}$

$\Delta G° = 2\ \text{mol NOBr} \times 82.4\ \text{kJ/mol} - 2\ \text{mol NO} \times 86.57\ \text{kJ/mol} - 1\ \text{mol}\ Br_2$
$\times\ 0\ \text{kJ/mol} = -8.3\ \text{kJ/mol}$

$\Delta G° = -RT \ln K_{eq}$

$$\ln K_{eq} = \frac{\Delta G°}{-RT} = \frac{\dfrac{-8.3\ \text{kJ}}{\text{mol}} \times \dfrac{10^3\ \text{J}}{\text{kJ}}}{\dfrac{-8.3145\ \text{J}}{\text{mol K}} \times 298.2\ \text{K}} = 3.3$$

$e^{\ln K_{eq}} = K_{eq} = e^{3.3} = 27$

$2\ NO + Br_2 \rightleftharpoons 2\ NOBr$

1 atm	--
1-X	X

$$27 = \frac{P^2_{NOBr}}{P^2_{NO}} = \frac{X^2}{(1-X)^2}$$

$$5.2 = \frac{X}{1-X}$$

$5.2 - 5.2\,X = X$

$X = 0.84$

$P_{NOBr} = 0.84\ \text{atm}$

$P_{NO} = 1.00 - 0.84 = 0.16\ \text{atm}$

17.10A $\Delta H°_{rxn} = \Sigma\Delta H°_f\ \text{product} - \Sigma\Delta H°_f\ \text{reactant}$

$\Delta H°_{rxn} = 1\ \text{mol}\ N_2O_4 \times 9.16\ \text{kJ /mol} - 2\ \text{mol}\ NO_2 \times 33.18\ \text{kJ /mol}$

$\Delta H°_{rxn} = -\ 57.20\ \text{kJ/mol}$

$$\ln\frac{K_2}{K_1} = \frac{\Delta H^\circ}{R}\left(\frac{1}{T_1}-\frac{1}{T_2}\right) = \frac{\frac{-57.20\,\text{kJ}}{\text{mol}}\times\frac{10^3\,\text{J}}{\text{kJ}}}{\frac{8.3145\,\text{J}}{\text{mol K}}}\left(\frac{1}{298\text{K}}-\frac{1}{338\text{K}}\right)$$

$$\ln\frac{K_2}{6.9} = -2.732$$

$$\frac{K_2}{6.9} = e^{-2.732} = 6.51\times10^{-2}$$

$K_2 = 0.45$

17.10B $\ln\frac{P_2}{P_1} = \frac{\Delta H^\circ}{R}\left(\frac{1}{T_1}-\frac{1}{T_2}\right)$

$$\ln\frac{P_2}{23.8\text{ mmHg}} = \frac{\frac{44.0\,\text{kJ}}{\text{mol}}\times\frac{10^3\,\text{J}}{\text{kJ}}}{8.3145\,\text{J/molK}}\left(\frac{1}{298.2\text{K}}-\frac{1}{313.2\text{K}}\right)$$

$$\ln\frac{P_2}{23.8\text{ mmHg}} == 0.850$$

Raise both sides of the equation to a power of e.

$$e^{\ln\frac{P_2}{23.8\text{ mmHg}}} = \frac{P_2}{23.8\text{ mmHg}} = e^{0.850} = 2.34$$

$P_2 = 55.7$ mmHg

From the table V.P. (40°) = 55.3 mmHg

Review Questions

1. (a) Spontaneous. Microorganisms that lead to its souring are present in the milk; no further intervention is needed.
   (b) Nonspontaneous. The extraction of copper metal from copper ores requires a great deal of external intervention.
   (c) Spontaneous. The corrosion of iron in moist air cannot be prevented without external intervention.

2. (c) The free energy change must be negative.

3. (a) Decrease in entropy. A liquid is converted to a solid.
   (b) Increase in entropy. A solid is converted to a gas.
   (c) Increase in entropy. A liquid combines with oxygen gas to produce an even greater amount of gaseous products.

9. NOF_3 has more atoms than NO_2F, more modes of vibration, and a greater entropy.

10. (b) the direction in which a net reactions occurs. For negative ΔG°, a reaction is spontaneous in the forward direction and nonspontaneous in the reverse direction.

11. Low temperature. At low temperatures, the negative ΔH term dominates in the Gibbs free energy equation and produces a $\Delta G < 0$.

13. The disorder increases when one molecule becomes two molecules or atoms. (d) $\Delta S > 0$

14. Because of its extensive hydrogen bonding, (c) NH_3 is not accurately described by Trouton's rule.

18. $\Delta G° = - RT \ln K_{eq}$
If $\Delta G° = 0$, $\ln K_{eq} = 0$ (d) $K_{eq} = 0$

Problems

19. (a) Increase. A gas is less ordered than a liquid.
(b) Increase. Production of a greater number of gaseous molecules leads to more disorder.
(c) One cannot tell, because there are the same number of gaseous molecules of about the
same complexity on both sides of the equation.
(d) Increase. A large amount of gas (a disordered phase) is produced from a solid (a more ordered phase).
(e) Decrease. A solid is more ordered than the liquid phase from which it is frozen.
(f) Indeterminate. The gases are all diatomic and the same number of moles of gas appear on each side of the equation.
(g) Increase. A liquid decomposes to produce a large amount of gas.
(h) Decrease. The conversion of a gas to an aqueous solution should produce a more ordered state.

21. Water is extensively hydrogen-bonded. To break all of the hydrogen bonds would require a large, positive ΔH that would not be offset by the increase in entropy, ΔS. Two separate liquids remain (octane floating on liquid water); there is very little mixing.

23. The correct statement would be: For a process to occur spontaneously, the total entropy or the entropy of the universe must increase; that is, the entropy of the system and of the surroundings together must increase.
Errors in other statements are: (a) Entropy of the system may increase in some cases and decrease in others; (b) Entropy of the surroundings may also increase or decrease; (c) Entropy of the system and surroundings need not both increase, as long as the increase in one exceeds the decrease in the other—that is, the total entropy increases.

25. From zero a sloped line goes to 194.5 K where a long vertical line jumps the standard molar entropy to a higher level. That vertical line corresponds to sublimation. Another sloped line begins at that point. The value at 298 K is 213.6 J/mol K. There is only one vertical line instead of two since, at 1 atm pressure, liquid carbond dioxide should not exist. (See Figure 11.12.)

27. The disintegration (oxidation) of aluminum is a spontaneous process, but because Al_2O_3 forms a protective coating on the exposed surface, which slows down further action of $O_2(g)$ and $H_2O(l)$, the complete disintegration takes a very long time. Spontaneity does not involve consideration of the speed of the reaction.

29. To use entropy change alone as a criterion for spontaneous change requires evaluating entropy changes in the system and in the surroundings. Free-energy change requires only measurements in the system: $\Delta G = \Delta H - T\Delta S$.

31.

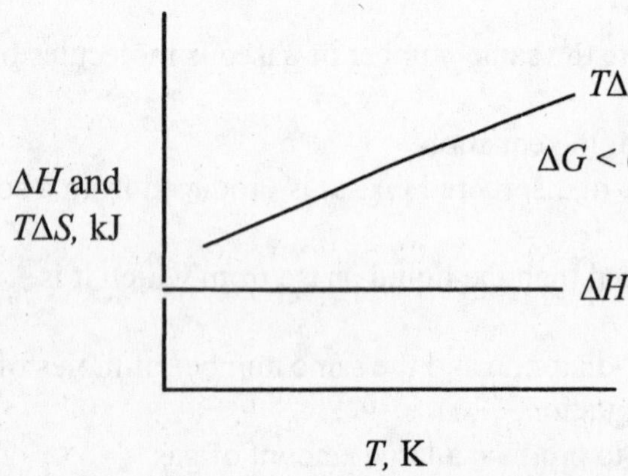

The $T\Delta S$ line does not cross the ΔH line. The $T\Delta S$ line is always above the ΔH line. ΔG is always negative, and the reaction is spontaneous.

33. (a) False. The change in *enthalpy* indicates whether a reaction is endothermic or exothermic.
(b) False. The entropy change indicates whether the reaction is accompanied by an increase or decrease in molecular disorder or in available energy levels.
(c) False. The free-energy change of a reaction indicates whether reactants at their stated conditions will spontaneously form products at their stated conditions. The correct statement would be that the free energy change indicates whether the forward reaction ($\Delta G < 0$), or reverse reaction ($\Delta G > 0$) is favored. If $\Delta G = 0$, the reversible reaction is at equilibrium. The value of ΔG depends on the relative values of $T\Delta S$ and ΔH.

35. (a) The melting of a solid is nonspontaneous below its melting point and spontaneous above its melting point. For water, this temperature is 0 °C.

(b) The condensation of a vapor at 1 atm pressure to liquid is spontaneous below the normal boiling point and nonspontaneous above the normal boiling point. For water, this temperature is 100 °C.

37. $\Delta S° < 0$ because one mole of gas is produced from two moles of gas. Because the reaction is spontaneous at room temperature, $\Delta H° < 0$. This corresponds to case 2 of Table 19.1. The reaction should become nonspontaneous at a higher temperature.

39. (a) $\Delta G° = 1 \text{ mol } H_2O(g) \times (-228.6 \text{ kJ/mol}) + 1 \text{ mol Fe} \times 0 - 1 \text{ mol FeO} \times (-251.5 \text{ kJ/mol}) - 1 \text{ mol } H_2 \times 0$

$\Delta G° = 22.9 \text{ kJ}$

(b) $\Delta G° = 1 \text{ mol } H_2O(l) \times (-237.2 \text{ kJ/mol}) + 1 \text{ mol } CdCl_2 \times (-344.0 \text{ kJ/mol}) - 1 \text{ mol CdO} \times (-228 \text{ kJ/mol}) - 2 \text{ mol HCl} \times (-95.30 \text{ kJ/mol})$

$\Delta G° = -163 \text{ kJ}$

41. $\Delta H° = 1 \text{ mol } CO_2 \times (-393.5 \text{ kJ/mol}) + 2 \text{ mol } SO_2 \times (-296.8 \text{ kJ/mol}) - 1 \text{ mol } CS_2 \times 89.70 \text{ kJ/mol} - 3 \text{ mol } O_2 \times 0$

$\Delta H° = -1076.8 \text{ kJ}$

$\Delta S° = 1 \text{ mol } CO_2 \times 213.6 \text{ J/mol K} + 2 \text{ mol } SO_2 \times (248.1 \text{ J/mol K}) - 1 \text{ mol } CS_2 \times 151.3 \text{ J/mol K} - 3 \text{ mol } O_2 \times 205.0 \text{ J/mol K}$

$\Delta S° = -56.5 \text{ J/K}$

$\Delta G° = \Delta H° - T\Delta S°$

$$\Delta G° = -1076.8 \text{ kJ} - 298 \text{ K} \times (-56.5 \text{ J/K}) \times \frac{\text{kJ}}{10^3 \text{ J}} = -1060.0 \text{ kJ}$$

$\Delta G° = 1 \text{ mol } CO_2 \times (-394.4 \text{ kJ/mol}) + 2 \text{ mol } SO_2 \times (-300.2 \text{ kJ/mol}) - 1 \text{ mol } CS_2 \times 65.27 \text{ kJ/mol} - 3 \text{ mol } O_2 \times 0 \text{ kJ/mol}$

$\Delta G° = -1060.1 \text{ kJ}$

Results of both methods cmpare very favorably.

43. With $\Delta G°$, K_{eq} can be evaluated: $\Delta G° = -RT \ln K_{eq}$. With K_{eq}, the equilibrium condition can be determined. To evaluate ΔG for nonstandard conditions, $\Delta G = \Delta G° + RT \ln Q$ (where Q is the reaction quotient). However, ΔG cannot be obtained without knowing $\Delta G°$.

45. $$\Delta S°_{vap} \approx 87 \text{ J/mol K} = \frac{\Delta H°_{vap}}{T_{bp}} = \frac{31.69 \text{ kJ/mol x} \frac{10^3 \text{ J}}{\text{kJ}}}{T_{bp}}$$

$T_{bp} = 364 \text{ K} - 273° = 91 \text{ °C}$

47. $$\Delta S°_{vap} \approx 87 \text{ J/K} = \frac{\Delta H°_{vap}}{T_{bp}} = \frac{\Delta H°_{vap} \times 10^3 \text{ J/kJ}}{(69.3 + 273.2)\text{K}}$$

$\Delta H^\circ_{vap} = 29.8$ kJ /mol
$SO_2Cl_2\ (l) \rightarrow SO_2Cl_2\ (g)$
$\Delta H^\circ_{vap} = 1 \times (-364.0 \text{ kJ /mol}) - 1 \times (-394.1 \text{ kJ /mol}) = 30.1$ kJ /mol
These values are in good agreement.

49. (a) $\Delta G = 0$ at equilibrium

(b) $$\ln K_p = \frac{\Delta G^\circ}{-RT} = \frac{-36.3\,\text{kJ/mol} \times 10^3\,\text{J/kJ}}{-8.3145\,\text{J/mol K} \times 850\,\text{K}}$$

$\ln K_p = 5.14$
$K_p = e^{5.14} = 171$

51. (a) $K_{eq} = P_{H_2O} P_{SO_2} = K_p$
(b) $K_{eq} = [Mg^{2+}][OH^-]^2 = K_{sp}$
(c) $$K_{eq} = \frac{[CH_3COOH][OH^-]}{[CH_3COO^-]} = \frac{K_w}{K_a}$$

53. (a) $\Delta G^\circ = 2 \text{ mol } SO_3 \times (-371.1 \text{ kJ /mol}) - 1 \text{ mol } O_2 \times 0$
$-2 \text{ mol } SO_2 \times (-300.2 \text{ kJ /mol}) = -141.8$ kJ

$\Delta G^\circ = -RT \ln K_p$

$$\ln K_p = \frac{-141.8 \text{ kJ /mol x } 10^3 \text{ J /kJ}}{-8.3145 \text{ J /mol K x 298 K}} = 57.2$$

$K_p = 7.2 \times 10^{24}$

(b) $\Delta G^\circ = 1 \text{ mol } CO_2 \times (-394.4 \text{ kJ /mol}) + 4 \text{ mol } H_2 \times 0$
$- 1 \text{ mol } CH_4 \times (-50.75 \text{ kJ /mol}) - 2 \text{ mol } H_2O(g) \times (-228.6 \text{ kJ /mol}) = 113.6$ kJ

$\Delta G^\circ = -RT \ln K_p$

$$\ln K_p = \frac{\Delta G^\circ}{-RT} = \frac{113.6 \text{ kJ x } 10^3 \text{ J /kJ}}{-8.3145 \text{ J /mol K x 298 K}}$$

$\ln K_p = -45.8$
$K_p = 1.3 \times 10^{-20}$

55. $\Delta G^\circ = \Delta H^\circ - T\Delta S^\circ$

$$\Delta G^\circ = 73.6 \text{ kJ /mol} - 298 \text{ K} \times 168.7 \text{ J /K mol} \times \frac{\text{kJ}}{10^3 \text{ J}}$$

$\Delta G^\circ = 23.3$ kJ /mol

$$\ln K_p = \frac{\Delta G^\circ}{-RT} = \frac{23.3 \text{ kJ /mol x } 10^3 \text{ J /kJ}}{-8.3145 \text{ J /mol K x 298 K}}$$

$\ln K_p = -9.40$

$K_p = 8.3 \times 10^{-5} = P_{C_{10}H_8}$

$$P_{C_{10}H_8} = 8.3 \times 10^{-5} \text{ atm} \times \frac{760 \text{ mmHg}}{\text{atm}} = 0.063 \text{ mmHg}$$

57. (a) $\dfrac{X_{CO_2}X_{H_2}}{X_{CO}X_{H_2O}} = \dfrac{\dfrac{P_{CO_2}}{P_T} \times \dfrac{P_{H2}}{P_T}}{\dfrac{P_{CO}}{P_T} \times \dfrac{P_{H_2O}}{P_T}} = \dfrac{P_{CO_2}P_{H_2}}{P_{CO}P_{H_2O}} = K_P$

$K_P = \dfrac{(0.320)^2}{0.0133 \text{ x } 0.347} = 22.2$

$\Delta G° = -RT \ln K_P = -8.3145 \text{ J/mol K} \times (345 + 273) \text{ K} \times \ln 22.2 \times \dfrac{\text{kJ}}{10^3 \text{ J}}$

$\Delta G° = -15.9 \text{ kJ/mol}$

(b) $Q_P = \dfrac{X_{CO_2}X_{H_2}}{X_{CO}X_{H_2O}} = \dfrac{\dfrac{\text{mol } CO_2}{\text{total mol}} \times \dfrac{\text{mol } H_2}{\text{total mol}}}{\dfrac{\text{mol CO}}{\text{total mol}} \times \dfrac{\text{mol } H_2O}{\text{total mol}}} = \dfrac{\text{mol } CO_2 \times \text{mol } H_2}{\text{mol CO} \times \text{mol } H_2O}$

$= \dfrac{0.145 \text{ mol } CO_2 \times 0.226 \text{ mol } H_2}{0.085 \text{ mol CO} \times 0.112 \text{ mol } H_2O} = 3.44$

$Q_P < K_P$ (22.2)
The reaction will go to the right.

(c)

	CO	+	H_2O	$\rightleftharpoons$	CO_2	+	H_2
I	0.085		0.112		0.145		0.226
C	$-X$		$-X$		$+X$		$+X$
E	$0.085-X$		$0.112-X$		$0.145+X$		$0.226+X$

$K = 22.2 = \dfrac{(0.145+X)(0.226+X)}{(0.085-X)(0.112-X)}$

$22.2 = \dfrac{0.0328 + 0.371X + X^2}{0.00952 - 0.197X + X^2}$

$22.2X^2 - 4.37X + 0.211 = X^2 + 0.371X + 0.0328$

$21.2X^2 - 4.74X + 0.178 = 0$

$X = \dfrac{4.74 \pm \sqrt{(4.74)^2 - 4 \text{ x } 21.1 \text{ x } 0.178}}{2 \text{ x } 21.2}$

$X = \dfrac{+4.74 \pm 2.73}{42.4}$

$X = 0.0474$ or 0.176 Too large, not valid.

? mol CO = 0.085 - 0.047 = 0.038 mol CO

? mol H_2O = 0.112 - 0.047 = 0.065 mol H_2O

? mol CO_2 = 0.145 + 0.047 = 0.192 mol CO_2

? mol H_2 = 0.226 + 0.047 = 0.273 mol H_2

Check: $\dfrac{0.192 \text{ mol } CO_2 \times 0.273 \text{ mol } H_2}{0.038 \text{ mol CO} \times 0.065 \text{ mol } H_2O} = 21.2$

59. (a) $C(graphite) + \frac{1}{2} O_2(g) \rightarrow CO(g)$

$\Delta G° = 1 \text{ mol CO} \times (-137.2 \text{ kJ/mol}) - 1 \text{ mol C} \times 0 - \frac{1}{2} \text{ mol } O_2 \times 0 = -137.2 \text{ kJ}$ No

(b) $1\ H_2(g) + \frac{1}{2} O_2(g) \rightarrow H_2O(g)$

$\Delta G° = 1 \text{ mol } H_2O \times (-228.6 \text{ kJ/mol}) - 1 \text{ mol } H_2 \times 0 - \frac{1}{2} \text{ mol } O_2 \times 0 = -228.6 \text{ kJ}$ No

(c) $CO(g) + \frac{1}{2} O_2(g) \rightarrow CO_2(g)$

$\Delta G° = 1 \text{ mol } CO_2 \times (-394.4 \text{ kJ/mol}) - 1 \text{ mol CO} \times (-137.2 \text{ kJ/mol}) - \frac{1}{2} \text{ mol } O_2 \times 0$

$= -257.2 \text{ kJ}$ Yes

61. $\Delta H°_{rxn} = 1 \text{ mol } CF_4 \times (-925 \text{ kJ/mol}) + 1 \text{ mol } SO_2 \times (-296.8 \text{ kJ/mol})$

$-1 \text{ mol } CO_2 \times (-393.5 \text{ kJ/mol}) - 1 \text{ mol } SF_4 \times (-763 \text{ kJ/mol})$

$\Delta H°_{rxn} = -65.3 \text{ kJ}$

$\Delta S°_{rxn} = 1 \text{ mol } CF_4 \times \frac{261.6 \text{ J}}{\text{mol K}} + 1 \text{ mol } SO_2 \times \frac{248.1 \text{ J}}{\text{mol K}} - 1 \text{ mol } CO_2 \times \frac{213.6 \text{ J}}{\text{mol K}}$

$- 1 \text{ mol } SF_4 \times \frac{299.6 \text{ J}}{\text{mol K}}$

$\Delta S° = \frac{-3.5 \text{ J}}{\text{K}}$

$\Delta G°318 = \Delta H°298 - T318\ \Delta S°298$

$\Delta G° = -65.3 \text{ kJ} - 318 \text{ K} \times \left(\frac{-3.5 \text{ J}}{\text{K}}\right) \times \frac{\text{kJ}}{10^3 \text{ J}}$

$\Delta G° = -64.2 \text{ kJ} = -RT \ln K_{eq}$

$$\ln K_{eq} = \frac{\frac{-64.2 \text{ kJ}}{\text{mol}} \times \frac{10^3 \text{ J}}{\text{kJ}}}{\frac{-8.3145 \text{ J}}{\text{K mol}} \times 318 \text{ K}} = 24.3$$

$e^{\ln Keq} = K_{eq} = e^{24.3} = 3.5 \times 10^{10}$

63. $$\ln\frac{P_2}{P_1} = \frac{\Delta H°}{R}\left(\frac{1}{T_1} - \frac{1}{T_2}\right)$$

$$\ln\frac{747 \text{ mmHg}}{760 \text{ mmHg}} = \frac{\frac{43.82 \text{ kJ}}{\text{mol}} \times \frac{10^3 \text{ J}}{\text{kJ}}}{\frac{8.3145 \text{ J}}{\text{mol K}}}\left(\frac{1}{(117.8 + 273.2)\text{K}} - \frac{1}{T_2}\right)$$

$$-0.0173 = 5.270 \times 10^3 \left(\frac{1}{391.0 \text{ K}} - \frac{1}{T_2}\right)$$

$$-3.28 \times 10^{-6} = 0.002558 \text{ K}^{-1} - \frac{1}{T_2}$$

$$\frac{1}{T_2} = 2.561 \times 10^{-3}\ \text{K}^{-1}$$
$$T_2 = 390.4\ \text{K} - 273.2 = 117.2\ °\text{C}$$

65. $\Delta H°_{rxn} = 2\ \text{mol NO}_2 \times 33.18\ \text{kJ/mol} - 2\ \text{mol NO} \times 90.25\ \text{kJ/mol} - 1\ \text{mol O}_2 \times 0 = -114.14\ \text{kJ}$

$$\Delta S°_{rxn} = 2\ \text{mol NO}_2 \times \frac{240.0\ \text{J}}{\text{mol K}} - 2\ \text{mol NO} \times \frac{210.6\ \text{J}}{\text{mol K}} - 1\ \text{mol O}_2 \times \frac{205.0\ \text{J}}{\text{mol K}} = \frac{-146.2\ \text{J}}{\text{K}}$$

$$\Delta G° = \Delta H° - T\Delta S° = -114.14\ \text{kJ} \times \frac{10^3\ \text{J}}{\text{kJ}} - (155\ °\text{C} + 273)\ \text{K} \times \left(\frac{-146.2\ \text{J}}{\text{K}}\right)$$
$$\Delta G° = -5.16 \times 10^4\ \text{J/mol} = -RT\ln K_{eq}$$
$$-5.16 \times 10^4\ \text{J/mol} = \frac{-8.3145\text{J}}{\text{K}} \times 428\ \text{K}\ln K$$
$$e^{\ln K} = K = e^{14.5} = 2.0 \times 10^6$$

67. $CCl_4(l) \rightarrow CCl_4(g)$

$$\Delta H°_{rxn} = \Delta H°_f(CCl_4\ (g)) - \Delta H°_f(CCl_4\ (l))$$
$$\Delta H°_{rxn} = \frac{-102.9\ \text{kJ}}{\text{mol}} - \left(\frac{-135.4\ \text{kJ}}{\text{mol}}\right) = \frac{32.5\ \text{kJ}}{\text{mol}}$$
$$\Delta G°_{rxn} = \Delta G°\ _f(CCl_4\ (g)) - \Delta G°\ _f(CCl_4\ (l))$$
$$\Delta G°_{rxn} = \frac{-60.63\ \text{kJ}}{\text{mol}} - \left(\frac{-65.27\ \text{kJ}}{\text{mol}}\right) = \frac{4.64\ \text{kJ}}{\text{mol}}$$
$$\Delta G° = \frac{4.64\ \text{kJ}}{\text{mol}} \times \frac{10^3\ \text{J}}{\text{kJ}} = -RT\ln K_p = \frac{-8.3145\ \text{J}}{\text{mol K}} \times 298.15\ \text{K} \times \ln K_p$$
$$\frac{4.64 \text{ x } 10^3\ \text{J}}{\text{mol}} = -\frac{2.48 \text{ x } 10^3\ \text{J}}{\text{mol}} \ln K_p$$
$$\ln K_p = -\ 1.87$$
$$e^{\ln K_p} = K_p = e^{-1.87} = 1.54 \times 10^{-1} = P_{CCl_4}\ \text{(in atm)}$$
$$\ln \frac{K_2}{K_1} = \frac{\Delta H°}{R}\left(\frac{1}{T_1} - \frac{1}{T_2}\right)$$

at the normal boiling point, $P_{CCl_4} = 1.00\ \text{atm} = K_1$

$$\ln \frac{0.154\ \text{atm}}{1.00\ \text{atm}} = \frac{\frac{32.5\ \text{kJ}}{\text{mol}} \text{ x } \frac{10^3\ \text{J}}{\text{kJ}}}{\frac{8.3145\ \text{J}}{\text{mol K}}} \times \left(\frac{1}{T_1} - \frac{1}{298.15\ \text{K}}\right)$$
$$-1.87 = 3.909 \times 10^3 \times \left(\frac{1}{T_1} - \frac{1}{298.15\ \text{K}}\right)$$
$$-\ 4.78 \times 10^{-4} = \left(\frac{1}{T_1} - \frac{1}{298.15\ \text{K}}\right)$$

$$\frac{1}{T_1} = 2.876 \times 10^{-3}$$

$T_1 = 347.7\text{ K} = 74.5\ °\text{C}$

69. $K_{eq} = P_{O_2} = 0.21\text{ atm}$

$$\Delta G° = -RT\ln K_{eq} = \frac{-8.3145\ JT}{\text{mol K}} \times \ln 0.21 = \frac{13.0\ JT}{\text{K}}$$

$\Delta H° = 1\text{ mol } O_2 \times 0\text{ kJ/mol} + 4\text{ mol Ag} \times 0\text{ kJ/mol} - 2\text{ mol } Ag_2O$
$\times (-31.0\text{ kJ/mol}) = 62.0\text{ kJ}$

$$\Delta S° = 1\text{ mol } O_2 \times \frac{205.0\text{ J}}{\text{mol K}} + 4\text{ mol Ag} \times \frac{42.55\text{ J}}{\text{mol K}} - 2\text{ mol } Ag_2O \times \frac{121\text{ J}}{\text{mol K}}$$
$= 133.2\text{ J/K}$

$$\Delta G° = 62.0\text{ kJ} \times \frac{10^3\text{ J}}{\text{kJ}} - T \times \frac{133.2\text{ J}}{\text{K}} = \frac{13.0\text{ J } T}{\text{K}}$$

$$6.20 \times 10^4\text{ J} = \frac{146.2\text{ J } T}{\text{K}}$$

$T = 424\text{ K} - 273$

$T = 151\ °\text{C}$

An alternate method is to determine K_p at 298.15 K and then use van't Hoff's equation to calculate the temperature at which $K_p = 0.21$.

71. The expansion in Figure 6.8 is not reversible because the process cannot be reversed by an infinitesimal change. The weights are too large; the change occurs too quickly.

74. In Chapter 14, it was seen that an endothermic reaction would be forced to the right by an increase in temperature. The effect of being forced to the right increases the K_{eq}. From the van't Hoff equation, for positive ΔH (endothermic) an increase in temperature increases the K_{eq}. That is, $\ln\frac{K_2}{K_1} = \frac{\Delta H°}{R}\left(\frac{1}{T_1} - \frac{1}{T_2}\right)$, and with $\Delta H > 0$ and $T_2 > T_1$, $K_2 > K_1$.

75. $\Delta H° = 1\text{ mole } CO_2 \times (-393.5\text{ kJ/mol}) + 1\text{ mol } H_2O \times (-241.8\text{ kJ/mol})$
$+ 1\text{ mol } Na_2CO_3 \times (-1131\text{ kJ/mol}) - 2\text{ mol } NaHCO_3 \times (-950.8\text{ kJ/mol})$

$\Delta H° = 135\text{ kJ/mol}$

$\Delta S° = 1\text{ mol } CO_2 \times 213.6\text{ J/mol K} + 1\text{ mol } H_2O \times 188.7\text{ J/mol K}$
$+1\text{ mol } Na_2CO_3 \times 135.0\text{ J/mol K} - 2\text{ mol } NaHCO_3 \times (102\text{ J/mol K})$

$\Delta S° = 333\text{ J/mol K}$

$\Delta G° = 135.3\text{ kJ} - 298.15\text{ K} \times 333\text{ J/K} \times \text{kJ}/10^3\text{ J}$

$$\ln K_p = \frac{\Delta G°}{-RT} = \frac{36.0\text{ kJ/mol} \times 10^3\text{ J/kJ}}{-8.3145\text{ J/mol K} \times 298.15\text{ K}}$$

$\ln K_p = -14.5$

$K_p = 5.0 \times 10^{-7}$

$P_{total} = 1.00 \text{ atm} = P_{CO_2} + P_{H_2O}$

by stoichiometry $P_{CO_2} = P_{H_2O} = 0.50$ atm

$K_p = P_{CO_2} \times P_{H_2O} = 0.50 \text{ atm} \times 0.50 \text{ atm} = 0.25$

$$\ln\frac{0.25}{5.0 \times 10^{-7}} = \frac{135 \text{ kJ/mol}}{8.3145 \text{ J/mol K}} \times \frac{10^3 \text{ J}}{\text{kJ}} \times \left(\frac{1}{298.15 \text{ K}} - \frac{1}{T}\right)$$

$$13.1 = 1.62 \times 10^4 \times \left(\frac{1}{298.15 \text{ K}} - \frac{1}{T}\right)$$

$$8.09 \times 10^{-4} = \frac{1}{298.15 \text{ K}} - \frac{1}{T}$$

$T = 392.9$ K

77. (a) $K_c = \dfrac{[CO][H_2O]}{[CO_2][H_2]} = \dfrac{0.224 \times 0.224}{0.276 \times 0.276} = 0.659$

$\Delta G^\circ = -RT \ln K_c$

$$\Delta G^\circ = \frac{-8.3145 \text{ J}}{\text{mol K}} \times 1000 \text{ K} \times \ln 0.659 \times \frac{\text{kJ}}{10^3 \text{ J}}$$

$\Delta G^\circ = 3.47$ kJ

(b) $CO_2 + H_2 \rightleftharpoons CO + H_2O$

0.0500 0.0700 0.0400 0.0850

$Q_c = \dfrac{0.0400 \times 0.0850}{0.0500 \times 0.0700} = 0.971$ $Q_c > K_c$

The reaction will go to the left.

(c) $+x$ $+x$ $-x$ $-x$

$0.0500 + x$ $0.0700 + x$ $0.0400 - x$ $0.0850 - x$

$$0.659 = \frac{(0.0400 - x)(0.0850 - x)}{(0.0500 + x)(0.0700 + x)}$$

$0.659x^2 + 0.0791x + 2.31 \times 10^{-3} = x^2 - 0.1250x + 3.40 \times 10^{-3}$

$0.341x^2 - 0.2041x + 1.09 \times 10^{-3} = 0$

$$x = \frac{0.2041 \pm \sqrt{(0.2041)^2 - 4 \times 0.341 \times 1.09 \times 10^{-3})}}{0.682}$$

$$x = \frac{0.2041 \pm 0.2004}{0.682}$$

$x = 0.593$ 5.4×10^{-3}

Too Large

Not Valid

? moles $CO_2 = 0.0500 + 0.0054 = 0.0554$ moles CO_2

? moles $H_2 = 0.0700 + 0.0054 = 0.0754$ moles H_2

? moles CO = 0.0400 – 0.0054 = 0.0346 moles CO
? moles H_2O = 0.0850 – 0.0054 = 0.0796 moles H_2O

$$\text{Check: } K_c = \frac{[CO][H_2O]}{[CO_2][H_2]} = \frac{0.0346\times 0.0796}{0.0554\times 0.0754} = 0.659$$

81. Hg(s) → Hg(l) $\quad \Delta H°_{fus} = 2.30$ kJ
Hg(l) → Hg(g) $\quad \Delta H°_{vapor} = 61.32$ kJ
Hg(s) → Hg(g) $\quad \Delta H°_{sub} = 63.62$ kJ

$$\Delta S°_{fus} = \frac{\Delta H°_{fus}}{T} = \frac{\Delta H°_{fus}}{(-38.86+273.15)\text{K}} = \frac{2.30\text{ kJ x }\frac{10^3\text{ J}}{\text{kJ}}}{(-38.86° + 273.15)} = \frac{9.82\text{ J}}{\text{K}}$$

$$\Delta S°_{vapor} = 1\text{ mol Hg(g)} \times \frac{174.9\text{ J}}{\text{mol K}} - 1\text{ mol Hg(l)} \times \frac{76.02\text{ J}}{\text{mol K}} = \frac{98.9\text{ J}}{\text{K}}$$

$$\Delta S°_{sub} = \Delta S°_{fus} + \Delta S°_{vap} = \frac{9.8\text{ J}}{\text{K}} + \frac{98.9\text{ J}}{\text{K}} = \frac{108.7\text{ J}}{\text{K}}$$

$$\Delta G°_{sub} = \Delta H°_{sub} - T\Delta S°_{sub} = 63.62\text{ kJ} \times \frac{10^3\text{ J}}{\text{kJ}} - (-78.5\text{ °C} + 273.2) \times \frac{108.7\text{ J}}{\text{K}}$$

$$= 4.246 \times 10^4\text{ J/mol} = -RT\ln K = \frac{-8.3145\text{ J}}{\text{mol K}} \times 194.7 \times \ln K$$

$-26.23 = \ln K$
$e^{\ln K} = K = e^{-26.23} = 4.06 \times 10^{-12}$

$$K = P_{Hg} = 4.06 \times 10^{-12}\text{ atm} \times \frac{760\text{ mmHg}}{\text{atm}} = 3.09 \times 10^{-9}\text{ mmHg}$$

83. $$\Delta H°_{rxn} = 1\text{ mole CO} \times \frac{-110.5\text{ kJ}}{\text{mol}} + 1\text{ mole } Cl_2 \times \frac{0\text{ kJ}}{\text{mol}} - 1\text{ mol } COCl_2 \times \frac{-220.9\text{ kJ}}{\text{mol}}$$
$$= 110.4\text{ kJ}$$

$$\Delta S°_{rxn} = 1\text{ mole CO} \times \frac{197.6\text{ J}}{\text{mole K}} + 1\text{ mole } Cl_2 \frac{223.0\text{ kJ}}{\text{mole}} - 1\text{ mol } COCl_2 \times \frac{283.8\text{ kJ}}{\text{mol}}$$
$$= \frac{136.8\text{ J}}{\text{K}}$$

$$\Delta G° = 110.4\text{ kJ} \times \frac{10^3\text{ J}}{\text{kJ}} - T \times \frac{136.8\text{ J}}{\text{K}}$$

$$COCl_2 \rightleftharpoons CO + Cl_2$$

$$K_p = \frac{\alpha^2 P}{1-\alpha^2} = \frac{(0.100)^2 \times 1.00\text{ atm}}{1-(0.100)^2}$$

$K_p = 1.01 \times 10^{-2}$

$$\Delta G° = -RT\ln K_p = \frac{-8.3145\text{ J}}{\text{mol K}} T\ln 1.01 \times 10^{-2} = \frac{3.82\text{ x }10^1\text{ J}}{\text{K}} T$$

$$\Delta G° = 1.104 \times 10^5\text{ J} - \frac{136.8\text{ J}}{\text{K}} T = \frac{3.82\text{ x }10^1\text{ J}}{\text{K}} T$$

$$\Delta G° = 1.104 \times 10^5 \text{ J} = \frac{175.0 \text{ J}}{\text{K}} T$$

$T = 631$ K - 273 = 358 °C

87. (a) False. $\Delta G° = 0$ at a single temperature at which products and reactants are at standard state conditions. $H_2O(g)$ is not at standard state.
(b) False. $H_2O(g)$ is not in its standard state, so the system is not at equilibrium.
(c) False. For $H_2O(g) = 1.5$ atm, $K_{eq} = 1.5$, $\Delta G° = -RT \ln K_{eq}$, and $\Delta G° < 0$.
(d) True. The process is nonspontaneous.
$\Delta G = 0$ applies to non-standard state conditions, and $\Delta G = 0$ is a condition for equilibrium. $\Delta G°$ applies only to standard-state conditions. If the standard-state conditions are also equilibrium conditions, $\Delta G° = 0$.

89. $PCl_5(g) \rightleftharpoons PCl_3(g) + Cl_2(g)$

$$\Delta G° = 1 \text{ mol } PCl_3 \times \left(\frac{-267.8 \text{ kJ}}{\text{mol}}\right) + 2 \text{ mol } Cl_2 \times 0 - 1 \text{ mol } PCl_5 \times \left(\frac{-305.0 \text{ kJ}}{\text{mol}}\right)$$
$$= 37.2 \text{ kJ}$$

$$\ln K = \frac{\Delta G°}{-RT} = \frac{\frac{37.2 \text{ kJ}}{\text{mol}} \times \frac{10^3 \text{ J}}{\text{kJ}}}{\frac{-8.3145 \text{ J}}{\text{mol K}} \times 298.15 \text{ K}} = -15.0$$

$e^{\ln K} = K = e^{-15.0} = 3.06 \times 10^{-7}$

$$\Delta H° = 1 \text{ mol } PCl_3 \times \left(\frac{-287.0 \text{ kJ}}{\text{mol}}\right) - 1 \text{ mol } Cl_2 \times 0 - 1 \text{ mol } PCl_5 \times \left(\frac{-374.9 \text{ kJ}}{\text{mol}}\right)$$
$$= 87.9 \text{ kJ}$$

$$\ln \frac{K_2}{K_1} = \frac{\Delta H°}{R}\left(\frac{1}{T_1} - \frac{1}{T_2}\right)$$

$$\ln \frac{K_2}{3.06 \times 10^{-7}} = \frac{\frac{87.9 \text{ kJ}}{\text{mol}} \times \frac{10^3 \text{ J}}{\text{kJ}}}{\frac{8.3145 \text{ J}}{\text{K mol}}}\left(\frac{1}{298.15 \text{ K}} - \frac{1}{500.15 \text{ K}}\right)$$

$$\ln \frac{K_2}{3.06 \times 10^{-7}} = 14.32$$

$$\frac{K_2}{3.06 \times 10^{-7}} = e^{14.32} = 1.66 \times 10^6$$

$K_2 = 0.508$

$$P_{Init} = \frac{nRT}{V} = \frac{0.100 \text{ mol} \times \frac{0.08206 \text{ L atm}}{\text{K mol}} \times 500 \text{ K}}{1.50 \text{ L}}$$

$P_{Init} = 2.74$ atm

$PCl_5 \rightleftharpoons PCl_3 + Cl_2$

2.74

$-x \quad x \quad x$

$2.74-x \quad x \quad x$

$$0.508 = \frac{x^2}{2.74 - x}$$

$$x^2 + 0.508\,x - 1.39 = 0$$

$$x = \frac{-0.508 \pm \sqrt{(0.508)^2 - 4 \times 1 \times (-1.39)}}{2}$$

$$x = \frac{-0.508 \pm 2.41}{2}$$

$x = -1.46$ Not Valid $\qquad$ 0.95 atm

$P_{PCl_5} = 2.74 - x = 1.79$ atm

$P_{Cl_2} = P_{PCl_3} = 0.95$ atm

$P_{total} = 3.69$ atm

Apply Your Knowledge

92. (a) $P_{O_2} = 0.25$ atm

$$(P_{O_2})^{\frac{1}{2}} = 0.50 = K_{eq2}$$

At 25 °C: $\Delta G^\circ = 58.56 \times 10^3$ J/mol $= -RT\ln K_{eq}$

58.56×10^3 J/mol $= -8.3145$ J/mol K $\times$ 298.15 K $\times \ln K_{eq}$

$\ln K_{eq} = -23.62$

$K_{eq1} = 5.52 \times 10^{-11}$

$$\ln\frac{K_2}{K_1} = \frac{-90.83 \times 10^3 \text{ J/mol}}{8.3145 \text{ J/mol K}}\left(\frac{1}{T_2} - \frac{1}{T_1}\right)$$

$$\ln\frac{0.50}{5.52 \times 10^{-11}} = \frac{-90.83 \times 10^3 \text{ J/mol}}{8.3145 \text{ J/mol K}}\left(\frac{1}{T_2} - \frac{1}{298.15\text{ K}}\right)$$

$$-2.10 \times 10^{-3}\text{ K}^{-1} = \left(\frac{1}{T_2} - \frac{1}{298.15\text{ K}}\right)$$

$$\frac{1}{T_2} = -2.10 \times 10^{-3}\text{ K}^{-1} + 3.354 \times 10^{-3}\text{ K}^{-1}$$

$T_2 = 797$ K $- 273. = 524$°C

(b) In order for the reaction

$HgO(s) \rightarrow Hg(l) + \frac{1}{2} O_2(g)$ $\qquad \Delta G^\circ = 58.56$ kJ/mol

to be coupled with another, the ΔG° for the sum of the two reactions must be negative. If the reaction to be coupled with the one above is based on ½ $O_2(g)$, $\Delta G^\circ_{tot} < 0$ or the coupling reaction must have $\Delta G^\circ < -58.56$ kJ/mol. If the coupling reaction is based on something other than the ½ $O_2(g)$, the two reactions must be combined as in the case for aluminum shown below. If the principal reactant in the coupling reaction is already in its highest oxidation state, that reactant cannot be oxidized further.

<u>Yes</u> $H_2(g) + \frac{1}{2} O_2(g) \rightarrow H_2O(l)$ $\Delta G° = -228.6$ kJ/mol
$H_2(g) + \frac{1}{2} O_2(g) \rightarrow H_2O(g)$ $\Delta G° = -237.2$ kJ/mol
<u>Yes</u> C (graphite) + ½ $O_2(g) \rightarrow CO(g)$ $\Delta G° = -137.2$ kJ/mol
<u>Yes</u> $CO(g) + \frac{1}{2} O_2(g) \rightarrow CO_2(g)$ $\Delta G° = -257.2$ kJ/mol
<u>No</u> $CO_2(g)$ (CO_2 cannot be oxidized.)
<u>Yes</u> $Cu_2O(s) + \frac{1}{2} O_2(g) \rightarrow 2\ CuO(s)$ $\Delta G° = -113.4$ kJ/mol
<u>No</u> CuO(s) (CuO cannot be oxidized.)
<u>No</u> N_2 (g) (All oxides of nitrogen have $\Delta G°_f > 0$.)
<u>No</u> $N_2O(g) + \frac{1}{2} O_2(g) \rightarrow 2NO(g)$ $\Delta G° = +68.94$ kJ/mol
<u>No</u> $NO(g) + \frac{1}{2} O_2(g) \rightarrow NO_2(g)$ $\Delta G° = -35.27$ kJ/mol
<u>Yes</u> $SO_2(g) + \frac{1}{2} O_2(g) \rightarrow SO_3(g)$ $\Delta G° = -70.9$ kJ/mol
<u>No</u> $SO_3(g)$ (SO_3 cannot be oxidized.)
<u>No</u> $Cl_2\ (g) + \frac{1}{2} O_2(g) \rightarrow Cl_2O(g)$ $\Delta G° = +97.49$ kJ/mol
<u>Yes</u> $Al(s) + 3/4\ O_2(g) \rightarrow \frac{1}{2} Al_2O_3(s)$ $\Delta G° = -791$ kJ/mol
$3\ HgO(s) \rightarrow 3\ Hg(l) + 3/2\ O_2(g)$ $\Delta G° = +175.7$ kJ/mol
$\Delta G_{tot} = -615$ kJ/mol
<u>No</u> $2\ Ag(g) + \frac{1}{2} O_2(g) \rightarrow Ag_2O(s)$ $\Delta G° = -11.2$ kJ/mol
<u>No</u> MgO(s) (MgO cannot be oxidized.)

94. $ATP + H_2O \rightarrow ADP + HPO_4^{2-}$ $\Delta G° = -30.5\ \frac{kJ}{mol}$

<u>HPO_4^{2-} + glucose → H_2O + glucose-6-phosphate</u> $\Delta G° = 13.9\ \frac{kJ}{mol}$

ATP + glucose → ADP + glucose-6-phosphate $\Delta G° = -16.6\ \frac{kJ}{mol}$

Normal body temperature = 37 °C = 310 K

$$\ln K = \frac{\Delta G°}{-RT} = \frac{\frac{-16.6\ kJ}{mol} \times \frac{10^3\ J}{kJ}}{\frac{-8.314\ J}{mol\ K} \times 310\ K} = 6.44$$

$e^{\ln K} = K = e^{6.44} = 627$

Concentrations of glucose-6-phosphate that is more than 627 times the glucose concentration will cause the reverse reaction to be spontaneous.

97. $\Delta G = \Delta G° + RT \ln Q$

$\Delta G = -RT \ln K + RT \ln Q$

$$\Delta G = \frac{-8.3145\ J}{mol\ K} \times 310\ K \times \ln 2.22 \times 10^5 + \frac{8.3145\ J}{mol\ K} \times 310\ K \times \ln \frac{6.0 \times 4.0}{3.0}$$

$$\Delta G = (-3.17 \times 10^4\ \text{J} + 5.36 \times 10^3\ \text{J}) \times \frac{\text{kJ}}{10^3\ \text{J}}$$

$$\Delta G = -26.3\ \text{kJ}$$

98. (a) $36\ \% = 100\ \% \times \dfrac{T_h - 314\,\text{K}}{T_h}$

$0.36\ T_h = T_h - 314\ \text{K}$

$0.64\ T_h = 314\ \text{K}$

$T_h = \dfrac{314}{0.64} = 491\ \text{K}$

The temperature of the steam is probably higher because some of the heat is lost to the surroundings.

(b) $\ln \dfrac{P_2}{P_1} = \dfrac{\Delta H^\circ_{vap}}{P}\left(\dfrac{1}{T_1} - \dfrac{1}{T_2}\right)$

$$\ln \frac{P_2}{1.00\,\text{atm}} = \frac{\dfrac{40\ \text{kJ}}{\text{mol}} \times \dfrac{10^3\ \text{J}}{\text{kJ}}}{\dfrac{8.3145\,\text{J}}{\text{mol K}}}\left(\frac{1}{373\,\text{K}} - \frac{1}{491\,\text{K}}\right)$$

$\ln P_2 = 3.10$

$e^{\ln P_2} = e^{3.10} = 22$

$P_2 = 22\ \text{atm}$

(c) You can't win, because you have to put energy into the engine to get work out, and you can't break even, because more energy has to be put in than work will be gotten out because of the loss of heat.

Chapter 18

Electrochemistry

Exercises

18.1A $Zn(s) \rightarrow Zn^{2+}$ $\qquad NO_3^- \rightarrow N_2O$

$2\ NO_3^- \rightarrow N_2O$

$2\ NO_3^- \rightarrow N_2O + 5\ H_2O$

$2\ NO_3^- + 10\ H^+ \rightarrow N_2O + 5\ H_2O$

$Zn(s) \rightarrow Zn^{2+} + 2\ e^-$ $\qquad 2\ NO_3^- + 10\ H^+ + 8\ e^- \rightarrow N_2O + 5\ H_2O$

$4\ Zn \rightarrow 4\ Zn^{2+} + 8\ e^-$

$4\ Zn(s) + 2\ NO_3^-(aq) + 10\ H^+(aq) \rightarrow 4\ Zn^{2+}(aq) + N_2O(g) + 5\ H_2O(l)$

18.1B $P_4 \rightarrow H_2PO_4^-$ $\qquad NO_3^- \rightarrow NO$

$P_4 \rightarrow 4\ H_2PO_4^-$

$P_4 + 16\ H_2O \rightarrow 4\ H_2PO_4^-$ $\qquad NO_3^- \rightarrow NO + 2\ H_2O$

$P_4 + 16\ H_2O \rightarrow 4\ H_2PO_4^- + 24\ H^+$ $\qquad NO_3^- + 4\ H^+ \rightarrow NO + 2\ H_2O$

$P_4 + 16\ H_2O \rightarrow 4\ H_2PO_4^- + 24\ H^+ + 20\ e^-$ $\qquad NO_3^- + 4\ H^+ + 3\ e^- \rightarrow NO + 2\ H_2O$

$3\ P_4 + 48\ H_2O \rightarrow 12\ H_2PO_4^- + 72\ H^+ + 60\ e^-$

$20\ NO_3^- + 80\ H^+ + 60\ e^- \rightarrow 20\ NO + 40\ H_2O$

$3\ P_4(s) + 20\ NO_3^-(aq) + 8\ H_2O(l) + 8\ H^+(aq) \rightarrow 12\ H_2PO_4^-(aq) + 20\ NO(g)$

18.2A $OCN^- \rightarrow CO_3^{2-} + N_2$ $\qquad OCl^- \rightarrow Cl^-$

$2\ OCN^- \rightarrow 2\ CO_3^{2-} + N_2$

$2\ OCN^- + 4\ H_2O \rightarrow 2\ CO_3^{2-} + N_2$ $\qquad OCl^- \rightarrow Cl^- + H_2O$

$2\ OCN^- + 4\ H_2O \rightarrow 2\ CO_3^{2-} + N_2 + 8\ H^+$ $\qquad OCl^- + 2H^+ \rightarrow Cl^- + H_2O$

$2\ OCN^- + 4\ H_2O \rightarrow 2\ CO_3^{2-} + N_2 + 8\ H^+ + 6\ e^-$

$OCl^- + 2H^+ + 2\ e^- \rightarrow Cl^- + H_2O$

$3\ OCl^- + 6\ H^+ + 6\ e^- \rightarrow 3\ Cl^- + 3\ H_2O$

$2\ OCN^- + 3\ OCl^- + H_2O \rightarrow 2\ CO_3^{2-} + N_2 + 2\ H^+ + 3\ Cl^-$

$2\ OCN^- + 3\ OCl^- + H_2O + 2\ OH^- \rightarrow 2\ CO_3^{2-} + N_2 + 2\ H_2O + 3\ Cl^-$

$2\ OCN^-(aq) + 3\ OCl^-(aq) + 2\ OH^-(aq)$
$\rightarrow 2\ CO_3^{2-}(aq) + N_2(g) + H_2O(l) + 3\ Cl^-(aq)$

18.2B $MnO_4^- \rightarrow MnO_2$

$CH_3CH_2OH \rightarrow CH_3CO_2^-$

$MnO_4^- \rightarrow MnO_2 + 2\ H_2O$

$C_2H_6O + H_2O \rightarrow C_2H_3O_2^-$

$MnO_4^- + 4\ H^+ \rightarrow MnO_2 + 2\ H_2O$

$C_2H_6O + H_2O \rightarrow C_2H_3O_2^- + 5\ H^+$

$MnO_4^- + 4\ H^+ + 3\ e^- \rightarrow MnO_2 + 2\ H_2O$

$C_2H_6O + H_2O \rightarrow C_2H_3O_2^- + 5\ H^+ + 4\ e^-$

$4\ MnO_4^- + 16\ H^+ + 12\ e^- \rightarrow 4\ MnO_2 + 8\ H_2O$

$3\ C_2H_6O + 3\ H_2O \rightarrow 3\ C_2H_3O_2^- + 15\ H^+ + 12\ e^-$

$4\ MnO_4^- + 3\ C_2H_6O + H^+ \rightarrow 4\ MnO_2 + 3\ C_2H_3O_2^- + 5\ H_2O$

$4\ MnO_4^- + 3\ C_2H_6O + H_2O \rightarrow 4\ MnO_2 + 3\ C_2H_3O_2^- + 5\ H_2O + OH^-$

$4\ MnO_4^-(aq) + 3\ C_2H_6O(aq)$
$\rightarrow 4\ MnO_2(s) + 3\ C_2H_3O_2^-(aq) + 4\ H_2O(l) + OH^-(aq)$

$4\ MnO_4^-(aq) + 3\ CH_3CH_2OH(aq)$
$\rightarrow 4\ MnO_2(s) + 3\ CH_3CO_2^-(aq) + 4\ H_2O(l) + OH^-(aq)$

18.3A anode $\{Al(s) \rightarrow Al^{3+} + 3\ e^-\} \times 2$
cathode $\underline{\{2\ H^+ + 2\ e^- \rightarrow H_2(g)\} \times 3}$
$2\ Al(s) + 6\ H^+(aq) \rightarrow 3\ H_2(g) + 2\ Al^{3+}(aq)$

18.3B anode $Cu(s) \rightarrow Cu^{2+} + 2\ e^-$
cathode $\{Ag^+ + e^- \rightarrow Ag(s)\} \times 2$
Cell reaction: $Cu(s) + 2\ Ag^+ \rightarrow Cu^{2+} + 2\ Ag(s)$
Cell diagram: $Cu(s)\ |\ Cu^{2+}(aq)\ \|\ Ag^+(aq)\ |\ Ag(s)$

18.4A $I_2 + 2\ e^- \rightarrow 2\ I^-$ $E°_{I_2/2I^-} = 0.535\ V$

$E°_{cell} = E°(right) - E°(left) = E°(cathode) - E°(anode) = E°_{I_2/2I^-} - E°_{Sm^{2+}/Sm}$

$3.21\ V = 0.535 - E°_{Sm^{2+}/Sm}$

$E°_{Sm^{2+}/Sm} = 0.535\ V - 3.21\ V = -2.67\ V$

18.4B $Cl_2 + 2\,e^- \rightarrow 2\,Cl^-$ $\quad E^\circ_{Cl_2/2Cl^-} = 1.358\text{ V}$

$E^\circ_{cell} = E^\circ(\text{right}) - E^\circ(\text{left}) = E^\circ(\text{cathode}) - E^\circ(\text{anode})$

$$= E^\circ_{Cl_2/2Cl^-} - E^\circ_{ClO_4^-/Cl_2}$$

$-0.034\text{ V} = 1.358\text{ V} - E^\circ_{ClO_4^-/Cl_2}$

$E^\circ_{ClO_4^-/Cl_2} = 1.392\text{ V}$

18.5A (a) $\{Al \rightarrow Al^{3+} + 3\,e^-\} \times 2$ oxidation anode

$[Cu^{2+} + 2\,e^- \rightarrow Cu\} \times 3$ reduction cathode

$E^\circ_{cell} = E^\circ(\text{cathode}) - E^\circ(\text{anode})$

$E^\circ_{cell} = E^\circ_{Cu^{2+}/Cu} - E^\circ_{Al^{3+}/Al}$

$E^\circ_{cell} = 0.340\text{ V} - (-1.676\text{ V})$

$E^\circ_{cell} = 2.016\text{ V}$

(b) $NO_3^- + 4\,H^+ + 3\,e^- \rightarrow NO + 2\,H_2O$ reduction cathode

$Pb^{2+} + 2\,H_2O \rightarrow PbO_2 + 4\,H^+ + 2\,e^-$ oxidation anode

$E^\circ_{cell} = E^\circ(\text{cathode}) - E^\circ(\text{anode})$

$E^\circ_{cell} = E^\circ_{NO_3^-/NO} - E^\circ_{PbO_2/Pb^{2+}}$

$E^\circ_{cell} = 0.956\text{ V} - 1.455\text{ V}$

$E^\circ_{cell} = -0.499\text{ V}$

18.5B $S_2O_8^{2-} + 2\,e^- \rightarrow 2\,SO_4^{2-}$ reduction cathode

$Mn^{2+} + 4\,H_2O \rightarrow MnO_4^- + 8\,H^+ + 5\,e^-$ oxidation anode

$2\,Mn^{2+}(aq) + 5\,S_2O_8^{2-}(aq) + 8\,H_2O(l)$

$$\rightarrow 2\,MnO_4^-(aq) + 10\,SO_4^{2-}(aq) + 16\,H^+(aq)$$

$E^\circ_{cell} = E^\circ(\text{cathode}) - E^\circ(\text{anode}) = E^\circ_{S_2O_8^{2-}/SO_4^{2-}} - E^\circ_{MnO_4^-/Mn^{2+}}$

$E^\circ_{cell} = 2.01\text{ V} - 1.51\text{ V}$

$E^\circ_{cell} = 0.50\text{ V}$

18.6A $Cu^{2+} + 2\ e^- \rightarrow Cu(s)$ $\qquad E^\circ_{Cu^{2+}/Cu} = 0.340\ V$

$\{Fe^{2+} \rightarrow Fe^{3+} + e^-\} \times 2$ $\qquad E^\circ_{Fe^{3+}/Fe^{2+}} = 0.771\ V$

$E^\circ_{cell} = E^\circ(\text{cathode}) - E^\circ(\text{anode})$

$E^\circ_{cell} = 0.340\ V - 0.771\ V = -0.431\ V$

Since E°_{cell} is negative, the reaction is not spontaneous in the forward direction.

18.6B $Mn^{2+} + 4\ H_2O \rightarrow MnO_4^- + 8\ H^+ + 5\ e^-$ oxidation anode

$2\ IO_3^- + 12\ H^+ + 10\ e^- \rightarrow I_2 + 6\ H_2O$ reduction cathode

$E^\circ_{cell} = E^\circ(\text{cathode}) - E^\circ(\text{anode})$

$E^\circ_{cell} = E^\circ_{IO_3^-/I_2} - E^\circ_{MnO_4^-/Mn^{2+}}$

$E^\circ_{cell} = 1.20\ V - 1.51\ V = -0.31\ V$

Since E°_{cell} is negative, the reverse direction is spontaneous.

18.7A Where the zinc strip touches the citric acid in the lemon, zinc metal atoms give up electrons (oxidation of Zn to Zn^{2+} occurs on the zinc electrode), because zinc is below SHE on Table 18.1 of standard electrode potentials. The electrons go through the zinc metal, the voltmeter, and the copper metal to where it touches the juice of the lemon. At that point, H^+ is changed to $H_2(g)$ (the reduction half-reaction is that of $H^+(aq)$ in citric acid to $H_2(g)$). The lemon juice supplies the H^+ and acts as an electrolyte. (Notice that both strips have to be in the same section of the lemon.) In part this reduction occurs directly on the zinc electrode, but also some electrons pass through the electric measuring circuit to the copper electrode, where reduction of $H^+(aq)$ also occurs.

18.7B (a) Acetic acid is a weak acid, bubbles will still form but not as rapidly.

(b) Nitric acid will provide the hydrogen ions to react with Zn, but nitrate will also react with copper to produce copper ions and NO gas. The copper ions will cause the solution to be blue in color.

(c) Result should be similar to Figure 18.11, but with the copious collection of $H_2(g)$ bubbles on the surface of the pool of mercury and a smaller amount on the zinc strip. The mercury will conduct an electric current the same as the zinc strip.

18.8A $\{Mg \rightarrow Mg^{2+} + 2\ e^-\} \times 3$ oxidation anode

$\{Al^{3+} + 3\ e^- \rightarrow Al\} \times 2$ reduction cathode

$E^\circ_{cell} = E^\circ(\text{cathode}) - E^\circ(\text{anode})$

$E^\circ_{cell} = E^\circ_{Al^{3+}/Al} - E^\circ_{Mg^{2+}/Mg} = -1.676\ V - (-2.356\ V)$

$E°cell = 0.680$ V

$\Delta G° = -nFE° = -6$ moles $e^- \times 96485$ C $(mol\ e^-)^{-1} \times 0.680$ V

$\Delta G° = -3.94 \times 10^5\ J \times \frac{kJ}{10^3\ J} = -394$ kJ

$E°cell = \frac{0.0592}{n} \log K_{eq}$ (The same results comes from $E°cell = \frac{0.02569}{n}$ ln K_{eq})

$\log K_{eq} = \frac{nE^{\circ}_{cell}}{0.0592} = \frac{6 \times 0.680}{0.0592} = 68.9$

Raise both sides of the equation to a power of e.

$10^{\log K_{eq}} = K_{eq} = 10^{68.9} = 8 \times 10^{68}$

18.8B $Ag \rightarrow Ag^+ + e^-$ oxidation anode

$NO_3^- + 4\ H^+ + 3\ e^- \rightarrow NO + 2\ H_2O$ reduction cathode

$3\ Ag(s) + NO_3^-\ (aq) + 4\ H^+(aq) + 3\ e^- \rightarrow 3\ Ag^+(aq) + NO(g) + 2\ H_2O(l)$

$E°cell = \frac{0.0592}{n} \log K_{eq}$

$E°cell = E°_{NO_3^-/NO} - E°_{Ag^+/Ag} = 0.956\ V - 0.800\ V = 0.156$ V

$E°cell = 0.156\ V = \frac{0.0592}{n} \log K_{eq}$

$\log K_{eq} = \frac{3 \times 0.156}{0.0592} = 7.91$

Raise both sides of the equation to a power of e.

$10^{\log K_{eq}} = K_{eq} = 10^{7.91} = 8.0 \times 10^7$

18.9A (a) $Zn\ (s) \rightarrow Zn^{2+} + 2\ e^-$ $E°_{Zn/Zn^{2+}} = -0.763$ V

$Cu^{2+} + 2\ e^- \rightarrow Cu\ (s)$ $E°_{Cu^{2+}/Cu} = 0.340$ V

$E°cell = E°(cathode) - E°(anode) = 0.340 - (-0.763\ V) = 1.103$ V

$E_{cell} = E°cell - \frac{0.0592}{n} \log \frac{[Zn^{2+}]}{[Cu^{2+}]}$

$E_{cell} = 1.103\ V - \frac{0.0592}{n} \log \frac{(2.0)}{(0.050)} = 1.103\ V - 0.047$ V

$E_{cell} = 1.056$ V

(b) $E_{cell} = E°cell - \frac{0.0592}{2} \log \frac{(0.050)}{(2.0)} = 1.103\ V + 0.047$ V

$E_{cell} = 1.150$ V

18.9B Cu (s) $\rightarrow$ Cu^{2+} + 2 e^- $\qquad E°_{Cu^{2+}/Cu}$ = 0.340 V

Cl_2 (g) + 2 e^- $\rightarrow$ 2 Cl^- $\qquad E°_{Cl_2/Cl^-}$ = 1.358 V

$E°_{cell} = E°(\text{cathode}) - E°(\text{anode}) = 1.358\text{ V} - 0.340\text{ V} = 1.018\text{ V}$

$$E_{cell} = E°_{cell} - \frac{0.0592}{2}\log\frac{(1.0)(0.25)^2}{(0.50)}$$

$E_{cell} = 1.018\text{ V} + 0.027\text{ V}$

$E_{cell} = 1.045\text{ V}$

18.10A The metal in the bend in the nail is more strained and thus more energetic, so it, along with the head and tail of the nail, is preferentially oxidized. Reduction occurs along the rest of the body of the nail.

18.10B In time there would be a formation of rust on the nail at the same points where the blue precipitate is seen in Figure 18.20.

18.11A The principal species in the solution are K^+ and Br^- ions and H_2O molecules.

K^+(aq) + e^- $\rightarrow$ K(s) $\qquad E°_{K^+/K}$ = –2.924 V

It is expected that the reduction half-reaction is:

2 H_2O(l) + 2 e^- $\rightarrow$ H_2(g) + 2 OH^-(aq) $\qquad E°_{H_2O/H_2}$ = –0.828 V

The two possibilities for oxidation are that of water molecules to O_2(g) and Br^-(aq) to Br_2(l). These oxidation half-reactions and their $E°_{ox}$ values are

2 Br^-(aq) $\rightarrow$ Br_2(l) + 2 e^- $\qquad E°_{Br^-/Br_2}$ = 1.065 V

2 H_2O(l) $\rightarrow$ O_2(g) + 4 H^+(aq) + 4 e^- $\qquad E°_{H_2O/O_2}$ = 1.229 V

We should expect the oxidation of Br^-(aq) to predominate and the net electrolysis reaction to be

2 Br^-(aq) + 2 H_2O(l) $\xrightarrow{\text{electrolysis}}$ Br_2(l) + H_2(g) + 2 OH^-(aq) $\qquad E°_{cell}$ = –1.893 V

18.11B Anode: Ag(s) $\rightarrow$ Ag^+(aq) + e^- $\qquad E°_{Ag^+/Ag}$ = – 0.800 V

Cathode: Cu^{2+}(aq) + 2 e^- $\rightarrow$ Cu(s) $\qquad E°_{Cu^{2+}/Cu}$ = 0.340 V

2 Ag(s) + Cu^{2+}(aq) $\longrightarrow$ Cu(s) +2 Ag^+(aq)

$E°_{cell} = E°(\text{cathode}) - E°(\text{anode}) = 0.340\text{ V} - 0.800\text{ V} = -0.460\text{ V}$

To force the electrolysis, 2 Ag(s) + Cu^{2+}(aq) $\longrightarrow$ Cu(s) +2 Ag^+(aq) requires the external voltage > 0.460 V. Note that without the external voltage, the solid copper will spontaneously deposit Ag(s) from $AgNO_3$(aq). (Recall Figure 18.1.)

18.12A Cell A $Zn^{2+} + Cu(s) \rightarrow Zn(s) + Cu^{2+}$
Cell B $Zn(s) + Cu^{2+} \rightarrow Zn^{2+} + Cu(s)$

18.12B The current will continue to flow only as long as the concentrations are unequal—that is, as long as there is a force to push the electrons.

18.13A $Cu^{2+}(aq) + 2e^- \rightarrow Cu(s)$

$$\frac{1.00\text{ g Cu}}{2.25\text{ A}} \times \frac{\text{mol}}{63.55\text{ g}} \times \frac{2\text{ mol e}^-}{\text{mol Cu}} \times \frac{96485\text{ C}}{\text{mol e}^-} \times \frac{1\text{ A x 1 s}}{1\text{ C}} \times \frac{\text{min}}{60\text{ s}} = 22.5\text{ min}$$

18.13B(a) $t = (21\text{ min} \times \frac{60\text{ s}}{\text{min}}) + 12\text{ s} = 1260\text{ s} + 12\text{ s} = 1272\text{ s}$

$$?\text{ C} = 2.175\text{ g Ag} \times \frac{\text{mol}}{107.87\text{ g}} \times \frac{\text{mol e}^-}{\text{mol Ag}} \times \frac{96485\text{ C}}{\text{mol e}^-} = 1.945 \times 10^3\text{ C}$$

(b) $?\text{ amp} = \frac{1.945 \times 10^3\text{ C}}{1272\text{ s}} = 1.529\text{ amp}$

18.14A $2\,Cl^-(aq) \rightarrow 2\,e^- + Cl_2(g)$ $E^\circ_{Cl_2/Cl^-} = 1.358\text{ V}$

$2\,H_2O(l) \rightarrow O_2(g) + 4\,H^+(aq) + 4\,e^-$ $E^\circ_{O_2/H_2O} = 1.229\text{ V}$

$2\,I^-(aq) \rightarrow 2\,e^- + I_2(s)$ $E^\circ_{I_2/I^-} = 0.535\text{ V}$

Nitrate ion cannot be oxidized, so the solution of magnesium nitrate will produce $O_2(g)$ at the anode. For each mole of electrons, there will be one-fourth of a mole of gas. $O_2(g)$ would be produced from a dilute solution of sodium chloride, $Cl_2(g)$ is produced from a concentrated solution of sodium chloride due to the over-voltage. For each mole of electrons, there will be one-half of a mole of chlorine gas produced but only one-fourth a mole of oxygen gas.

$I_2(s)$ would be produced at the anode from a potassium iodide solution. That is a solid—no gas is produced.

The sodium chloride solution would produce the most gas for the same amount of charge unless the solution is a very dilute solution, then it will produce the same as the magnesium nitrate.

18.14B (a) $?\text{ sec} = \frac{1.00\text{ g}}{2.50\text{ A}} \times \frac{\text{As}}{\text{C}} \times \frac{96500\text{ C}}{\text{mol e}^-} \times \frac{\text{mol e}^-}{\text{mol Na}} \times \frac{\text{mol Na}}{22.99\text{ g}} = 1.68 \times 10^3\text{ s}$

(b) $?\text{ sec} = \frac{1.00\text{ g}}{2.50\text{ A}} \times \frac{\text{As}}{\text{C}} \times \frac{96500\text{ C}}{\text{mol e}^-} \times \frac{2\text{ mol e}^-}{\text{mol Cu}} \times \frac{\text{mol Cu}}{63.5\text{ g}} = 1.21 \times 10^3\text{ s}$

(c) $?\text{ sec} = \frac{1.00\text{ g}}{2.50\text{ A}} \times \frac{\text{As}}{\text{C}} \times \frac{96500\text{ C}}{\text{mol e}^-} \times \frac{\text{mol e}^-}{\text{mol Ag}} \times \frac{\text{mol Ag}}{107.9\text{ g}} = 358\text{ s}$

(d) $? \text{ sec} = \frac{1.00 \text{ g}}{2.50 \text{ A}} \times \frac{\text{As}}{\text{C}} \times \frac{96500 \text{ C}}{\text{mol } e^-} \times \frac{2 \text{ mol } e^-}{\text{mol Ni}} \times \frac{\text{mol Ni}}{58.69 \text{ g}} = 1.32 \times 10^3 \text{ s}$

(c) is the shortest time.

Review Questions

5. Np in NpO_2^+ is a +5 oxidation state so $NpO_2^+ \rightarrow Np^{4+}$ is a (c) reduction reaction.

6. (d) Reduction occurs at the cathode. Electrons move from the anode to the cathode.

8. Standard electrode potentials are based on an arbitrary value of zero assigned to the half-reaction, $2 H^+(1 M) + 2 e^- \rightarrow H_2(g, 1 atm)$. If the reduction most easily achieved is assigned $E° = 0$, all others will have $E° < 0$. If the reduction most difficult to achieve is assigned $E° = 0$, all others will have $E° > 0$.

9. $E = E^o - \frac{0.0592 \text{ V}}{n} \log \frac{[Zn^{2+}]}{[Pb^{2+}]}$

$E = 0.66 \text{ V} - \frac{0.0592 \text{ V}}{2} \log \frac{0.1}{0.01} = 0.66 \text{ V} - 0.030 \text{ V}$

(d) $E = 0.63$ V

11. Au and Ag have positive reduction potentials and do not react with HCl (aq).

12. Redox reactions outside of an electrochemical cell are still reactions involving an oxidation and a reduction. The sign of $E°_{cell}$ of those reactions is still applicable. The prediction of the direction of spontaneous change based on the sign of E_{cell} is essentially a prediction based on the sign of ΔG ($\Delta G = -nFE_{cell}$). Because ΔG is a state function, the prediction of spontaneous change is independent of the path of the reaction, whether in a voltaic cell or directly between the reactants.
The electrode potentials can be used to predict spontaneity for all redox reactions, but they are not applicable to other types, for example, acid-base reactions. .

13. $E°_{cell} < 0$ means that a reaction will not occur spontaneously with reactants and products at standard-state conditions. Often it can be made to occur under other conditions.

21. Standard electrode potentials are used to calculate the minimum voltage required to cause electrolysis.

22. (a) Since $E^o = -1.02$ v, the reaction can only occur when forced by an outside current.

23. The metal needs to be the cathode, so that metal ions in solution are reduced to metal atoms, which become attached to the plated metal.

24. $4.5\text{ g Al} \times \frac{\text{mol Al}}{26.98\text{ g}} \times \frac{3\text{ mol e}^-}{\text{mol Al}} \times \frac{\text{mol H}_2}{2\text{ mol e}^-} \times \frac{22.4\text{ L}}{\text{mol H}_2} = 5.6\text{ L H}_2$ (d)

$2H^+ + 2e^- \rightarrow H_2(g)$

Problems

25. (a) $HNO_2(aq) \rightarrow NO_2(g) + H^+(aq) + e^-$ oxidation
(b) $PbO_2(s) + 2H^+(aq) + 2e^- \rightarrow PbO(s) + H_2O(l)$ reduction
(c) $CH_3CH_2OH + 3\ H_2O \rightarrow 2\ CO_2 + 12\ H^+ + 12\ e^-$ oxidation
$CH_3CH_2OH + 3\ H_2O + 12\ OH^- \rightarrow 2\ CO_2 + 12\ H^+ + 12\ OH^- + 12\ e^-$
$CH_3CH_2OH(aq) + 12\ OH^-(aq) \rightarrow 2\ CO_2(g) + 9\ H_2O(l) + 12\ e^-$

27. The solution to the balancing of redox equations is listed in this order: First, the two balanced half-reactions are listed side by side. These are followed by the final balanced equation for acidic solution. For basic solutions, two more equations are listed. One is to show the addition of the OH^- ions, and the other, the final balanced equation in basic solution.
(a) $Fe^{2+} \rightarrow Fe^{3+} + e^-$ $Cr_2O_7^{2-} + 14\ H^+ + 6\ e^- \rightarrow 2\ Cr^{3+} + 7\ H_2O$
$Cr_2O_7^{2-}(aq) + 6\ Fe^{2+}(aq) + 14\ H^+(aq) \rightarrow 2\ Cr^{3+}(aq) + Fe^{3+}(aq) + 7\ H_2O(l)$
(b) $S_8 + 32\ H_2O \rightarrow 8\ SO_4^{2-} + 64\ H^+ + 48\ e^-$ $O_2 + 4\ H^+ + 4\ e^- \rightarrow 2\ H_2O$
$S_8(s) + 12\ O_2(g) + 8\ H_2O(l) \rightarrow 8\ SO_4^{2-}(aq) + 16\ H^+(aq)$
(c) $Fe^{3+} + e^- \rightarrow Fe^{2+}$ $2\ NH_2OH_2^+ \rightarrow N_2O + H_2O + 6\ H^+ + 4\ e^-$
$4\ Fe^{3+}(aq) + 2\ NH_2OH_2^+(aq) \rightarrow 4\ Fe^{2+}(aq) + N_2O(g) + H_2O(l) + 6\ H^+(aq)$

29. (a) $Fe(OH)_2 + H_2O \rightarrow Fe(OH)_3 + H^+ + e^-$ $O_2 + 4\ H^+ + 4\ e^- \rightarrow 2\ H_2O$
$4\ Fe(OH)_2(s) + 2\ H_2O(l) + O_2(g) \rightarrow 4\ Fe(OH)_3(s)$
(b) $S_8 + 12\ H_2O \rightarrow 4\ S_2O_3^{2-} + 24\ H^+ + 16\ e^-$ $S_8 + 16\ e^- \rightarrow 8\ S^{2-}$
$2\ S_8 + 12\ H_2O \rightarrow 4\ S_2O_3^{2-} + 8\ S^{2-} + 24\ H^+$
$2\ S_8 + 12\ H_2O + 24\ OH^- \rightarrow 4\ S_2O_3^{2-} + 8\ S^{2-} + 24\ H^+ + 24\ OH^-$
$S_8(s) + 12\ OH^-(aq) \rightarrow 2\ S_2O_3^{2-}(aq) + 4\ S^{2-}(aq) + 6\ H_2O(l)$
(c) $CrI_3 + 16\ H_2O \rightarrow CrO_4^{2-} + 3\ IO_4^- + 32\ H^+ + 27\ e^-$
$H_2O_2 + 2\ H^+ + 2\ e^- \rightarrow 2\ H_2O$
$2\ CrI_3 + 27\ H_2O_2 \rightarrow 2\ CrO_4^{2-} + 6\ IO_4^- + 10\ H^+ + 22\ H_2O$
$2\ CrI_3 + 27\ H_2O_2 + 10\ OH^- \rightarrow 2\ CrO_4^{2-} + 6\ IO_4^- + 10\ H^+ + 10\ OH^- + 32\ H_2O$
$2\ CrI_3 + 27\ H_2O_2 + 10\ OH^- \rightarrow 2\ CrO_4^{2-} + 6\ IO_4^- + 32\ H_2O$

31. (a) $MnO_4^- + 8\ H^+ + 5\ e^- \rightarrow Mn^{2+} + 4\ H_2O$ $C_2H_2O_4 \rightarrow 2\ CO_2 + 2\ H^+ + 2\ e^-$
$2\ MnO_4^-(aq) + 5\ C_2H_2O_4(aq) + 6\ H^+(aq)$
$\rightarrow 2\ Mn^{2+}(aq) + 10\ CO_2(g) + 8\ H_2O(l)$

(b) $Cr_2O_7^{2-} + 14 H^+ + 6 e^- \rightarrow 2 Cr^{3+} + 7 H_2O$

$UO^{2+} + H_2O \rightarrow UO_2^{2+} + 2 H^+ + 2 e^-$

$Cr_2O_7^{2-}(aq) + 3 UO^{2+}(aq) + 8 H^+(aq)$

$\rightarrow 2 Cr^{3+}(aq) + 3 UO_2^{2+}(aq) + 4 H_2O(l)$

(c) $Zn \rightarrow Zn^{2+} + 2 e^-$ $NO_3^- + 9 H^+ + 8 e^- \rightarrow NH_3(g) + 3 H_2O$

$4 Zn + NO_3^- + 9 H^+ \rightarrow 4 Zn^{2+} + NH_3(g) + 3 H_2O$

$4 Zn + NO_3^- + 9 H^+ + 9 OH^- \rightarrow 4 Zn^{2+} + NH_3(g) + 3 H_2O + 9 OH^-$

$4 Zn(s) + NO_3^-(aq) + 6 H_2O(l) \rightarrow 4 Zn^{2+}(aq) + NH_3(g) + 9 OH^-(aq)$

33. (a) $V^{2+} \rightarrow V^{3+} + e^-$ oxidation

$Sn^{2+} + 2 e^- \rightarrow Sn(s)$ reduction

$2 V^{2+}(aq) + Sn^{2+}(aq) \rightarrow 2 V^{3+}(aq) + Sn(s)$

$E°_{cell} = E°(cathode) - E°(anode) = E°(reduction) - E°(oxidation)$

$E°_{cell} = E°_{Sn^{2+}/Sn} - E°_{V^{3+}/V^{2+}} = -0.137 V - (-0.255 V)$

$E°_{cell} = 0.118 V$ Spontaneous, so reaction as written.

(b) $V^{2+} \rightarrow V^{3+} + e^-$ oxidation

$Zn^{2+} + 2 e^- \rightarrow Zn$ reduction

$2 V^{2+} + Zn^{2+} \rightarrow 2 V^{3+} + Zn(s)$

$E°_{cell} = E°(cathode) - E°(anode) = E°(reduction) - E°(oxidation)$

$E°_{cell} = E°_{Zn^{2+}/Zn} - E°_{V^{3+}/V^{2+}} = -0.763 V - (-0.255 V)$

$E°_{cell} = -0.508 V$ Nonspontaneous, the reverse reaction is spontaneous.

(c) $V^{2+} \rightarrow V^{3+} + e^-$ oxidation

$Cu^{2+} + 2 e^- \rightarrow Cu$ reduction

$2 V^{2+}(aq) + Cu^{2+}(aq) \rightarrow 2 V^{3+}(aq) + Cu(s)$

$E°_{cell} = E°(cathode) - E°(anode) = E°(reduction) - E°(oxidation)$

$E°_{cell} = E°_{Cu^{2+}/Cu} - E°_{V^{3+}/V^{2+}} = 0.340 V - (-0.255 V)$

$E°_{cell} = 0.595 V$ Spontaneous, so reaction as written.

35. $\{SbO^+ + 2 H^+ + 3 e^- \rightarrow Sb + H_2O\} \times 2$ $E°_{cathode} = ?$

$\{V(s) \rightarrow V^{2+} + 2 e^-\} \times 3$ $E°_{anode} = -1.13 V$

$E°_{cell} = E°_{cathode} - E°_{anode}$

$1.334 V = E°_{cathode} - (-1.13 V)$

$E°_{cathode} = 1.334 V - 1.13 V = 0.20 V$

37. $V(s) \rightarrow V^{2+} + 2 e^-$ $E°_{anode} = ?$

$Cu^{2+} + 2\ e^- \rightarrow Cu$ $\qquad E°_{cathode} = 0.340$ V

$E°_{cell} = E°_{cathode} - E°_{anode} = 1.47\ V = 0.340\ V - E°_{anode}$

$E°_{anode} = -1.13\ V = E°_{V^{2+}/V}$

39. (a) $\{Fe^{2+}(aq) \rightarrow Fe^{3+}(aq) + e^-\} \times 6$ $\qquad E°_{anode} = 0.771$ V

$Cr_2O_7^{2-}(aq) + 14\ H^+ + 6\ e^- \rightarrow 2\ Cr^{3+}(aq) + 7\ H_2O(l)$ $\qquad E°_{cathode} = 1.33$ V

$6\ Fe^{2+}(aq) + Cr_2O_7^{2-}(aq) + 14\ H^+(aq) \rightarrow 6\ Fe^{3+}(aq) + 2\ Cr^{3+}(aq) + 7\ H_2O(l)$

$E°_{cell} = E°_{cathode} - E°_{anode} = 1.33\ V - 0.771\ V = 0.56$ V

(b) $NO + 2\ H_2O \rightarrow NO_3^- + 4\ H^+ + 3\ e^-$ $\qquad E°_{anode} = -0.956$ V

$H_2O_2 + 2\ H^+ + 2\ e^- \rightarrow 2\ H_2O$ $\qquad E°_{cathode} = 1.763$ V

$2\ NO(g) + 3\ H_2O_2(aq) \rightarrow 2\ NO_3^-(aq) + 2\ H^+(aq) + 2\ H_2O(l)$

$E°_{cell} = E°_{cathode} - E°_{anode} = 1.763\ V - (-0.956\ V) = 2.719$ V

41. (a) $\{Fe^{3+} + e^- \rightarrow Fe^{2+}\} \times 2$ $\qquad E°_{cathode} = 0.771$ V

$Sn^{2+} \rightarrow Sn^{4+} + 2\ e^-$ $\qquad E°_{anode} = 0.154$ V

$2\ Fe^{3+}(aq) + Sn^{2+}(aq) \rightarrow Sn^{4+}(aq) + 2\ Fe^{2+}(aq)$

$Pt\ |\ Sn^{2+}(aq), Sn^{4+}(aq)\ ||\ Fe^{2+}(aq), Fe^{3+}(aq)\ |\ Pt$

$E°_{cell} = E°_{cathode} - E°_{anode} = 0.771\ V - 0.154\ V = 0.617$ V

(b) $\{Cu(s) \rightarrow Cu^{2+} + 2\ e^-\} \times 3$ $\qquad E°_{anode} = 0.340$ V

$\{4\ H^+ + NO_3^- + 3\ e^- \rightarrow 2\ H_2O + NO(g)\} \times 2$ $\qquad E°_{cathode} = 0.956$ V

$3\ Cu(s) + 8\ H^+(aq) + 2\ NO_3^-(aq) \rightarrow 3\ Cu^{2+}(aq) + 4\ H_2O(l) + 2\ NO(g)$

$Cu(s)\ |\ Cu^{2+}(aq)\ ||\ H^+(aq), NO_3^-(aq), NO(g)\ |\ Pt(s)$

$E°_{cell} = E°_{cathode} - E°_{anode} = 0.956\ V - 0.340\ V = 0.616$ V

43. (a) $Sn^{4+} + 2\ e^- \rightarrow Sn^{2+}$ $\qquad E°_{cathode} = 0.154$ V

$2\ I^- \rightarrow I_2 + 2e^-$ $\qquad E°_{anode} = 0.535$ V

$E°_{cell} = E°_{cathode} - E°_{anode} = 0.154\ V - 0.535\ V = -0.381$ V

not spontaneous

(b) $ClO^- + H_2O + 2\ e^- \rightarrow Cl^- + 2\ OH^-$ $\qquad E°_{cathode} = 0.890$ V

$MnO_2(s) + 4\ OH^- \rightarrow MnO_4^- + 2H_2O + 3e^-$ $\qquad E°_{anode} = 0.60$ V

$E°_{cell} = E°_{cathode} - E°_{anode} = 0.890\ V - 0.60\ V = 0.29$ V

spontaneous

45. (a) $Sn^{4+} + 2\ e^- \rightarrow Sn^{2+}$ $\qquad E°_{cathode} = 0.154$ V

$Cu(s) \rightarrow Cu^{2+} + 2\ e^-$ $\qquad E°_{anode} = 0.340$ V

$E°_{cell} = E°_{cathode} - E°_{anode} = 0.154\ V - 0.340\ V = -0.186$ V not spontaneous

(b) $I_2 + 6\,H_2O \rightarrow 2\,IO_3^{-} + 12\,H^+ + 10\,e^-$ $E°_{anode} = 1.20$ V

$O_3 + 2\,H^+ + 2\,e^- \rightarrow O_2 + H_2O$ $E°_{cathode} = 2.075$ V

$E°_{cell} = E°_{cathode} - E°_{anode} = 2.075\text{ V} - 1.20\text{ V} = 0.88\text{ V}$

spontaneous

(c) $Cr(OH)_3$ is soluble in alkaline solution and forms $Cr(OH)_4^-$. $H_2O_2 + OH^-$ reacts to form $HO_2^- + H_2O$.

$Cr(OH)_4^- + 4\,OH^- \rightarrow CrO_4^{2-} + 4\,H_2O + 3\,e^-$ $E°_{anode} = -0.13$ V

$HO_2^- + H_2O + 2\,e^- \rightarrow 3\,OH^-$ $E°_{cathode} = 0.88$ V

$E°_{cell} = E°_{cathode} - E°_{anode} = 0.88\text{ V} - (-0.13\text{ V}) = 1.01\text{ V}$

spontaneous

47. (a) Silver does not react with HCl(aq) because $H^+(aq)$ is not a good enough oxidizing agent to oxidize Ag(s) to $Ag^+(aq)$. $E°_{cell}$ for the reaction is – 0.800 V. For HNO_3, NO_3^- acts as the oxidizing agent to oxidize Ag(s).

(b) Nitrate ion in acidic solution is a good enough oxidizing agent to oxidize Ag (s) to Ag+ (aq). $E°_{cell}$ for the reaction is 0.156 V.

$3\,Ag(s) + 4\,H^+(aq) + NO_3^-(aq) \rightarrow 3\,Ag^+(aq) + NO(g) + 2\,H_2O(l)$

49. Because Ag displaces $Pd^{2+}(aq)$, $E°_{cathode}$ for the half-reaction $Pd^{2+} + 2\,e^- \rightarrow Pd(s)$ must be greater than 0.800 V. Because Pd(s) is oxidized to $Pd^{2+}(aq)$ by $HNO_3(aq)$, $E°_{anode}$ for $Pd(s) \rightarrow Pd^{2+} + 2\,e^-$ must be less posiitve than 0.956 V. Thus, for the half-reaction $Pd^{2+} + 2\,e^- \rightarrow Pd\,(s)$ $0.800\text{ V} < E°_{Pd^{2+}/Pd} < 0.956\text{ V}$.

51. (a) $O_2 + 4\,H^+ + 4\,e^- \rightarrow 2\,H_2O$ $E°_{cathode} = 1.229$ V

$\{2\,I^- \rightarrow + I_2 + 2\,e^-\} \times 2$ $E°_{anode} = 0.535$ V

$E°_{cell} = E°_{cathode} - E°_{anode} = 1.229\text{ V} - 0.535\text{ V} = 0.694\text{ V}$

$\Delta G° = -nFE° = -4\text{ moles e}^- \times 96485\text{ C/mol e}^- \times 0.694\text{ V}$

$\Delta G° = -2.68 \times 10^5\text{ J} = -268\text{ kJ}$

(b) $Cr_2O_7^{2-} + 14\,H^+ + 6\,e^- \rightarrow 2\,Cr^{3+} + 7\,H_2O$ $E°_{cathode} = 1.33$ V

$\{Cu\,(s) \rightarrow Cu^{2+} + 2\,e^-\} \times 3$ $E°_{anode} = 0.340$ V

$E°_{cell} = E°_{cathode} - E°_{anode} = 1.33\text{ V} - 0.337\text{ V} = 0.99\text{ V}$

$\Delta G° = -nFE° = -6\text{ mol e}^- \times 96485\text{ C/mol e}^- \times 0.99\text{ V}$

$\Delta G° = -5.7 \times 10^5\text{ J} = -5.7 \times 10^2\text{ kJ}$

53. (a) $PbO_2 + 4\,H^+ + 2\,e^- \rightarrow Pb^{2+} + 2\,H_2O$ $E°_{cathode} = 1.455$ V

$2\,Cl^- \rightarrow Cl_2 + 2\,e^-$ $E°_{anode} = 1.358$ V

$E°_{cell} = E°_{cathode} - E°_{anode} = 1.455\text{ V} - 1.358\text{ V} = 0.097\text{ V}$

$$K_{eq} = \frac{[Pb^{2+}]P_{Cl_2}}{[H^+]^4[Cl^-]^2}$$

$$E°_{cell} = 0.097\text{ V} = \frac{0.0592}{n}\log K_{eq} = \frac{0.0592}{2}\log K_{eq}$$

$$\log K_{eq} = \frac{2 \times 0.097}{0.0592} = 3.28$$

$$10^{\log K_{eq}} = K_{eq} = 10^{3.28} = 2 \times 10^3$$

(b) $\{O_2 + 2\,H_2O + 4\,e^- \rightarrow 4\,OH^-\} \times 3$ $\qquad E°_{cathode} = 0.401\text{ V}$

$\{Br^- + 6\,OH^- \rightarrow BrO_3^- + 6\,e^- + 3\,H_2O\} \times 2$ $\qquad E°_{anode} = 0.584\text{ V}$

$E°_{cell} = E°_{cathode} - E°_{anode} = 0.401\text{ V} - 0.584\text{ V} = -0.183\text{ V}$

$$K_{eq} = \frac{[BrO_3^-]^2}{[Br^-]^2 P_{O_2}^3}$$

$$E°_{cell} = -0.183\text{ V} = \frac{0.0592}{12}\log K_{eq}$$

$$\log K_{eq} = \frac{12 \times (-0.183)}{0.0592} = -37.09$$

$$10^{\log K_{eq}} = K_{eq} = 10^{-37.09} = 8 \times 10^{-38}$$

55. $2\,Ag^+(aq) + Cu(s) \rightarrow 2\,Ag(s) + Cu^{2+}(aq)$

$\{Ag^+(aq) + e^- \rightarrow Ag(s)\} \times 2$ $\qquad E°_{cathode} = 0.800\text{ V}$

$Cu(s) \rightarrow Cu^{2+} + e^-$ $\qquad E°_{anode} = 0.340\text{ V}$

$E°_{cell} = E°_{cathode} - E°_{anode} = 0.800\text{ V} - 0.340\text{ V} = 0.460\text{ V}$

$$E°_{cell} = \frac{0.0592}{n}\log K_{eq}$$

$$\log K_{eq} = \frac{2 \times 0.460}{0.0592} = 15.54$$

$$10^{\log K_{eq}} = K_{eq} = 10^{15.54} = 3.5 \times 10^{15}$$

Assume that the reaction initially goes to completion and then approaches equilibrium from the right.

	$Cu(s) + 2\,Ag^+$	$\rightarrow$	$Cu^{2+} + 2\,Ag(s)$
init	—		0.500 M
change	$+X$		$-0.50\,X$
eq.	X		$0.500 - 0.50\,X$

$$\frac{[Cu^{2+}]}{[Ag^+]^2} = \frac{0.500 - 0.50\,X}{X^2} = 3.5 \times 10^{15}$$

Assume $X << 0.50$

$$X^2 = \frac{0.50}{3.5\times10^{15}} = 1.4\times10^{-16}$$

$X = 1.2\times10^{-8}$ M = $[Ag^+]$ $\qquad$ $[Cu^{2+}] = 0.50$ M

57. (a) $E_{cell} = E°_{cell} - \frac{0.0592}{n}\log Q$

$Pb(s) \rightarrow Pb^{2+} + 2\,e^-$ $\qquad$ $E°_{anode} = -0.125$ V

$2\,H^+ + 2\,e^- \rightarrow H_2$ $\qquad$ $E°_{cathode} = 0.000$ V

$E°_{cell} = E°_{cathode} - E°_{anode} = 0.000\text{ V} - (-0.125\text{ V}) = 0.125$ V

$n = 2$

$$Q = \frac{[Pb^{2+}]P_{H_2}}{[H^+]^2}$$

$$E_{cell} = 0.125\text{ V} - \frac{0.0592}{2}\log\frac{(0.85)(0.95)}{(0.0025)^2}$$

$E_{cell} = 0.125\text{ V} - 0.151\text{ V} = -0.026$ V

(b) $E_{cell} = E°_{cell} - \frac{0.0592}{n}\log Q$

$\{Mn^{2+} + 2\,H_2O \rightarrow MnO_2 + 4\,H^+ + 2\,e^-\}\times3$ $\qquad$ $E°_{anode} = 1.23$ V

$ClO_3^- + 6\,H^+ + 6\,e^- \rightarrow Cl^- + 3\,H_2O$ $\qquad$ $E°_{cathode} = 1.450$ V

$E°_{cell} = E°_{cathode} - E°_{anode} = 1.450\text{ V} - 1.23\text{ V} = 0.22$ V

$n = 6$

$$Q = \frac{[Cl^-][H^+]^6}{[ClO_3^-][Mn^{2+}]^3}$$

$$E_{cell} = 0.22\text{ V} - \frac{0.0592}{6}\log\frac{(1.50)(1.25)^6}{(0.65)(0.25)^3}$$

$E_{cell} = 0.22\text{ V} - 0.027\text{ V} = 0.19$ V

59. $H_2(g, 1\text{ atm}) + 2\,H^+(1\text{ M}) \rightarrow 2\,H^+(0.0025\text{ M}) + H_2(g, 1\text{ atm})$

$$E_{cell} = E°_{cell} - \frac{0.0592}{n}\log Q$$

$$E_{cell} = 0.00 - \frac{0.0592}{2}\log\frac{[H^+]^2P_{H_2}}{[H^+]^2P_{H_2}}$$

$$E_{cell} = 0.00 - \frac{0.0592}{2}\log\frac{(0.0025)^2\,(1)}{(1)^2\,(1)} = 0.15\text{ V}$$

OR

concentration cell: $2\ H^+\ (1M) \rightarrow 2\ H^+\ (0.0025\ M)$

$$E_{cell} = -\frac{0.0592}{n}\log\frac{[H^+]^2 prod}{[H^+]^2 reac}$$

$$E_{cell} = -\frac{0.0592}{2}\log\frac{(0.0025)^2}{(1)^2} = 0.15\ V$$

61. $H_2(g, 1\ atm) + 2\ H^+(1\ M) \rightarrow 2\ H^+(X\ M) + H_2(g, 1\ atm)$

$$Q = \frac{[H^+]^2 P_{H_2}}{[H^+]^2 P_{H_2}}$$

$n = 2$

$$E_{cell} = E^\circ_{cell} - \frac{0.0592}{2}\log\frac{[H^+]^2\ (1)}{(1)^2\ (1)}$$

$$0.108 = 0 - \frac{0.0592}{2}\log[H^+]^2$$

$-3.65 = \log[H^+]^2$ or $-3.65 = 2\log[H^+]$

$[H^+]^2 = 10^{-3.65}$ $\quad -1.83 = \log[H^+]$

$[H^+]^2 = 2.24 \times 10^{-4}$ $\quad [H^+] = 10^{-1.83}$

$[H^+] = 1.50 \times 10^{-2}$ $\quad pH = -\log[H^+] = 1.83$

$pH = 1.83$

63. $Cu \rightarrow Cu^{2+} + 2\ e^-$ $\quad E^\circ_{anode} = 0.340\ V$

$\underline{Ag^+ + e^- \rightarrow Ag}$ $\quad E^\circ_{cathode} = 0.800\ V$

$Cu(s) + 2\ Ag^+(aq) \rightarrow Cu^{2+}(aq) + 2\ Ag(s)$

$E^\circ_{cell} = E^\circ_{cathode} - E^\circ_{anode} = 0.800\ V - 0.340\ V = 0.460\ V$

$n = 2$

$$Q = \frac{[Cu^{2+}]}{[Ag^+]^2}$$

$$E_{cell} = E^\circ_{cell} - \frac{0.0592}{n}\log\frac{[Cu^{2+}]}{[Ag^+]^2} = 0.460 - \frac{0.0592}{2}\log\frac{[Cu^{2+}]}{[Ag^+]^2}$$

The higher cell voltage will be the cell with the lower $[Cu^{2+}]/[Ag^+]^2$ ratio.

(a) $\frac{1.25}{(0.55)^2} = 4.13$ $\quad E_{cell} = 0.442$

(b) $\frac{0.12}{(0.60)^2} = 0.33$ $\quad E_{cell} = 0.474$

In this case the $[Ag^+]$ is about the same. So the lower value of $[Cu^{2+}]$ in cell (b) means a greater tendency for oxidation to occur there and the higher cell voltage.

65. $\{Mg(s) \rightarrow Mg^{2+} + 2\,e^-\} \times 2$ $E°_{anode} = -2.356$ V
$\underline{O_2(g) + 2\,H_2O + 4\,e^- \rightarrow 4\,OH^-(aq)}$ $E°_{cathode} = 0.401$ V
$2\,Mg(s) + O_2(g) + 2\,H_2O(l) \rightarrow 2\,Mg(OH)_2(s)$
$E°_{cell} = E°_{cathode} - E°_{anode} = 0.401\text{ V} -(-2.356\text{ V}) = 2.757\text{ V}$

67. anode: $Zn(s) + 2\,OH^- \rightarrow ZnO(s) + H_2O + 2\,e^-$
cathode: $\underline{Ag_2O + H_2O + 2\,e^- \rightarrow 2\,Ag + 2\,OH^-}$
$Zn(s) + Ag_2O(s) \rightarrow ZnO(s) + 2\,Ag(s)$

69. Oxygen is the oxidizing agent required to oxidize Fe (s) to Fe^{2+} and then to Fe^{3+}. Water is a reactant in the reduction half-reaction, in which O_2 (g) is reduced to OH^- (aq). Water is also a reactant in the conversion of $Fe(OH)_2$ to $Fe_2O_3{\cdot}\underline{x}H_2O$ (rust). The electrolyte completes the electrical circuit between the cathodic and anodic areas.

71. A sacrificial anode is used up—that is, the anode metal is oxidized to metal ions at the same time that a cathodic half-reaction occurs on the protected metal.

73. Zinc protects the iron from corrosion. Zinc is oxidized instead of the iron. The faint white precipitate is zinc ferricyanide.

75. (a) anode $Cu(s) \rightarrow Cu^{2+} + 2\,e^-$
cathode $Cu^{2+} + 2\,e^- \rightarrow Cu(s)$
$Cu(s,anode) \rightarrow Cu(s,cathode)$
This is similar to the electrorefining of copper.
(b) anode $2\,H_2O \rightarrow 4\,H^+ + O_2 + 4\,e^-$
cathode $Cu^{2+} + 2\,e^- \rightarrow Cu(s)$
$2\,H_2O(l) + 2\,Cu^{2+}(aq) \rightarrow 4\,H^+(aq) + O_2(g) + 2\,Cu(s)$
Cu(s) is deposited on an iron core.
(c) anode $2\,H_2O \rightarrow 4\,H^+ + O_2 + 4\,e^-$
cathode $Cu^{2+} + 2\,e^- \rightarrow Cu(s)$
$2\,H_2O(l) + 2\,Cu^{2+}(aq) \rightarrow 2\,Cu(s) + 4\,H^+(aq) + O_2(g)$
Cu(s) is deposited on a platinum core.

77. (a) anode $2\,Cl^- \rightarrow Cl_2 + 2\,e^-$ $E°_{anode} = 1.358$ V
cathode $Ba^{2+} + 2\,e^- \rightarrow Ba(l)$ $E°_{cathode} = -2.92$ V
$BaCl_2(l) \xrightarrow{\text{electrolysis}} Ba(l) + Cl_2(g)$
probable products Ba(l) and $Cl_2(g)$
$E°_{cell}\text{ (voltaic)} = E°_{cathode} - E°_{anode} = -2.92\text{ V} - 1.358\text{ V} = -4.28\text{ V}$
$E°_{cell}\text{ (electrolytic)} > 4.28\text{ V}$
(b) anode $2\,Br^- \rightarrow Br_2(l) + 2\,e^-$ $E°_{anode} = 1.065$ V

cathode $2\,H^+ + 2\,e^- \rightarrow H_2(g)$ $E°_{cathode} = 0.00$ V

net $2\,Br^-(aq) + 2\,H^+(aq) \rightarrow Br_2(l) + H_2(g)$

probable products $Br_2(l)$ and $H_2(g)$

$E°_{cell}$ (voltaic) = $E°_{cathode} - E°_{anode} = 0.00\text{ V} - 1.065\text{ V} = -1.065\text{ V}$

$E°_{cell}$ (electrolytic) > 1.065 V

(c) anode $2\,H_2O \rightarrow 4\,H^+ + O_2(g) + 4\,e^-$ $E°_{anode} = 1.229$ V

cathode $\{2\,H_2O + 2\,e^- \rightarrow H_2(g) + 2\,OH^-\} \times 2$ $E°_{cathode} = -0.828$ V

net $2\,H_2O(l) \rightarrow 2\,H_2(g) + O_2(g)$

probable products $O_2(g)$ and $2\,H_2(g)$

$E°_{cell}$ (voltaic) = $E°_{cathode} - E°_{anode} = -0.828\text{ V} - 1.229\text{ V} = -2.057\text{ V}$

$E°_{cell}$ (electrolytic) > 2.057 V

79. $Ag^+ + e^- \rightarrow Ag(s)$

$$? \text{ g Ag} = 1.73\text{ A} \times \frac{\text{C}}{\text{A s}} \times 2.05\text{ h} \times \frac{3600\text{ s}}{\text{h}} \times \frac{\text{mol } e^-}{96485\text{ C}} \times \frac{\text{mol Ag}}{\text{mol } e^-} \times \frac{107.9\text{ g Ag}}{\text{mol Ag}} = 14.3\text{ g Ag}$$

81. $Cu^{2+} + 2\,e^- \rightarrow Cu$

$$? \text{ C} = 25.0\text{ g Cu} \times \frac{\text{mol Cu}}{63.55\text{ g Cu}} \times \frac{2\text{ mol } e^-}{\text{mol Cu}} \times \frac{96485\text{ C}}{\text{mol } e^-} = 7.59 \times 10^4\text{ C}$$

83. Na (s) does not electrodeposit from $NaNO_3$ (aq). Of the remaining solutions, $AgNO_3$ (aq) yields the greatest number of moles of deposit. One mole of silver is formed for every mole of electrons ($Ag^+ + e^- \rightarrow Ag$), whereas only one-half mole of copper and zinc is formed ($M^{2+} + 2\,e^- \rightarrow M$). Also, given that the atomic weight of Ag is greater than those of Cu and Zn, $AgNO_3$(aq) (solution b) yields the greatest mass of metal deposit.

$$1.00\text{ A} \times 1\text{ h} \times \frac{3600\text{s}}{\text{h}} \times \frac{\text{C}}{\text{A s}} \times \frac{\text{mol } e^-}{96485\text{ C}} \times \frac{\text{mol M}}{n\text{ mol } e^-} \times \frac{\text{molar mass g}}{\text{mol M}} =$$

Additional Problems

86. $CN^- + H_2O \rightarrow OCN^- + 2\,H^+ + 2\,e^-$ $Cl_2 + 2\,e^- \rightarrow 2\,Cl^-$

$CN^- + 2\,OH^- \rightarrow OCN^- + H_2O + 2\,e^-$

NaCNaq) + 2 NaOH(aq) + Cl_2(g) → NaOCN(aq) + 2 NaCl(aq) + H_2O(l)

$2\,OCN^- + 4\,H_2O \rightarrow 2\,HCO_3^- + N_2 + 6\,H^+ + 6\,e^-$ $Cl_2 + 2\,e^- \rightarrow 2\,Cl^-$

$2\,OCN^- + 6\,OH^- \rightarrow 2\,HCO_3^- + N_2 + 2\,H_2O + 6\,e^-$

2 NaOCN(aq) + 6 NaOH(aq) + 3 Cl_2(g)

→ 2 $NaHCO_3$(aq) + N_2(g) + 2 H_2O(l) + 6 NaCl(aq)

88. $NO + 5\,H^+ + 5\,e^- \rightarrow NH_3 + H_2O$ $H_2 \rightarrow 2\,H^+ + 2\,e^-$

$2\,NO(g) + 5\,H_2(g) \rightarrow 2\,NH_3(g) + 2\,H_2O(g)$

The method works because the half-reactions involved do not have to be in aqueous solutions. The electrons in the oxidation have to equal the electrons in the reduction whether in solution or not. Moreover, as the case here, ionic species appearing in the half-reactions cancel out in the overall reaction.

91. $Ag^+ + e^- \rightarrow Ag\,(s)$

$2\,O^{2-} + 4\,e^- \rightarrow O_2$

$$P_{total} = P_{O_2} + VP_{H_2O}$$

$$P_{O_2} = 761.5\text{ mmHg} - 17.5\text{ mmHg}$$

$$P_{O_2} = 744.0\text{ mmHg}$$

$$?\text{ mL } O_2 = 1.02\text{ g Ag} \times \frac{\text{mol Ag}}{107.9\text{ g Ag}} \times \frac{\text{mol e}^-}{\text{mol Ag}} \times \frac{\text{mol } O_2}{4\text{ mol e}^-} \times \frac{\frac{62.364\text{ L mmHg}}{\text{K mol}} \times 293.2\text{ K}}{744.0\text{ mmHg}} \times \frac{\text{mL}}{10^{-3}\text{ L}} = 58.1\text{ mL } O_2$$

92. $$?\text{ mol Ag} = 1.75\text{ A} \times 25.0\text{ min} \times \frac{60\text{ s}}{\text{min}} \times \frac{\text{C}}{\text{A s}} \times \frac{\text{mol e}^-}{96485\text{ C}} \times \frac{\text{mol Ag}}{\text{mol e}} = 0.0272\text{ mol Ag}$$

? mol = 0.100 L × 0.785 M = 0.0785 mol initial
<u>−0.0272 mol electrolyzed</u>
0.0513 mol remaining

$$?\text{ M } AgNO_3 = \frac{0.0513\text{ mol}}{0.100\text{ L}} = 0.513\text{ M } AgNO_3$$

Water is oxidized to $O_2(g)$, but there is very little volume change.

$$?\text{ mol } H_2O = 1.75\text{ A} \times 25.0\text{ min} \times \frac{60\text{ s}}{\text{min}} \times \frac{\text{C}}{\text{A s}} \times \frac{\text{mol e}^-}{96485\text{ C}} \times \frac{\text{mol } H_2O}{2\text{ mol e}^-} = 0.0136\text{ mol } H_2O$$

$$?\text{ L } H_2O = 0.0136\text{ mol } H_2O \times \frac{18.02\text{ g}}{\text{mol}} \times \frac{\text{mL}}{1.0\text{ g}} \times \frac{10^{-3}\text{ L}}{\text{mL}} = 0.00025\text{ L}$$

96. $H_2\,(g) \rightarrow 2\,H^+ + 2\,e^-$ $\qquad E^\circ_{anode} = 0.00\text{ V}$

$2\,H^+ + 2\,e^- \rightarrow H_2\,(g)$ $\qquad E^\circ_{cathode} = 0.00\text{ V}$

$$E_{cell} = E^\circ_{cell} - \frac{0.0592}{2}\log\frac{[H^+]^2\text{ anode}}{[H^+]^2\text{ cathode}}$$

The buffer equation can be used to calculate $[H^+]$. The buffer equation is obtained as follows:

$$NH_4^+ + H_2O \rightleftharpoons H_3O^+ + NH_3 \qquad K_a = \frac{K_w}{K_b}$$

$$\frac{[H_3O^+][NH_3]}{[NH_4^+]} = \frac{K_w}{K_b}$$

$$[H_3O^+] = \frac{K_w}{K_b} \times \frac{[\text{acid}]}{[\text{base}]} = \frac{K_w}{K_b} \times \frac{[NH_4^+]}{[NH_3]} = \frac{10^{-14} \times (0.15)}{1.8\times10^{-5} \times (0.45)} = 1.85 \times 10^{-10}\text{ M}$$

$$E_{cell} = 0.00 - \frac{0.0592}{2}\log\frac{(0.010)^2}{(1.85\times10^{-10})^2}$$

$$E_{cell} = -0.457\text{ V}$$

99. $Cu^{2+} + 2\,e^- \rightarrow Cu\,(s)$ cathode

$Cu\,(s) \rightarrow Cu^{2+} + 2\,e^-$ anode

$$E_{cell} = E°_{cell} - \frac{0.0592}{2}\log\frac{[Cu^{2+}]\text{ anode}}{[Cu^{2+}]\text{ cathode}}$$

$$E_{cell} = 0.00 - \frac{0.0592}{2}\log\frac{0.025}{1.50}$$

$$E_{cell} = 0.0526\text{ V}$$

As the cell operates, the concentration of the 1.50 M solution will decrease, and that of the 0.025 M solution will increase, causing the voltage to decrease. When the concentrations of the solutions become equal (0.76 M in each half-cell), E_{cell} will fall to zero.

102. $Ag(s) \rightarrow Ag^+ + e^-$ $E°(\text{anode}) = 0.800\text{ V}$

$\underline{Ag^+ + e^- \rightarrow Ag(s)}$ $E°(\text{cathode}) = 0.800\text{ V}$

$Ag^+ \rightarrow Ag^+$

$Ag(s)|AgI(\text{satd})||Ag^+(0.100\text{ M})|Ag(s)$

$E°_{cell} = E°(\text{cathode}) - E°(\text{anode}) = 0.800\text{ V} - 0.800\text{ V} = 0.00\text{ V}$

$$E = E°_{cell} - \frac{0.0592}{n}\log Q = 0 - \frac{0.0592}{1}\log\frac{\text{anode}}{\text{cathode}}$$

$$0.417 = 0 - \frac{0.0592}{1}\log\frac{[Ag^+]}{(0.100)}$$

$$\frac{0.417\times(-1)}{0.0592} = \log\frac{[Ag^+]}{(0.100)}$$

$$10^{-7.04} = 9.12 \times 10^{-8} = 10^{\log\frac{[Ag^+]}{0.100}} = \frac{[Ag^+]}{(0.100)}$$

$$[Ag^+] = 9.12 \times 10^{-9}$$

$$K_{sp} = [Ag^+]^2 = (9.12 \times 10^{-9})^2 = 8.3 \times 10^{-17}$$

104. $Cu^{2+} + 2\,e^- \rightarrow Cu$

$$? \text{ mol } Cu^{2+}_{initial} = 250.0 \text{ mL} \times \frac{10^{-3} \text{ L}}{\text{mL}} \times 0.1000 \text{ M} = 2.500 \times 10^{-2} \text{ mol } Cu^{2+}$$

$$\text{mol } Cu^{2+} \text{ electrolyzed} = 3.512 \text{ A} \times 1368 \text{ s} \times \frac{\text{C}}{\text{A s}} \times \frac{\text{mol e}^-}{96485 \text{C}} \times \frac{\text{mol } Cu^{2+}}{2 \text{ mol e}^-}$$

$$= 2.490 \times 10^{-2} \text{ mol } Cu^{2+}$$

$$\text{mol } Cu^{2+} \text{ in solution} = 2.500 \times 10^{-2} \text{ mol} - 2.490 \times 10^{-2} \text{ mol} = 0.010 \times 10^{-2} \text{ mol}$$

$$= 1.0 \times 10^{-4} \text{ mol}$$

$$[Cu^{2+}]_{init} = \frac{1.0 \times 10^{-4} \text{ mol}}{250.0 \text{ mL}} \times \frac{\text{ml}}{10^{-3} \text{ L}} = 4.0 \times 10^{-4} \text{ M}$$

To simplify the calculation, it is assumed that the formation reaction goes to completion. Then $[Cu^{2+}]$ formed in the reverse reaction is determined to see if $[Cu^{2+}]$ is indeed negligible.

	Cu^{2+}	+	$4\ NH_3$	$\rightleftharpoons$	$[Cu(NH_3)_4]^{2+}$
Init			0.10 M		4×10^{-4} M
Change	$+X$		no change		$-X$
Eq	X		0.10 M		$4 \times 10^{-4}\ \text{M} - X$

$$K_f = 1.1 \times 10^{13} = \frac{4.0 \times 10^{-4} - X}{(0.10)^4 \times X}$$

Assume $X << 4.0 \times 10^{-4}$

$$X = \frac{4.0 \times 10^{-4}}{(0.10)^4 \times 1.1 \times 10^{13}} = 3.6 \times 10^{-13}$$

$$[Cu(NH_3)_4]^{2+} = 4.0 \times 10^{-4} - 3.6 \times 10^{-13} = 4.0 \times 10^{-4} \text{ M}$$

Yes, the blue color will be seen.

106. $Hg^{2+} + 2\ e^- \rightarrow Hg\ (l)$ $\qquad E°_{cathode} = 0.854$ V

$[Fe^{2+} \rightarrow Fe^{3+} + e^-\} \times 2$ $\qquad E°_{anode} = 0.771$ V

$Hg^{2+} + 2\ Fe^{2+} \rightarrow 2\ Fe^{3+} + Hg\ (l)$

$E°_{cell} = E°_{cathode} - E°_{anode} = 0.854\ \text{V} - (0.771\ \text{V}) = 0.083\ \text{V}$

At equilibrium, $E_{cell} = 0$

$$0 = 0.083 - \frac{0.0592}{2} \log \frac{[Fe^{3+}]^2}{[Fe^{2+}]^2[Hg^{2+}]}$$

$$6.4 \times 10^2 = K = \frac{[Fe^{3+}]^2}{[Fe^{2+}]^2[Hg^{2+}]}$$

	Hg^{2+}	+	$2\ Fe^{2+}$	$\rightarrow$	$2\ Fe^{3+}$	+	Hg(l)
Init	0.250 M		0.180 M		0.210 M		
Change	$-X$		$-2X$		$+2X$		
Eq	$0.250 - X$		$0.180 - 2X$		$0.210 + 2X$		

$$6.4 \times 10^2 = \frac{(0.210 + 2X)^2}{(0.250-X)(0.180-2X)^2}$$

The maximum value of X must be less than 0.090.

Assume $X = 0.050$, $\frac{(0.310)^2}{(0.200)(0.080)^2} = 75 < 640$

Assume $X = 0.060$, $\frac{(0.330)^2}{(0.190)(0.060)^2} = 159 < 640$

Assume $X = 0.080$, $\frac{(0.370)^2}{(0.170)(0.020)^2} = 2013 > 640$

Assume $X = 0.070$, $\frac{(0.350)^2}{(0.180)(0.040)^2} = 425 < 640$

Assume $X = 0.075$, $\frac{(0.360)^2}{(0.175)(0.030)^2} = 823 > 640$

Assume $X = 0.071$, $\frac{(0.352)^2}{(0.179)(0.038)^2} = 479 < 640$

Assume $X = 0.072$, $\frac{(0.354)^2}{(0.178)(0.036)^2} = 543 < 640$

Assume $X = 0.073$, $\frac{(0.356)^2}{(0.177)(0.034)^2} = 619 < 640$

$X = 0.073$

$[Hg^{2+}] = 0.250 - X = 0.177$ M

$[Fe^{2+}] = 0.180 - 2X = 0.034$ M

$[Fe^{3+}] = 0.210 + 2X = 0.356$ M

109. $? \text{ moles } Cd^{2+} = 1 \text{ pA} \times \frac{10^{-12}\text{ A}}{\text{pA}}\ 1 \text{ ms} \times \frac{10^{-3}\text{ s}}{\text{ms}} \times \frac{\text{C}}{\text{A s}} \times \frac{\text{mole e}^-}{96485\text{ C}} \times \frac{\text{mole } Cd^{2+}}{2 \text{ mole e}^-}$

$= 5.18 \times 10^{-21}$ moles Cd^{2+}

$? \text{ ions } Cd^{2+} = 5.18 \times 10^{-21} \text{ moles} \times \frac{6.022 \times 10^{23}\text{ ions}}{\text{mole}} = 3.12 \times 10^3$ ions

112. (a) $O_2\,(g) + 2\,H_2O + 4e^- \rightarrow 4\,OH^-$ cathode

$(H_2\,(g) + 2\,OH^- \rightarrow 2H_2O + 2\,e^-) \times 2$ anode

$2\,H_2 + O_2 \rightarrow 2\,H_2O$

$E°_{cell} = E°(\text{cathode}) - E°(\text{anode}) = 0.401\text{ V} - (-0.828\text{ V}) = 1.229\text{ V}$

See page 778. A higher temperature causing a lower voltage would change this and all following values.

(b) $E = E°_{cell} - \frac{0.0592}{n}\log\frac{1}{P_{O_2} P_{H_2}^{\ 2}}$

$$E = 1.229 - \frac{0.0592}{4}\log\frac{1}{(2.0)(2.0)^2}$$

$E = 1.242$ V

(c) $? \text{ kJ} = \dfrac{205 \text{ atm} \times 225 \text{ L}}{\dfrac{0.08206 \text{ L atm}}{\text{K mol}} \times 298 \text{ K}} \times \dfrac{4 \text{ mol e}^-}{\text{mol O}_2} \times \dfrac{96485 \text{C}}{\text{mol e}^-} \times 1.242 \text{ V} \times \dfrac{\text{J}}{\text{VC}} \times \dfrac{\text{kJ}}{10^3 \text{ J}}$

$\times\, 0.82 = 7.4 \times 10^5$ kJ

(d) $? \text{ watts} = ?\dfrac{\text{J}}{\text{s}} = \dfrac{7.41 \times 10^5 \text{ kJ}}{10 \text{ days}} \times \dfrac{\text{d}}{24 \text{ h}} \times \dfrac{\text{h}}{3600 \text{ s}} \times \dfrac{10^3 \text{ J}}{\text{kJ}} = 8.6 \times 10^2 \dfrac{\text{J}}{\text{s}}$

$= 8.6 \times 10^2$ watts

(e) The potential is an average potential; it may change especially as the pressure of gas decreases. Ideal gas pressure is assumed; cell is nonstandard.

Apply Your Knowledge

115. $\text{g cyclohexene} = 9.65 \text{ mA} \times \dfrac{10^{-3} \text{ A}}{\text{mA}} \times 235 \text{ s} \times \dfrac{\text{C}}{\text{A s}} \times \dfrac{\text{mol e}^-}{96485 \text{C}} \times \dfrac{\text{mol Br}_2}{2 \text{ mol e}^-}$

$\times \dfrac{\text{mole C}_6\text{H}_{10}}{\text{mole Br}_2} \times \dfrac{82.14 \text{ g}}{\text{mole C}_6\text{H}_{10}} = 9.65 \times 10^{-4}$ g

116. (a) $C_3H_5(ONO_2)_3(aq) + H^+(aq) + 2e^- \rightarrow C_3H_5OH(ONO_2)_2(aq) + NO_2^-(aq)$

$NO_2^-(aq) + 2H^+(aq) + e^- \rightarrow NO(g) + H_2O(l)$

(b) $C_3H_5(ONO_2)_3(aq) + NADH(aq)$

$\rightarrow C_3H_5OH(ONO_2)_2(aq) + NAD^+(aq) + NO_2^-(aq)$

$2NO_2^-(aq) + NADH(aq) + 3H^+(aq) \rightarrow 2NO(g) + 2H_2O(l) + NAD^+(aq)$

Chapter 19

Nuclear Chemistry

Exercises

19.1A (a) $^{212}_{86}Rn \rightarrow ^{4}_{2}He + ?$

$212 = 4 + A \quad A = 208$

$86 = 2 + Z \quad Z = 84$ Polonium

$^{212}_{86}Rn \rightarrow ^{4}_{2}He + ^{208}_{84}Po$

(b) $^{37}_{18}Ar + ^{0}_{-1}e \rightarrow ?$

$37 + 0 = A \quad A = 37$

$18 - 1 = Z \quad Z = 17$

$^{37}_{18}Ar + ^{0}_{-1}e \rightarrow ^{37}_{17}Cl$

(c) $^{60}_{28}Ni^{*} \rightarrow \gamma + ^{60}_{28}Ni$

19.1B (a) The sum of the mass numbers of $^{4}_{2}He$ and $^{214}_{82}Pb$ equals the mass number of the starting nucleus (218). The sum of the atomic numbers of He (2) and Pb(82) equals the atomic number of the parent nucleus (84): $^{218}_{84}Po \rightarrow ^{214}_{82}Pb + ^{4}_{2}He$.

(b) The sum of the mass numbers of $^{36}_{16}S$ and $^{0}_{1}e$ equals the mass number of the starting nucleus (36). The sum of the atomic numbers of S (16) and a positron (+1) equals the parent nucleus's atomic number (17): $^{36}_{17}Cl \rightarrow ^{36}_{16}S + ^{0}_{1}e$

19.2A $\ln \dfrac{A_t}{A_o} = -\lambda t \qquad t_{1/2} = 14.3\text{ d} = \dfrac{0.693}{\lambda}$

$$\lambda = \frac{0.693}{t_{1/2}} = \frac{0.693}{14.3\text{ d}} = 0.0485\text{ d}^{-1}$$

$$\ln \frac{A_t}{2.50x10^{10}\ \frac{\text{atom}}{\text{sec}}} = -0.0485\text{ d}^{-1} \times 365\text{ d} = -17.7$$

$$e^{\ln \frac{A_t}{2.50x10^{10}\ \frac{\text{atom}}{\text{sec}}}} = \frac{A_t}{2.50x10^{10}\ \frac{\text{atom}}{\text{sec}}} = e^{-17.7} = 2.06 \times 10^{-8}$$

$$A_t = 515\ \frac{\text{atoms}}{\text{sec}}$$

19.2B $\ln \frac{A_t}{A_o} = -\lambda t$ $\qquad \lambda = \frac{0.693}{t_{1/2}} = \frac{0.693}{2.411 \times 10^4\ y} = \frac{2.874 \times 10^{-5}}{y}$

$\ln \frac{1}{100} = \frac{-2.874 \times 10^{-5}}{y}\ t = -4.61$

$t = 1.60 \times 10^5\ y$

19.3A ? Fraction $= \frac{1}{(2)^{30}} = 9.3 \times 10^{-10}$

$? \text{ d} = \frac{8.040 \text{ d}}{\text{half - life}} \times 30 \text{ half-life} = 241.2 \text{ days}$

19.3B (b). There are about 10 times more ^{24}Na atoms than ^{11}C atoms, but ^{11}C atoms disintegrate about 50 times as fast. There are about $10^4 - 10^5$ times as many ^{238}U atoms as ^{11}C atoms, but $t_{1/2}$ of ^{238}U exceeds that of ^{11}C by a factor of more than 10^9.

An alternate approach to the problem using activity follows.

$A = \lambda N \qquad \lambda = \frac{0.693}{t_{\frac{1}{2}}}$

(a) $^{24}Na\ A = \frac{0.693}{14.659 \text{ h}} \times \frac{\text{h}}{60 \text{ min}} \times 1\ \mu\text{mol} \times \frac{1 \times 10^{-6} \text{ mol}}{\mu\text{mol}} \times \frac{6.02 \times 10^{23} \text{ atoms}}{\text{mol}}$

(b) $^{11}C\ A = \frac{0.693}{20.39 \text{ min}} \times 1 \times 10^{-6} \text{ g} \times \frac{1 \text{ mol}}{11 \text{ g}} \times \frac{6.02 \times 10^{23} \text{ atoms}}{\text{mol}}$

(c) $^{238}U\ A = \frac{0.693}{4.51 \times 10^9 \text{ y}} \times \frac{\text{y}}{365 \text{ d}} \times \frac{\text{d}}{24 \text{ h}} \times \frac{\text{h}}{60 \text{ min}} \times 1 \text{ g} \times \frac{1 \text{ mol}}{238 \text{ g}} \times \frac{6.02 \times 10^{23} \text{ atoms}}{\text{mol}}$

By ignoring the factors that are the same, such as 0.693, 10^{-6}, and 6.02×10^{23}, it is evident that C-11 has the greatest rate of decay because (b) $\frac{1}{20 \times 11}$ is more than (a) $\frac{1}{15 \times 60}$ and much more than (c) $\frac{1}{10^9 \times 10^2 \times 10^1 \times 10^1 \times 10^2}$.

Actual $^{24}Na \quad A = \frac{4.7 \times 10^{14} \text{ atoms}}{\text{min}}$

$^{11}C \quad A = \frac{1.9 \times 10^{15} \text{ atoms}}{\text{min}}$

$^{238}U \quad A = \frac{7.4 \times 10^5 \text{ atoms}}{\text{min}}$

19.4A $\ln\frac{A_t}{A_o} = -\lambda t = \ln\frac{A_t}{\frac{7.2\text{ dis}}{\text{min}}} = -(1.21 \times 10^{-4}\text{ y}^{-1}) \times 1500\text{ y} = -0.182$

$$e^{\ln\frac{A_t}{A_o}} = \frac{A_t}{A_o} = \frac{A_t}{\frac{7.2\text{ dis}}{\text{min}}} = e^{-0.182} = 0.834$$

$$A_t = \frac{6.0\text{ dis}}{\text{min}}$$

19.4B The activity falls to one-half its initial value in one half-life period—$t_{1/2}$ = 12.26 y. The brandy is 12.26 y old, not 25 years. The claimed age is not authentic.

19.5A The sum of the mass numbers of $^{35}_{17}$Cl and $^{1}_{0}$n equals the sum of the mass numbers of $^{35}_{16}$S and the product (1). The sum of the atomic numbers of Cl (17) and ^{1}n (0) equals the sum of the atomic number of S (16) and the product (1). The product must be a hydrogen nucleus. $^{35}_{17}\text{Cl} + ^{1}_{0}\text{n} \rightarrow ^{35}_{16}\text{S} + ^{1}_{1}\text{H}$

19.5B $^{249}_{98}\text{Cf} + ^{15}_{7}\text{N} \rightarrow ^{260}_{105}\text{Db} + ?^{1}_{0}\text{n}$

$249 + 15 = 260 + A \quad A = 4$

$98 + 7 = 105$

$^{249}_{98}\text{Cf} + ^{15}_{7}\text{N} \rightarrow ^{260}_{105}\text{Db} + 4\,^{1}_{0}\text{n}$

19.6A Yes, Zinc-74 should be radioactive because it has a higher neutron to proton ratio than the "belt of stability". It should decay by Beta emission.

19.6B $^{40}_{20}$Ca is not radioactive because it is an even nuclide, and 20 is a magic number. $^{48}_{20}$Ca is not radioactive because it is an even-even nuclide, and 20 and 28 are magic numbers. $^{39}_{20}$Ca is radioactive because it is an odd-even nuclide that is below the ratio of stability. $\frac{n}{p+} = 0.95 < 1.0$.

19.7A To bring $^{84}_{39}$Y closer to the belt of stability, the ratio of neutrons to protons needs to increase ($\frac{45}{39}$ = 1.15 stable Y is $\frac{50}{39}$ = 1.28). This means $^{84}_{39}$Y either decays by positron emission or electron capture. The resultant isotope would be $^{84}_{38}$Sr ($\frac{46}{38}$ = 1.21). The product is barely inside the belt of stability and actually is stable.

19.7B β^- emission effectively changes a neutron into a proton and thus decreases the neutron/proton ratio. β^+ emission increases the neutron/proton ratio. The isotope with the larger neutron/proton ratio should decay by β^- emission; this is $^{22}_{9}F$. $^{17}_{9}F$ decays by β^+ emission.

$^{22}_{9}F \rightarrow {}^{0}_{-1}e + {}^{22}_{10}Ne$

$^{17}_{9}F \rightarrow {}^{0}_{1}e + {}^{17}_{8}O$

19.8A $^{237}_{93}Np \rightarrow {}^{4}_{2}He + {}^{233}_{91}Pa$

$\Delta E = 232.9901\text{ u} + 4.0015\text{ u} - 236.9970\text{ u} = -0.0054\text{ u}$

$?\text{MeV} = -0.0054\text{ u} \times \dfrac{931.5\text{ MeV}}{\text{u}} = -5.0\text{ MeV}$

19.8B Mass $e^- = 9.1093897 \times 10^{-31}\text{ kg} \times \dfrac{\text{u}}{1.6605402 \times 10^{-27}\text{ kg}} = 5.485799 \times 10^{-4}\text{ u}$

nuclear masses

$^{27}_{13}Al$ $\quad 26.9815\text{ u} - 13 \times 5.485799 \times 10^{-4}\text{ u} = 26.9744\text{ u}$

$^{4}_{2}He$ $\quad 4.0026\text{ u} - 2 \times 5.485799 \times 10^{-4}\text{ u} = 4.0015\text{ u}$

$^{30}_{15}P$ $\quad 29.9783\text{ u} - 15 \times 5.485799 \times 10^{-4}\text{ u} = 29.9701\text{ u}$

$\Delta E = 29.9701\text{ u} + 1.0087\text{ u} - 26.9744\text{ u} - 4.0015\text{ u} = 0.0029\text{ u}$

$\Delta E = 0.0029\text{ u} \times \dfrac{931.5\text{ MeV}}{\text{u}} = 2.7\text{ MeV}$

ΔE from atomic masses

$\Delta E = 29.9783\text{ u} + 1.0087\text{ u} - 26.9815\text{ u} - 4.0026\text{ u} = 0.0029\text{ u}$

$\Delta E = 0.0029\text{ u} \times \dfrac{931.5\text{ MeV}}{\text{u}} = 2.7\text{ MeV}$

Both calculations are the same because the difference is the difference in masses of nuclei. Electrons are not involved in the reaction.

Review Questions

2. (b) β^- emission increases the atomic number by 1.

3. In this set of answers, only (c) β^- rays are charged and will be deflected by a magnetic field.

5. The activity is equal to the decay constant times the amount of material.
$A = \lambda N$
The most radioactive will have the largest (d) decay constant.

7. $\ln \dfrac{1}{100} = -\lambda t$

$$\lambda = \frac{0.693}{11.4\,\text{d}} = 0.0608\ \text{d}^{-1}$$

$$t = \frac{\ln \frac{1}{100}}{-\lambda} = \frac{\ln \frac{1}{100}}{-0.0608\ \text{d}^{-1}} = 75.7\ \text{d}$$

10. (d) Silver-108 is most likely to be radioactive as it is an odd-odd isotope. It has an odd number of both neutrons and protons.

11. Low neutron to proton ratios are more likely to decay by β^+ emission. (a) ^{59}Cu is the answer.

13. The binding energy curve peaks at iron. (c) $^{56}_{26}\text{Fe}$ is the answer.

15. The critical mass is the minimum mass of a fissionable isotope (such as uranium-235) that must be brought into a small volume for a self-sustaining nuclear fission reaction to occur.

Problems

21. (a) $^{219}_{86}\text{Rn} \rightarrow {}^{215}_{84}\text{Po} + {}^{4}_{2}\text{He}$

(b) $^{167}_{69}\text{Tm} + {}^{0}_{-1}\text{e} \rightarrow {}^{167}_{68}\text{Er}$

(c) $^{90}_{38}\text{Sr} \rightarrow {}^{0}_{-1}\text{e} + {}^{90}_{39}\text{Y}$

(d) $^{79}_{36}\text{Kr} \rightarrow {}^{0}_{+1}\text{e} + {}^{79}_{35}\text{Br}$

23. (a) $^{123}_{49}\text{In} + {}^{0}_{-1}\text{e} \rightarrow {}^{123}_{48}\text{Cd}$

(b) $^{103}_{42}\text{Mo} \rightarrow {}^{0}_{-1}\text{e} + {}^{103}_{43}\text{Tc}$

$^{103}_{43}\text{Tc} \rightarrow {}^{0}_{-1}\text{e} + {}^{103}_{44}\text{Ru}$

$^{103}_{44}\text{Ru} \rightarrow {}^{0}_{-1}\text{e} + {}^{103}_{45}\text{Rh}$

(c) $^{217}_{85}\text{At} \rightarrow {}^{4}_{2}\text{He} + {}^{213}_{83}\text{Bi}$

$^{213}_{83}\text{Bi} \rightarrow {}^{4}_{2}\text{He} + {}^{209}_{81}\text{Tl}$

$^{209}_{81}\text{Tl} \rightarrow {}^{0}_{-1}\text{e} + {}^{209}_{82}\text{Pb}$

$^{209}_{82}\text{Pb} \rightarrow {}^{0}_{-1}\text{e} + {}^{209}_{83}\text{Bi}$

25. (a) ${}^{7}_{4}\text{Be} + {}^{0}_{-1}\text{e} \rightarrow {}^{7}_{3}\text{Li}$

(b) ${}^{242}_{94}\text{Pu} \rightarrow {}^{4}_{2}\text{He} + {}^{238}_{92}\text{U}$

27.

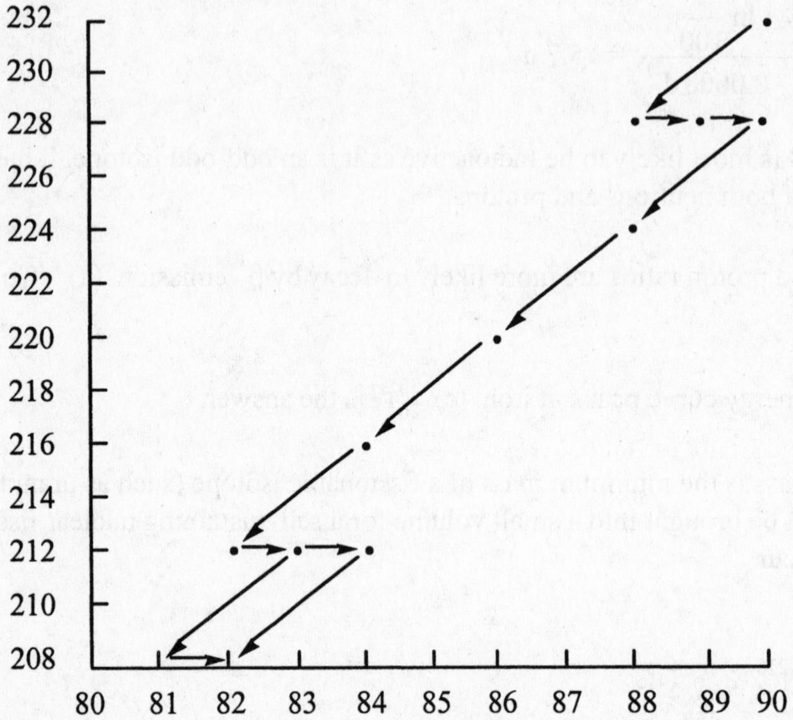

29. Chemical reactions are based on collision and a series of elementary steps that can lead to reactions of different orders. Radioactive decay is a nuclear process unrelated to chemical reactions. Although the decay may involve different types of particles (α, β, and γ rays), all radioactive decay follows the same rate law, which is first-order.

31. (a) $A = \lambda N = \dfrac{0.693}{t_{1/2}} N$

The shortest half-life would be the greatest activity. ^{15}O.

(b) 25% left means two half-lives. If two days is two half-lives, then one half-life is one day. The half-life of ^{28}Mg is close to one day.

33. For the same mass, there are about four times as many atoms of ^{32}P as of ^{131}I and the decay constant (λ) of ^{32}P is about one-half that of ^{131}I. The activity is proportional to the number of atoms by the factor of the decay constant: $A = \lambda \cdot N$. The activity of ^{32}P is about twice that of ^{131}I.

$$\frac{A_{\text{P-32}}}{A_{\text{I-131}}} = \frac{\lambda_{\text{P-32}}}{\lambda_{\text{I-131}}} \times \frac{N_{\text{P-32}}}{N_{\text{I-131}}} = \tfrac{1}{2} \times 4 = 2$$

35. (a) $A = \lambda N$

$$\lambda = \frac{0.693}{11.5\ \text{h}} = 0.0603\ \text{h}^{-1}$$

$$A = 0.0603\ \text{h}^{-1} \times 10^6\ \text{atoms} = 6.03 \times 10^4\ \text{decay/h}$$

(b) $$\lambda = \frac{0.693}{20.39\ \text{min}} \times \frac{60\ \text{min}}{\text{h}} = 2.04\ \text{h}^{-1}$$

$$A = 2.04\ \text{h}^{-1} \times 10^6\ \text{atoms} = 2.04 \times 10^6\ \text{decay/h}$$

37. $$\ln\frac{118}{138} = -\lambda \times 20.0\ \text{d}$$

$$-0.157 = -\lambda \times 20.0\ \text{d}$$

$$\lambda = 7.83 \times 10^{-3}/\text{d}$$

$$t_{1/2} = \frac{0.693}{7.83 \times 10^{-3}/\text{d}} = 88.5\ \text{d}$$

39. (a) 24.0 h is not quite 3 half-lives.

$$884 \xrightarrow{1} 442 \xrightarrow{2} 221 \xrightarrow{3} 110$$

A little more than 110 ~ 115 $\frac{\text{dis}}{\text{min}}$.

$$\ln\frac{A}{884\frac{\text{dis}}{\text{min}}} = -\lambda t = \frac{-0.693}{8.28\ \text{hr}} \times 24.0\ \text{hr} = -2.01$$

$$e^{\ln\frac{A}{884\frac{\text{dis}}{\text{min}}}} = \frac{A}{884\frac{\text{dis}}{\text{min}}} = e^{-2.01} = 0.134$$

$$A = 119\frac{\text{dis}}{\text{min}}$$

41. $$\ln\frac{A_t}{A_0} = -\lambda t = -\frac{0.693}{5730\ \text{y}} \times 2.80 \times 10^3\ \text{y} = -0.339$$

$$e^{\ln\frac{A_t}{A_0}} = \frac{A_t}{A_0} = \frac{A_t}{15\frac{\text{dis}}{\text{min g}}} = e^{-0.339} = 0.71$$

$$A_t = 11\frac{\text{dis}}{\text{min g}}$$

43. (a) ${}^{9}_{4}\text{Be} + {}^{4}_{2}\text{He} \rightarrow {}^{1}_{0}\text{n} + {}^{12}_{6}\text{C}$

(b) ${}^{130}_{52}\text{Te} + {}^{2}_{1}\text{H} \rightarrow {}^{130}_{53}\text{I} + 2\,{}^{1}_{0}\text{n}$

(c) ${}^{10}_{5}\text{B} + {}^{1}_{0}\text{n} \rightarrow {}^{7}_{3}\text{Li} + {}^{4}_{2}\text{He}$

45. (a) $2\,{}^{2}_{1}H \rightarrow {}^{3}_{2}He + {}^{1}_{0}n$

(b) ${}^{241}_{95}Am + {}^{4}_{2}He \rightarrow {}^{243}_{97}Bk + 2\,{}^{1}_{0}n$

(c) ${}^{238}_{92}U + {}^{4}_{2}He \rightarrow {}^{239}_{94}Pu + 3\,{}^{1}_{0}n$

47. ${}^{99}_{43}Tc + {}^{1}_{0}n \rightarrow {}^{100}_{43}Tc \rightarrow {}^{100}_{44}Ru + {}^{0}_{-1}e$

49. mass of protons 1.0073 × 26 = 26.1898 u
mass of neutrons 1.0087 × (56-26) = <u>30.2610 u</u>
56.4508 u

mass defect = 56.4508 - 55.92068 = 0.5301 u

51. mass of protons 1.0073 × 16 = 16.1168 u
mass of neutrons 1.0087 × 16 = <u>16.1392 u</u>
32.2560 u

mass defect = 32.2560 u - 31.96329 u = 0.2927 u

$$\frac{BE}{nucleon} = \frac{0.2927\,u}{32\text{ nucleons}} \times \frac{931.5\,MeV}{u} = 8.520\text{ MeV/nucleon}$$

53. ${}^{32}P \rightarrow {}^{0}_{-1}e + {}^{32}S$

energy = (31.9633 u + 0.0005486 u - 31.9657 u)

$$= -0.0019\,u \times \frac{931.5\,MeV}{u} = -1.8\,MeV$$

55. (a) ${}^{228}_{90}Th \rightarrow {}^{4}_{2}He + {}^{224}_{88}Ra$

$$\text{mass defect} = -5.50\,MeV \times \frac{u}{931.5\,MeV} = -0.00590$$

223.9719 u + 4.00150 u - mass Th 228 = – 0.00590
nuclear mass Th 228 = 227.9793 u

(b) atomic mass = 227.9793 u + 90 × 5.486 × 10^{-4} u = 228.0287 u

57. $\text{energy} = (235.0439\,u + 1.0087\,u - 4.0026\,u - 232.0382\,u) \times \frac{931.5\,MeV}{u}$

$= 11.0\,MeV$

59. We expect the neutron/proton ratio to be about 1:1 for nuclides of low atomic number, which these four are. The two with neutron/proton ratio closest to unity should not be radioactive:(b) ${}^{15}_{7}N$ and (c) ${}^{30}_{14}Si$. The other two should be radioactive:(a) ${}^{10}_{6}C$ and (d) ${}^{36}_{15}P$. Also, ${}^{36}_{15}P$ has odd numbers for both Z and N.

Only four stable nuclides exist with this combination, and phosphorus is not included among them. (a) $^{10}_{6}C$ has a neutron to proton ratio less than 1 so is radioactive.

61. Cu–64, Cl–36, and Ga–70, isotopes with mass numbers close to the weighted-average atomic mass do not occur naturally because they are odd-odd isotopes. Isotopes on each side of the weighted-average atomic mass are stable and average to the tabulated atomic mass.

63. Calcium-40 has a doubly magic number and is in the region where nuclides with the same number of neutrons and protons are the most stable.

65. $\Delta E = (138.8891\text{ u} + 94.8985\text{ u} + 2 \times (1.0087\text{ u}) - 234.9934\text{ u} - 1.0087\text{ u})$

$$= -0.1971\text{u} \times \frac{931.5\text{ MeV}}{\text{u}} = -183.6\text{ MeV}$$

67. A short half-life nuclide will decay to a smaller number of radioactive atoms quickly and will no longer be a hazard. A long half-life has a low activity and is less of a hazard than a nuclide of intermediate half-life. The intermediate half-life nuclide will have a fairly strong activity and will last for a considerable length of time.

69. First determine the decay constant. $\lambda = \dfrac{0.69315}{8.040\text{ d}} = 0.08621\text{ d}^{-1}$

Activity $= A = \lambda N$

Now determine the number of ^{131}I atoms to produce one curie.

$$3.70 \times 10^{10}\text{ s}^{-1} = 0.08621\text{ d}^{-1} \times \frac{1\text{ d}}{24\text{ h}} \times \frac{1\text{ h}}{3600\text{ s}} \times N$$

$$N = \frac{3.70 \times 10^{10}\text{ s}^{-1}}{9.978 \times 10^{-7}\text{ s}^{-1}} = 3.71 \times 10^{16}\text{ I-131 atoms/curie}$$

$$^{131}\text{I mass} = 40 \times 10^{6}\text{ curie} \times \frac{3.71 \times 10^{16}\ ^{131}\text{I atoms}}{\text{curie}} \times \frac{\text{mol}\ ^{131}\text{I}}{6.022 \times 10^{23}\text{ atoms}} \times \frac{131\text{ g I-131}}{1\text{ mol}\ ^{131}\text{I}} = 320\text{ g}\ ^{131}\text{I minimum.}$$

$$^{131}\text{I mass} = 50 \times 10^{6}\text{ curie} \times \frac{3.71 \times 10^{16}\ ^{131}\text{I atoms}}{\text{curie}} \times \frac{\text{mol}\ ^{131}\text{I}}{6.022 \times 10^{23}\text{ atoms}} \times \frac{131\text{ g I-131}}{1\text{ mol}\ ^{131}\text{I}} = 400\text{ g}\ ^{131}\text{I maximum.}$$

The amount of ^{131}I released ranges from 320 to 400 g

71. Since the positron reacts with an electron, the stopping distance would be very short, maybe a few micrometers. The positron would react with the first electron it encounters and there are a lot of electrons in matter.

73. Adding tritium to the hydrogen supply will cause Beta radiation at the leak which can be detected by a Geiger counter. The Beta rays will not get through the supply line at other places. An inert radioactive gas such as ^{37}Ar or ^{41}Ar can be added to the stream and produce the same effect, radiation at the leak.

Additional Problems

78. We convert the mass defect into MeV.

$$\Delta\text{energy} = (234.0436\text{ u} + 4.0026\text{ u} - 238.0508\text{ u}) \times \frac{931.5\text{ MeV}}{1\text{ u}} = -4.3\text{ MeV}$$

79. First determine the energy per photon in joules, then the wavelength.

$$0.050\text{ MeV} \times \frac{1.6022\times10^{-13}\text{ J}}{\text{MeV}} = 8.0\times10^{-15}\text{ J} = \frac{hc}{\lambda}$$

$$\lambda = \frac{hc}{8.0\times10^{-15}\text{ J}} = \frac{6.626\times10^{-34}\text{ J}\cdot\text{s}\times3.00\times10^{8}\text{ m/s}}{8.0\times10^{-15}\text{ J}} \times \frac{\text{pm}}{10^{-12}\text{ m}} = 25\text{ pm}$$

80. (a) $\frac{214}{4} = 53.5$ 0.5 is $\frac{2}{4}$ $4n + 2$ series

(b) $\frac{216}{4} = 54$ $4n$ series

(c) $\frac{215}{4} = 53.75$ 0.75 is $\frac{3}{4}$ $4n + 3$ series

(d) $\frac{235}{4} = 58.75$ 0.75 is $\frac{3}{4}$ $4n + 3$ series

81. $\Delta E = (74.9225\text{ u} + 2\times1.0087\text{ u} - 74.9216\text{ u} - 2.0140\text{ u})$

$$= 4.3\times10^{-3}\text{ u}\times931.5\frac{\text{MeV}}{\text{u}} = 4.01\text{ MeV}$$

$$= 4.01\text{ MeV}\times\frac{1.6022\times10^{-13}\text{ J}}{\text{MeV}} = 6.42\times10^{-13}\text{ J}$$

$E = ½mv^2$

$$v = \sqrt{\frac{2E}{m}} = \sqrt{\frac{2\times6.42\times10^{-13}\text{ J}}{2.0140\text{ u}}\times\frac{\text{u}}{1.6605\times10^{-27}\text{ kg}}\times\frac{\text{kg m}^2\text{ s}^{-2}}{\text{J}}}$$

$$v = 1.96\times10^{7}\frac{\text{m}}{\text{s}}$$

82. $C + O_2 \rightarrow CO_2$

$$\Delta H^\circ_{rxn} = \Delta H^\circ_f[CO_2(g)] = -393.5 \frac{kJ}{mol}$$

$$1.00 \text{ kg C} \times \frac{10^3 \text{ g}}{\text{kg}} \times \frac{\text{mol}}{12.01 \text{ g}} \times \frac{393.5 \text{ kJ}}{\text{mol}} \times \frac{10^3 \text{ J}}{\text{kJ}} \times \frac{\text{eV}}{1.602 \times 10^{-19} \text{ J}} \times \frac{\text{MeV}}{10^6 \text{ eV}}$$

$$\times \frac{\text{u}}{931.5 \text{MeV}} \times \frac{1.661 \times 10^{-24} \text{ g}}{\text{u}} = 3.65 \times 10^{-7} \text{ g}$$

86. (a) Rate = λN

$$\lambda = \frac{\text{Rate}}{N} = \frac{2.40 \times 10^2 \text{ dis/s}}{40 \text{ ng}} \times \frac{\text{ng}}{10^{-9} \text{ g}} \times \frac{226 \text{ g}}{\text{mole}} \times \frac{\text{mole}}{6.022 \times 10^{23} \text{ atom}} =$$

$$2.25 \times 10^{-12} \text{ s}^{-1}$$

$$\ln \frac{N}{N_0} = \ln \frac{A}{A_0} = \ln \frac{A}{240 \text{ s}^{-1}} = -\lambda t$$

$$= -2.25 \times 10^{-12} \text{ s}^{-1} \times 50 \text{ y} \times \frac{365 \text{ d}}{\text{y}} \times \frac{24 \text{ hr}}{\text{d}} \times \frac{3600 \text{ s}}{\text{hr}} = -0.00355$$

$$e^{\ln \frac{A}{240 \text{ s}^{-1}}} = \frac{A}{240 \text{ s}^{-1}} = e^{-0.00355} = 0.996$$

$A_t = 0.996 \times 240 \text{ s}^{-1} = 239 \text{ s}^{-1}$

At 50 years the rate is 2.39×10^2 dis s^{-1}.

(b) $$\ln \frac{N}{N_0} = \ln \frac{A}{A_0} = \ln \frac{A}{240 \text{ s}^{-1}} = -\lambda t$$

$$= -2.25 \times 10^{-12} \text{ s}^{-1} \times 100 \text{ y} \times \frac{365 \text{ d}}{\text{y}} \times \frac{24 \text{ hr}}{\text{d}} \times \frac{3600 \text{ s}}{\text{hr}} = -0.00710$$

$$e^{\ln \frac{A}{240 \text{ s}^{-1}}} = \frac{A}{240 \text{ s}^{-1}} = e^{-0.00710} = 0.993$$

$A_t = 0.993 \times 240 \text{ s}^{-1} = 238 \text{ s}^{-1}$

At 100 years the rate is 2.38×10^2 dis s^{-1}.

89. $$? \text{ kJ} = -3.2 \times 10^{-11} \frac{\text{J}}{\text{atom}} \times \frac{6.022 \times 10^{23} \text{ atom}}{\text{mol}} \times \frac{\text{mol}}{235 \text{ g}} \times \frac{10^{-3} \text{ g}}{\text{mg}}$$

$$\times 1.00 \text{ mg} \times \frac{\text{kJ}}{10^3 \text{ J}} = -8.2 \times 10^4 \text{ kJ (from fission)}$$

$CH_4(g) + 2\ O_2(g) \rightarrow CO_2(g) + 2\ H_2O(l)$

$\Delta H^\circ_{rxn} = \Sigma \Delta H^\circ_f \text{(products)} - \Sigma \Delta H^\circ_f \text{(reactants)}$

$$\Delta H^\circ_{rxn} = 1 \text{ mol } CO_2 \times -393.5 \frac{kJ}{mol} + 2 \text{ mol } H_2O \times -285.8 \frac{kJ}{mol}$$

$$- 1 \text{ mol } CH_4 \times -74.81 \frac{kJ}{mol} - 2 \text{ mol } O_2 \times 0 = -890.3 \text{ kJ/mol } CH_4$$

$2\ C_2H_6(g) + 7\ O_2(g) \rightarrow 4\ CO_2(g) + 6\ H_2O(l)$

$$\Delta H^\circ \text{ rxn} = 4 \text{ mol } CO_2 \times -393.5 \frac{kJ}{mol} + 6 \text{ mol } H_2O \times -285.8 \frac{kJ}{mol}$$

$$- 2 \text{ mol } C_2H_6 \times -84.68 \frac{kJ}{mol} - 7 \text{ mol } O_2 \times 0 = \frac{-3119.4 \text{ kJ}}{2 \text{ mol } C_2H_6}$$

$$\Delta H^\circ \text{ rxn} = -1559.7 \frac{kJ}{\text{mol } C_2H_6}$$

$$\text{Energy per mole gas} = 0.920 \times -890.2 \text{ kJ} + 0.080 \times -1559.7 \text{ kJ} = -943.8 \frac{kJ}{\text{mol gas}}$$

$$? \text{ L} = -8.2 \times 10^4 \text{ kJ} \times \frac{\text{mol gas}}{-943.8 \text{ kJ}} \times \frac{\frac{62.4 \text{ mmHg}}{\text{K mol}} \text{ x } (22^\circ + 273)\text{K}}{744 \text{ mmHg}}$$

$$= 2.1 \times 10^3 \text{ L natural gas}$$

92. Co-57 – $1s^12s^22p^63s^23p^64s^23d^7$
Co-57 - $1s^22s^12p^63s^23p^64s^23d^7$
$\lambda = 0.179$ nm

$$\nu = \frac{3.00 \times 10^8 \text{ m/s}}{0.179 \text{ nm}} \times \frac{\text{nm}}{10^{-9} \text{ m}} = 1.68 \times 10^{18} \text{ s}^{-1}$$

$$E = h\nu = \frac{6.626 \times 10^{-34} \text{ Js}}{\text{atom}} \times 1.68 \times 10^{18} \text{ s}^{-1} \times \frac{6.022 \times 10^{23} \text{ atom}}{\text{mol}} \times \frac{kJ}{10^3 \text{ J}}$$

$$= \frac{6.70 \times 10^5 \text{ kJ}}{\text{mol}}$$

93. Copper-63 and copper-65 are stable. Copper-64 is an odd-odd isotope so it is radioactive. Both zinc-64 and nickel-64 are even-even isotopes, close to the belt of stability, and are likely to be stable. Copper-64 can thus decay in either direction, by beta decay to zinc-64 or by either electron capture or positron emission to nickel-64.

Apply Your Knowledge

94. Ra– emanation $\xleftarrow{\alpha}$ radium $\xleftarrow{\alpha}$ ionium
radon-222 radium-226 thorium-230

$\xleftarrow{\alpha}$ uranium II $\xleftarrow{\beta}$ uranium x_2 $\xleftarrow{\beta}$ uranium x_1
uranium-234 protactinium-234 thorium-234

$\xleftarrow{\alpha}$ uranium I
uranium-238

95. $? \frac{J}{kg} = \frac{1000\,rem}{1} \times \frac{1\,rad \times Q}{1\,rem} \times \frac{0.01\,Gy}{rad} \times \frac{J}{Gy\ kg} = \frac{10\,J}{kg}$

Q =1 for beta emission

$? J = 66\ kg \times \frac{10\,J}{kg} = 660\ J$

$? J = 660\ J \times \frac{43\ kg\ O}{66\ kg\ body} = 430\ J$

$? \frac{J}{g} = \frac{1314\,kJ}{mol} \times \frac{mol}{15.999\,g} \times \frac{10^3\,J}{kJ} = 8.213 \times 10^4\ \frac{J}{g}$

$? g = 430\ J \times \frac{g}{8.213 \times 10^4\ J} = 5.24 \times 10^{-3}\ g\ O$

$? \text{fraction} = \frac{5.24 \times 10^{-3}\ g}{43\ kg} \times \frac{kg}{10^3\ g} = 1.22 \times 10^{-7}$ or 1.22×10^{-5}% O ionized

$? J = 660\ J \times \frac{16\ kg\ C}{66\ kg\ body} = 160\ J$

$? \frac{J}{g} = \frac{1086\,kJ}{mol} \times \frac{mol}{12.011\,g} \times \frac{10^3\,J}{kJ} = 9.04 \times 10^4\ \frac{J}{g}$

$? g = 160\ J \times \frac{g}{9.04 \times 10^4\ J} = 1.77 \times 10^{-3}\ g\ C$

$? \text{fraction} = \frac{1.77 \times 10^{-3}\ g}{16\ kg} \times \frac{kg}{10^3\ g} = 1.11 \times 10^{-7}$ or 1.11×10^{-5}% C ionized

$? J = 660\ J \times \frac{7\ kg\ H}{66\ kg\ body} = 70\ J$

$? \frac{J}{g} = \frac{1312\,kJ}{mol} \times \frac{mol}{1.0079\,g} \times \frac{10^3\,J}{kJ} = 1.302 \times 10^6\ \frac{J}{g}$

$? g = 70\ J \times \frac{g}{1.302 \times 10^6\ J} = 5.4 \times 10^{-5}\ g\ H$

$? \text{fraction} = \frac{5.34 \times 10^{-5}\ g}{7\ kg} \times \frac{kg}{10^3\ g} = 7.62 \times 10^{-9}$ or 7.62×10^{-7}% H ionized

$$? \text{total fraction} = \frac{5.24 \times 10^{-3}\ g\ O + 1.77 \times 10^{-3}\ g\ C + 5.4 \times 10^{-5}\ g\ H}{66\ kg \times \frac{10^3\ g}{kg}}$$

$$= 1.1 \times 10^{-7} \times 100\% = 1.1 \times 10^{-5}\%$$

99.

t	cpm	–background of 27 cpm	ln A
0	1784	1757	7.47
1	1232	1205	7.09
2.33	880	853	6.75

3.5	656	629	6.44
4.5	554	527	6.27
5.5	342	315	5.75
6.5	266	239	5.48

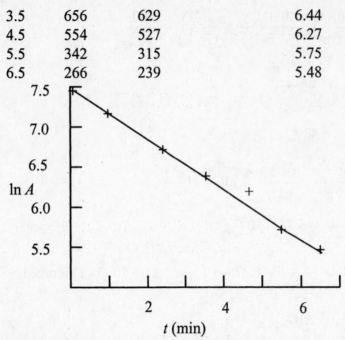

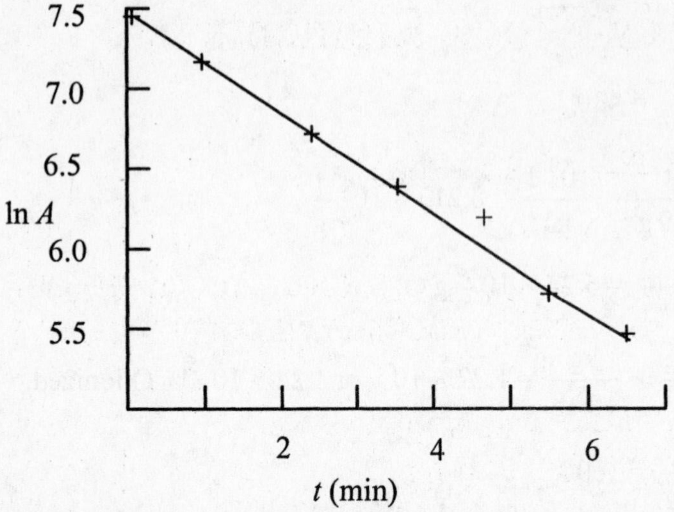

In a research lab, the data would be fed into a computer to give the slope of the best straight time. For this problem, use the third and last points to get the slope. Notice that we do not use the outlying point.

$$\text{slope} = -\lambda = \frac{5.48 - 6.75}{6.5 - 2.33} = -0.305$$

$\lambda = 0.305 \text{ min}^{-1}$

$$t_{1/2} = \frac{0.693}{0.305 \text{ min}^{-1}} = 2.27 \text{ min}$$

Note that from 0 to 2.33 min, the cpm declines by slightly more than half. One computer program gives $\lambda = 0.297 \text{ min}^{-1}$ and $t_{1/2} = 2.33$ min.

Chapter 20

The *s*-Block Elements

Exercises

20.1A (a) $2\ NaCl(l) \xrightarrow{\text{electrolysis}} 2\ Na(l) + Cl_2(g)$

$2\ Na(s) + H_2(g) \xrightarrow{\Delta} 2\ NaH(s)$

(b) $2\ NaCl(aq) + 2\ H_2O(l) \xrightarrow{\text{electrolysis}} 2\ NaOH(aq) + Cl_2(g) + H_2(g)$

$2\ NaOH(aq) + Cl_2(g) \rightarrow NaCl(aq) + NaOCl(aq) + H_2O(l)$

20.1B $Na(s) \rightarrow NaOH(aq) \xrightarrow{SO_2} Na_2SO_3(aq)$

$Na(s) + 2\ H_2O(l) \rightarrow 2\ NaOH(aq) + H_2(g)$

$2\ NaOH\ (aq) + SO_2\ (g) \rightarrow Na_2SO_3(aq) + H_2O(l)$

20.2A (a) If the sole product were MgO, the mass of the product would be

$$? \text{ g MgO} = 1.000 \text{ g Mg} \times \frac{1 \text{ mol Mg}}{24.305 \text{ g Mg}} \times \frac{2 \text{ mol MgO}}{2 \text{ mol Mg}} \times \frac{40.304 \text{ g MgO}}{1 \text{ mol MgO}}$$
$$= 1.658 \text{ g MgO maximum mass}$$

(b) If the sole product were Mg_3N_2, the mass of the product would be

$$? \text{ g } Mg_3N_2 = 1.000 \text{ g Mg} \times \frac{1 \text{ mol Mg}}{24.305 \text{ g Mg}} \times \frac{1 \text{ mol } Mg_3N_2}{3 \text{ mol Mg}}$$
$$\times \frac{100.93 \text{ g } Mg_3N_2}{1 \text{ mol } Mg_3N_2} = 1.384 \text{ g } Mg_3N_2 \text{ minimum mass}$$

20.2B Slightly less than the maximum might indicate mostly MgO is formed with a little Mg_3N_2. Since N_2 is much more abundant in the air than is O_2, the rate of reaction must be much greater with O_2 than N_2.

Review Questions

1. Sodium is the most abundant alkali metal, and calcium is the most abundant alkaline earth metal.
The alkali and alkaline earth metals occur principally as carbonates, chlorides, and sulfates.

4. If only a small quantity of water is electrolyzed, H^+ produced at the anode is removed at the cathode and the concentration of sulfuric acid is essentially unchanged. However, if a larger quantity is electrolyzed, the concentration of the sulfuric acid increases slowly. This slow increase results from the gradual decrease on the amount of water.

5. Helium does not resemble the Group 2A elements physically or chemically. It has a filled shell of electrons in the configuration $1s^2$ and it resembles the other noble gases. The other 2A elements have an empty *p* subshell in the valence shell.

7. (c) is incorrect. CaH_2 is an ionic hydride. A molecular hydride is hydrogen with a nonmetal.

13. (c) the electrolysis of H_2O is not involved in the Dow process.

Problems

21. Na_2CO_3 is most soluble because 1A salts are more soluble than 2A salts. Li_2CO_3 is next in solubility because lithium is like magnesium in solubility of its salt, even though it is a 1A element. $MgCO_3$ is least soluble because it is a group 2A salt. Interionic attractions are strongest between the small highly charged Mg^{2+} and $CO_3{}^{2-}$ and are greater for Li_2CO_3 than for Na_2CO_3 because Li^+ is a smaller ion than Na^+.

23. Density, electrode potential, electrical conductivity, and hardness. The regular periodic trends usually depends on one changing property. The irregular trends depend on more than one property which are conflicting.

25. (a) potassium peroxide
(b) calcium hydrogen carbonate (calcium bicarbonate)
(c) $Ba_3(PO_4)_2$
(d) $Sr(NO_3)_2 \cdot 4H_2O$

27. (a) calcium oxide, CaO
(b) sodium sulfate decahydrate, $Na_2SO_4 \cdot 10H_2O$
(c) calcium sulfate dihydrate, $CaSO_4 \cdot 2H_2O$
(d) $CaCO_3 \cdot MgCO_3$

29.

$$MgCO_3 \xrightarrow{HCl} MgCl_2(aq) \xrightarrow{\text{evaporation}} MgCl_2(s) \xrightarrow{\text{electrolysis}} Mg(l)$$

$$MgCO_3(s) + 2\,HCl \rightarrow MgCl_2(aq) + H_2O + CO_2(g)$$

$$MgCl_2(s) \xrightarrow{\text{electrolysis}} Mg(l) + Cl_2(g)$$

31. Glauber's salt is sodium sulfate decahydrate $Na_2SO_4 \cdot 10\,H_2O$.

$$H_2SO_4(\text{conc aq}) + 2\,NaCl(s) \xrightarrow{\Delta} Na_2SO_4(s) + 2\,HCl(g).$$

$$Na_2SO_4(s) + 10\,H_2O(l) \rightarrow Na_2SO_4 \cdot 10H_2O(s)$$

A volatile acid is produced by heating one of its salts (NaCl) with a nonvolatile acid (H_2SO_4).

33. NaOH(s) has to be prepared from NaCl(aq) by electrolysis. To obtain Na from NaOH requires an additional electrolysis. This is a more expensive route than to electrolyze NaCl(l), despite the requirement of a higher temperature.

35.

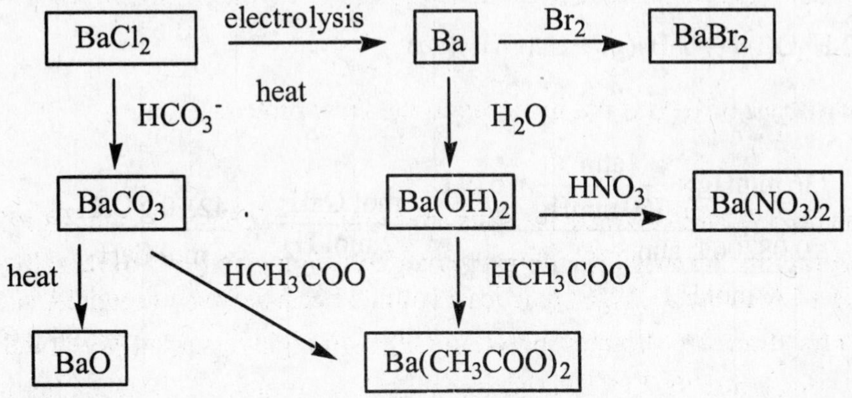

37. $Cl_2 + 2e^- \rightarrow 2Cl^-$ $E° = 1.358$ V
$O_2 + 4H^+ + 4e^- \rightarrow 2H_2O$ $E° = 1.229$ V
$2H^+ + 2e^- \rightarrow H_2$ $E° = 0.00$ V
$Mg^{2+} + 2e^- \rightarrow Mg$ (s) $E° = -2.356$ V
Electrolysis of $MgCl_2$ (aq) will produce H_2 (g) and O_2 (g). Electrolysis without the water produces Mg (s) and Cl_2 (g).

39. (a) $Ca(s) + 2\ HCl(aq) \rightarrow H_2(g) + CaCl_2(aq)$
(b) $KHCO_3(s) + HCl(aq) \rightarrow KCl(aq) + H_2O(l) + CO_2(g)$
(c) $2\ NaF(s) + H_2SO_4(conc.\ aq) \xrightarrow{\Delta} Na_2SO_4(s) + 2\ HF(g)$

41. (a) $2\ Li(s) + Cl_2(g) \rightarrow 2\ LiCl(s)$
(b) $2\ K(s) + 2\ H_2O(l) \rightarrow 2\ KOH(aq) + H_2(g)$
(c) $2\ Cs(s) + Br_2(l) \rightarrow 2\ CsBr(s)$
(d) $K(s) + O_2(g) \rightarrow KO_2(s)$

43. (a) $BeF_2(s) + 2\ Na(l) \xrightarrow{\Delta} Be(s) + 2\ NaF(s)$
(b) $Ca(s) + 2\ CH_3COOH(aq) \rightarrow Ca(CH_3COO)_2(aq) + H_2(g)$
(c) $PuO_2(s) + 2\ Ca(s) \xrightarrow{\Delta} Pu(s) + 2\ CaO(s)$
(d) $CaCO_3{\cdot}MgCO_3(s) \xrightarrow{\Delta} CaO(s){\cdot}MgO(s) + 2\ CO_2(g)$

45. (a) $MgCO_3(s) + 2\ HCl(aq) \rightarrow MgCl_2(aq) + H_2O(l) + CO_2(g)$
(b) $CO_2(g) + 2\ KOH(aq) \rightarrow K_2CO_3(aq) + H_2O(l)$
(c) $2\ KCl(s) + H_2SO_4(conc.\ aq) \xrightarrow{\Delta} K_2SO_4(s) + 2\ HCl(g)$

47. (a) $2\ NaCl(s) + H_2SO_4(conc.\ aq) \xrightarrow{\Delta} Na_2SO_4(s) + 2\ HCl(g)$

(b) $Mg(HCO_3)_2(aq) \xrightarrow{\Delta} MgCO_3(s) + H_2O(l) + CO_2(g)$

$MgCO_3(s) \xrightarrow{\Delta} MgO(s) + CO_2(g)$

49. $CaH_2(s) + 2\ H_2O(l) \rightarrow 2\ H_2(g) + Ca(OH)_2(aq)$

$$n = \frac{PV}{RT}$$

$$?\ g\ CaH_2 = \frac{736\ mmHg \times \frac{atm}{760\ mmHg} \times 679\ L}{\frac{0.08206\ L\ atm}{K\ mol\ H_2} \times (11+273)\ K} \times \frac{mol\ CaH_2}{2\ mol\ H_2} \times \frac{42.10\ g\ CaH_2}{mol\ CaH_2}$$

$$= 594\ g\ CaH_2$$

51. $CaCO_3{\cdot}MgCO_3(s) \rightarrow CaO{\cdot}MgO(s) + 2\ CO_2(g)$

$$1.00 \times 10^3\ kg \times \frac{10^3\ g}{kg} \times \frac{mol\ CaCO_3{\cdot}MgCO_3}{184.4\ g} \times \frac{2\ mol\ CO_2}{mol} \times \frac{0.08206\ atm}{K\ mol}$$

$$\times \frac{295\ K}{748\ mmHg \times \frac{atm}{760\ mmHg}} \times \frac{1\ m^3}{10^3\ L} = 267\ m^3\ CO_2$$

53. $$1.00\ kg\ Mg \times \frac{10^3\ g}{kg} \times \frac{ton\ seawater}{1270\ g\ Mg} \times \frac{2000\ lb}{ton} \times \frac{454\ g}{lb} \times \frac{mL}{1.03\ g} \times \frac{10^{-3}\ L}{mL}$$

$$= 694\ L\ seawater$$

The actual volume required is greater than 694 L because not every step in the overall process has a 100% yield.

55.

					Actual
(a)	$\frac{8 \times 18 \times 100\%}{137 + 32 + 2 + (8 \times 18)}$	≈	$\frac{144 \times 100\%}{170 + 144}$	< 50%	46%
(b)	$\frac{6 \times 18 \times 100\%}{40 + (2 \times 35.5) + (6 \times 18)}$	≈	$\frac{108 \times 100\%}{111 + 108}$	≈ 50%	49%
(c)	$\frac{7 \times 18 \times 100\%}{24 + 32 + 64 + (7 \times 18)}$	≈	$\frac{126 \times 100\%}{120 + 126}$	> 50%	51%

(c) has the greatest mass percent water. It is the only one that is larger than 50%.

57. The base NH_3 reacts with HCO_3^- to form CO_3^{2-}, which precipitates Mg^{2+}, Ca^{2+} and Fe^{2+}.

$NH_3 + HCO_3^- \rightarrow NH_4^+ + CO_3^{2-}$

$M^{2+} + CO_3^{2-} \rightarrow MCO_3(s)$ $\qquad M^{2+} = Mg^{2+}, Ca^{2+}, Fe^{2+}$

Alternatively, think of the reactions as

$NH_3 + H_2O \rightarrow NH_4^+ + OH^-$

$OH^{-} + HCO_3^{-} \rightarrow CO_3^{2-} + H_2O$
$M^{2+} + CO_3^{2-} \rightarrow MCO_3(s)$ $\qquad M^{2+} = Mg^{2+}, Ca^{2+}, Fe^{2+}$

59. $Ca(HCO_3)_2(aq) \rightarrow CaCO_3(s) + H_2O(g) + CO_2(g)$

$$725 \text{ mL water} \times \frac{10^{-3}\text{ L}}{\text{mL}} \times \frac{115 \text{ g } HCO_3^-}{10^3 \text{ L water}} \times \frac{\text{mol } HCO_3^-}{61.02 \text{ g}} \times \frac{\text{mol } CaCO_3}{2 \text{ mol } HCO_3^-} \times \frac{100.09 \text{ g } CaCO_3}{\text{mol } CaCO_3} = 0.0684 \text{ g "scale"}$$

61. $Ca(HCO_3)_2(aq)$ drips from a cave roof. When the drop evaporates, it leaves $CaCO_3(s)$, which builds up from the floor, making stalagmites. If the drop evaporates before falling, stalactites are produced. Columns are formed where a stalactite meets a stalagmite. The caves themselves are formed when calcium carbonate is dissolved by water charged with $CO_2(g)$.The water is "hard water" after calcium carbonate dissolves in it. That is, the formation of a cave or of "hard water" occurs through the reaction: $CaCO_3(s) + H_2O(l) + CO_2(aq) \rightarrow Ca(HCO_3)_2(aq)$. Stalactites, stalagmites, and "boiler scale" (which forms when hard water is boiled) form in the reverse of this reaction.

63. The water is first run through a cation-exchange resin to exchange all cations for H^+. Then an anion-exchange resin will exchange the anions for OH^-. The OH^- and H^+ combine to form water.
$Ca^{2+}(aq) + \text{resin - 2 H(s)} \rightarrow \text{resin - Ca(s)} + 2\,H^+(aq)$
$Fe^{3+}(aq) + \text{resin - 3 H(s)} \rightarrow \text{resin - Fe(s)} + 3\,H^+(aq)$
$SO_4^{2-}(aq) + \text{resin - 2 OH(s)} \rightarrow \text{resin - } SO_4(s) + 2\,OH^-(aq)$
$Cl^-(aq) + \text{resin - OH(s)} \rightarrow \text{resin - Cl(s)} + OH^-(aq)$
$H^+(aq) + OH^-(aq) \rightarrow H_2O(l)$

Additional Problems

65. (a) $4\,Li(s) + O_2(g) \rightarrow 2\,Li_2O(s)$
(b) $6\,Li(s) + N_2(g) \rightarrow 2\,Li_3N(s)$
(c) $Li_2CO_3(s) \rightarrow Li_2O(s) + CO_2(g)$

66. $2\,Na_2O_2(aq) + 2\,CO_2(g) \rightarrow 2\,Na_2CO_3(aq) + O_2(g)$

67. $CaO(s) + CO_2(g) \rightarrow CaCO_3(s)$. Carbon dioxide reacts with the CaO or water vapor reacts with the CaO.
$CaO(s) + H_2O(g) \rightarrow Ca(OH)_2(s)$ followed by:
$Ca(OH)_2(s) + CO_2(g) \rightarrow CaCO_3(s) + H_2O(g)$.

71. Let X be the mass of Mg that reacts to form MgO and $1.000 - X$, the mass of Mg that reacts to form Mg_3N_2.

$$?\text{g MgO} = X\text{ g Mg} \times \frac{1\text{ mol Mg}}{24.305\text{ g Mg}} \times \frac{2\text{ mol MgO}}{2\text{ mol Mg}} \times \frac{40.304\text{ g MgO}}{\text{mol MgO}} = 1.658\,X\text{ gMgO}$$

$$?\text{g Mg}_3\text{N}_2 = (1.000\text{-}X)\text{ g Mg} \times \frac{1\text{ mol Mg}}{24.305\text{ g Mg}} \times \frac{1\text{ mol Mg}_3\text{N}_2}{3\text{ mol Mg}} \times \frac{100.93\text{ g Mg}_3\text{N}_2}{\text{mol Mg}_3\text{N}_2}$$

$$= 1.384 \times (1.000 - X) = (1.384 - 1.384\,X)\text{gMg}_3\text{N}_2$$

The total mass of product is 1.537g. That is,
$1.658\,X + (1.384 - 1.384\,X) = 1.537$
$0.274\,X = 0.153$
$X = 0.558$
Mass of MgO = $1.658\,X = 1.658 \times 0.558 = 0.925$ g

$$\text{Mass \% MgO} = \frac{0.925\text{ g MgO} \times 100\%}{1.537\text{ g products}} = 60.2\%\text{ MgO.}$$

74. (a) $Ca^{2+}(aq) + Na_2CO_3(aq) \rightarrow CaCO_3(s) + 2\,Na^+(aq)$

(b) $$?\text{ g Na}_2\text{CO}_3 = 162\text{ L water} \times \frac{117\text{ g SO}_4^{2-}}{10^3\text{ L water}} \times \frac{\text{mol SO}_4^{2-}}{96.06\text{ g}} \times \frac{\text{mol Ca}^{2+}}{\text{mol SO}_4^{2-}}$$

$$\times \frac{\text{mol Na}_2\text{CO}_3}{\text{mol Ca}^{2+}} \times \frac{105.99\text{ g}}{\text{mol}} = 20.9\text{ g Na}_2\text{CO}_3$$

80. (a) The size of the paint particles is in the colloidal range.
surface area of sphere = $4\pi r^2$

$$?\frac{\text{ions}}{\text{particle}} = \frac{4\,\pi \times (500\text{ mm})^2}{\text{particle}} \times \frac{\text{ion}}{0.16\text{ mm}^2} = 1.96 \times 10^7\,\frac{\text{ions}}{\text{particle}}$$

volume of sphere = $4\pi\, r^3/3$

$$?\,\frac{\text{g}}{\text{particle}} = \frac{4\,\pi}{3} \times \left(500\text{ nm} \times \frac{10^{-9}\text{ m}}{\text{nm}} \times \frac{\text{cm}}{10^{-2}\text{ m}}\right)^3 \times \frac{\text{mL}}{\text{cm}^3} \times \frac{4.26\text{ g}}{\text{mL}}$$

$$= 2.23 \times 10^{-12}\,\frac{\text{g}}{\text{particle}}$$

$$?\text{ ions} = 1.00\text{ kg paint} \times \frac{10^3\text{ g}}{\text{kg}} \times \frac{40.0\text{ g TiO}_2}{100.0\text{ g paint}} \times \frac{\text{particle}}{2.23 \times 10^{-12}\text{ g}}$$

$$\times\, 1.96 \times 10^7\,\frac{\text{ions}}{\text{particle}} = 3.51 \times 10^{21}\text{ ions}$$

$$?\text{ g sodium dodecyl sulfate} = 3.51 \times 10^{21}\text{ ions} \times \frac{\text{mol}}{6.022 \times 10^{23}} \times \frac{288.4\text{ g}}{\text{mol}} = 1.7\text{ g}$$

(b) ? C = 1.96×10^7 ion $\times\ 1.602 \times 10^{-19}$ C/ion = 3.1×10^{-12} C
The charged colloidal particles repel one another, making it easier for them to slide past one another, that is, imparting less resistance to flow or lowering viscosity.

(c) In oil-based paint, the surfactant (a salt) will not dissolve well in the oil. In latex paint, water molecules are attracted to ions of the surfactant. The surfactant dissolves in the solvent and can easily be carried to the particles.

83. Spectroscopic methods should reveal whether the emitted radiation is that of hydrogen or of other elements.

Apply Your Knowledge

84.
$$\frac{15.17\text{ mL} \times \frac{10^{-3}\text{ L}}{\text{mL}} \times 0.02650\text{ M NaOH}}{100.0\text{ mL solution} \times \frac{10^{-3}\text{ L}}{\text{mL}}} \times \frac{1\text{ mol H}^+}{1\text{ mol OH}^-} \times \frac{1\text{ mol Ca}^{2+}}{2\text{ mol H}^+}$$
$$\times \frac{40.08\text{ g Ca}^{2+}}{\text{mol Ca}^{2+}} \times \frac{\text{ppm}}{\text{g}/10^3\text{ L}} = 80.56\text{ ppm Ca}^{2+}$$

$(1\text{ ppm} = \text{g}/10^3\text{ L},\ 10^3\text{ L} = 10^6\text{ g})$

85.
$$?\text{ years} = \frac{3.0\text{ V} \times 0.5\text{ A h}}{5\,\mu\text{W}} \times \frac{\mu\text{ W}}{10^{-6}\text{ W}} \times \frac{\text{W}}{\text{V A}} \times \frac{\text{d}}{24\text{h}} \times \frac{\text{yr}}{365\text{ d}} = 34\text{ years}$$
$$?\text{ g Li} = 0.5\text{ A h} \times \frac{3600\text{ s}}{\text{h}} \times \frac{\text{C}}{\text{A s}} \times \frac{\text{mol e-}}{96485\text{ C}} \times \frac{\text{mol Li}}{\text{mol e}} \times \frac{6.94\text{ g Li}}{\text{mol Li}} = 0.13\text{ g Li}$$

89.
$$?\frac{\text{kg CO}_2}{\text{d}\cdot\text{person}} = \frac{1.4\times10^4\text{ L air}}{\text{d}\cdot\text{person}} \times \frac{4.0\text{ L CO}_2}{100\text{ L air}} \times \frac{1.00\text{ atm}}{\frac{0.08206\text{ L atm}}{\text{K mol}} \times 298\text{ K}} \times \frac{44.01\text{ g}}{\text{mol CO}_2}$$
$$\times \frac{\text{kg}}{10^3\text{ g}} = 1.0\frac{\text{kg CO}_2}{\text{d}\cdot\text{person}}$$

(a) $2\text{ LiOH} + CO_2 \rightarrow Li_2CO_3 + H_2O$

$$?\text{ kg LiOH} = 1.0\frac{\text{kg CO}_2}{\text{d}\cdot\text{person}} \times 3\text{ people} \times 7\text{ d} \times \frac{\text{mol CO}_2}{44.01\text{ g CO}_2} \times \frac{2\text{ mol LiOH}}{\text{mol CO}_2}$$
$$\times \frac{23.95\text{ g LiOH}}{\text{mol LiOH}} = 23\text{ kg LiOH}$$

(b) $2\text{ NaOH} + CO_2 \rightarrow Na_2CO_3 + H_2O$

$$?\text{ kg NaOH} = 1.0\frac{\text{kg CO}_2}{\text{d}\cdot\text{person}} \times 3\text{ people} \times 7\text{ d} \times \frac{\text{mol CO}_2}{44.01\text{ g CO}_2} \times \frac{2\text{ mol NaOH}}{\text{mol CO}_2}$$
$$\times \frac{40.00\text{ g NaOH}}{\text{mol NaOH}} = 38\text{ kg NaOH}$$

$$?\$ = (38 - 23\ )\text{ kg} \times \frac{\$20{,}000}{\text{kg}} = \$300{,}000$$

91. $? \text{ mol linolenic} = 13600 \text{ kg/day} \times \frac{\text{mol}}{278.42 \text{ g}} = 48.847 \frac{\text{kmol}}{\text{day}}$

$$? \frac{\text{kg } H_2}{\text{day}} \text{ for steric} = \frac{24.424 \text{ kmol}}{\text{day}} \times \frac{3 \text{ mol } H_2}{\text{mol linolenic}} \times \frac{2.0158 \text{ g}}{\text{mol}} = 147.70 \frac{\text{kg}}{\text{day}}$$

$$? \frac{\text{kg } H_2}{\text{day}} \text{ for oleic} = \frac{24.424 \text{ kmol}}{\text{day}} \times \frac{2 \text{ mol } H_2}{\text{mol linolenic}} \times \frac{2.0158 \text{ g}}{\text{mol}} = 98.460 \frac{\text{kg}}{\text{day}}$$

$$? \frac{\text{kg } H_2}{\text{day}} \text{ total} = 147.70 \frac{\text{kg}}{\text{day}} + 98.460 \frac{\text{kg}}{\text{day}} = 246.17 \frac{\text{kg}}{\text{day}}$$

Chapter 21

The *p*-Block Elements

Exercise

21.1A $2\ BCl_3(g) + 3\ H_2(g) \rightarrow 2\ B(s) + 6\ HCl(g)$

21.1B $[B_2O_4(OH)_4]^{2-}(aq) + H_2O(l) \rightarrow H_2O_2(aq) + [B(OH)_4]^-(aq)$

$[B_2O_4(OH)_4]^{2-} \rightarrow 2\ [B(OH)_4]^-$ $\qquad$ $2\ H_2O \rightarrow H_2O_2$

$+ 4\ H^+ + 4e^-$ $\qquad$ $+ 2H^+ + 2e^-$

$[B_2O_4(OH)_4]^{2-}(aq) + 4\ H_2O(l) \rightarrow 2\ H_2O_2(aq) + 2\ [B(OH)_4]^-(aq)$

21.2A $2\ Br^-(aq) \rightarrow Br_2(l) + 2\ e^-$ $\qquad$ $E°_{Br^2/Br^-} = 1.065\ V$

$SO_4^{2-}(aq) + 4\ H^+(aq) + 2\ e^- \rightarrow 2\ H_2O + SO_2(g)$ $\qquad$ $E°_{SO_4^{2-}/SO_2} = 0.17\ V$

$E°_{cell} = E°_{cathode} - E°_{anode} = E°_{SO_4^{2-}/SO_2} - E°_{Br^2/Br^-} = -0.90\ V$

$\Delta G° = -nFE°_{cell}$

$\Delta G° = -2\ mol\ e^- \times \frac{96485\ C}{mol\ e^-} \times -0.90\ V \times \frac{J}{C\ V} \times \frac{kJ}{1000\ J} = 174\ kJ$

The positive $\Delta G°$ indicates that the reaction is nonspontaneous, just as was concluded in Example 21.2. The $\Delta G°$ values do not agree because the states of the reactants and products differ, for example, $H_2SO_4(l)$ and NaBr(s) in Example 21.2 and $[Br^-] = 1M$, $[H^+] = 1\ M$ and $[SO_4^{2-}] = 1\ M$ in this exercise.

21.2B $2\ NaBr(s) + 2\ H_2SO_4(l) \rightarrow Na_2SO_4(s) + 2\ H_2O(g) + SO_2(g) + Br_2(g)$

from Example 21.2

−349.0	−690.1	−1270	−228.6	−300.2	3.14

$\Delta G°_{rxn} = 54\ kJ$

The two problems are similar except for NaI instead of NaBr and I_2 instead of Br_2.

NaI(s)	$I_2(s)$
−286.1	19.36

$\Delta G°_{rxn} = 54 + (19.4-3.1) - (2 \times (-286.1) - (2 \times(-349)))$

$\Delta G°_{rxn} = -56\ kJ$

The $\Delta G°_{rxn}$ is negative, so the reaction is spontaneous at 25°C and the reaction is spontaneous at a lower temperature than the reaction in Example 21.2.

Review Questions

3. Oxygen, silicon, and aluminum are the three most abundant elements in Earth's crust.

4. Aluminum is the most active metal, and fluorine is the most active nonmetal in the *p* block of elements.

8. Because aluminum is less dense than copper, the transmission lines can be made lighter than if copper were used. Even if a larger cross section wire is used to equal the current carrying capacity of a copper cable, the aluminum cable is lighter and easier to support.

15. Water has extensive hydrogen bonding between molecules, which makes it persist as a liquid at much higher temperatures than is the case for H_2S, in which hydrogen bonding is unimportant.

21. (a) silver azide (b) potassium thiocyanate
(c) astatine oxide (d) telluric acid

22. (a) H_2SeO_4 (b) $H_2Te(aq)$ (c) $Pb(N_3)_2$ (d) AgAt

23. bauxite mixture of hydrates of Al_2O_3, boric oxide B_2O_3, corundum Al_2O_3, cyanogen $(CN)_2$, hydrazine NH_2NH_2, silica SiO_2

Problems

25. $3\,Mg(s) + B_2O_3(s) \xrightarrow{\Delta} 2\,B(s) + 3\,MgO(s)$

27. H_3BO_3 is an acid because it accepts an OH^- from water forming both $[B(OH)_4]^-$ and H^+ or H_3O^+. The molecule $B(OH)_3$ is only able to accept one OH^- through a lone pair of electrons on the OH^- ion. Therefore it is a monoprotic acid. H_3PO_4 is a triprotic acid because it has the structure $(HO)_3PO$ and can lose three H^+ ions from the molecule.

29. $Na_2B_4O_7{\cdot}10H_2O(s) + H_2SO_4(aq) \rightarrow 4\,B(OH)_3(s) + Na_2SO_4(aq) + 5\,H_2O(l)$
$2\,B(OH)_3(s) \xrightarrow{\Delta} B_2O_3(s) + 3\,H_2O(g)$
$B_2O_3(s) + 3\,C(s) + 3\,Cl_2(g) \xrightarrow{\Delta} 2\,BCl_3(s) + 3\,CO(g)$
$2\,BCl_3(s) + 3\,H_2(g) \xrightarrow{\Delta} 2\,B(s) + 6\,HCl(g)$

31. $2\,BCl_3 + 6\,LiAlH_4 \rightarrow B_2H_6 + 6\,Li^+ + 6\,Cl^- + 6\,AlH_3$

33. $12\,NaHCO_3(s) + 4\,KAl(SO_4)_2(s) \xrightarrow{\Delta} 2\,K_2SO_4(s) + 6\,Na_2SO_4(s) + 4\,Al(OH)_3(s) + 12\,CO_2(g)$

35. Aluminum does oxidize but only a thin outside layer. That layer protects the inside aluminum from further oxidation or corrosion. The iron oxide layer is more permeable and catalyzes more iron to oxidize.

37. $\Delta H^\circ_{rxn} = \Sigma\Delta H^\circ_f \text{ products} - \Sigma\Delta H^\circ_f \text{ reactants}$
$\Delta H^\circ_{rxn} = \Delta H^\circ_f(Al_2O_3) - \Delta H^\circ_f(Fe_2O_3) = -1676 \text{ kJ/mol} - (-824.2 \text{ kJ/mol})$
$\Delta H^\circ_{rxn} \approx -852 \text{ kJ/mol}$
This result is only approximate because it is based on data at 298 K, whereas the reaction occurs at a very high temperature. Also, the estimate assumes $\Delta H^\circ_f = 0$ for Fe(s), even though at the temperature of the reaction the stable form of iron is Fe(l).

39. Hydrolysis of Al^{3+}(aq) produces $Al(OH)_3$(s). At a lower pH, the H^+ reacts with the hydroxide from $Al(OH)_3$ and the compound dissolves, producing Al^{3+}(aq). At a high pH, $Al(OH)_4^-$ is formed and again the gelatinous $Al(OH)_3$ dissolves. $Al(OH)_3$ (s) exists as a suspended solid in water only if the pH is between 5 and 8.

41.

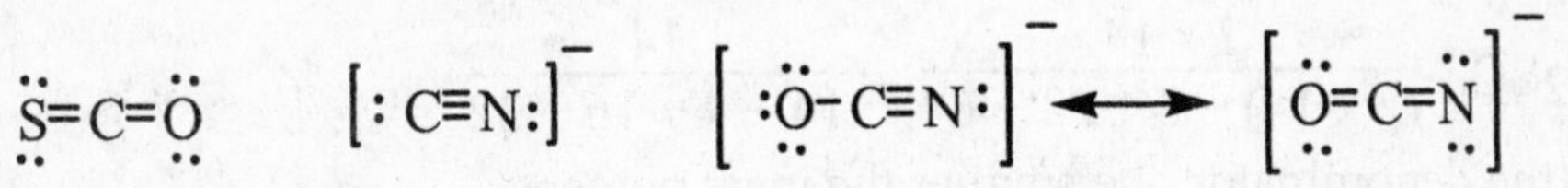

43. $Al_4C_3(s) + 12\ H_2O(l) \rightarrow 3\ CH_4(g) + 4\ Al(OH)_3(s)$

45.

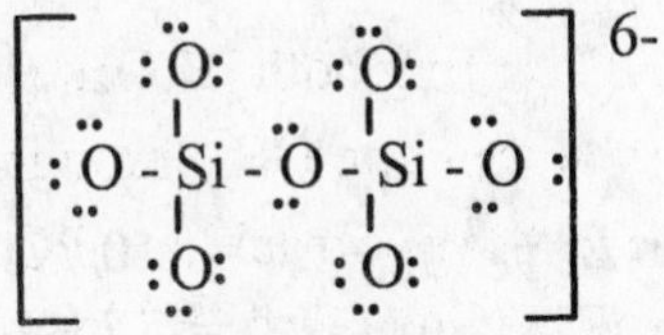

This would be twc tetrahedra with a common corner.

47. $\overset{+1}{K}\overset{+3}{Al_2}(\overset{+3}{Al}\overset{+4}{Si_3}\overset{-2}{O_{10}})^{-5}(\overset{-2}{O}\overset{+1}{H})_2^{-1}$ $\quad 1 + 2(+3) + (-5) + 2(-1) = 0$

49. (a) $Si(CH_3)_4$ (b) $SiCl_2(CH_3)_2$ (c) $SiH(C_2H_5)_3$

51. $Sn^{2+} \rightarrow Sn^{4+} + 2\ e^-$ $\qquad E^\circ_{Sn^{4+}/Sn^{2+}} = 0.154$ V
(a) $Cu^{2+} + e^- \rightarrow Cu^+$ $\qquad E^\circ_{Cu^{2+}/Cu^+} = 0.159$ V
Yes: $E^\circ_{cell} = E^\circ_{Cu^{2+}/Cu^+} - E^\circ_{Sn^{4+}/Sn^{2+}} = 0.005$ V
although the reaction is not likely to go to completion.
(b) $V^{3+} + e^- \rightarrow V^{2+}$ $\qquad E^\circ_{V^{3+}/V^{2+}} = -0.255$ V
No: $E^\circ_{cell} = E^\circ_{V^{3+}/V^{2+}} - E^\circ_{Sn^{4+}/Sn^{2+}} = -0.409$ V
(c) $Ag^+ + e^- \rightarrow Ag$ $\qquad E^\circ_{Ag^+/Ag} = 0.800$ V
Yes: $E^\circ_{cell} = E^\circ_{Ag^+/Ag} - E^\circ_{Sn^{4+}/Sn^{2+}} = 0.646$ V
(d) $Cr^{3+} + e^- \rightarrow Cr^{2+}$ $\qquad E^\circ_{Cr^{3+}/Cr^{2+}} = -0.424$ V

No; $E°_{cell} = E°_{Cr^{3+}/Cr^{2+}} - E°_{Sn^{4+}/Sn^{2+}} = -0.578$ V

53. (a) $Sn(s) + 2\ HCl(aq) \rightarrow SnCl_2(aq) + H_2(g)$
(b) $SnCl_2(s) + Cl_2(g) \rightarrow SnCl_4(g)$
(c) $SnCl_4(aq) + 4\ NH_3(aq) + 2\ H_2O(l) \rightarrow SnO_2(s) + 4\ NH_4^+(aq) + 4Cl^-(aq)$

55. (a) $N_2O\ \frac{2 \times 14}{(2 \times 14) + 16} = \frac{14}{14 + 8}$
(b) $NH_3\ \frac{14}{14 + 3}$
(c) $NO\ \frac{14}{14 + 16}$
(d) $NH_4Cl\ \frac{14}{14 + 4 + 35}$
(e) $H_2NNH_2\ \frac{2 \times 14}{(2 \times 14) + 4} = \frac{14}{14 + 2}$
(f) $(NH_4)_2SO_4\ \frac{2 \times 14}{(2 \times 14) + 8 + 32 + 64} = \frac{14}{14 + 4 + 16 + 32}$
The larger the denominator, the smaller the mass percent.
In increaing order, f < d < c < a < b < e.

57. (a) $3\ NO_2(g) + H_2O(l) \rightarrow 2\ HNO_3(aq) + NO(g)$
(b) $4\ NH_3(g) + 5\ O_2(g) \rightarrow 4\ NO(g) + 6\ H_2O(g)$
(c) $NH_4NO_3(s) \xrightarrow{200\,°C} N_2O(g) + 2\ H_2O(g)$

59. $4(Fe^{3+}(aq) + e^- \rightarrow Fe^{2+}(aq))$ reduction $E°_{Fe^{3+}/Fe^{2+}} = 0.771$ V
$\underline{NH_2NH_3^+(aq) \rightarrow N_2(g) + 5\ H^+(aq) + 4\ e^-}$ oxidation $E°_{N_2/NH_2NH_3^+} = ?$ V
$4\ Fe^{3+}(aq) + NH_2NH_3^+(aq) \rightarrow N_2(g) + 4\ Fe^{2+}(aq) + 5\ H^+(aq)$
$E°_{cell} = E°_{Fe^{3+}/Fe^{2+}} - E°_{N_2/NH_2NH_3^+} = 0.771\ V - E°_{N_2/NH_2NH_3^+} = 1.00$ V
$E°_{N_2/NH_2NH_3^+} = 0.771\ V - 1.00\ V = -0.23$ V

61. The principal allotropes of phosphorus are white P and red P, with white phosphorus being the more reactive. The molecular structure of white phosphorus consists of individual P_4 tetrahedra. In red phosphorus, the P_4 tetrahedra are joined into long chains.

63. $P_4(s) + 3\ KOH(aq) + 3\ H_2O(l) \rightarrow 3\ KH_2PO_2(aq) + PH_3(g)$

65. (a) $4\ Al(s) + 3\ O_2(g) \rightarrow 2\ Al_2O_3(s)$
(b) $2\ KClO_3(s) \rightarrow 2\ KCl(s) + 3\ O_2(g)$
(c) $2\ Na_2O_2(s) + 2\ H_2O(l) \rightarrow 4\ NaOH(aq) + O_2(g)$
(d) $Pb^{2+}(aq) + O_3(g) + H_2O(l) \rightarrow PbO_2(s) + O_2(g) + 2\ H^+(aq)$

67. Sulfur will melt at 119 °C. Water is heated under pressure to produce steam at about 160 °C. The super-heated steam melts the sulfur underground. Sulfur neither reacts with hot water nor dissolves in it, so the liquid sulfur can be brought to the surface in a pure condition by compressed air. (see Figure 21.13.)

69. reduction: $2\ SO_3^{2-} + 3\ H_2O + 4\ e^- \rightarrow S_2O_3^{2-} + 6\ OH^-$

oxidation: $\underline{2\ S(s) + 6\ OH^- \rightarrow S_2O_3^{2-} + 3\ H_2O + 4\ e^-}$

$2\ SO_3^{2-} + 2\ S(s) \rightarrow 2\ S_2O_3^{2-}$

$SO_3^{2-}(aq) + S(s) \rightarrow S_2O_3^{2-}(aq)$

71. $Br_2 + 2\ e^- \rightarrow 2\ Br^-$ $\quad E° = 1.065$ V

$I_2 + 2\ e^- \rightarrow 2\ I^-$ $\quad E° = 0.535$ V

$F_2 + 2\ e^- \rightarrow 2\ F^-$ $\quad E° = 2.866$ V

$Cl_2 + 2\ e^- \rightarrow 2\ Cl^-$ $\quad E° = 1.358$ V

To displace Br_2 from an aqueous solution of Br^- requires oxidation of Br^-.

$2\ Br^-(aq) \rightarrow Br_2(l) + 2\ e^-$ $\quad E°_{anode} = 1.065$ V

$E°_{cell} = E°(\text{reduction}) - E°_{Br_2/Br^-} = E°(\text{reduction}) - 1.065$ V

Only a reduction half-reaction with $E° > 1.065$ will work, and this must be $Cl_2(g) + 2\ e^- \rightarrow 2\ Cl^-(aq)$ $\quad E° = 1.358$ V. I_2 is too poor an oxidizing agent to work, and I^-, Cl^-, and F^- can only be reducing agents, not oxidizing agents.

73. $I^-(aq) + 3\ Cl_2(g) + 3\ H_2O(l) \rightarrow IO_3^-(aq) + 6\ Cl^-(aq) + 6\ H^+(aq)$

$2\ Br^-(aq) + Cl_2(g) \rightarrow Br_2(l) + 2\ Cl^-(aq)$

When the products of the reaction are treated with $CS_2(l)$, the Br_2 dissolves in $CS_2(l)$ and the other products remain in the aqueous solution.

75. $MgBr_2$ is a halide.

$NaClO_3$ is a halate salt.

BrF_3 is an interhalogen compound and fits neither of the two categories above.

77. (a) $2\ KI(s) + 2\ H_2SO_4(\text{conc.}) \rightarrow I_2(s) + K_2SO_4(s) + 2\ H_2O(l) + SO_2(g)$

(b) $KI(s) + H_3PO_4(\text{conc.}) \rightarrow KH_2PO_4(s) + HI(g)$

79. (a)

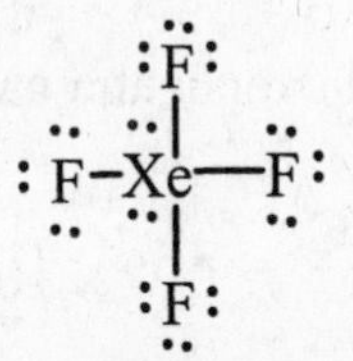

4 bonded atoms

2 lone pairs

AX_4E_2

square planar

(b)

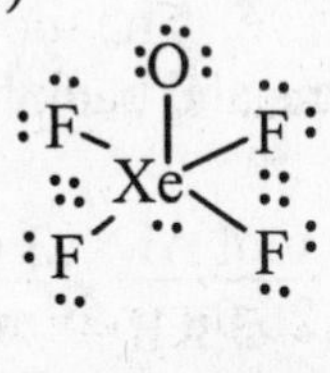

5 bonded atoms

1 lone pair

AX_5E

square pyramidal

81. The nucleus $^{4}_{2}He$ is very stable and is thus easily formed and emitted as an alpha particle from many nuclei. Argon is not emitted from other nuclei; it is formed only by the nucleus of $^{40}_{19}K$ absorbing an inner shell electron to become argon.

Additional Problems

83.

(a) 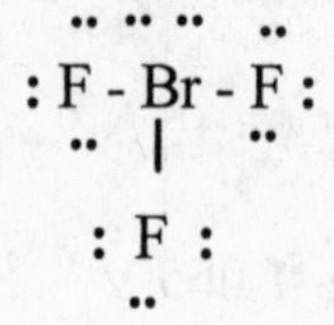

(b) 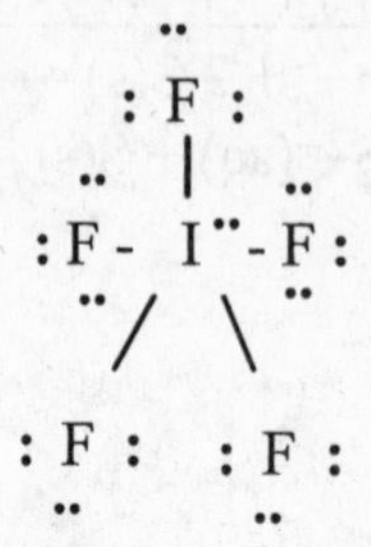

(a)	(b)
3 bonded atoms	5 bonded atoms
2 L.P.	1 L.P.
AX_3E_2	AX_5E
T-shape	square pyramidal
trigonal bipyramidal e^- geometry	octahedral e^- geometry

86. $2\ Al(s) + 2\ KOH(aq) + 6\ H_2O(l) \rightarrow 2\ KAl(OH)_4(aq) + 3\ H_2(g)$

$KAl(OH)_4(aq) + 2\ H_2SO_4(aq) + 8\ H_2O(l) \rightarrow KAl(SO_4)_2{\cdot}12H_2O(s)$

91. $$2.50\ L \times 1.75\ M\ Na_2SO_3 \times \frac{2\ mol\ H_2S}{mol\ SO_3^{2-}} \times \frac{100\ mol\ air}{1.5\ mol\ H_2S} \times \frac{\frac{62.4\ mm\ Hg\ L}{K\ mol} \times 298\ K}{755\ mm\ Hg} = 1.4 \times 10^4\ L\ air$$

92. $$1.00 \times 10^3\ kg\ Al \times \frac{1000\ g\ Al}{1\ kg\ Al} \times \frac{mol\ Al}{26.98\ g} \times \frac{3\ mol\ C}{4\ mol\ Al} \times \frac{12.01\ g\ C}{mol\ C} \times \frac{1\ kg\ C}{1000\ g\ C} = 334\ kg\ C$$

$$1.00 \times 10^3\ kg\ Al \times \frac{1000\ g\ Al}{1\ kg\ Al} \times \frac{mol\ Al}{26.98\ g} \times \frac{mol\ Al_2O_3}{2\ mol\ Al} \times \frac{101.96\ g\ Al_2O_3}{mol\ Al_2O_3} \times \frac{1\ kg\ Al}{1000\ g\ Al} = 1.89 \times 10^3\ kg\ Al_2O_3$$

(Note that the conversion to grams and back to kilograms will cancel and can be left out.)

$$1.00 \times 10^3\ kg\ Al \times \frac{1000\ g\ Al}{1\ kg\ Al} \times \frac{mol\ Al}{26.98\ g} \times \frac{3\ mol\ e^-}{mol\ Al} \times \frac{96485\ C}{mol\ e^-} = 1.07 \times 10^{10}\ C$$

95. $Mg + 2\ H_2O \rightarrow Mg(OH)_2 + 2\ H^+ + 2\ e^-$

$NO_3^- + 9\ H^+ + 8\ e^- \rightarrow NH_3 + 3\ H_2O$

$4\ Mg + 5\ H_2O + NO_3^- + H^+ \rightarrow 4\ Mg(OH)_2 + NH_3$

$4\ Mg + 6\ H_2O + NO_3^- \rightarrow 4\ Mg(OH)_2 + NH_3 + OH^-$

$NH_3 + HCl \rightarrow NH_4Cl$

$HCl + NaOH \rightarrow Na^+ + Cl^- + H_2O$

$$[NO_3^-] = ((50.00\text{ mL} \times 0.1500\text{ M HCl} - 32.10\text{ mL} \times 0.1000\text{ M NaOH} \times \frac{\text{mol HCl}}{\text{mol NaOH}}) \times \frac{\text{mol}NH_3}{\text{mol}HCl} \times \frac{\text{mol }NO_3^-}{\text{mol }NH_3}) \times \frac{1}{25.00\text{ mL}} = 0.1716\text{ M }NO_3^-$$

98. (a) $1/2\ Cl_2(g) + 1/2\ F_2(g) \rightarrow ClF(g)$

$\Delta H = \text{BE(react)} - \text{BE(prod)}$

$$\Delta H = 1/2\text{ mol }Cl_2 \times \frac{243\text{ kJ}}{\text{mol}} + 1/2\text{ mol }F_2 \times \frac{159\text{ kJ}}{\text{mol}} - 1\text{ mol ClF} \times \frac{251\text{ kJ}}{\text{mol}} = \frac{-50\text{ kJ}}{\text{mol}}$$

(b) $1/2\ O_2(g) + F_2(g) \rightarrow OF_2(g)$

$$\Delta H = 1/2\text{ mol }O_2 \times \frac{498\text{ kJ}}{\text{mol}} + 1\text{ mol }F_2 \times \frac{159\text{ kJ}}{\text{mol}} - 2\text{ mol OF} \times \frac{187\text{ kJ}}{\text{mol}} = \frac{34\text{ kJ}}{\text{mol}}$$

(c) $Cl_2(g) + 1/2\ O_2(g) \rightarrow Cl_2O(g)$

$$\Delta H = 1\text{ mol }Cl_2 \times \frac{243\text{ kJ}}{\text{mol}} + 1/2\text{ mol }O_2 \times \frac{498\text{ kJ}}{\text{mol}} - 2\text{ mol ClO} \times \frac{205\text{ kJ}}{\text{mol}} = \frac{82\text{ kJ}}{\text{mol}}$$

(d) $1/2\ N_2(g) + 3/2\ F_2(g) \rightarrow NF_3(g)$

$$\Delta H = 1/2\text{ mol }N_2 \times \frac{946\text{ kJ}}{\text{mol}} + 3/2\text{ mol }F_2 \times \frac{159\text{ kJ}}{\text{mol}} - 3\text{ mol NF} \times \frac{280\text{ kJ}}{\text{mol}} = \frac{-129\text{ kJ}}{\text{mol}}$$

Apply Your Knowledge

105. Area of Earth's surface $= 4\pi r^2 = 4 \times \pi \times (4000\text{ mi})^2 \times \left(\frac{5280\text{ ft}}{\text{mi}}\right)^2 \times \left(\frac{12\text{ in}}{\text{ft}}\right)^2 \times \left(\frac{2.54\text{ cm}}{\text{in}}\right)^2 = 5.21 \times 10^{18}\text{ cm}^2$

$$V_{O_3} = \text{area} \times \text{thickness} = 5.21 \times 10^{18}\text{ cm}^2 \times 0.3\text{ cm} = 2 \times 10^{18}\text{ cm}^3 = 2 \times 10^{18}\text{ mL}$$

$$?\ O_3\text{ molecules} = 2 \times 10^{18}\text{ mL} \times \frac{10^{-3}\text{ L}}{\text{mL}} \times \frac{\text{mol}}{22.4\text{ L}} \times \frac{6.02 \times 10^{23}}{\text{mol}} = 5 \times 10^{37}\ O_3\text{ molecules}$$

Chapter 22

The *d*-Block Elements and Coordination Chemistry

Exercises

22.1A Essentially nothing. The match used in an attempt to ignite the dichromate salt might flare a bit initially, with the dichromate, an oxidizing agent, promoting the oxidation of the match. However, a reaction could not be sustained, because there is no reducing agent in the solid.

22.1B $4\ FeCr_2O_4(s) + 16\ NaOH(l) + 7\ O_2(g)$

$$\xrightarrow{\Delta} 8\ Na_2CrO_4(s) + 4\ Fe(OH)_3(s) + 2\ H_2O(g)$$

22.2A $MnO_4^-(aq) + 2\ H_2O(l) + 3\ e^- \rightarrow MnO_2(s) + 4\ OH^-(aq)$ $\quad E° = 0.60$ V

$MnO_4^-(aq)$ in basic solution will oxidize any species for which $E° < 0.60$ V. This includes $Br^-(aq)$ to $BrO_3^-(aq)$, $Br_2(l)$ to $BrO^-(aq)$, Ag(s) to $Ag_2O(s)$, $NO_2^-(aq)$ to $NO_3^-(aq)$, S(s) to $SO_3^{2-}(aq)$, and so on.

22.2B (a) $BiO_3^- + 4\ H^+ + 2\ e^- \rightarrow BiO^+ + 2\ H_2O$

$$Mn^{2+} + 4\ H_2O \rightarrow MnO_4^- + 8\ H^+ + 5\ e^-$$

$$2\ Mn^{2+}(aq) + 5\ BiO_3^-(aq) + 4\ H^+(aq) \rightarrow 2\ MnO_4^-(aq) + 5\ BiO^+(aq) + 2\ H_2O(l)$$

(b) The standard electrode potential for the BiO_3^- reduction to BiO^+ must exceed that of the reduction of MnO_4^- to Mn^{2+}, which is 1.51 V.

22.3A (a) Coordination number = 6

$ox\#_{Co}$ + anionic charge$_{SO_4^{2-}}$ = species charge

$ox\#_{Co} + (-2) = +1$

$ox\#_{Co} = +3$

(b) Coordination number = 6

$ox\#_{Fe} + 6 \times$ anionic charge$_{CN^-}$ = species charge

$ox\#_{Co} + 6 \times (-1) = -4$

$ox\#_{Fe} = +2$

22.3B $[CrCl_2(H_2O)_4]Cl$

22.4A (a) hexaamminecobalt(II) ion

(b) pentaamminebromocobalt(III) bromide

22.4B (a) tetrachloroaurate(III) ion
(b) aquabromobis(ethylenediamine)cobalt(III) tetracyanonickelate(II)

22.5A (a) $[Coox(en)_2]^{3+}$ (b) $[CrCl_4(NH_3)_2]^-$

22.5B (a) $[Cr(NH_3)_6][Co(CN)_6]$
(b) $[PtCl_2(en)_2]SO_4$

22.6A (a) These are isomers. The atom donating electrons is S in the first and N in the second.
(b) These are isomers. The ligands, NH_3, are bonded to Zn in one structure and Cu in the other. Similarly, the Cl^- ligands are bonded to Cu in one structure and Zn in the other.
(c) These are not isomers. There are two NH_3 in one structure and 4 NH_3 in the other.

22.6B (a) Different structural isomers are not possible. The H_2O can bond at any one of the four equivalent sites with NH_3 in the other three sites.
(b) $[Pt(NH_3)_4][CuCl_4]$ is another isomer, as are other combinations of NH_3 and Cl^- on Cu^{2+} and Pt^{2+} such as $[CuCl(NH_3)_3][PtCl_3(NH_3)]$ and $[PtCl(NH_3)_3][CuCl_3(NH_3)]$.

22.7A Assuming that the ethylenediamine molecule can link only to adjacent coordination sites, there is only one form of the molecule and no geometric or optical isomers.

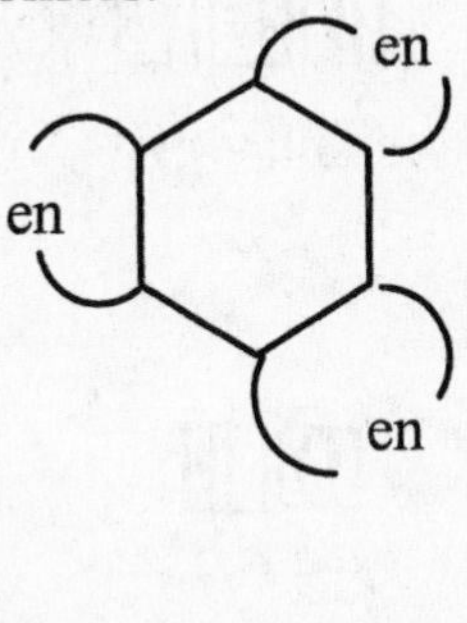

22.7B

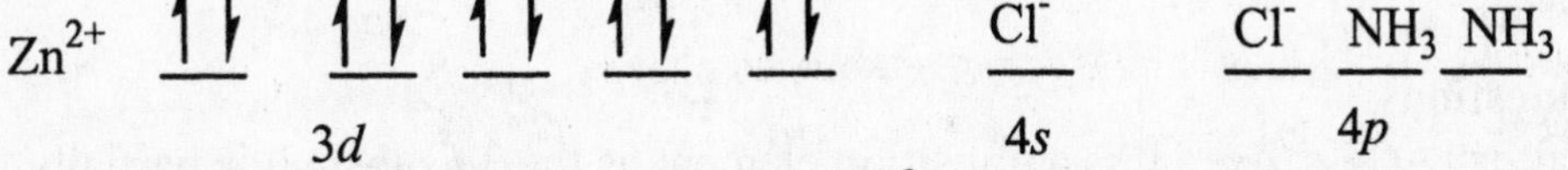

$[ZnCl_2(NH_3)_2]$ is a tetrahedral complex (sp^3 hybrid)and has no cis-trans isomerism. All four vertices are equivalent.

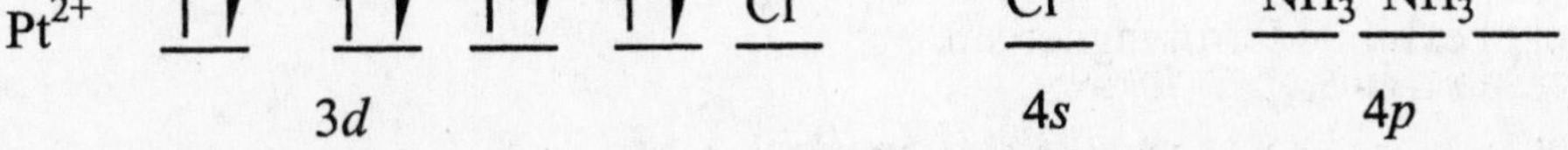

$[PtCl_2(NH_3)_2]$ is a square planar complex (*dsp*^{2}hybrid) and can have cis-trans isomerism. There is a cis isomer with Cl^- next to each other. The trans isomer has the two Cl^- ions are at opposite corners.

22.8A

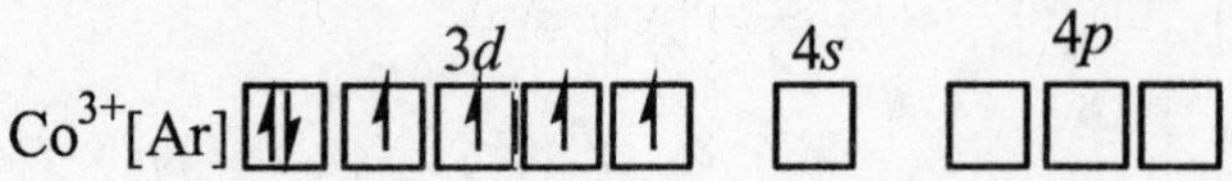

For a strong field ligand (CN^-), the electrons are paired in the lower set of orbitals.

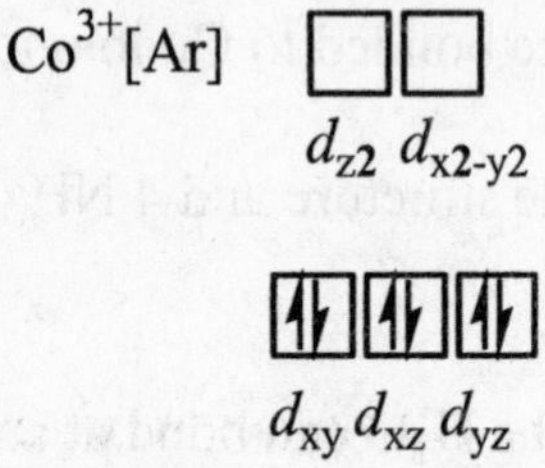

There are no unpaired electrons.

22.8B

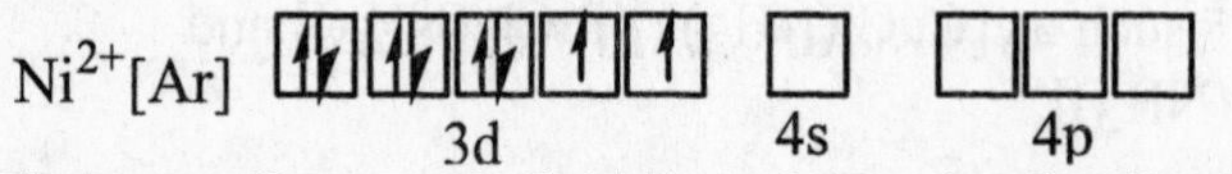

Cl^- is usually a weak-field ligand, but the distribution of the eight 3*d* electrons would be the same even if it were strong field. There are two unpaired electrons.

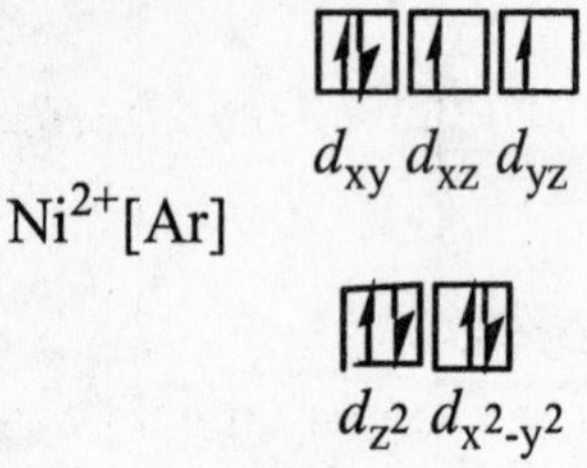

weak field

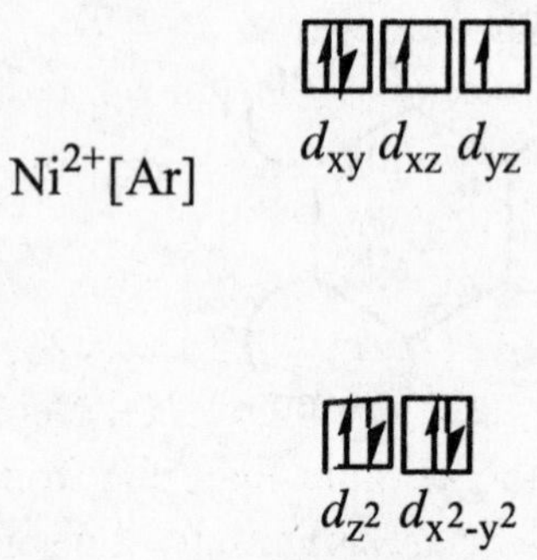

strong field

Review Questions

2. Cobalt is [Ar] $3d^7 4s^2$. It is a transition element as the 3*d* subshell is partially filled. [Kr] $4d^{10}5s^2 5p^1$ is a representative element, indium. The 4*d* subshell is filled, and the 5*p* subshell is partially filled.

5. (c) Nickel exhibits ferromagnetism.

6. (a) Au can have a +3 oxidation state. The oxidation states of (b) Ag is +1, (c) Ni is +2, and (d) Cu is +1 and +2.

7. Fe, Co and Ni lose the two 4*s* electrons to form 2+ ions. Iron loses the two 4*s* electrons and one 3*d* electron to leave a half-filled 3*d* subshell, which is an especially stable electron configuration.

8. (a) scandium chloride (b) iron(II) silicate
(c) sodium manganate (d) chromium (VI) oxide

12. Ni is (c) +2.

13. Pt is (d) 6.

18. The central metal atom is part of a complex anion.

20. (b) $[Pt(en)Cl_2]^{2+}$ can exhibit cis-trans isomerism.

Problems

23. In both cases electrons are lost to produce the electron configuration of Ar. With calcium this means the two 4*s* electrons, and the ion Ca^{2+} is formed. With scandium, the 3*d* electron is lost as well, producing Sc^{3+}.

25. Cr might be expected to have a +1 oxidation state because of its one s electron.

27. (a) The Ca atom is smaller than the K atom because it has a higher nuclear charge (+20 compared to +19), but the same number of electrons in its noble-gas core (18), coupled with the fact that the two 4*s* electrons are not effective in screening one another.
(b) The Mn atom is smaller than the Ca atom because it has a higher nuclear charge (+25 compared to +20) and the same number of valence electrons in the same configuration ($4s^2$).
(c) The Mn and Fe atoms are about the same size because they have about the same nuclear charge (+25 and +26, respectively), the same number of valence electrons ($4s^2$), and inner shell electrons that are about equally effective in shielding the valence electrons from the nucleus. The effective nuclear charges of Mn and Fe are essentially the same and roughly equal to {25 – (18 + 5)} and {26 – (18 + 6)}.

29. (a) $2\ Sc(s) + 6\ HCl(aq) \rightarrow 2\ ScCl_3(aq) + 3\ H_2(g)$
(b) $Sc(OH)_3(s) + 3\ HCl(aq) \rightarrow ScCl_3(aq) + 3\ H_2O(l)$
(c) $Sc(OH)_3(s) + 3\ Na^+(aq) + 3\ OH^-(aq) \rightarrow [Sc(OH)_6]^{3-}(aq) + 3\ Na^+(aq)$

31. (a) $K_2Cr_2O_7(aq) + 14\ HCl(aq) \rightarrow 3\ Cl_2(g) + 2\ CrCl_3(aq) + 2\ KCl(aq) + 7\ H_2O(l)$
Although the reaction has a negative $E°$ (−0.028 V), the reaction can be forced by exccess HCl and the removal of $Cl_2(g)$. Separate $CrCl_3(aq)$ and 2 KCl(aq) by fractional crystallization.

or
$K_2Cr_2O_7(aq) + 14\ HBr(aq) \rightarrow 3\ Br_2(l) + 2\ CrBr_3(aq) + 2\ KBr(aq) + 7\ H_2O(l)$
This reaction has a positive $E°$ (0.26 V). The products can be separated by fractional crystallization.

(b) $2\ KMnO_4(aq) + 5\ H_2C_2O_4(aq) + 6\ HCl(aq) \rightarrow$
$10\ CO_2(g) + 2\ MnCl_2(aq) + 2\ KCl(aq) + 8\ H_2O(l)$
Treatment of the reaction products with K_2CO_3.
$MnCl_2(aq) + K_2CO_3(aq) \rightarrow MnCO_3(s) + 2\ KCl(aq)$

33. (a) $3\ MnO_2(s) + KClO_3(aq) + 6\ KOH(aq) \rightarrow 3\ K_2MnO_4(aq) + KCl(aq) + 3\ H_2O(l)$
(b) $2\ (MnO_4^{2-} \rightarrow MnO_4^- + e^-)$
$Cl_2 + 2\ e^- \rightarrow 2\ Cl^-$
$2\ MnO_4^{2-} + Cl_2 \rightarrow 2\ MnO_4^- + 2\ Cl^-$
Add four K^+ ions to both sides of the equation.
$2\ K_2MnO_4(aq) + Cl_2(l) \rightarrow 2\ KMnO_4(aq) + 2\ KCl(aq)$
Separate by fractionla crystallization.

35. $[CrO_4^{2-}]$ in the reversible reaction
$Cr_2O_7^{2-} + H_2O \rightleftharpoons 2\ CrO_4^{2-} + 2\ H^+$
is large enough that K_{sp} of $PbCrO_4(s)$ is exceeded and a precipitate of $PbCrO_4(s)$ is formed.

37. Fe(s) is the best reducing agent. The reducing agent is oxidized, and $E° = -0.440$ V for the half-reaction: $Fe^{2+}(aq) + 2\ e^- \rightarrow Fe(s)$. For the reaction of $Co^{2+}(aq) + 2\ e^- \rightarrow Co(s)$, $E° = -0.277$ V, and for the half reaction, $Fe^{3+}(aq) + e^- \rightarrow Fe^{2+}(aq)$, $E°$ = 0.771 V, which means that $Fe^{3+}(aq)$ is readily reduced, making it an oxidizing agent. $Co^{3+}(aq)$ is an oxidizing agent (reduced to $Co^{2+}(aq)$), not a reducing agent.

39. $Co^{3+}(aq) + e^- \rightarrow Co^{2+}(aq)$ $\qquad E°_{cathode} = 1.92$ V
$\underline{2\ H_2O(l) \rightarrow O_2(g) + 4\ H^+(aq) + 4\ e^-}$ $\qquad E°_{anode} = 1.23$ V
$4\ Co^{3+}(aq) + 2\ H_2O(l) \rightarrow O_2(g) + 4\ H^+(aq) + 4\ Co^{2+}(aq)$
$E°_{cell} = E°_{cathode} - E°_{anode} = 1.92\ V - 1.23\ V = 0.69\ V$
In water, $Co^{3+}(aq)$ spontaneously converts to $Co^{2+}(aq)$.

41. (a) $\{Ag(s) \rightarrow Ag^+ + e^-\} \times 2$
$\underline{SO_4^{2-} + 4\ H^+ + 2\ e^- \rightarrow SO_2 + 2\ H_2O}$
$2\ Ag(s) + SO_4^{2-}(aq) + 4\ H^+(aq)$
$\rightarrow SO_2(g) + 2\ H_2O(l) + 2\ Ag^+(aq)$ net ionic equation
(b) $NO_3^- + 4\ H^+ + 3\ e^- \rightarrow NO(g) + 2\ H_2O$
$3\ Ag(s) + NO_3^- + 4\ H^+(aq)$
$\rightarrow NO(g) + 2\ H_2O(l) + 3\ Ag^+(aq)$ net ionic equation

43. $NO_3^- + 4\ H^+ + 3\ e^- \rightarrow NO(g) + 2\ H_2O$ $\qquad E°_{cathode} = 0.956$ V
$Ag \rightarrow Ag^+ + e^-$ $\qquad E°_{anode} = 0.800$ V
$E°_{cell}$ for $NO_3^- + Ag$ is 0.156 V, the Ag reaction is spontaneous.

$Au \rightarrow Au^{3+} + 3\ e^-$ $\qquad E°_{anode} = 1.52\ V$
$E°_{cell}$ for $NO_3^- + Au$ is -0.56 V, the gold reaction will not occur spontaneously.

45. Group 2B elements form complexes like transition elements. The 2A elements don't form many complexes. Group 2B atomic radii are smaller than Group 2A. Compared to Group 2A, the 2B elements have higher ionization energies and $E°$ values that are less negative.

47. $Cd^{2+} + 2\ e^- \rightarrow Cd(s)$ $\quad -0.403$
$Fe^{2+} + 2\ e^- \rightarrow Fe(s)$ $\quad -0.440$
$Zn^{2+} + 2\ e^- \rightarrow Zn(s)$ $\quad -0.763$
No, cadmium will not protect iron as zinc does. Iron will react before cadmium but after zinc.

49. (a) $3(Hg(l) \rightarrow Hg^{2+} + 2\ e^-)$
$\underline{2(NO_3^- + 4\ H^+ + 3\ e^- \rightarrow NO(g) + 2\ H_2O)}$
$3\ Hg(l) + 2\ NO_3^-(aq) + 8\ H^+(aq) \rightarrow 2\ NO(g) + 4\ H_2O(l) + 3\ Hg^{2+}(aq)$
(b) $ZnO(s) + 2\ CH_3CO_2H(aq) \rightarrow Zn^{2+}(aq) + 2\ CH_3CO_2^-(aq) + H_2O(l)$

51. (a) 6 (b) 2 (c) 4

53. (a) 6 (b) 3 (c) 3 (d) 1

55. (a) +2 (b) +3 (c) +4 (d) 0

57. (a) hexaamminezinc(II) ion
(b) tetrachloroferrate(II) ion
(c) tetraamminedichloroplatinum(IV) ion
(d) tris(ethylenediamine)cobalt(III) ion

59. (a) $[CoF_6]^{3-}$ (b) $[CrCl_2(H_2O)_4]^+$ (c) $[CoBr_2(en)_2]^+$

61. (a) potassium hexacyanochromate(II) (b) potassium tris(oxalato)chromate(III)

63. (a) $[Cu(NH_3)_4]Cl_2$ (b) $[Cr(en)_3][Co(CN)_6]$ (c) $[Pt(NH_3)_4][PtCl_4]$

65. (a) It is a cation, so the name ends as copper(II): tetraaquacopper(II) ion.
(b) The oxidation state is not listed. The correct name could be a neutral complex, pentaamminesulfatocobalt(II), but is more likely a complex ion, pentaamminesulfatocobalt(III) ion.

67. (a) Yes, these are structural isomers. The NCS ion is bonded through the N atom in one isomer and through the S atom in the other.
(b) Yes, these are isomers. They have identical overall compositions:

Pt$_3$Cl$_6\cdot$6NH$_3$, but they differ in the compositions of the complex cations and anions.

(c) No. These are different compounds; they do not have the same number of K atoms.

(d) Yes, these are structural isomers. Specifically, they differ in one of the ligands in the complex cations.

69. The en can only bond at adjacent coordination sites. Since the other four sites are all filled by Cl$^-$, there exists only one form. All others can be made by a rotation of the first. *Cis-trans* isomerism is not possible.

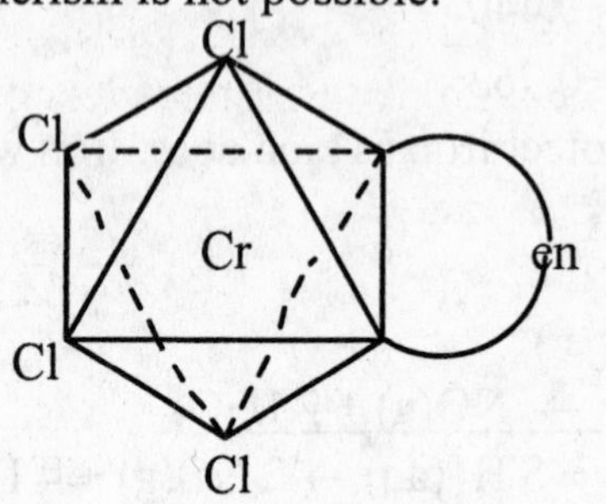

71.

Yes.

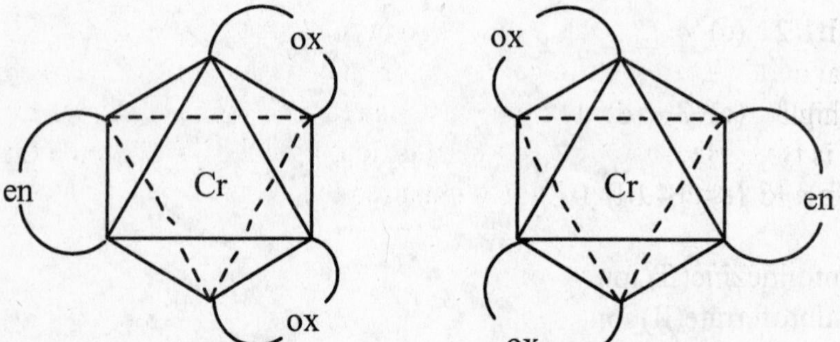

These mirror images are not superimposable.

73. Mirror image tetrahedral isomers which are not superimposable are found by using four different ligands. There are only two ways to organize the four different ligands. Those two different arrangements are mirror images of each other.

75. (a)

$$\boxed{}\,\boxed{}$$
$d_{z^2}\ d_{x^2-y^2}$

Cr^{3+} [Ar]

$$\boxed{\uparrow}\,\boxed{\uparrow}\,\boxed{\uparrow}$$
$d_{xy}\ d_{xz}\ d_{yz}$

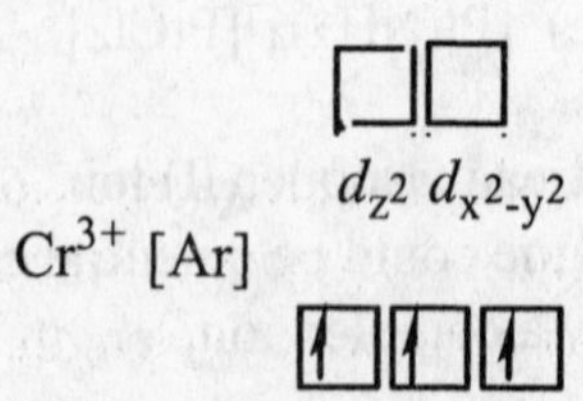

Paramagnetic. Cr^{3+}- [Ar] $3d^3$. with only three electrons, it doesn't matter whether it is low-spin or high-spin; they are the same. There are three unpaired electrons.

(b)

Fe^{3+} [Ar]

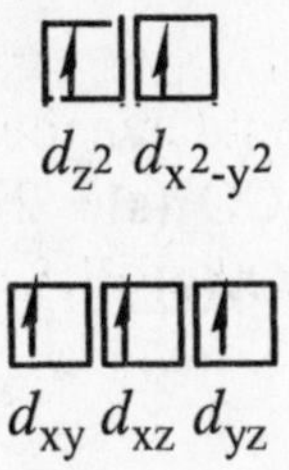

weak field: high-spin

Fe^{3+} [Ar]

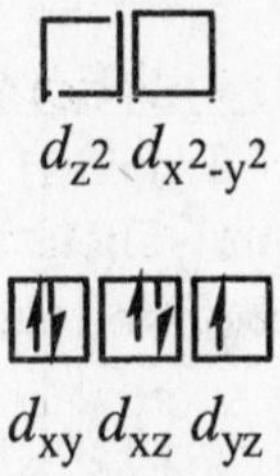

strong field: low-spin

Paramagnetic. Fe^{3+}-[Ar]$3d^5$. There is an odd number of electrons so it must be paramagnetic. Because Cl^- is a weak-field ligand it is expected that the high-spin state will exist with five unpaired electrons.

(c)

Mn^{3+} [Ar]

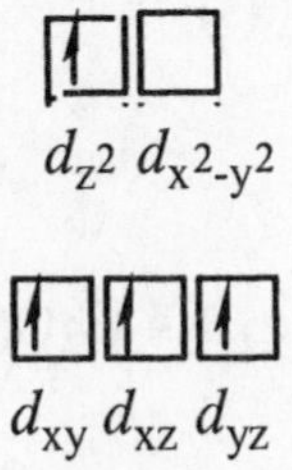

weak field: high-spin

Mn^{3+} [Ar]

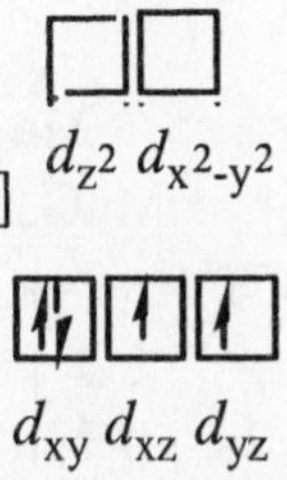

strong field: low-spin

Paramagnetic. The electron configuraton of Mn^{3+} is [Ar]$3d^4$. If the complex is of the high-spin type, the number of unpaired electrons is four; for the low-spin type it is two. Because CN^- is a strong-field ligand, it is expected that the ion is in a low-spin state with two unpaired e^-.

(d)

Co^{2+} [Ar]

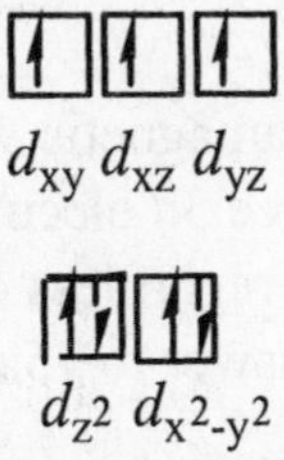

weak field or strong field

Paramagnetic. The electron configuration of Co^{2+} is [Ar]$3d^7$. Cl^- is usually a weak ligand, but in a strong field or a weak field, there are three unpaired electrons.

77. $[Cr(H_2O)_6]^{3+}$ is violet and $[Cr(NH_3)_6]^{3+}$ is yellow. NH_3 is a stronger field ligand, causing a greater Δ value, and requiring light of a higher frequency to stimulate an electronic transition. Light of a higher frequency has a shorter wavelength. Violet is a shorter wavelength than yellow. $[Cr(NH_3)_6]^{3+}$ should absorb violet and transmit yellow light.

Additional Problems

80. $\{Cu(s) \rightarrow Cu^{2+} + 2\ e^-\} \times 2$

$O_2(g) + 2\ H_2O(l) + 4\ e^- \rightarrow 4\ OH^-(aq)$

$\underline{CO_2(g) + H_2O(l) \rightarrow 2\ H^+(aq) + CO_3^{2-}(aq)}$

$2Cu(s) + O_2(g) + 3\ H_2O(l) + CO_2(g) \rightarrow Cu_2(OH)_2CO_3(s) + 2\ H_2O(l)$

$2Cu(s) + O_2(g) + H_2O(l) + CO_2(g) \rightarrow Cu_2(OH)_2CO_3(s)$

An alternate method follows.

$\{Cu(s) \rightarrow Cu^{2+} + 2\ e^-\} \times 2$

$\underline{O_2(g) + 2\ H_2O(l) + 4\ e^- \rightarrow 4\ OH^-(aq)}$

$2\ Cu(s) + O_2(g) + 2\ H_2O(l) \rightarrow 2\ Cu^{2+} + 4\ OH^-(aq)$

$\underline{2\ Cu^{2+} + 4\ OH^-(aq) \rightarrow 2\ Cu(OH)_2(s)}$

$2\ Cu(s) + O_2(g) + 2\ H_2O(l) \rightarrow 2\ Cu(OH)_2(s)$

$\underline{2\ Cu(OH)_2(s) + CO_2(g) \rightarrow Cu_2(OH)_2CO_3(s) + H_2O(l)}$

$2Cu(s) + O_2(g) + H_2O(l) + CO_2(g) \rightarrow Cu_2(OH)_2CO_3(s)$

81. $MnO_4^{2-} + 8\ H^+ + 4\ e^- \rightarrow Mn^{2+} + 4\ H_2O$

$\underline{\{MnO_4^{2-} \rightarrow MnO_4^- + e^-\} \times 4}$

$5\ MnO_4^{2-}(aq) + 8\ H^+(aq) \rightarrow Mn^{2+}(aq) + 4\ MnO_4^-(aq) + 4\ H_2O(l)$

OR

$MnO_4^{2-} + 4\ H^+ + 2\ e^- \rightarrow MnO_2(s) + 2\ H_2O$

$\underline{\{MnO_4^{2-} \rightarrow MnO_4^- + e^-\} \times 2}$

$3\ MnO_4^{2-}(aq) + 4\ H^+(aq) \rightarrow MnO_2(s) + 2\ MnO_4^-(aq) + 2\ H_2O(l)$

85. $CrO_3 + 6\ H^+ + 6\ e^- \rightarrow Cr\ (s) + 3\ H_2O$

$$2.57\ A \times 3.25\ h \times \frac{3600\ s}{h} \times \frac{C}{A\ s} \times \frac{mol\ e^-}{96485\ C} \times \frac{mol\ Cr}{6\ mol\ e^-} \times \frac{52.00\ g\ Cr}{mol\ Cr} = 2.70\ g\ Cr$$

89. Cr^{3+} has an odd number of electrons so it will be paramagnetic with any ligand. Cr^{3+} has the electron configuration $[Ar]3d^3$. The three 3*d* electrons will remain unpaired in the lower-energy set of three 3*d* orbitals, regardless of whether the ligands (L) are strong field or weak field (see also, answer to 75a).

93. Structures (a) and (c) are identical. (a) and (b) [or (a) and (d)] are geometric isomers and (b) and (d) are optical isomers.

Apply Your Knowledge

97. $CoCl_2 \cdot 6\ H_2O + 2\ en + HCl \rightarrow [CoCl_2(en)_2]Cl + e^- + 6\ H_2O + H^+$

$H_2O_2 + 2\ H^+ + 2\ e^- \rightarrow 2\ H_2O$

$2\ CoCl_2 \cdot 6\ H_2O + 4\ en + 2\ HCl + H_2O_2 \rightarrow 2\ [CoCl_2(en)_2]Cl + 14\ H_2O$

$$\frac{?\ mol}{coeff} = 5.0\ g\ CoCl_2 \cdot 6\ H_2O \times \frac{mol}{238\ g} \times \frac{2mol[CoCl_2(en)_2]Cl}{2\ mol\ CoCl_2 \cdot 6\ H_2O}$$

$$= 0.021\ mol\ product$$

$$\frac{?\,\text{mol}}{\text{coeff}} = 19\text{ mL en} \times \frac{1.00\text{ g}}{\text{mL}} \times \frac{10.0\text{ g en}}{100\text{ g solution}} \times \frac{\text{mol}}{60.1\text{ g}} \times \frac{2\text{mol}[\text{CoCl}_2(\text{en})_2]\text{Cl}}{4\text{ mol en}}$$
$$= 0.016\text{ mol product}$$

$$\frac{?\,\text{mol}}{\text{coeff}} = 2.0\text{ mL H}_2\text{O}_2 \times \frac{1.44\text{ g}}{\text{mL}} \times \frac{30.0\text{ g H}_2\text{O}_2}{100\text{ g solution}} \times \frac{\text{mol}}{34.0\text{ g}} \times \frac{2\text{mol}[\text{CoCl}_2(\text{en})_2]\text{Cl}}{1\text{ mol H}_2\text{O}_2}$$
$$= 0.051\text{ mol product}$$

$$\frac{?\,\text{mol}}{\text{coeff}} = 12.0\text{ mL HCl} \times 12\text{ M} \times \frac{10^{-3}\text{ L}}{\text{mL}} \times \frac{2\text{mol}[\text{CoCl}_2(\text{en})_2]\text{Cl}}{2\text{ mol HCl}}$$
$$= 0.14\text{ mol product}$$

ethylene diamine is limiting.

$$?\text{ g complex} = 19\text{ mL en} \times \frac{1.00\text{ g}}{\text{mL}} \times \frac{10.0\text{ g en}}{100\text{ g solution}} \times \frac{\text{mol}}{60.1\text{ g}}$$
$$\times \frac{2\text{mol}[\text{CoCl}_2(\text{en})_2]\text{Cl}}{4\text{ mol en}} \times \frac{285\text{ g}[\text{CoCl}_2(\text{en})_2]\text{Cl}}{\text{mol}[\text{CoCl}_2(\text{en})_2]\text{Cl}} = 4.5\text{ g }[\text{CoCl}_2(\text{en})_2]\text{Cl}$$

$$?\,\%\text{ yield} = \frac{2.77\text{ g}}{4.5\text{ g}} \times 100\% = 62\%$$

99. The unipositive cation must be $K^+$and all the other species must be found in the complex cation: $K[PtCl_3(C_2H_4)]$.

100. $? \text{ M Ni}^{2+} = (25.0\text{ mL} \times 0.0200\text{ M EDTA} - 3.32\text{ mL} \times 0.0347\text{ M})$
$$\times \frac{\text{mmol Ni}^{2+}}{\text{mmol EDTA}} \times \frac{1}{25.0\text{ mL}} = 0.0154\text{ M Ni}^{2+}$$

The results would have been the same because EDTA reacts with all polyvalent metal ions in a one to one ratio regardless of charge.

Chapter 23

Chemistry and Life: More on Organic, Biological, and Medicinal Chemistry

Exercises

23.1A (a) Aliphatic. The compound has a conjugated bonding system, but it does not extend completely around the ring. It also has only eight π electrons instead of the 10 $(4 \times 2 + 2)$ π electrons needed.

(b) Aliphatic. The compound has a conjugated bonding system, but it does not extend completely around the ring. It does have the six π electrons $[(4 \times 1) + 2]$ needed to be aromatic.

(c) Aliphatic. The compound has a conjugated bonding system, that does extend completely around the ring. It does not have the six π electrons $[(4 \times 1) + 2]$ or ten π electrons $[(4 \times 2) + 2]$ needed to be aromatic; it has eight π electrons.

(d) Aromatic. The compound has a conjugated bonding system that does extend completely around the ring. It does have the 14 π electrons $[(4 \times 3) + 2]$ needed to be aromatic.

23.1B Pyrrole has 6 π electrons. The N (sp^2 hybrid) in the ring supplies a pair of electrons in an unhybridized p orbital to contribute to the π bonding around the ring. Pyrrole is cyclic and planar.

23.2A The sign of the optical rotation cannot be predicted from the structure alone.

(a) Gucose has six carbon atoms and an aldehyde group. In the Fisher projection, the OH^- on the bottom chiral carbon is on the left –the L configuration. It is an L–aldohexose.

(b) Erythrulose has four carbon atoms and a ketone group. In the Fisher projection, the OH^- on the bottom chiral carbon is on the left—the L configuration. It is an L–ketotetrose.

23.2B No, D – (+) –fructose cannot exist. The direction of rotation (–) is determined by experiment for each molecule and for D-fructose the direction is (–).

Review Questions

3. To make methyl ethyl ketone, oxidize 2-butanol.

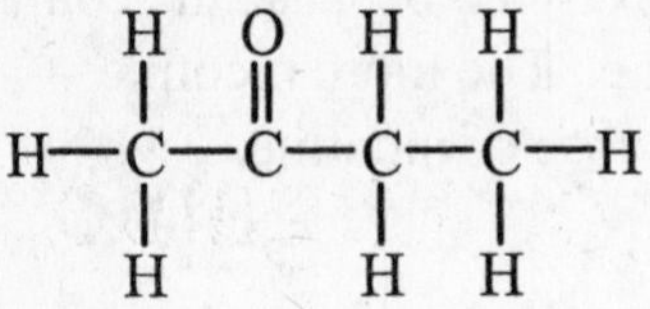

methyl ethyl ketone

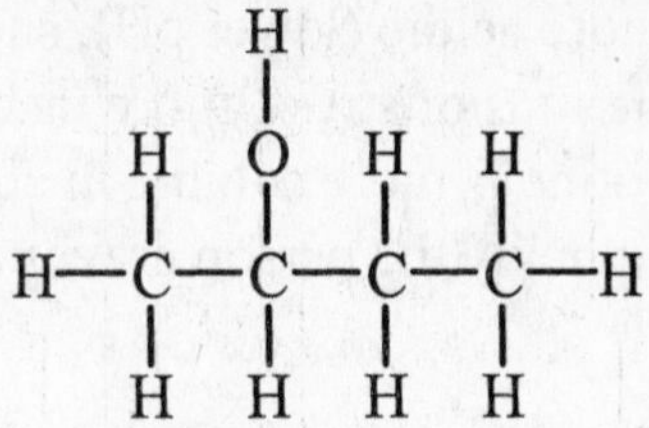

2-butanol

6. (a) Aromatic. The compound has a conjugated π bond system completely around the ring and has six π electrons [(4 × 1) +2].
   (b) Aliphatic. The conjugated bond system does not extend completely around the ring.
   (c) Aliphatic. The π bond system is not conjugated.
   (d) The ring is aromatic; the side-chain is aliphatic. The side chain has no double bonds. The ring has a conjugated π bond system completely around the ring and has six π electrons. The compound is aromatic.

8. The fatty acid, linoleic acid, is unsaturated so glycerol trilinoleate is (a) a fat.

13. (c) Glyceral is not part of a nucleic acid chain.

15. DNA is (b) a double helix.

18. A conjugated system of double bonds refers to alternating single and double bonds. Three examples are 1,3-butadiene, 1,3,5-hexatriene, and 1,3,5,7-octatetraene.

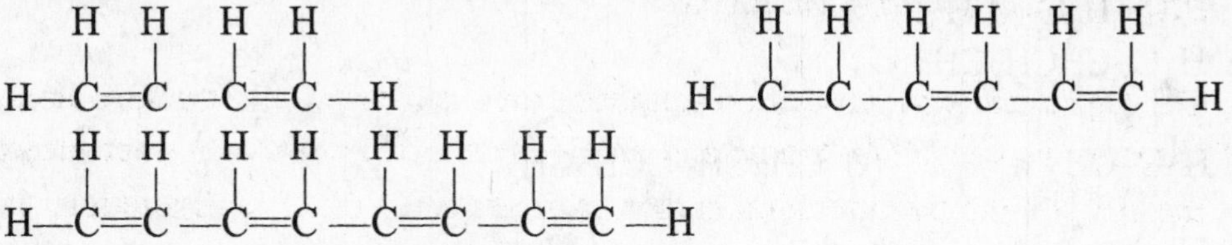

19. Absorption of infrared radiation causes molecules to vibrate more rapidly, to stretch and to bend, and also to rotate more rapidly. Absorption of UV-VIS radiation raises electrons to higher energy levels, especially π electrons.

20. Of those listed, UV-spectroscopy uses radiation of the highest energy content; NMR absorbs radiation of the smallest energy content or longest wavelength.

21. We have two relationships: $\nu = c/\lambda$ and $\tilde{\nu} = 1/\lambda$ with λ in cm. Thus, $\nu = c\tilde{\nu}$, with $c = 2.9979 \times 10^{10}$ cm/sec.

25. Penicillin molecules have a β lactam, a four-member ring with one nitrogen and three carbons, one an amide. The amide carbon in penicillin attaches to the enzyme that catalyzes the synthesis of bacterial cell walls. This deactivates the enzyme. Human cells are not harmed, because they do not have cell walls.

32. When the urine is more acidic (lower pH), such as after a meal, after drinking alcohol, or during periods of stress, more nicotine is excreted because the conjugate acid of the base nicotine is more soluble in acidic urine. The more nicotine excreted, the more quickly the person craves another dose of nicotine.

Problems

33. (a)
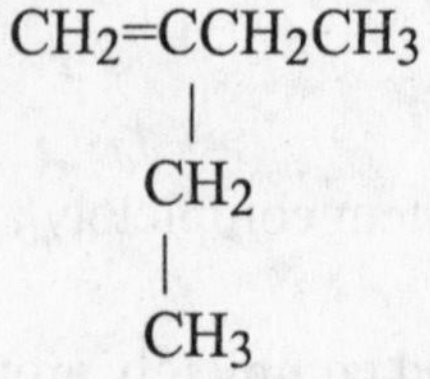

(b)
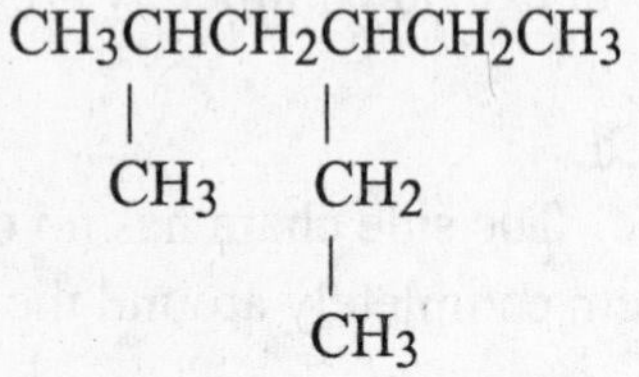

(c)
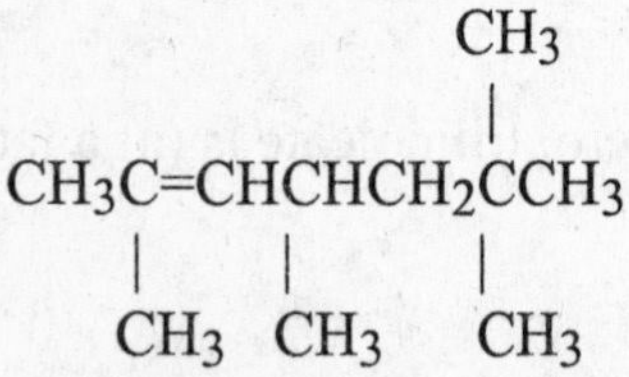

(d)
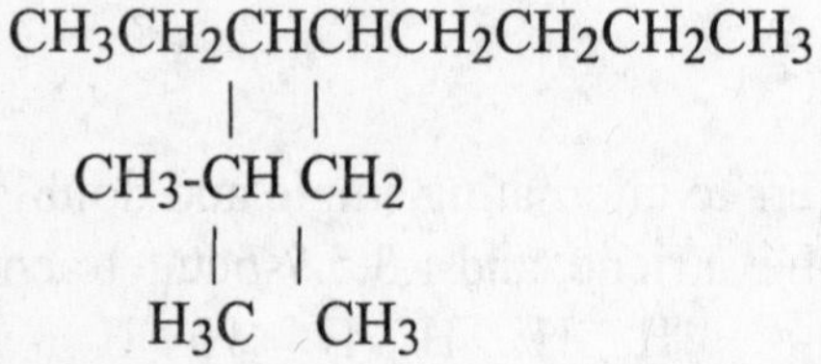

(e)
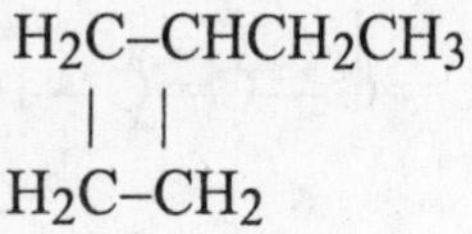

(f)
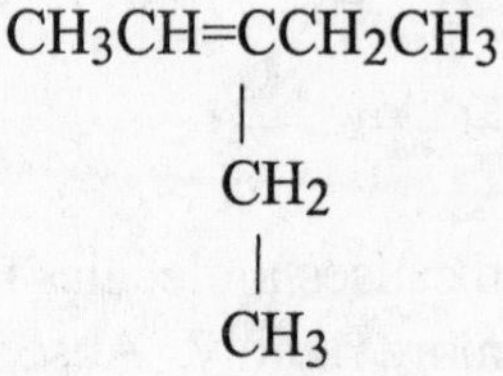

35. (a) 2-methyl-2-butene
(b) 1,1,6-trichloro-3-heptyne
(c) 2,2,4-trimethyl-3-hexene
(d) 2-methyl-1-pentene

37. (a) 2,2-dimethyl-3-pentanol
(b) 2-methyl-1-propanol
(c) 2-chloro-4-iodobenzoic acid
(d) 4-chloro-2-nitroaniline

39. (a) $CH_3CH_2CHCH_2CH_2CH_3$

|

OH

(b) CH_3

|

CH_3CHCCH_3

| |

OH CH_3

(c) $CH_3CH_2CHCHCHCH_2CH_3$

| | |

OH CH_3 CH_3

(d) $HOCH_2CHCH_2CH_3$

|

CH_2

|

CH_3

(e)

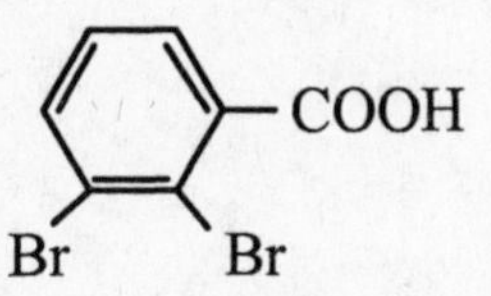

(f)

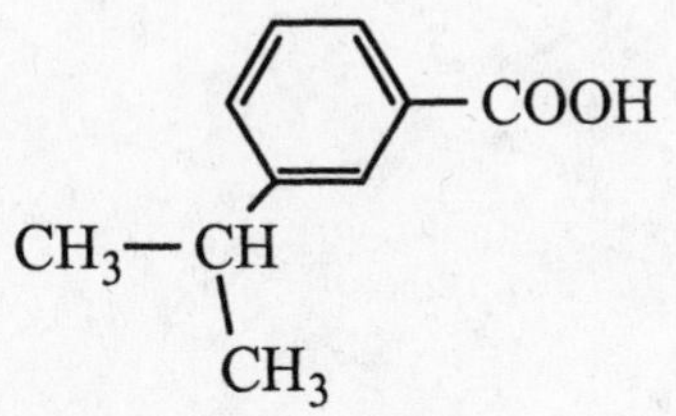

41. (a) chloro, carboxylic acid
(b) alkene
(c) aromatic, carboxylic acid, bromo

43. (a) 3-chloropentanoic acid, $CH_3CH_2CHClCH_2COOH$
(b) 2-ethyl-1-pentene, $CH_2{=}C(CH_2CH_3)CH_2CH_2CH_3$
(c) 2,4-dibromobenzoic acid, $(2,4\text{-}Br_2C_6H_3COOH)$

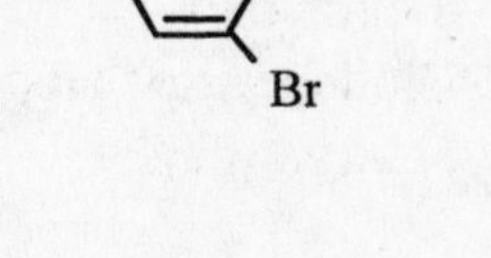

45. (a) oxidation
(b) hydration

47. (a) $CH_3CH_2CH_2CH_3$
(b) $(CH_3)_2CCH(CH_3)_2$
|
O
H

49. (a) $HOCH_2CH_2CH_3 \xrightarrow{H_2SO_4 \text{ and } \Delta} CH_2{=}CHCH_3 + H_2O$
(b) $2\ HOCH_2CH_2CH_3 \xrightarrow{H_2SO_4} CH_3CH_2CH_2OCH_2CH_2CH_3$

51. (a) $CH_2{=}CHCH_3$ (b) $CH_2{=}CCH_3$
|
CH_3

53. (a) $CH_3CH_2CH_2COOH + NaHCO_3 \rightarrow CH_3\ CH_2CH_2COO^-Na^+ + H_2O + CO_2$
(b) $CH_3COOCH_2CH_3 + H_2O \xrightarrow{H_2SO_4} CH_3COOH + HOCH_2CH_3$

55. (a) saturated 8C
(b) unsaturated 10C
(c) unsaturated 18C (3 double bonds)

57. (a) $CH_3(CH_2)_{16}COOH$
(b) $CH_3(CH_2)_{12}COO^-K^+$
(c) $CH_2OCO(CH_2)_7CH{=}CH(CH_2)_7CH_3$
|
$CHOCO(CH_2)_7CH{=}CH(CH_2)_7CH_3$
|
$CH_2OCO(CH_2)_7CH{=}CH(CH_2)_7CH_3$
(d) $CH_2OCO(CH_2)_{14}CH_3$
|
$CHOCO(CH_2)_{14}CH_3$
|
$CH_2OCO(CH_2)_{14}CH_3$

59. (a) aldose, triose
(b) aldose, pentose
(c) aldose, pentose
(d) ketose, hexose
(e) aldose, hexose

61. (a) D-sugar
(b) L-sugar
(c) D-sugar

63. (a)

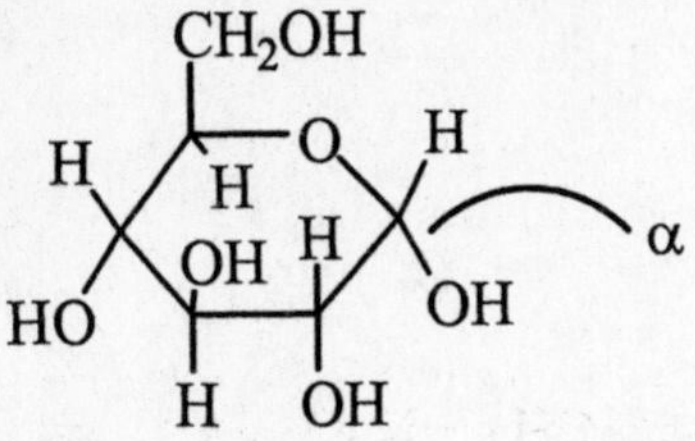

CH_2OH

H   O   H

H

OH   H   α

HO   OH

H   OH

(b)

CH_2OH O   OH   β

H   HO CH_2OH

OH   H

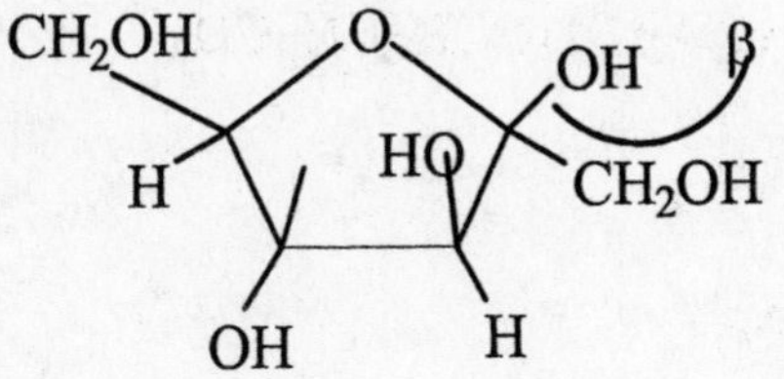

65. (a) β
     (b) α

67. (a) $-CH_2COOH$ (b) $-CH_2SH$ (c) $-CHCH_3$
                                                          |
                                                         CH_3

     (d) $-CH_2OH$ (e)

     $-CH_2-$⬡

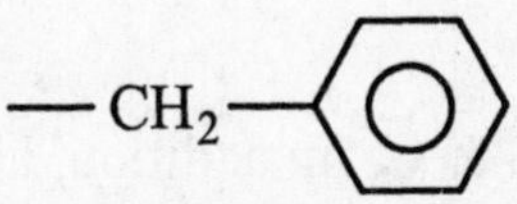

69. These are three amino acids with basic side chains:

lysine,   arginine,
$NH_2CHCOOH$ $NH_2CHCOOH$
      | |
$(CH_2)_4NH_2$ $(CH_2)_3NHCNH_2$
                                     ||
                                     NH

and histidine
$NH_2CHCOOH$
      |
    CH_2

    NH

  N

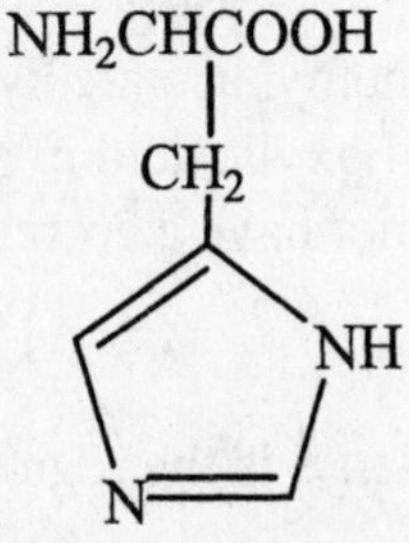

71.

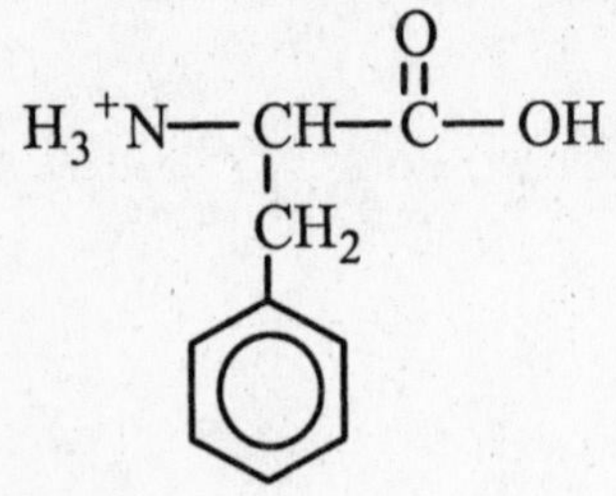

73. (a) pH < pI
(b) pH > pI
(c) at pH equal to the isoelectric point.

75. The sugar is ribose; the heterocyclic base is the pyrimidine base cytosine.

77. This molecule is not a nucleotide, since it has no phosphate group. A nucleotide which is missing a phosphate group is called a nucleoside. The base is a pyrimidine since it has but one ring; it is thymine. Since the sugar is deoxyribose, this compound would be incorporated into DNA.

79. The complementary DNA strand is CTAATGT

81. The RNA strand would be CUAAUGU

83.. The pairing between RNA and DNA is G to C. In addition, U on RNA corresponds to A on DNA, while A on RNA corresponds to T on DNA. Thus, the DNA strand that produced mRNA with the sequence UCCGAU is AGGCTA.

85. The pairing in RNA is always A to U and G to C.
(a) AAA is the base triplet taht pairs with (anticodon) UUU.
(b) GUA is the base triplet taht pairs with (anticodon)CAU.
(c) UCG is the base triplet taht pairs with (anticodon)AGC.
(d) GGC is the base triplet taht pairs with (anticodon)CCG.

87. We should be able to distinguish between acetic acid and methyl acetate by detecting an absorption peak for the O—H group of the carboxylic acid in acetic acid in the range 2500 to 3000 cm^{-1}. Methyl acetate does not have a corresponding peak.

89. $3250 - 3450\ cm^{-1}$ is the region for alcohols, $1680 - 1750\ cm^{-1}$ is the region for ketone.
It would be better if both regions were checked as other groups also absorb at different wavelengths.

91. Yes, we would expect absorption in the infrared because the organic molecule will likely have one or more of the structural features noted in Table 23.6.

93. A substance that appears yellow absorbs the complementary color, violet light with wavelength of about 425 nm (see Table 23.7).

Additional Problems

96. (a) $CH_3CH_2CH_2OH$ (b) $HOCH_2CH_2OH$
(c) HCH_2OH (d) $(CH_3)_2CHCH_2CH_2OH$

99.

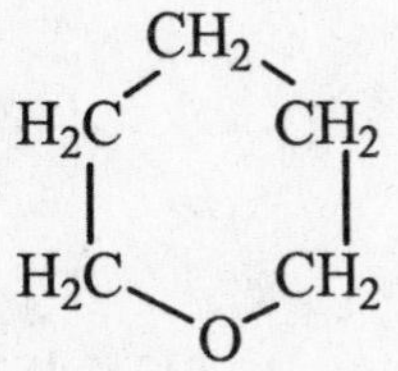

102. One would expect *p*-hydroxyphenol (HOC_6H_4OH) to be colorless; its conjugated system is no more extensive than that of benzene. The conjugated system of azobenzene[$(C_6H_5N)_2$] extends over 14 atoms and thus should produce a colored (orange-red) compound.

103. (a) $C_6H_5—NH_2$ will show aromatic C—H at ≈ 3030 cm^{-1} and N—H at 3300-3500 cm^{-1}. $C_6H_5—NH—C(=O)—CH_3$ will show those absorptions plus an amide C=O at 1630-1690 cm^{-1} and alkyl C—H at 2850-2960 cm^{-1}.
(b) The alcohol will show alkyl C—H at 2850-2960 cm^{-1} and the alcohol O—H at 3250-3450 cm^{-1}, while the ketone will show alkyl C—H at 2850-2960 cm^{-1} and ketone C=O at 1680-1750 cm^{-1}.

113.

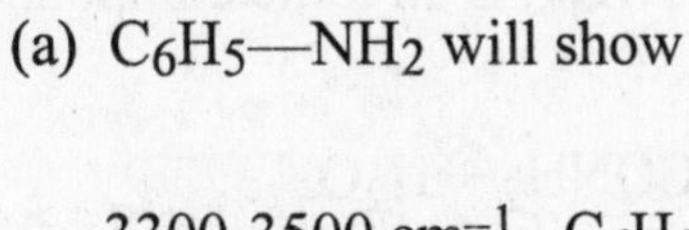

C_1 and C_5 are equivalent and are 1°, C_4 is 1°, C_3 is 2°, and C_2 is 3°.
Chlorine on C_1 or C_5 makes the same isomer 1, $ClCH_2CH(CH_3)CH_2CH_3$
Chlorine on C_4 makes the isomer 2, $CH_3CH(CH_3)CH_2CH_2Cl$
Chlorine on C_2 makes the isomer 3, $CH_3CCl(CH_3)CH_2CH_3$
Chlorine on C_3 makes the isomer 4, $CH_3CH(CH_3)CHClCH_3$
There are four isomers.
There are 12 possible replacement sites (12 H sites). One site on C_2, 2 sites on C_3, and 3 sites on C_1, C_4, and C_5.

$$\text{H site} \times \text{reactivity} = \text{\# of molecules}$$

$$
\begin{array}{lccccll}
C_2 & 1 & \times & 4.3 & = & 4.3 & \text{isomer 3} \\
C_3 & 2 & \times & 3 & = & 6 & \text{isomer 4} \\
C_1 & 3 & \times & 1 & = & 3 & \text{isomer 1} \\
C_4 & 3 & \times & 1 & = & 3 & \text{isomer 2} \\
C_5 & 3 & \times & 1 & = & \underline{3} & \text{isomer 1} \\
& & & & & 19.3 &
\end{array}
$$

$$\text{isomer 1} = \frac{6}{19.3} \times 100\,\% = 31.1\,\%$$

$$\text{isomer 2} = \frac{3}{19.3} \times 100\,\% = 15.5\,\%$$

$$\text{isomer 3} = \frac{4.3}{19.3}\, 100\,\% = 22.3\,\%$$

$$\text{isomer 4} = \frac{6}{19.3} \times 100\,\% = 31.1\,\%$$

114. (a) $CH_3CH_2CH_2COOCOCH_2CH_2CH_3 + H_2O \rightarrow 2\ CH_3CH_2CH_2COOH$

$CH_3CH_2CH_2COOCOCH_2CH_2CH_3 + 2\ NH_3 + H_2O \rightarrow$

$$2\ CH_3CH_2CH_2CO_2{}^-NH_4{}^+$$

(b) The product from NH_3, i.e., $CH_3CH_2CH_2CO_2{}^-NH_4{}^+$, is an ionic compound and would have a higher melting point.

(c) $CH_3CH_2CH_2CO_2{}^-NH_4{}^+ \xrightarrow{\Delta} CH_3CH_2CH_2CONH_2 + H_2O$

Apply Your Knowledge

115. $? \text{ moles acid} = 125\ mL \times \dfrac{10^{-3}\ L}{mL} \times 0.400\ M\ NaOH \times \dfrac{mol\ acid}{mol\ NaOH}$

$$= 5.00 \times 10^{-2}\ \text{moles}$$

$? \text{ molar mass} = \dfrac{5.10\ g}{5.00 \times 10^{-2}\ mol} = 102\ \dfrac{g}{mol}$

COOH is $12.01 + 2 \times 16.00 + 1.01 = 45.02$

CH_3 is $12.01 + 3 \times 1.01 = 15.04$

CH_2 is $12.01 + 2 \times 1.01 = 14.03$

$102 - 45 - 15 = 42$

$\dfrac{42}{14} = 3 \qquad CH_3(CH_2)_3COOH$

4 possible structural formulas

$CH_3(CH_2)_3COOH$
   pentanoic acid

COOH
|
$CH_3CH_2CHCH_3$
2-methylbutanoic acid

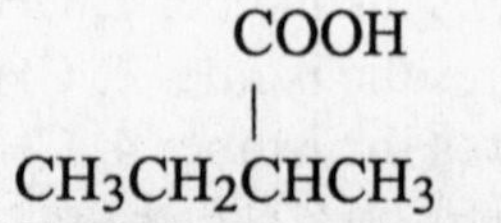

CH_3CHCH_2COOH
|
CH_3 3-methylbutanoic acid

CH_3
|
CH_3CCOOH
|
CH_3
2,2-dimethylpropanoic acid

117. Acetone is polar.
Carbon disulfide is nonpolar.
Carbon tetrachloride is nonpolar.
Diethyl ether is polar.
Chloroform is polar.
Carbon disulfide and carbon tetrachloride are useful solvents for infrared spectroscopy. Carbon disulfide and carbon tetrachloride are useful solvents for NMR because they have no hydrogens to produce an interfering signal.
Inorganic salts have ionic bonds whereas organic molecules have covalent bonds. The scan of vibrational frequencies usually don't include frequencies that produce signals from inorganic salts.

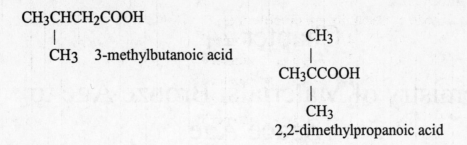

Chapter 24

Chemistry of Materials: Bronze Age to Space Age

Exercise

24.1A The main reason for using zinc for the displacement reaction is that if a more active metal such as aluminum were used, this metal would displace Zn(s) as well as other, less active metals. This would reduce the yield of Zn(s) obtained in the electrolysis.

24.1B $Ag_2S(s) + 4\,CN^-(aq) + 2\,O_2(g) \rightarrow 2\,Ag(CN)_2^-(aq) + SO_4^{2-}(aq)$

24.2A The error in the given structure is that bonding does not occur through the -CH_3 group; the carbon atom of the methyl group would form five bonds—an impossibility for carbon.

24.2B In addition polymerization, usually another molecule attaches at each end of the double bond that is broken. In this case one or two propadiene molecules can bond to the center atom as well as the ends producing a branched polymer.

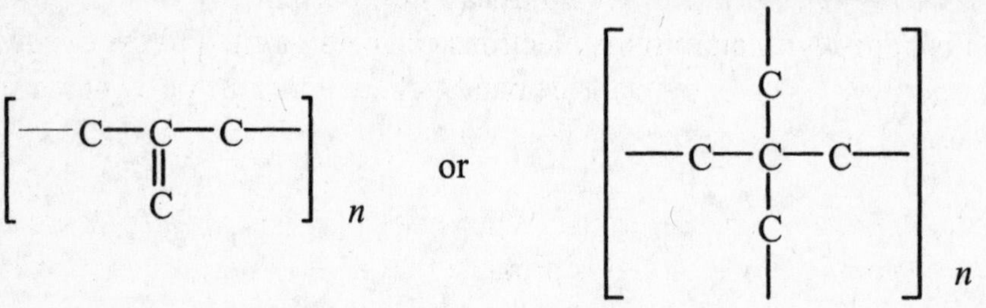

24.3A $n\,HOCH_2CH_2COOH \rightarrow \!\left[O{-}CH_2{-}CH_2{-}\overset{\overset{\displaystyle O}{\|}}{C} \right]_n + n\,H_2O$

24.3B Some amino acids can polymerize through their sidechains as well as through the amine group and the carboxylic acid group. Serine, threonine, and tyrosine have alcohol groups which can form polyesters. Tryptophan, asparagine, glutamine, proline, lysine, arginine, and histidine have nitrogen atoms in the sidechains which can form other polyamides. Aspartic acid and glutamic acid have a second carboxylic acid group on the sidechain which can form other polyamides.

Review Questions

1. (c) copper. The free metals must not be very reactive, (below H_2 in the activity series of Figure 4.13) or they would not be found free (unreacted).

2. (b) sulfates Oxides, sulfides, carbonates, and chlorides have economically feasible extraction methods already developed.

5. Reduction is the process of obtaining the free metal from its oxide. Thus, this is the process of removing oxygen from a metal compound. The preferred or most widely used metallurgical reducing agent is carbon, as coke. Often the CO(g) produced in a furnace is the actual reducing agent. Carbon and CO are especially effective in reducing metal oxides. There are a few metals that can be extracted by heating an ore to a high temperature and thus do not require a reducing agent.

7. (b) An alloy is a metal with one or more elements added to improve properties.

13. One model of metallic bonding states that the valence electrons are free to move around the remaining atomic cores. It is this electron gas(comprised of valence electrons only) or electron sea comprised of all the valence electrons in the metal crystal that bonds metal atoms together.

16. When metal atoms are added to a semiconductor that have either empty atomic orbitals near the bottom of the band gap or full atomic orbitals near the top of the band gap, that process is called doping. Either addition creates a slightly impure semiconductor that conducts electricity more readily. If the dopant has an extra valence electron, the resulting semiconductor is *n*-type, since negative electrons carry the current. If the dopant has one fewer valence electron, the resulting semiconductor is *p*-type, since positive holes carry the current.
When silicon is doped with aluminum, electrons can be promoted to the conduction band, leaving positive holes in the valence band. These holes move by being filled with an electron. The electron leaves a hole where it moved from.

17. (c) P and As have five valence electrons. When trace amounts of P or As are added to Si, an electron conducting material is produced. It takes less energy to free electrons than in an insulator. Thus, (c) is the incorrect statement.

20. (d) CH_3COOH does not have a double bond suitable for addition polymerization and does not have two functional groups for condensation polymerization.

23. (b) Polyethylene is a polymer, comprised of only one material.

PROBLEMS

25. An ore containing a high percentage of a metal may not be the best industrial source of the metal because the metal may be in a compound that is chemically difficult to reduce. For example, many clays contain high percentages of aluminum, but the metal is difficult to extract from them. Another difficulty can be the by-products produced during the extraction process. If these by-products complicate the purification process or are hazardous or difficult to dispose of, their formation will make the use of the ore uneconomical.

27. It is reasonable to find cadmium as an impurity since cadmium and zinc are quite similar chemically, as indicated by their adjacent positions in Group 2B of the periodic table. Cadmium can be removed either by fractional distillation of liquid zinc or by adding powdered zinc to displace the less active cadmium before refining the zinc electrolytically.

29. When HgS(s) is roasted, any HgO(s) that would form would immediately decompose to the elements: $2\ HgO(s) \xrightarrow{\Delta} 2\ Hg(l) + O_2(g)$. Decomposition of HgO(s) is one of the ways that Joseph Priestly, the discoverer of $O_2(g)$, prepared the element.

31. $2\ [Ag(CN)_2]^-(aq) + Zn(s) \rightarrow 2\ Ag(s) + [Zn(CN)_4]^{2-}(aq)$ summarizes the displacement of silver from its cyano complex by zinc.

33. Dissolving: $ZnO(s) + H_2SO_4(aq) \rightarrow ZnSO_4(aq) + H_2O(l)$
Displacement: $Zn(s) + CdSO_4(aq) \rightarrow Cd(s) + ZnSO_4(aq)$
Electrolysis: $2\ ZnSO_4(aq) + 2\ H_2O(l) \xrightarrow{\text{electrolysis}} 2\ Zn(s) + 2\ H_2SO_4(aq) + O_2(g)$
$Zn^{2+}(aq) + 2e^- \rightarrow Zn(s)$ cathode
$2H_2O(l) \rightarrow O_2(g) + 4e^- + 4H^+(aq)$ anode
The $H_2SO_4(aq)$ solution is all that remains after electrolysis, and it then can be recycled to dissolve more ZnO(s).

35. Chlorination: $Sn(s) + 2\ Cl_2(g) \rightarrow SnCl_4(l)$
Hydrolysis: $SnCl_4(l) + (2+x)H_2O(l) \rightarrow SnO_2{\cdot}xH_2O(s) + 4\ HCl(aq)$
Dehydration: $SnO_2{\cdot}xH_2O(s) \xrightarrow{\Delta} SnO_2(s) + x\ H_2O(g)$
Reduction: $SnO_2(s) + 2\ C(s) \xrightarrow{\Delta} Sn(l) + 2\ CO(g)$

37. $$\text{mass of ore} = 1.0 \times 10^7 \text{ kg pig iron} \times \frac{95 \text{ kg Fe}}{100 \text{ kg pig iron}} \times \frac{1 \text{ kmol Fe}}{55.85 \text{ kg Fe}} \times \frac{1 \text{ kmol } Fe_2O_3\text{(hematite)}}{2 \text{ kmol Fe}} \times \frac{159.7 \text{ kg } Fe_2O_3}{1 \text{ kmol } Fe_2O_3} \times \frac{100 \text{ kg ore}}{82 \text{ kg hematite}} = 1.7 \times 10^7 \text{ kg ore}$$

39. $2\ PbS(s) + 3\ O_2(g) \rightarrow 2\ PbO(s) + 2\ SO_2(g)$
$PbO(s) + C(s) \rightarrow Pb(l) + CO(g)$
$PbO(s) + CO(g) \rightarrow Pb(l) + CO_2(g)$

41. Calcium is a better conductor of electricity than potassium because calcium has twice as many valence electrons. These valence electrons are free to move through the 4*p* band in calcium, which overlaps the 4*s* band. Potassium metal has half as many electrons, mostly confined to the 4*s* band.

43. The lustrous appearance of metals results from their reflecting all incident visible light. Since there are energy levels at all spacings in a band, photons of all energies can be absorbed. When they are re-emitted, we see lustre. Because of the greater number of transitions of some energies, colored metals reflect some wavelengths of light to a greater extent than others.

45. $$? \text{ energy levels} = 55.5 \text{ mg Mg} \times \frac{10^{-3}\text{ g}}{\text{mg}} \times \frac{\text{mol Mg}}{24.31 \text{ g Mg}} \times \frac{6.022 \text{ x } 10^{23} \text{ atoms}}{\text{mol Mg}} \times \frac{1 \text{ energy level}}{\text{atom}} = 1.37 \times 10^{21} \text{ energy levels}$$

$$? \text{ e}^- = 55.5 \text{ mg Mg} \times \frac{10^{-3}\text{ g}}{\text{mg}} \times \frac{\text{mol Mg}}{24.31 \text{ g Mg}} \times \frac{6.022 \text{ x } 10^{23} \text{ atoms}}{\text{mol Mg}} \times \frac{2 \text{ e}^-}{\text{atom}} = 2.75 \times 10^{21} \text{ e}^-$$

47. In a semiconductor, there is an energy gap between the valence and conduction bands. In a metallic conductor, either the valence band is itself a conduction band or it overlaps one; there is no energy gap.

49. (a) *n*-type
(b) *p*-type

51. In *n*-type semiconductors, conduction electrons and positive holes are present in the same number, but the positive holes are not able to move, since they are in the atomic orbitals of the dopant, such as in P^+ in phosphorus-doped silicon. (The dopant's electrons have moved into the conduction band and are the freely moving conduction electrons.) In *p*-type semiconductors, the number of holes equals the number of electrons that are now trapped in the valence orbitals of the dopant, such as in Al^- in aluminum-doped silicon. But these electrons are not conduction electrons; they are not free to move throughout the metal. Positive holes predominate as charge carriers.

53. Metals become superconductors near 0 K because all resistance to electron flow disappears. Semiconductors usually do not become superconductors at low temperatures as the band gap remains as an energy barrier to be overcome, and the ability of electrons to jump from the valence to the conduction band *decreases* as the temperature is lowered.

55. We obtain the number of valence electrons (v.e.), outer-shell electrons, from the electron configuration of each element.

(a) [Cu] = [Ar] $3d^{10}\,4s^1$ 1 v.e. [S] = [Ne] $3s^2\,3p^4$ 6 v.e.
average = 3.5 v.e.

(b) [Zn] = [Ar] $3d^{10}\,4s^2$ 2 v.e. [Se] = [Ar] $3d^{10}\,4s^2\,4p^4$ 6 v.e.
average = 4 v.e.

(c) [Pb] = [Xe] $4f^{14}\,5d^{10}\,6s^2\,6p^2$ 4 v.e. [O] = $1s^2\,2s^2\,2p^4$ 6 v.e.
average = 5 v.e.

(d) $[Ga] = [Ar]\ 3d^{10}\ 4s^2\ 4p^1$ 3 v.e $[P] = [Ne]\ 3s^2\ 3p^3$ 5 v.e.
average = 4 v.e.
ZnSe and GaP meet the average number of valence electrons criterion for being semiconductors.

57. (a) Cellophane is cellulose. It is made into a final product by forcing a viscous solution through a narrow slit to form a film.
(b) LDPE, low density polyethylene, has the formula $\left[CH_2-CH_2 \right]_n$. It is made by the free-radical polymerization of ethylene, $CH_2{=}CH_2$. Its chains are branched and pack together poorly.

59. Rubber is elastic because its coiled polymer chains are straightened out during stretching but return to their coiled state when relaxed. Vulcanization forms crosslinks between chains which more readily pull the chains back into their original shape after stretching.

61.

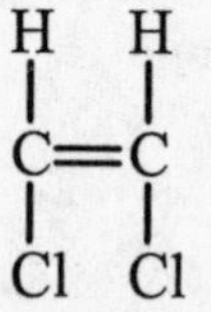

$\left[CHCl-CHCl \right]_n$ is the condensed formula for polydichloroethene,

or 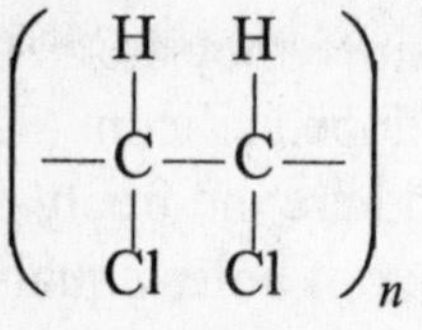

63. (a) polystyrene

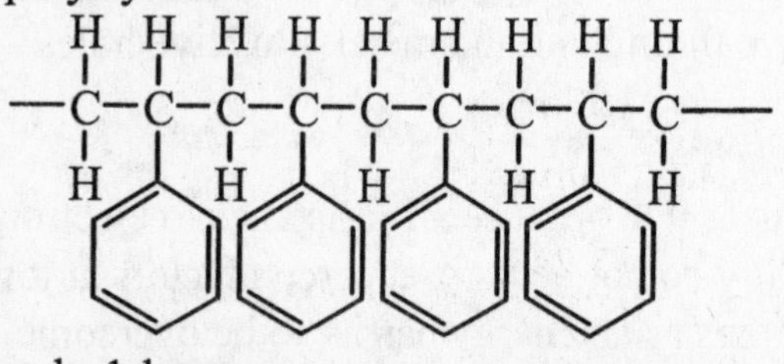

(b) poly-1-hexene

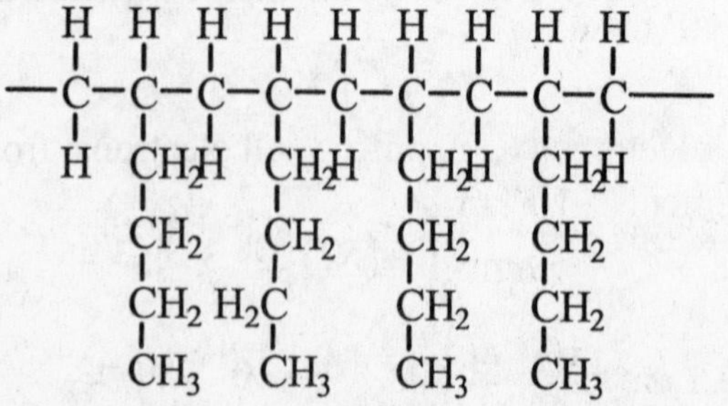

65. (a) polylactic acid

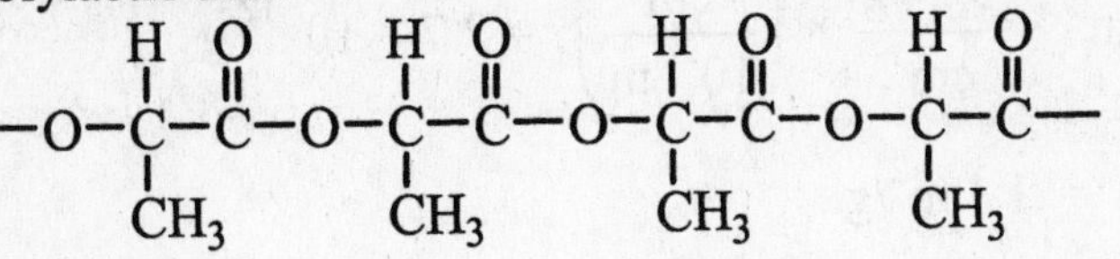

$$-O-\overset{\overset{\displaystyle H}{|}}{\underset{\underset{\displaystyle CH_3}{|}}{C}}-\overset{\overset{\displaystyle O}{\|}}{C}-O-\overset{\overset{\displaystyle H}{|}}{\underset{\underset{\displaystyle CH_3}{|}}{C}}-\overset{\overset{\displaystyle O}{\|}}{C}-O-\overset{\overset{\displaystyle H}{|}}{\underset{\underset{\displaystyle CH_3}{|}}{C}}-\overset{\overset{\displaystyle O}{\|}}{C}-O-\overset{\overset{\displaystyle H}{|}}{\underset{\underset{\displaystyle CH_3}{|}}{C}}-\overset{\overset{\displaystyle O}{\|}}{C}-$$

(b) nylon-46

$$-N-(H_2C)_4-\underset{\underset{\displaystyle H}{|}}{N}-\overset{\overset{\displaystyle O}{\|}}{C}-(CH_2)_4-\overset{\overset{\displaystyle O}{\|}}{C}-\underset{\underset{\displaystyle H}{|}}{N}-(CH_2)_4-\underset{\underset{\displaystyle H}{|}}{N}-\overset{\overset{\displaystyle O}{\|}}{C}-(CH_2)_4-\overset{\overset{\displaystyle O}{\|}}{C}-$$

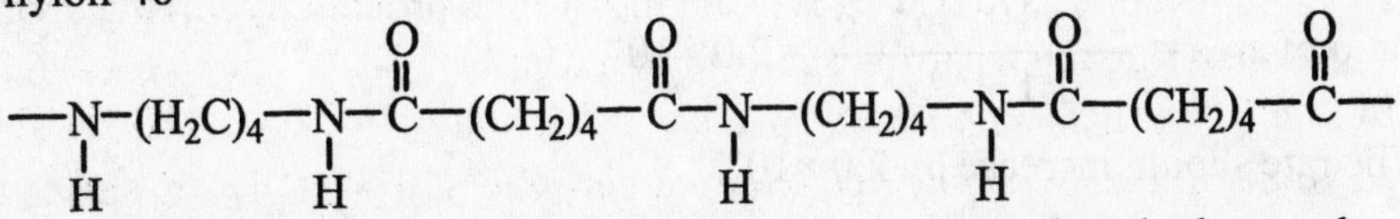

This is a condensation polymer; a molecule of water is produced when each monomer is added to the polymer. Because of the size of the units, only two repeating units are shown.

67. (a)

$$H-O-\overset{\overset{\displaystyle O}{\|}}{C}-\bigcirc-\overset{\overset{\displaystyle O}{\|}}{C}-O-H \quad \text{and} \quad NH_2-\bigcirc-NH_2$$

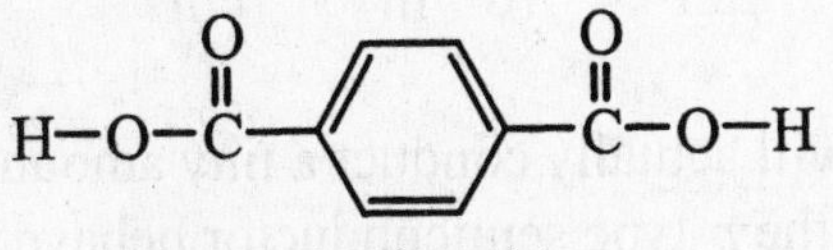

terephthalic acid   1,4 phenylenediamine

(b) polyamide

69. Tempered iron is more dense because fcc has a greater packing efficiency than bcc. (See problem 89, chapter 11.) All tempered metals need not follow this pattern. For example, if a metal underwent transition from fcc to bcc as the temperature was increased, a decrease in density would be seen.

71. for 5.0 nm catalyst

$$\text{surface area} = 4\pi r^2 = 4 \times \pi \times \left(2.5\,\text{nm} \times \frac{10^{-9}\,\text{m}}{\text{nm}}\right)^2 = 7.9 \times 10^{-17}\,\text{m}^2$$

$$\text{volume} = \frac{4}{3}\pi r^3 = \frac{4}{3} \times \pi \times \left(2.5\,\text{nm} \times \frac{10^{-9}\,\text{m}}{\text{nm}}\right)^3 = 6.6 \times 10^{-26}\,\text{m}^3$$

$$\frac{\text{mass}}{\text{sphere}} = 6.6 \times 10^{-26}\,\text{m}^3 \times \left(\frac{\text{cm}}{10^{-2}\,\text{m}}\right)^3 \times \frac{5.24\,\text{g}}{\text{cm}^3} = 3.4 \times 10^{-19}\,\text{g}$$

$$\frac{\text{area}}{\text{mass}} = \frac{7.9 \times 10^{-17}\,\text{m}^2}{3.4 \times 10^{-19}\,\text{g}} = 2.3 \times 10^2\,\text{m}^2/\text{g}$$

for 1.0 μm catalyst

$$\text{surface area} = 4\pi r^2 = 4 \times \pi \times \left(0.5\,\mu\text{m} \times \frac{10^{-6}\,\text{m}}{\mu\text{m}}\right)^2 = 3.1 \times 10^{-12}\,\text{m}^2$$

$$\text{volume} = \frac{4}{3}\pi r^3 = \frac{4}{3} \times \pi \times \left(0.5\,\mu\text{m} \times \frac{10^{-6}\,\text{m}}{\mu\text{m}}\right)^3 = 5.2 \times 10^{-19}\,\text{m}^3$$

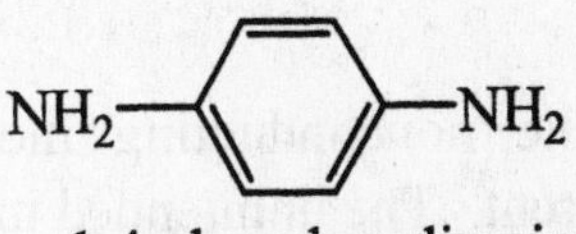

$$\frac{\text{mass}}{\text{sphere}} = 5.2 \times 10^{-19}\ \text{m}^3 \times \frac{5.24\ \text{g}}{\text{cm}^3} \times \left(\frac{\text{cm}}{10^{-2}\ \text{m}}\right)^3 = 2.7 \times 10^{-12}\ \text{g}$$

$$\frac{\text{area}}{\text{mass}} = \frac{3.1\times10^{-12}\ \text{m}^2}{2.7\times10^{-12}\ \text{g}} = 1.1\ \text{m}^2/\text{g}$$

$$\text{ratio of area} = \frac{2.3\times10^2\ \text{m}^2/g}{1.1\ \text{m}^2/g} = 2.0\times10^2$$

The rate should increase by 2.0×10^2.

Additional Problems

73. ? g = area × depth × density

$$?\ \text{g} = \pi \times \left(100\ \text{in} \times \frac{2.54\ \text{cm}}{\text{in}}\right)^2 \times 1.0\ \mu\text{m} \times \frac{10^{-6}\ \text{m}}{\mu\text{m}} \times \frac{\text{cm}}{10^{-2}\ \text{m}} \times \frac{2.698\ \text{g}}{\text{cm}^3} = 55\ \text{g}$$

75. In the "nonconducting" mode, the diode will actually conduct a tiny amount of current. The unintended impurity makes the *p*-type semiconductor behave as *n*-type to a very limited extent, and vice versa.

77. First determine the energy per photon.

$$E = \frac{110\ \text{kJ}}{\text{mol}} \times \frac{1000\ \text{J}}{1\ \text{kJ}} \times \frac{1\ \text{mol}}{6.022\times10^{23}\ \text{photon}} = 1.83\times10^{-19}\ \text{J}$$

Then determine the wavelength.

$$\lambda = \frac{hc}{E} = \frac{6.626\times10^{-34}\ \text{J}\cdot\text{s}\times3.00\times10^{8}\ \text{m/s}}{1.83\times10^{-19}\ \text{J}} \times \frac{10^6\ \mu\text{m}}{1\ \text{m}} = 1.09\ \mu\text{m}$$

This is infrared radiation. Silicon should have sufficient energy levels in its conduction band so that it absorbs all visible light.

81. 1,2-ethanediol has two functional groups (—OH) but 1,2,3-propanetriol has <u>three</u> functional groups (also —OH). The polymer of 1,2,3-propanetriol and 1,2-benzenedicarboxylic acid has extensive cross-linking and is a more rigid polymer.

84. number average molecular mass = (0.28 × 786 u) + (0.25 × 702 u) + (0.15 × 814 u) + (0.32 × 758 u) = 760 u

Apply Your Knowledge

88. First, write an equation for the titration reaction.

oxidation: $5\ \{Fe^{2+}(aq) \rightarrow Fe^{3+}(aq) + e^-\}$

reduction: $MnO_4^-(aq) + 8\ H^+(aq) + 5\ e^- \rightarrow Mn^{2+}(aq) + 4\ H_2O(l)$

net: $MnO_4^-(aq) + 5\ Fe^{2+}(aq) + 8\ H^+(aq) \rightarrow Mn^{2+}(aq) + 4\ H_2O(l) + 5\ Fe^{3+}(aq)$

Now determine $[MnO_4^-]$.

$$[MnO_4^-] = \frac{\text{mmol } MnO_4^-}{\text{mL soln}}$$

$$[MnO_4^-] = \frac{17.66\,\text{mL} \times \frac{0.0826\,\text{mmol Fe}^{2+}}{1\,\text{mL}} \times \frac{1\,\text{mmol MnO}_4^-}{5\,\text{mmol Fe}^{2+}}}{10.00\,\text{mL}} = 0.0292\ \text{M}$$

Now determine the mass of Mn in the 100.0 mL of solution.

$$\text{Mn mass} = 100.0\ \text{mL} \times \frac{0.0292\ \text{mmol MnO}_4^-}{1\ \text{mL soln}} \times \frac{1\ \text{mmol Mn}}{1\ \text{mmol MnO}_4^-} \times \frac{54.94\ \text{mg Mn}}{1\ \text{mmol Mn}} \times \frac{1\ \text{g}}{1000\ \text{mg}} = 0.160\ \text{g Mn}$$

$$\text{Mn \%} = \frac{0.160\ \text{g Mn}}{1.250\ \text{g sample}} \times 100\% = 12.8\%\ \text{Mn}$$

91. $$?\ \text{g Ni} = 0.0906\ \text{g} \times \frac{\text{mole}}{288.9\ \text{g}} \times \frac{\text{mol Ni}}{\text{mole complex}} \times \frac{58.69\ \text{g}}{\text{mole Ni}} \times \frac{250.0\ \text{mL}}{25.00\ \text{mL}} = 0.184\ \text{g Ni}$$

$$?\ \text{\% Ni} = \frac{0.184\ \text{g Ni}}{5.108\ \text{g sample}} \times 100\% = 3.60\ \%\ \text{Ni}$$

Chapter 25

Environmental Chemistry

Exercise

25.1A $? \text{ mmHg} = \dfrac{38.5\%}{100\%} \times 17.5 \text{ mmHg} = 6.74 \text{ mmHg}$

25.1B $? \text{ mmHg} = \dfrac{73.1\%}{100\%} \times 17.5 \text{ mmHg} = 12.8 \text{ mmHg}$

$$? \text{ g} = \frac{PVM}{RT} = \frac{12.3 \text{ mmHg} \times 10.0 \text{ L} \times 18.01 \text{ g/mol}}{62.36 \dfrac{\text{L mmHg}}{\text{K mol}} \times 293.2 \text{ K}} = 0.126 \text{ g}$$

Review Questions

1. (a) troposphere
   (b) stratosphere

6. (a) carbon monoxide and nitrogen monoxide
   (b) methane, ozone, nitrous oxide, CFCs
   (c) nitric acid and sulfuric acid

10. A fuel mixture that is rich in fuel and lean in oxygen will produce more carbon monoxide.
    Carbon monoxide ties up the iron atom in hemoglobin, preventing the transport of oxygen to the cells of the body.

13. (a) The electrostatic precipitator collects particulates out of flue gases.
    (b) A catalytic converter changes hydrocarbons and carbon monoxide to carbon dioxide and water. A second converter converts nitrogen monoxide to nitrogen.

19. Cholera, typhoid fever, and dysentery. Industralized nations clean the water supply using Cl_2, O_3, or other such oxidizing agents.

22. BOD is biochemical oxygen demand or the amount of oxygen necessary to decompose aerobically the organic matter in water. A high BOD means that there is a lot of organic material in the water. Usually that means that the oxygen in the water is used up and living organisms will die.

25. Chlorinated hydrocarbons are stable; they do not easily react with other compounds.

Problems

47. If 99% of the mass is within 30 km, then only 1% is outside 30 km, and this can support a mercury column that is only about 1% × 760 mmHg = 7.6 mmHg or ≈ 10 mmHg.

49. The elements are separated by molar mass with the lowest molar mass at the highest attitude and highest molar mass at the lowest altitude.

51. $? \text{ mol \%} = \frac{2.00 \text{ mmHg}}{760 \text{ mmHg}} \times 100\% = 0.263 \text{ mol \% } H_2O$

$$\frac{0.263 \text{ mol } H_2O}{100 \text{ mol air}} \times \frac{10^4}{10^4} = \times\times \frac{2.36\times10^3 \text{ mol } H_2O}{10^6 \text{ mol air}} = 2.63 \times 10^3 \text{ ppm } H_2O$$

53. $\frac{75.5\%}{100\%} \times 17.5 \text{ mmHg} = 13.2 \text{ mmHg}$

55. The dewpoint is about 20.4 °C, the temperature at which the V.P. of water is 18.0 mmHg.

57. Combustion of a hydrocarbon is an oxidation-reduction reaction in which the oxidation state of carbon atoms can increase to either +2 (CO) or +4 (CO_2). The reaction of an acid with a metal carbonate is an acid-base reaction. Only CO_2 can form. The oxidation state of C is +4 in the metal carbonate and remains the same in CO_2. A reduction would be required to produce CO, but there is no accompanying oxidation.

59. $2\ C_6H_{14} + 19\ O_2 \rightarrow 12\ CO_2 + 14\ H_2O$
It is impossible because no one can know the ratio of CO to CO_2 produced at any one time. The ratio can even change during the reaction.

61. (a) $CH_4 + 2\ O_2 \rightarrow CO_2 + 2\ H_2O$

$$? \text{ metric tons } CH_4 = 19.8\ t \times \frac{10^3 \text{ kg}}{1\ t} \times \frac{10^3 \text{ g}}{1 \text{ kg}} \times \frac{\text{mol } CO_2}{44.01 \text{ g } CO_2} \times \frac{\text{mol } CH_4}{\text{mol } CO_2}$$
$$\times \frac{16.04 \text{ g } CH_4}{\text{mol } CH_4} \times \frac{10^{-3} \text{ kg}}{1 \text{ g}} \times \frac{10^{-3}\ t}{1 \text{ kg}} = 7.22 \text{ metric tons } CH_4$$

(b) $2\ C_8H_{18} + 25\ O_2 \rightarrow 16\ CO_2 + 18\ H_2O$

$$? \text{ metric tons } C_8H_{18} = 19.8\ t \times \frac{10^3 \text{ kg}}{1\ t} \times \frac{10^3 \text{ g}}{1 \text{ kg}} \times \frac{\text{mol } CO_2}{44.01 \text{ g } CO_2} \times \frac{2 \text{ mol } C_8H_{18}}{16 \text{ mol } CO_2}$$
$$\times \frac{114.22 \text{ g } C_8H_{18}}{\text{mol } C_8H_{18}} \times \frac{10^{-3} \text{ kg}}{1 \text{ g}} \times \frac{10^{-3}\ t}{1 \text{ kg}} = 6.42 \text{ metric tons } C_8H_{18}$$

(c) $C + O_2 \rightarrow CO_2$

$$? \text{ metric tons coal} = 19.8 \text{ t} \frac{10^3 \text{ kg}}{1 \ t} \times \frac{10^3 \text{ g}}{1 \text{ kg}} \times \frac{\text{mol CO}_2}{44.01 \text{ g CO}_2} \times \frac{\text{mol C}}{\text{mol CO}_2}$$

$$\times \frac{12.01 \text{ g C}}{\text{mol C}} \times \frac{100 \text{ g coal}}{94.1 \text{ g C}} \times \frac{10^{-3} \text{ kg}}{1 \text{ g}} \times \frac{10^{-3} \ t}{1 \text{ kg}} = 5.74 \text{ metric tons coal}$$

63. Nitrogen oxides are periodical when air (nitrogen and oxygen) is exposed to flame.

65. (a) $S_8(s) + 8\ O_2(g) \rightarrow 8\ SO_2(g)$
(b) $2\ ZnS(s) + 3\ O_2(g) \rightarrow 2\ ZnO(s) + 2\ SO_2(g)$
(c) $2\ SO_2(g) + O_2(g) \rightarrow 2\ SO_3(g)$
(d) $SO_3(g) + H_2O(l) \rightarrow H_2SO_4(aq)$
(e) $H_2SO_4(aq) + 2\ NH_3(aq) \rightarrow (NH_4)_2SO_4(aq)$

67. (a) Wave action causes droplets of salt water to be sprayed into the air. When the water evaporates, a salt particle is left in the air.
(b) $S(\text{in coal}) + O_2(g) \rightarrow SO_2(g)$
$2\ SO_2(g) + O_2(g) \rightarrow 2\ SO_3(g)$
$SO_3(g) + H_2O(l) \rightarrow H_2SO_4(aq)$
$2\ NH_3(g) + H_2SO_4(aq) \rightarrow (NH_4)_2SO_4(s)$

69. $CaCO_3(s) + 2H^+(aq) \rightarrow Ca^{2+}(aq) + H_2CO_3(aq)$
$H_2CO_3(aq) \rightarrow H_2O(l) + CO_2(g)$

71. $H_2SO_4(aq) + Ca(OH)_2(aq) \rightarrow CaSO_4(aq) + 2\ H_2O(l)$

$$? \text{ kg} = (300\text{m} \times 200\text{m} \times 5.0\text{m}) \times \left(\frac{\text{cm}}{10^{-2} \text{ m}}\right)^3 \times \frac{\text{mL}}{\text{cm}^3} \times \frac{10^{-3} \text{ L}}{\text{mL}} \times 2.0 \times 10^{-4} \text{ M}$$

$$\times \frac{\text{mol Ca(OH)}_2}{\text{mol H}_2\text{SO}_4} \times \frac{74.10 \text{ g}}{\text{mol}} \times \frac{\text{kg}}{10^3 \text{ g}} = 4.4 \times 10^3 \text{ kg}$$

Additional Problems

73. $$V_{\text{particle}} = \frac{4}{3}\pi r^3 = \frac{4}{3}\pi(0.5\ \mu\text{m})^3 \times \left(\frac{10^{-6} \text{ m}}{\mu m}\right)^3 \times \left(\frac{\text{cm}}{10^{-2} \text{ m}}\right)^3 = \frac{4}{3}\pi(0.5 \times 10^{-4} \text{ cm})^3$$

$$= 5.2 \times 10^{-13} \text{ cm}^3\text{/particle}$$

$$? \text{ particles/cm}^3 \text{ air} = \frac{100\ \mu\text{g particles}}{\text{m}^3 \text{ air}} \times \left(\frac{10^{-2} \text{ m}}{\text{cm}}\right)^3 \times \frac{10^{-6} \text{ g particle}}{\mu\text{g particle}}$$

$$\times \frac{\text{cm}^3 \text{ particle}}{1 \text{g particle}} \times \frac{\text{particle}}{5.2 \times 10^{-13} \text{ cm}^3} = 2 \times 10^2 \text{ particles/cm}^3 \text{ air}$$

75. $$\text{R.H.} = \frac{5.67 \text{ mmHg}}{17.5 \text{ mmHg}} \times 100\% = 32.4\%$$

It does take some moisture out of the air, but only if the initial relative humidity is greater than 32.4%.

78. (a) Surface area of a sphere = $4\pi r^2$

$$V_{gas} = \text{depth} \times \text{area} = 0.3\text{ cm} \times 4 \times 3.1416 \times (4000\text{ miles})^2 \times \left(\frac{5280\text{ ft}}{\text{mile}}\right)^2$$

$$\times \left(\frac{12\text{ in}}{\text{ft}}\right)^2 \times \left(\frac{2.54\text{ cm}}{\text{in}}\right)^2 \times \frac{\text{mL}}{\text{cm}^3} \times \frac{10^{-3}\text{ L}}{\text{mL}} = 1.6 \times 10^{15}\text{ L}$$

$$?\text{ molecules O}_3 = 1.6 \times 10^{15}\text{ L} \times \frac{\text{mol}}{22.4\text{ L}} \times \frac{6.022 \times 10^{23}\text{ molecules}}{\text{mol}}$$

$$= 4 \times 10^{37}\text{ molecules O}_3$$

(b) There is no easy way to collect the ozone molecules or to transport them. Moreover, even and if they were collected, they would quickly react to make O_2.

79. C_8H_{18} is less dense than water; that is, d < 1.00 g/mL. Let's assume about 0.80 g/mL.

$$1\text{ L} \times \frac{\text{mL}}{10^{-3}\text{ L}} \times \frac{0.80\text{ g}}{\text{mL}} \times \frac{\text{mol}}{114.22\text{ g}} \times \frac{8\text{ mol CO}}{\text{mol C}_8\text{H}_{18}} \approx 56\text{ mol CO}$$

Assume a molar volume of air of about 25 L/mol at the prevailing T and P.

$$95\text{ m} \times 38\text{ m} \times 16\text{ m} \times \left(\frac{\text{cm}}{10^{-2}\text{ m}}\right)^3 \times \frac{\text{mL}}{\text{cm}^3} \times \frac{10^{-3}\text{ L}}{\text{mL}} \times \frac{1\text{ mol air}}{25\text{ L air}} \approx 2.3 \times 10^6\text{ mol}$$

$$\frac{56\text{ mol CO}}{2.3 \times 10^6\text{ mol air}} \approx 24\text{ ppm}$$

The limit of 35 ppm would not be exceeded.

85. $C_3H_8O + \frac{9}{2}\ O_2 \rightarrow 3\ CO_2 + 4\ H_2O$

$$? = \frac{\text{mg O}_2}{\text{L}} = \frac{875\text{ kg C}_3\text{H}_8\text{O}}{1.8 \times 10^8\text{ L}} \times \frac{10^3\text{ g}}{\text{kg}} \times \frac{\text{mol C}_3\text{H}_8\text{O}}{60.09\text{ g C}_3\text{H}_8\text{O}} \times \frac{9\text{ mol O}_2}{2\text{ mol C}_3\text{H}_8\text{O}}$$

$$\times \frac{32.00\text{ g O}_2}{\text{mol O}_2} \times \frac{\text{mg}}{10^{-3}\text{ g}} = 12\ \frac{\text{mg O}_2}{\text{L}} = \text{increase in BOD}$$

90. (a) $\frac{P_1}{T_1} = \frac{P_2}{T_2}$ $\qquad \frac{9.2\text{ Torr}}{283\text{K}} = \frac{P_2}{294\text{K}}$

$$P_2 = \frac{9.2\text{ Torr} \times 294\text{K}}{283\text{K}} = 9.6\text{ Torr}$$

$$\text{relative humidity at 21 °C} = \frac{\text{partial pressure } H_2O}{VP} \times 100\,\% = \frac{9.6\text{ Torr}}{18.7\text{ Torr}} \times 100\,\%$$

$$= 51\,\%$$

(b) $P_{H_2O} = 0.85 \times 31.8\text{ Torr} = 27.0\text{ Torr at 30 °C}$

$$8.0\text{ hr} \times \frac{60\text{ min}}{\text{hr}} \times \frac{100\text{ ft}^3}{\text{min}} \times \left(\frac{12\text{ in}}{\text{ft}}\right)^3 \times \left(\frac{2.54\text{ cm}}{\text{in}}\right)^3 \times \frac{\text{mL}}{\text{cm}^3} \times \frac{10^{-3}\text{ L}}{\text{mL}}$$

$$\times \frac{\dfrac{(27.0-9.2)\text{ Torr}}{\dfrac{760\text{ Torr}}{\text{atm}}}}{} \times \frac{1}{\dfrac{0.08206\text{ L atm}}{\text{K mole}}} \times \frac{1}{283\text{ K}} \times \frac{18.0\text{ g}}{\text{mole}} = 2.47 \times 10^4\text{ g}$$

(c) $\Delta H = n\Delta H_{cond} = 2.47 \times 10^4\text{ g} \times \dfrac{44.0\text{ kJ}}{\text{mole}} \times \dfrac{\text{mole}}{18.01\text{ g}} = 6.03 \times 10^4\text{ kJ}$

Apply Your Knowledge

94. $C + O_2 \rightarrow CO_2$ $\quad\quad \Delta H_f = -393.5\text{ kJ/mol}$

$$8.7 \times 10^8\text{ ton} \times \frac{2000\text{ lb}}{\text{ton}} \times \frac{454\text{ g}}{\text{lb}} \times \frac{\text{mol}}{12.01\text{ g}} \times \frac{393.5\text{ kJ}}{\text{mol}} \times \frac{2\text{ mg } SO_2}{\text{kJ}} \times \frac{10^{-3}\text{ g}}{\text{mg}}$$

$$\times \frac{\text{mol } SO_2}{64.07\text{ g}} \times \frac{\text{mol } H_2SO_4}{\text{mol } SO_2} \times \frac{98.09\text{ g}}{\text{mol } H_2SO_4} \times \frac{\text{lb}}{454\text{ g}} = 1.75 \times 10^{11}\text{ lb } H_2SO_4.$$

This mass of H_2SO_4 is about twice the typical U.S. annual production.

97. (a) $? \dfrac{\text{mol C}}{\text{L}} = \dfrac{505\text{ mg}}{\text{L}} \times \dfrac{10^{-3}\text{ g}}{\text{mg}} \times \dfrac{\text{mol}}{12.01\text{ g}} = \dfrac{4.2 \times 10^{-2}\text{ mol C}}{\text{L}}$

$? \dfrac{\text{mol N}}{\text{L}} = \dfrac{92\text{ mg}}{\text{L}} \times \dfrac{10^{-3}\text{ g}}{\text{mg}} \times \dfrac{\text{mol}}{14.01\text{ g}} = \dfrac{6.6 \times 10^{-3}\text{ mol N}}{\text{L}}$

$? \dfrac{\text{mol P}}{\text{L}} = \dfrac{14\text{ mg}}{\text{L}} \times \dfrac{10^{-3}\text{ g}}{\text{mg}} \times \dfrac{\text{mol}}{30.97\text{ g}} = \dfrac{4.5 \times 10^{-4}\text{ mol P}}{\text{L}}$

$\dfrac{4.20 \times 10^{-2}}{4.5 \times 10^{-4}} = 93.3 \quad\quad \dfrac{6.6 \times 10^{-3}}{4.5 \times 10^{-4}} = 14.7$

ratio 93.3 C : 14.7 N : 1 P

$\dfrac{93.3}{14.7} = 6.3 \quad\quad \dfrac{106}{16} = 6.6$

C is the limiting nutrient.

(b) Begin with either $(NH_4)_2HPO_4$ or $NH_4H_2PO_4$. If $(NH_4)_2HPO_4$ is chosen, it provides the P and 2 moles N. Add 2.5 moles $CO(NH_2)_2$ for the N and 13.83 moles $CH_3CHOHCOOH$ for the rest of the C.

$1\text{ mole } (NH_4)_2HPO_4 \times \dfrac{132.06\text{ g}}{\text{mole}} = 132\text{ g}$

$$2.5 \text{ mole } CO(NH_2)_2 \times \frac{60.05 \text{ g}}{\text{mole}} = 150 \text{ g}$$

$$13.83 \text{ mole } CH_3CHOHCOOH \times \frac{90.08 \text{ g}}{\text{mole}} = 1246$$

$$\frac{150 \text{ g}}{132 \text{ g}} = 1.136 \qquad 1528 \text{ g}$$

$$\frac{1246 \text{ g}}{132 \text{ g}} = 9.439$$

The mass ratio of $CH_3CHOHCOOH$ to $CO(NH_2)_2$ to $(NH_4)_2HPO_4$ is 9.439:1.136:1.000.

If instead one begins with 1 mole $NH_4H_2PO_4$. That provides the P and one mole of N; add 3 moles of $CO(NH_2)_2$ for the N and 3 C and 13.66 moles of $CH_3CHOHCOOH$ for the rest of the C.

$$1 \text{ mole } NH_4H_2PO_4 \times \frac{115.09 \text{ g}}{\text{mole}} = 115$$

$$3 \text{ mole } CO(NH_2)_2 \times \frac{60.05 \text{ g}}{\text{mole}} = 180$$

$$13.66 \text{ moles } CH_3CHOHCOOH \times \frac{90.08 \text{ g}}{\text{mole}} = \underline{1230}$$

$$\frac{180 \text{ g}}{115 \text{ g}} = 1.565 \qquad 1525 \text{ g}$$

$$\frac{1230}{115} = 10.70$$

The mass ratio of $CH_3CHOHCOOH$ to $CO(NH_2)_2$ to $NH_4H_2PO_4$ is 10.70:1.565:1.0000

NH_4NO_3 was not used. It would provide the N but not the P and would thus increase the total mass.

98. (a) $? \text{ L in pond} = 7.91 \text{ acre} \times \frac{(\text{mile})^2}{640 \text{ acre}} \times \left(\frac{5280 \text{ ft}}{\text{mile}}\right)^2 \times 5.25 \text{ ft} \times \left(\frac{12 \text{ in}}{\text{ft}}\right)^3$

$$\times \left(\frac{2.54 \text{ cm}}{\text{in}}\right)^3 \times \frac{\text{mL}}{\text{cm}^3} \times \frac{10^{-3} \text{ L}}{\text{mL}} = 5.12 \times 10^7 \text{ L}$$

$$? \text{ ppm} = \frac{2.50 \times 10^5 \text{ g}}{5.12 \times 10^7 \text{ L}} \times \frac{10^{-3} \text{ L}}{\text{mL}} \times \frac{\text{mL}}{1 \text{ g}} \times 10^6 \text{ ppm} = 4.88 \text{ ppm}$$

$$? \text{ molecules} = 2.50 \times 10^2 \text{ kg} \times \frac{10^3 \text{ g}}{\text{kg}} \times \frac{\text{mole}}{172.0 \text{ g}} \times \frac{6.022 \times 10^{23} \text{ molecules}}{\text{mole}}$$

$$= 8.75 \times 10^{26} \text{ molecules}$$

$$? \frac{\text{molecules}}{\text{L}} = \frac{8.75 \times 10^{26} \text{ molecules}}{5.12 \times 10^7 \text{ L}} = 1.71 \times 10^{19} \text{ molecules/L}$$

(b) $\ln \frac{N}{N_0} = -\lambda t$ $\qquad \lambda = \frac{0.693}{t_{1/2}} = \frac{0.693}{1.00\ \text{y}} = 0.693$

$$t = \frac{\ln \frac{1.00\ \text{ppm}}{4.88\ \text{ppm}}}{-0.693} = 2.29\ \text{y}$$

(c) $\ln \frac{N}{N_0} = -\lambda t$

before 2nd application

$$\ln \frac{N}{4.88\ \text{ppm}} = -0.693(1/2)$$

$$\frac{N}{4.88\ \text{ppm}} = e^{-\frac{0.693}{2}} = 0.707$$

$N = 3.45$ ppm

before 3rd application, 3.45 ppm + 4.88 ppm = 8.33 ppm

$$\ln \frac{N}{8.33\ \text{ppm}} = -0.693(1/2)$$

$N = 0.707 \times 8.33$ ppm = 5.89 ppm

before 4th application, 5.89 ppm + 4.88 ppm = 10.77 ppm

$$\ln \frac{N}{10.77\ \text{ppm}} = -0.693(1/2)$$

$N = 0.707 \times 10.77$ ppm = 7.61 ppm

before 5th application, 7.61 ppm + 4.88 ppm = 12.49 ppm

$$\ln \frac{N}{12.49\ \text{ppm}} = -0.693(1/2)$$

$N = 0.707 \times 12.49$ ppm = 8.83 ppm

before 6th application, 8.83 ppm + 4.88 ppm = 13.71 ppm

$$\ln \frac{N}{13.71\ \text{ppm}} = -0.693(1/2)$$

$N = 0.707 \times 13.71 = 9.69$ ppm

(d) The increase becomes smaller each time, so that the solubility limit is probably never reached. Each time, the increase is 70.7% of the previous increase. It only requires 12 times before the increase is less than 0.1. At that point the concentration is only about 13.07 ppm, less than the saturation limit 18 ppm. Increases 2.44 + 1.72 + 1.22 + 0.86 + 0.61 + 0.43 + 0.30 + 0.22 + 0.15 + 0.11 + 0.076 + 0.054 = 8.19 ppm.

Total = 8.19 + 3.45 = 13.07ppm after 12th application, and 13.11 ppm

$$?\ \text{ppm} = \frac{18\ \text{mg dichlob}}{\text{L}} \times \frac{10^{-3}\ \text{g}}{\text{mg}} \times \frac{10^{-3}\ \text{L}}{\text{mL}} \times \frac{\text{mL}}{1.0\ \text{g}} \times \frac{\text{ppm}\ 10^{6}\ \text{g solution}}{\text{g dichlob}} = 18\ \text{ppm}$$